Suzanne's Diary for Nicholas

Suzanne's Diary for Nicholas

James Patterson

Thorndike Press • Chivers Press
Waterville, Maine USA Bath, England

This Large Print edition is published by Thorndike Press, USA and by Chivers Press, England.

Published in 2001 in the U.S. by arrangement with Little, Brown and Company, Inc.

Published in 2001 in the U.K. by arrangement with Headline Book Publishing Ltd.

U.S. Hardcover 0-7862-3297-8 (Basic Series Edition)
U.S. Softcover 0-7862-3298-6
U.K. Hardcover 0-7540-1707-9 (Windsor Large Print)
U.K. Softcover 0-7540-9106-6 (Paragon Large Print)

The text of this Large Print edition is unabridged.
Other aspects of the book may vary from the original edition.

Set in 16 pt. Plantin by Christina S. Huff.

Printed in the United States on permanent paper.

British Library Cataloguing-in-Publication Data available

Library of Congress Cataloging-in-Publication Data

Patterson, James, 1947–
    Suzanne's diary for Nicholas : a novel / James Patterson.
        p.   cm.
    ISBN 0-7862-3297-8 (lg. print : hc : alk. paper)
    ISBN 0-7862-3298-6 (lg. print : sc : alk. paper)
    1. Mothers and sons — Fiction.   2. Loss (Psychology) —
    Fiction.   3. Infants — Fiction.   4. Large type books.
    I. Title.
PS3566.A822 S8 2001b
    813′.54—dc21   2001027094

Katie

For those who have loved, and lost,
and loved again

For Robin Schwarz, whose valuable assistance
and big heart are much appreciated

Also, thanks for the help —
Mary, Fern, Barbara, Irene, Maria, Darcy,
Mary Ellen, and Carole Anne

Most of all, for Suzie and Jack; and for Jane

Katie Wilkinson sat in warm bathwater in the weird but wonderful old-fashioned porcelain tub in her New York apartment. The apartment exuded "old" and "worn" in ways that practitioners of shabby chic couldn't begin to imagine. Katie's Persian cat, Guinevere, looking like a favorite gray wool sweater, was perched on the sink. Her black Labrador, Merlin, sat in the doorway leading to the bedroom. They watched Katie as if they were afraid for her.

She lowered her head when she finished reading the diary and set the leatherbound book on the wooden stool beside the tub. Her body shivered.

Then she started to sob, and Katie saw that her hands were shaking. She was losing it, and she didn't lose it often. She was a strong person, and always had been. Katie whispered words she'd once heard in her father's church in Asheboro, North Carolina. "Oh, Lord, oh, Lord, are you *anywhere*, my Lord?"

9

She could never have imagined that this small volume would have such a disturbing effect on her. Of course, it wasn't just the diary that had forced her into this state of confusion and duress.

No, it wasn't just Suzanne's diary for Nicholas.

She visualized Suzanne in her mind. Katie *saw* her at her quaint cottage on Beach Road on Martha's Vineyard.

Then little Nicholas. Twelve months old, with the most brilliant blue eyes.

And finally, Matt.

Nicholas's daddy.

Suzanne's husband.

And Katie's former lover.

What did she think of Matt now? Could she ever forgive him? She wasn't sure. But at least she finally understood some of what had happened. The diary had told her bits and pieces of what she needed to know, as well as deep, painful secrets that maybe she didn't need to know.

Katie slipped down farther into the water, and found herself thinking back to the day she had received the diary — July 19.

Remembering the day started her crying again.

10

On the morning of the nineteenth, Katie had felt drawn to the Hudson River, and then to the Circle Line, the boat ride around Manhattan Island that she and Matt had first taken as a total goof but had enjoyed so much that they kept coming back.

She boarded the first boat of the day. She was feeling sad, but also angry. Oh, God, she didn't know what she was feeling.

The early boat wasn't too crowded with tourists. She took a seat near the rail of the upper deck and watched New York from the unique vantage point of the brooding waterways surrounding it.

A few people noticed her sitting there alone — especially the men.

Katie usually stood out in a crowd. She was tall — almost six feet, with warm, friendly blue eyes. She had always thought of herself as gawky and felt that people were staring at her for all the wrong reasons. Her friends begged to differ; they said she was close to breathtaking, stunning in her strength. Katie always responded, "Uh-huh, sure, don't I wish." She didn't see herself that way and knew she never would. She was an ordinary, regular person. A North Carolina farm girl at heart.

She often wore her brunette hair in a long braid, and had since she was eight years old. It used to look tomboyish, but now it was supposed to be big-city cool. She guessed she'd finally caught up with the times. The only makeup she ever wore was a little mascara and sometimes lipstick. Today she wore neither. She definitely didn't look breathtaking.

Sitting there on the top deck, she remembered a favorite line from the movie *The African Queen*: "Head up, chin out, hair blowing in the wind, the living picture of the hero-eyne," Bogart had teased Hepburn. It cheered her a bit — a *titch*, as her mother liked to say back home in Asheboro.

She had been crying for hours, and her eyes were puffy. The night before, the man she loved had suddenly and inexplicably ended their relationship. She'd been completely sucker punched. She hadn't seen it coming. It almost didn't seem possible that Matt had left her.

Damn him! How could he? Had he been lying to me all this time — months and months?

Of course he had! The bastard. The total creep.

She wanted to think about Matt, about what had happened to separate them, but she wound up thinking of times they'd shared, mostly good times.

Begrudgingly, she had to admit that she had always been able to talk to him freely and easily about anything. She could talk to Matt the way she talked to her women friends. Even her girlfriends, who could be catty and generally had terrible luck with men, liked Matt. *So what happened between us?* That's what she desperately wanted to know.

He *was* thoughtful — at least he *had been*. Her birthday was in June, and he had sent her a single rose every day of what he called "your birthday month." He always seemed to notice whether he'd seen her in a certain blouse or sweater before, her shoes, her moods — the good, the bad, and occasionally the stressed-out ugly.

He liked a lot of the same things Katie did, or so he said. *Ally McBeal, The Practice, Memoirs of a Geisha, The Girl with the Pearl Earring.* Dinner, *then* drinks at the bar at One if by Land, Two if by Sea. Waterloo in the West Village; Coup in the East; Bubby's on Hudson Street. Foreign movies at the Lincoln Plaza Cinema. Vintage black-and-white photos, oil paintings that they found at flea markets. Trips to NoLita (North of Little Italy) and Williamsburg (the new SoHo).

He went to church with her on Sundays, where she taught a Bible class of pre-schoolers. They both treasured Sunday afternoons at her apartment — with Katie reading the *Times* from cover to cover, and Matt revising his poems, which he spread out on her bed and on the bedroom floor and even on the butcher-block kitchen table.

Tracy Chapman or Macy Gray, maybe Sarah Vaughan, would be playing softly in the background. Delicious. Perfect in every way.

He made her feel at peace with herself, completed her circle, did *something* that was good and right. No one else had ever made her feel that way before. Completely, blissfully at peace.

What could beat being in love with Matt?

Nothing that Katie knew of.

One night they had stopped at a little juke bar on Avenue A. They danced, and Matt sang "All Shook Up" in her ear, doing a funny but improbably good Elvis impersonation. Then Matt did an even better Al Green, which completely blew her away.

She had wanted to be with him all the time. *Corny, but true.*

When he was away on Martha's Vineyard, where he lived and worked, they would talk for hours every night on the phone — or send each other funny e-mails. They called it their "long-distance love affair." He had always stopped Katie from actually visiting him on the Vineyard, though. Maybe *that* should have been her early-warning signal?

Somehow, it had worked — for eleven glorious months that seemed to go by in an instant. Katie had expected him to propose soon. She was sure of it. She had even told her mother. But, of course, she had been so wrong that it was pathetic. She felt like a fool — and she hated herself for it.

How could she have been so stupefyingly wrong about him? About everything? It wasn't like her to be this out of touch with her instincts. They were usually *good;* she was *smart;* she didn't do really *dumb things.*

Until now. And, boy, had she made a doozy of a mistake this time.

Katie suddenly realized that she was sobbing and that everyone around her on the deck of the boat was staring at her.

"I'm sorry," she said, and motioned for them to please look away. She blushed. She was embarrassed and felt like such an idiot. "I'm okay."

But she wasn't okay.

Katie had never been so hurt in her life. Nothing came close to this. She had lost the only man she had ever loved; God, how she loved Matt.

Katie couldn't bear to go in to work that day. She couldn't face the people at her office. Or even strangers on a city bus. She'd gotten enough curious looks on the boat to last a lifetime.

When she got back to her apartment after her trip on the Circle Line, a package was propped up against the front door.

She thought it was a manuscript from the office. She cursed work under her breath. Couldn't they leave her alone for a single day? She was entitled to a personal day now and then. God, she worked so hard for them. They knew how passionate she was about her books. They knew how much Katie cared.

She was a senior editor at a highly thought of, collegial, very pleasant New York publishing house that specialized in literary novels and poetry. She loved her job. It was where she had met Matt. She had enthusiastically bought his first volume of poetry from a small literary agency in Boston about a year before.

The two of them hit it off right away, *really* hit it off. Just weeks later they had fallen in love — or so she had believed with her heart, soul, body, mind, woman's intuition.

How could she have been so wrong? What had happened? Why?

As she reached down for the package, she recognized the handwriting. *It was Matt's.* There was no doubt about it.

She wanted to hurl the package away with all the power and strength in her body, and nearly dropped it.

She didn't. Too much self-control — that was her problem. *One* of her problems. Katie stared at the package for some time. Finally, she took a deep breath and tore away the brown paper wrapping.

What she found inside was a small antique-looking diary. Katie frowned. She didn't understand. Then she felt her stomach begin to knot.

Suzanne's Diary for Nicholas was handwritten on its front cover — handwritten, but it wasn't Matt's handwriting.

Suzanne's?

Suddenly Katie's head was reeling and she could barely catch a breath. She couldn't think straight, either. Matt had always been closemouthed and secretive about his past. One of the things she had found out was that his wife's name was Suzanne. That much had slipped out one night after they had drunk two bottles of wine. But then Matt hadn't wanted to talk about Suzanne.

The only arguments they'd ever had were over the silence about his past. Katie had insisted on knowing more, which only made Matt quieter and more mysterious. It was so unlike him. After they actually had a fight about it, he'd told her that he wasn't married to Suzanne anymore; he swore it, but that was all he was going to say on the subject.

Who was Nicholas? And why had Matt sent her this diary? Why now? She was completely puzzled, and more than a little upset.

Katie's fingers were trembling as she opened the diary to its first page. A note from Matt was affixed. Her eyes began to well up, and she angrily wiped the tears away. She read what he'd written.

Dear Katie,

No words or actions could begin to tell you what I'm feeling now. I'm so sorry about what I allowed to happen between us. It was all my fault, of course. I take all the blame.

19

You are perfect, wonderful, beautiful. It's not you. It's me.

Maybe this diary will explain things better than I ever could. If you have the heart, read it.

It's about my wife and son, and me.

I will warn you, though, there will be parts that may be hard for you to read.

I never expected to fall in love with you, but I did.

Matt

Katie turned the page.

The Diary

Dear Nicholas, my little prince —

There were years and years when I wondered if I would ever be a mother.

During this time, I had a recurring daydream that it would be so wonderful and wise to make a videotape every year for my children and tell them who I was, what I thought about, how much I loved them, what I worried about, the things that thrilled me, made me laugh or cry, made me think in new ways. And, of course, all my most personal secrets.

I would have treasured such videotapes if my mother and father had recorded them each year, to tell me who they were, what they felt about me and the world.

As it turned out, I don't know who they are, and that's a little sad. No, it's a lot sad.

So, I am going to make a videotape for you every year — but there's something else I want to do for you, sweet boy.

23

I want to keep a diary, *this* diary, and I promise to be faithful about writing in it.

As I write this very first entry, you are two weeks old. But I want to start by telling you about some things that happened before you were born. I want to start *before* the beginning, so to speak.

This is for your eyes only, Nick.

This is what happened to Nicholas, Suzanne, and Matt.

Let me start the story on a warm and fragrant spring evening in Boston.

I was working at Massachusetts General Hospital at the time. I had been a physician for eight years. There were moments that I absolutely loved, cherished: seeing patients get well, and even being with some when it was clear they wouldn't recover. Then there were the bureaucracy and the hopeless inadequacy of our country's current health-care program. There were my own inadequacies as well.

I had just come off a twenty-four-hour rotation and I was tired beyond anything you can imagine. I was out walking my trusted and faithful golden retriever, Gustavus, a.k.a. Gus.

I suppose I should give you a little snapshot of myself back then. I had long blond hair, stood about five foot five, not exactly beautiful but nice enough to look at, a friendly smile most of the time, for most of the human race. Not *too* caught up in appearances.

It was a late Friday afternoon, and I remember that the weather was so nice, the air was sweet and as clear as crystal. It was the kind of day that I live for.

I can see it all as if it just happened.

Gus had sprinted off to harass and chase a poor, defenseless city duck that had wandered away from the safety of the pond. We were in the Boston Public Garden, by the swan boats. This was our usual walk, especially if Michael, my boyfriend, was working, as he was that night.

Gus had broken from his lead, and I ran after him. He is a gifted retriever, who lives to retrieve anything: balls, Frisbees, paper wrappers, soap bubbles, reflections on the windows of my apartment.

As I ran after Gus, I was suddenly struck by the worst pain I have ever felt in my life. *Jesus, what is this?*

It was so intense that I fell to my hands and knees.

Then it got worse. Razor-sharp knives were shooting up and down my arm, across my back, and into my jaw. I gasped. I couldn't catch my breath. I couldn't focus on anything in the Public Garden. Everything was a blur. I couldn't actually be sure of what was happening to me, but something told me *heart*.

What was wrong with me?

I wanted to cry out for help, but even a few words were beyond me. The tree-laden Garden was spinning like a whirligig. Concerned people began crowding around, then hovering over me.

Gus had come skulking back. I heard him barking over my head. Then he was licking my cheek, but I barely felt his tongue.

I was flat on my back, holding my chest.

Heart? My God, I am only thirty-five years old.

"Get an ambulance," someone cried. "She's in trouble. I think she's dying."

I am not! I wanted to shout. *I can't be dying.*

My breathing was becoming shallower and I was fading to black, to nothingness. *Oh, God,* I thought. *Stay alive, breathe, keep conscious, Suzanne.*

That's when I remember reaching out for a stone that was near me in the dirt. *Hang on to this stone,* I thought, *hang on tight.* I believed it was the only thing that would keep me attached to the earth at that scary moment. I wanted to call out for Michael, but I knew it wouldn't help.

Suddenly I realized what was happening to me. I must have passed out for several minutes. When I came to, I was being lifted into an ambulance. Tears streamed down my face. My body was soaked with sweat.

The EMT woman kept saying, "You're gonna be fine. You're all right, ma'am." But I knew I wasn't.

I looked at her with whatever strength I could muster and whispered, "Don't let me die."

All the while I was holding the small stone tightly in my hand. The last thing I recall is an oxygen mask being slipped over my face, a deathly weakness spreading through my body, and the stone finally dropping from my hand.

So, Nicky,

I was only thirty-five when I had the heart attack in Boston. The following day I had a coronary bypass at Mass. General. It put me out of action, out of circulation for almost two months, and it was during my recuperation that I had time to think, really think, maybe for the first time in my life.

I thoroughly, painfully examined my life in Boston, just how hectic it had become, with rounds, research, overtime, overwork, and double shifts. I thought about how I'd been feeling just before this awful thing happened. I also dealt with my own denial. My grandmother had died of heart failure. My family had a history of heart disease. And still I hadn't been as careful as I should have been.

It was while I was recuperating that a doctor friend told me the story of the five balls. You should never forget this one,

29

Nicky. This is terribly important.

It goes like this.

Imagine life is a game in which you are juggling five balls. The balls are called work, family, health, friends, and integrity. And you're keeping all of them in the air. But one day you finally come to understand that work is a rubber ball. If you drop it, it will bounce back. The other four balls — family, health, friends, integrity — are made of glass. If you drop one of these, it will be irrevocably scuffed, nicked, perhaps even shattered. And once you truly understand the lesson of the five balls, you will have the beginnings of balance in your life.

Nicky, *I finally understood.*

Nick —

As you can probably tell, this is all pre-Daddy, pre-Matt.

Let me tell you about Dr. Michael Bernstein.

I met Michael in 1996 at the wedding reception for John Kennedy and Carolyn Bessette on Cumberland Island, Georgia. I must admit that both of us had led pretty charmed lives up until then. My parents had died when I was two, but I was fortunate enough to have been raised with great love and patience by my grandparents in Cornwall, New York. I went to Lawrenceville Academy in New Jersey, then Duke, and finally Harvard Medical School.

I felt incredibly lucky to be at each of the three schools, and I couldn't have gotten a better education — except that nowhere did I learn the lesson of the five balls.

Michael also went to Harvard Medical

31

School, but he had graduated four years before I got there. We didn't meet until the Kennedy wedding. I was a guest of Carolyn's; Michael was a guest of John's. The wedding itself was magical, full of hope and promise. Maybe that was part of what drew Michael and me together.

What kept us together for the next four years was a little more complicated. Part of it was pure physical attraction, and at some point I want to talk to you about that — but not now. Michael was — *is* — tall and dashing, with a radiant smile. We had a lot of mutual interests. I loved his stories, always so droll, laconic, biting; I loved to listen to him play the piano and sing anything from Sinatra to Sting. Also, we were both workaholics — me at Mass. General, Michael at Children's Hospital in Boston.

But none of these things are what love is really about, Nicholas. Trust me on that.

About four weeks after my heart attack, I woke up one morning at eight o'clock. The apartment where we lived was quiet, and I luxuriated in the peacefulness for a few moments. It seemed to have a healing quality. Finally I got up and went to the kitchen to make myself breakfast before I went off to rehab.

I jumped back when I heard a noise, the

scratch of a chair leg against the floor. Nervously, I went to see who was out there.

It was Michael. I was surprised to see him still home, as he was almost always out of the house by seven. He was sitting at the small pine table in the breakfast nook.

"You almost gave me a heart attack," I said, making what I thought was a pretty decent joke.

Michael didn't laugh. He patted the chair next to him at the table.

Then, with the calmness and self-reverence I was used to from him, he told me the three main reasons why he was leaving me: he said he couldn't talk or relate to me the way he could with his male friends; he didn't think that I could have a baby now, because of my heart attack; he had fallen for someone else already.

I ran out of the kitchen, and then out of the house. That morning the pain I felt was even worse than the heart attack. Nothing was right with my life; I had gotten it all wrong so far. *Everything!!!*

I did love being a doctor, but I was trying to do it in a large, somewhat bureaucratic, big-city hospital, which just wasn't right for me.

I was working so hard — because there was nothing else of value in my life. I earned about

$120,000 a year, but I was spending it on dinners in town, getaway weekends, clothes that I didn't need or even like that much.

I had wanted children all my life, yet here I was without a significant other, without a child, without a plan, and no prospects to change any of it.

Here's what I did, little boy.

I began to *live* the lesson of the five balls.

I left my job at Mass. General. I left Boston. I left my murderous schedule and commitments behind. I moved to the one place in the world where I had always been happy. I went there, truly, to mend a broken heart.

I was turning endlessly around and around like a hamster on a wheel in a tiny cage. My life was stretched to the limit, and something was bound to give. Unfortunately, it had been my heart.

This wasn't a small change, Nicky; I had decided to change everything.

Nicky,

I arrived on the island of Martha's Vineyard like an awkward tourist, lugging the baggage of my past, not knowing what to do with it yet. I would spend the first couple of months filling cupboards with wholesome, farm-fresh foods, throwing out old magazines that had followed me to my new home, and I would also settle into a new job.

From the time I was five until I was seventeen, I had spent summers with my grandparents on Martha's Vineyard. My grandfather was an architect, as my father had been as well, and he could work from his home. My grandmother Isabelle was a homemaker, and she was gifted at making our living space the most comfortable and loving place I could begin to imagine.

I loved being back on the Vineyard, loved everything about it. Gus and I often went to the beach in the early evening, and we sat out there until the light of day was gone. We played ball, or sometimes with a Frisbee for the first hour or so. Then we huddled together on a blanket until the sun went down.

I had negotiated for the practice of a general practitioner who was moving to Illinois. We were switching lives in some ways. He was going to Chicago just when I was exiting city life. My office was one of five doctors' offices in a white clapboard house in Vineyard Haven. The house was more than a hundred years old and had four beautiful antique rockers on the front porch. I even had a rocker at the desk where I worked.

Country doctor resonated with a wonderful sound for me, like recess bells of an old country school. I was inspired to hang out a shingle that said as much: SUZANNE BEDFORD — COUNTRY DOCTOR — IN.

I began to see a few patients in my second month on Martha's Vineyard.

Emily Howe, seventy, part-time librarian, honored member of the Daughters of the American Revolution, hard, steadfast, and against everything that had occurred since about 1900. Diagnosis: bronchitis; Prognosis: good.

Dorris Lathem, ninety-three, had already outlived three husbands, eleven dogs, and a house fire. Healthy as a horse. Diagnosis:

old gal; Prognosis: will live forever.

Earl Chapman, Presbyterian minister. General Outlook — always his own. Diagnosis: acute diarrhea; Prognosis: possible recurrence of what the Lord might call getting even.

My first patient list read like a who's who of a William Carlos Williams poem. I imagined Dr. Williams walking the streets of the Vineyard on his appointed rounds, an icy wind blowing from the distant hills, milk frozen on every landing, the famous wheelbarrow soldered into the winter mud. There he'd be, making a late-afternoon call on the boy who fell off his sled and broke an arm along with his pride.

This was for me. I was experiencing a fantasy that was a million miles away when I lived in Boston.

But, in fact, it was just down Route 3 and across the water.

I felt I had come home.

Nicholas,

I had no idea that the love of my life was here — just waiting for me. If I had, I would have run straight into Daddy's arms. In a heartbeat.

When I first arrived on Martha's Vineyard, I was unsure about everything, but especially where to settle. I drove around looking for something that said "home," "you'll be okay here," "look no further."

There are so many parts of our island that are beautiful, and even though I knew it in some ways, it sang out differently to me this time.

Everything was different because *I* felt different. Up Island was always special to me, because this is where I had spent so many glorious summers. It lay like a child's picture book of farms and fences, dirt roads, and cliffs. Down Island was a whirl of widow's walks, gazebos, lighthouses, and harbors.

It was a turn-of-the-century boathouse

that finally stole my heart. And still does. This truly was home.

It needed to be fixed up, but it was winterized, and I loved it at first sight, first smell, first touch. Old beams — which had once supported stored boats — crisscrossed the ceiling. Upstairs I eventually put in corner portholes to let the sun come in in hoops of light. The walls *had* to be painted robin's-egg blue because the whole downstairs opened to a view of the sea. Big barnlike doors slid port and starboard to bring everything that was once outside, *inside*.

Can you imagine, Nicky, living practically right on the beach, like that? Every part of me, body and soul, knew I'd made the right decision. Even my sensible side was in agreement. I now lived between Vineyard Haven and Oak Bluffs. Sometimes I'd be working out of my home or making house calls, but the rest of the time I'd be at Martha's Vineyard Hospital or the Vineyard Walk-In Medical Center in Vineyard Haven. I was also doing some cardiology rehab at the Medical Center.

I was alone, except for Gus, living a solitary life, but I was content for the most part.

Maybe it was because I had no idea what I was missing at the time: *your daddy and you.*

Nicholas,

I was driving home from the hospital when I heard a funny noise. What's that? *Shhhhh . . . bump shhhh . . . bump shhhh . . . bump.*

I had to pull over onto the shoulder of the road. I got out of my Jeep to take a look.

Shitfire and save matches. The right wheel was as flat as a pancake. I could have, and I would have, changed the tire if I hadn't taken out the spare in order to make room for all my other stuff when I was moving.

I called the gas station from my cell phone, mad at myself for having to call a garage. A guy answered and condescended to me a little; *another* guy would come to fix the flat. It made me feel like "such a girl," and I hated that. I knew how to change a tire perfectly well. I pride myself on self-sufficiency and independence. And good old-fashioned stubbornness.

I was standing against the passenger-side

door, pretending to admire the beautiful landscape and making it seem to passing cars that I had pulled over for that reason, when a car pulled up right in back of mine.

Clearly it wasn't from the gas station.

Not unless they'd sent a forest green Jaguar convertible.

"You need some help?" a man asked. He was already walking slowly toward my car, and honestly, I couldn't take my eyes off him.

"No, thanks . . . I called the Shell station in town. They'll be here soon. Thanks, anyway."

There was something familiar about this guy. I wondered if I had met him in one of the stores around the island. Or maybe at the hospital.

But he was tall and good-looking, and I thought that I'd have remembered him. He had a nice, easy smile and he was kind of laid-back.

"I can change the tire," he offered, and somehow managed *not* to be condescending when he said it. "I know I drive a fancy car, but I'm not really a fancy person."

"Thanks, but I took my spare out to make room for more important things like my stereo and my antique candlestick collection."

He laughed . . . and he was *so familiar.*

41

Who was he? Where did I know him from?

"I'm flattered, though," I continued. "A man in a shiny convertible willing to change a tire."

He laughed again — a nice laugh. *So familiar.*

"Hey, I'm vast. . . . I contain multitudes."

"Walt Whitman!" I said — and then I remembered who this was. "You used to say that *all the time*. You quoted Walt Whitman. *Matt?*"

"Suzanne Bedford!" he said. "I was almost sure it was you."

He was so surprised — bumping into me like this after such a long time. It must have been almost twenty years.

Matt Wolfe was even handsomer than I remembered him. At thirty-seven, he had grown up very nicely. He was slender, with closely cropped brown hair and an endearing smile. He looked in great shape. We talked on the side of the road. He had become a lawyer for the Environmental Protection Agency as well as a fine-arts dealer. I had to laugh when he told me that. Matt used to joke that he would never become an *entremanure,* as he called businesspeople back then.

He wasn't surprised to learn that I was a doctor. What surprised Matt was that I wasn't with someone, that I had come back to Martha's Vineyard *alone*.

We continued to catch up on each other's life. He was funny, easy to talk to. When I had dated Matt, he was eighteen, I was sixteen. That was the last year my grandparents had rented for the summer on the Vineyard — but obviously, I never forgot the island or its many treasures. I'd been having dreams about the ocean and the beaches on the Vineyard ever since I could remember.

I think we were both a little disappointed to see the bright yellow Shell tow truck pull in behind us. I know that I was. Just before I turned to go, Matt mumbled a few words about how nice this was — my flat tire. Then he asked me what I was doing Saturday night.

I think I blushed. I know I did. "You mean a date?"

"Yes, Suzanne, a date. Now that I've seen you again, I'd like to see you *again*."

I told Matt I would love to see him on Saturday. My heart was pounding a little, and I took that to be a very good sign.

Nick,

Who the heck was sitting on my porch? As I drove up late that same afternoon, I couldn't really tell.

It couldn't be the electric guy, or the phone guy, or the cable guy — I'd seen all of them the day before.

Nope, it was the painting guy, the one who was going to help me with everything around the cottage that needed a ladder or an outlet or a finish.

We walked around the cottage as I pointed out several of the problems I'd inherited: windows that wouldn't close, floors that buckled at the door, a leak in the bathroom, a broken pump, a cracked gutter, and a whole cottage that needed scraping and painting.

What this house had in cute, it lacked in practical.

But this guy was great, took notes, asked

pertinent questions, and told me he could fix everything by the millennium. The next millennium. We struck a deal on the spot (which gave me the distinct feeling I'd made out pretty good).

Suddenly life was looking a lot better to me. I had a new practice that I loved, I had a housepainter with a good reputation, and I had a hot date with Matt.

When I was finally alone in my little cottage by the sea, I threw up both arms and shouted hooray.

Then I said, "Matt Wolfe. Hmmm. Imagine that. How terrific. How very cool."

Nick,

Just about everybody has an occasional fantasy about somebody they really liked in high school, or maybe even grade school, coming back into their life. For me, that person was Matt.

Who knows, maybe he was a small part of what drew me back to Martha's Vineyard. Probably not, but who can tell about these things?

Nevertheless, I was nearly an hour late for our date on Saturday night. I had to get a patient admitted, run home and feed Gustavus, get pretty, and find my beeper all before I left. Plus — I must confess — I can be a bit disorganized at times. My grandfather used to say, "Suzie, you have a lot *in* your mind."

When I entered Lola's, which is a neat spot on the beach between Vineyard Haven and Oak Bluffs, Matt was waiting with a bottle of pinot noir. He looked relaxed, and I liked that.

Also handsome. I liked that just fine, too.

"Matt, I'm so, so sorry," I said. "This is one of the negatives about dating a doctor."

He laughed. "After twenty years . . . what's twenty minutes? Or fifty? And besides, you look beautiful, Suzanne. You're worth the wait."

I was flattered, and a little embarrassed. It had been a while since someone had paid me a compliment, even as a joke. But I liked it. And I eased smoothly into the evening like someone slipping into satin sheets.

"So, you're back on the Vineyard for good?" Matt asked after I told him some, but not all, of the events that had led up to my decision. I didn't tell him about the heart attack. I would, but not yet.

"I love it here. Always have. I feel like I've come home," I said. "Yes, I'm back here for good."

"How are your grandparents?" he asked. "I remember them both."

"My grandfather's still alive, and he's doing great. Grandmother died six years ago. Her heart."

Matt and I talked and talked — about work, summers on the Vineyard, college, our twenties, thirties, successes, disappointments. He had spent his twenties living all over the world: Positano, Madrid, London,

New York. He'd gotten into New York University Law School when he was twenty-eight, moved back to the Vineyard two years ago. Loved it. It felt so good to talk to him again; it was such a nice trip down memory lane.

After dinner Matt followed me home in his Jag. He was just being thoughtful. We both got out in the driveway and talked some more under a beautiful full moon. I was really enjoying myself.

He started to laugh. "Remember our first date?"

Actually, I did. There had been a wicked thunderstorm and it knocked out the electricity in my house. I had to get dressed in the dark. By mistake, I picked up a can of Lysol instead of hair spray. I smelled of disinfectant all night.

Matt grimaced and asked, "Do you remember the first time I got my nerve up to kiss you? Probably not. I was scared."

That surprised me a little. "I couldn't tell. As I remember it, you were always pretty confident."

"My *lips* were shaking, my teeth hitting together. I had the biggest crush on you. I wasn't the only one."

I laughed. This was silly, but it sure was fun. In a way, seeing Matt again was a fantasy come true. "I don't believe any of this, but I love hearing it."

"Suzanne, could I kiss you?" he asked in a gentle voice.

Now *I* was shaking a little. I was out of practice at this. "That would be okay. That would be good, actually."

Matt leaned over and, in the sweetest way, kissed me. A kiss, just one. But it was really something after all these years.

Dear Nicky,

Bizarre! That's the only word I can use to describe life sometimes. Just freaking bizarre.

Remember the housepainter I told you about? Well, he was over here the morning after my date with Matt, giving the joint a face-lift. I know this because he left me a bouquet of the most beautiful wildflowers.

There they were — pinks, reds, yellows, blues, and purples, sitting pretty in a mason jar by the front door.

Very sweet, very nice, and unexpectedly touching.

At first I thought they were from Matt, but damn it, they weren't.

There was also a note. *Dear Suzanne, The lights are still out in your kitchen, but I hope these will brighten your day some. Maybe we can get together sometime and do whatever you want to do, whenever you want to, wherever you want to.* He signed himself

Picasso — more readily known as your house-painter.

I was blown away. Until the night before, I hadn't had a date since I left Boston; I hadn't wanted to date since Michael Bernstein left me.

Anyway, I heard the painter-maintenance man hammering something somewhere, and I went outside. There he was, perched like a gull on the steep slanted roof.

"Picasso," I yelled, "thank you so much for the beautiful flowers. What a nice present. A nice thought."

"Oh, you're welcome. They just reminded me of you, and I couldn't resist."

"Well, you guessed right; they're all my favorites."

"What do you think, Suzanne? Maybe we could grab a bite sometime, go for a ride, catch a movie, play Scrabble. Did I leave anything out?"

I smiled in spite of myself.

"It's kind of a crazy time for me right now, with patients and all. I just have to make that a priority for the time being. But it was really nice of you to ask."

He took the rejection in stride. He smiled down at me. But then he ran his hand through his hair and said, "I understand. Of course you realize if you don't go out with

51

me just once, I'll have no choice but to raise your rates."

I called back to him, "No, I didn't know that."

"Yeah. It's absolutely despicable, a totally unfair business practice. But what can you do? It's the way of the world."

I laughed, and told him I'd take that under serious consideration. "Hey, by the way, what do I owe you for the extra work you've already done over the garage?" I asked.

"That? That's nothing . . . nothing at all. No charge."

I shrugged, smiled, waved. What he'd said was nice to hear — maybe because it *wasn't* the way of the world.

"Hey, thanks, Picasso."

"Hey, no problem, Suzanne."

And he resumed his task of putting a roof over my head.

Dear Nicholas,

I am watching over you as I write this, and you are absolutely gorgeous.

Sometimes I look at you and just can't believe you're mine. You have your father's chin, but you definitely have my smile.

There's a little toy that hangs over your crib and when you pull on it, it plays "Whistle a Happy Tune." This makes you laugh immediately. I think Daddy and I love to hear that song as much as you do.

Sometimes at night, if I'm driving home late or taking a walk, I'll hear that little melody in my head, and I'll feel such longing for you.

Right now, I just want to pick you up out of your sleep and hold you as close as I can.

The other thing that always makes you laugh is "One Potato, Two Potato." I don't know why. Maybe it's the sound of it, the silly lyrical bounce of the words. Maybe it's the part of you that's Irish. All I know is, the word *potato* can send you into fits and wiggles of happiness.

Sometimes I can't imagine your being any other age than the one you are this second. But I think all mothers tend to hold their children frozen in time, or maybe pressed like flowers, forever perfect, forever eternal. Sometimes when I rock you, I feel as if I were holding a little bit of heaven in my arms. I have a sense that there are protective angels all around you, all around us.

I believe in angels now. Just looking at you, sweet baby boy, I would have to.

I'm thinking about how much I loved you when you were in mommy's tummy. I loved you *the moment we met.* Seeing you for the first time, you looked right at Daddy and me. The look in your eyes said "Hey I'm here, hi!"

You were incredibly alert, checking everything out. Finally, Daddy and I could see you after nine months of imagining what you would be like. I took your head and pulled it gently to my chest. You were six pounds three ounces of sheer happiness.

After I held you, Daddy held you next. He couldn't believe how a baby, just minutes old, could be looking back at him.

Matt's little boy.

Our beautiful little Nicholas.

Katie

Matt's little boy.

Our beautiful little Nicholas.

Katie Wilkinson put down the diary, sighed, and took a deep breath. Her throat felt raw and sore. She ran her fingers through Guinevere's soft gray fur, and the cat purred gently. She blew her nose into a tissue. She hadn't been ready for this. She definitely hadn't been ready for Suzanne.

Or Nicholas.

And especially not Nicholas, Suzanne, and Matt.

"This is so crazy and so bad, Guinny," she said to the cat. "I've gotten myself into such a mess. God, what a disaster."

Katie got up and wandered around her apartment. She had always been so proud of it. She had done much of the work herself, and liked nothing better than to throw on a T-shirt, cutoffs, and work boots, then build and hang her own cabinets and bookcases. Her place was filled with authentic

57

antique pine, old hooked rugs, small water-colors like the one of the Pisgah Bridge, just south of Asheboro.

Her grandmother's jelly cabinet was in her study, and the interior planks still held the aroma of homemade molasses and jellies. Several vellum-paged, hand-sewn board books were displayed in the jelly cabinet. Katie had made them herself. She'd learned bookbinding at the Penland School of Crafts in North Carolina.

There was a phrase she loved, and also lived by — Hands to work, hearts to God.

She had so many questions right now, but no one to answer them. No, that wasn't completely true, was it? She had the diary.

Suzanne.

She liked her. Damn it, she liked Suzanne. She hadn't wanted to — but there it was. Under different circumstances they might have been friends. She *had* friends like Suzanne in New York and back home in North Carolina. Laurie, Robin, Susan, Gilda, Lynn — lots of really good friends.

Suzanne had been gutsy and brave to get out of Boston and move to Martha's Vineyard. She had chased her dream to be the kind of doctor, the kind of woman, she needed to be. She had learned from her near-fatal heart attack: she'd learned to treasure every moment as a *gift*.

And what about Matt? What had Katie meant to him? Was she just another woman in a doomed affair? God, she felt as if she should be wearing the Scarlet Letter. Suddenly she was ashamed. Her father used to ask her a question all the time when she was growing up: "Are you right with God, Katie?" She wasn't sure now. She didn't know if she was right with anyone. She had never felt that way before, and she didn't like it.

"Jerk," she whispered. "You creep. Not *you*, Guinevere. I'm talking about Matt! Damn him!"

Why didn't he just tell her the truth? Had he been cheating on his perfect wife? Why hadn't he wanted to talk about Suzanne? Or Nicholas?

How could she have allowed Matt to seal off his past from her? She hadn't pushed as much as she could have. Why? Because it wasn't her style to be pushy. Because she didn't like being pushed herself. She certainly didn't like confrontations.

But the most compelling reason had been the look in Matt's eyes whenever they started to talk about his past. There was such sadness — but also intimations of anger. And Matt had *sworn* to her that he was no longer married.

Katie kept remembering the horrible night Matt left her. She was still trying to make sense of it. Had she been a fool to trust *someone she thought she loved?*

On the night of July 18, she had prepared a special dinner. She was a good cook, though she seldom had the time to do this kind of elaborate affair. She'd set the wrought-iron table on her small terrace with her beautiful Royal Crown Derby china and her grandmother's silver. She'd bought a dozen roses, a mixture of red and white. She had Toni Braxton, Anita Baker, Whitney, and Eric Clapton on the CD player.

When Matt arrived, she had the best, the most wonderful surprise waiting for him. It was really great: the first copy of the book of poems he'd written, which she had edited at the publishing house where she worked. It had been a labor of love. She also gave him the news that the printing was 11,500 copies — very large for a collection of poems. "You're on your way. Don't forget your friends when you get to the top," she'd said.

Less than an hour later, Katie found herself in tears, shaking all over, and feeling as if she were living a horrible nightmare that couldn't possibly be real. Matt had barely come in the door when she knew something was wrong. She could see it in his eyes, hear it in the tone of his voice. Matt had finally told her, "Katie, I have to break this off. I can't see you again. I won't be coming to New York anymore. I know how awful that sounds, how unexpected. I'm sorry. I had to tell you in person. That's why I came here tonight."

No, he had no idea how awful it sounded, or was. Her heart was broken. It *still* was broken. She had trusted him. She'd left herself completely open to hurt. She'd never done that before.

And she had wanted to talk to him that night — she'd had important things to tell him.

Katie just never got the chance.

After he left her apartment, she opened a drawer in the antique dresser near the door leading to the terrace.

There was another present for Matt hidden inside.

A special present.

Katie held it in her hand, and she began to shake again. Her lips quivered, then her teeth started to chatter. She couldn't help herself, couldn't make it stop. She pulled away the wrapping paper and ribbon, and then she opened the small oblong box.

Oh, God!

Katie started to cry as she peered inside. The tears streamed from her eyes. The hurt she felt was almost unbearable.

She'd had something so important and so wonderful to share with Matt that night.

Inside the box was a beautiful silver baby rattle.

She was pregnant.

The Diary

Nicholas:

This is the rhythm of my life, and it is as regular and comforting as the Atlantic tides I see from the house. It is so natural, and good, and right. I know in my heart that this is where I am supposed to be.

I get up at six and take Gus for a long romp down past the Rowe farm. It opens to a field of ponies, which Gus regards with a certain laissez-faire. I think he believes they're giant golden retrievers. We eventually come out to a stretch of beach rimmed with eight- and ten-foot-high dunes and waving sea grass. Sometimes I wave back. I can be such a kook that it's embarrassing.

The route is somewhat varied, but usually we end up cutting through Mike Straw's property that has a lane of noble oaks. If it's hot or raining, the old trees act as a canopy. Gus seems to like this time of the day almost as much as I do.

What I especially like about the walks is the peaceful, easy feeling I have inside. I think a lot of it is due to the fact that I've taken back my life, reclaimed myself.

Remember the five balls, Nicky — always remember the five balls.

That is my exact thought as I start down the long road that leads home.

Just before I turn in to my driveway, I pass the Bone house next door. Melanie Bone was amazingly gracious and generous when I first moved in, supplying me with everything from helpful phone numbers to hammers, nails, paint, use of her phone, and cold, tangy lemonade, depending on the requirement. In fact, that's how I got my housepainter's number. Melanie recommended Picasso to me.

She is my age and already has four kids, God love her. I'm always in awe of anybody who can do that. All mothers are amazing. Just keeping extracurricular activities straight is like trying to run Camp Kippewa. Melanie is small, just a little over five feet, with jet black hair, and the loveliest, most welcoming smile.

Did I mention that the Bone kids are all girls? Ages one through four! I've always been bad with names, so I keep them organized by calling them by their ages. "Is Two sleeping?" "Is that Four outside on the swings?" "I think this will fit Three."

The Bones all giggle when I do this, and they think it's so silly, they've inducted Gus as honorary number five. Lord, if anyone ever overheard my system, they'd never come to see Dr. Bedford.

But they do come, Nicky, and I heal, and I am healing myself.

Now listen to what happens next. I had another date with Matt. I was invited to a party at his house.

My little man,

The house outside Vineyard Haven was beautiful, tasteful, impressive, and very expensive. I couldn't help but be impressed. As I looked around, the men and the women, even the children, arranged themselves into one demographic group: successful. It was Matt's world. It was as if the whole Upper East and West Sides of Manhattan, some smatterings of TriBeCa, and all of SoHo had been transplanted to the Vineyard. Party-goers were spread across the decks, the stone walkways, and the various gorgeously furnished rooms that opened to endless views of the sea.

The house was definitely not *me*, but I could still appreciate its beauty, even the love that had gone into making it what it was.

Matt took my arm and introduced me to his friends. Still, I felt out of place. I don't know exactly why. I had attended more than my share of events like this in Boston. Ribbon cuttings for new hospital wings, large and small cocktail soirees, the endless invitations to whatever was newsworthy in Boston.

But I really felt uncomfortable, and I didn't want to tell Matt, to spoil the night for him. My recent stint on Martha's Vineyard had been more down-home. Growing vegetables, hanging shutters, waterproofing porch floors.

At one crazy point, I actually looked down to see if I'd gotten all the white paint off my hands before I came.

You know what it was like, Nick? Sometimes when we hang together, and it's just the two of us, I'll talk Nicky-talk with you. That's the special language of made-up words; strange, funny noises; and other indecipherable codes and signals that only the two of us understand.

Then an adult will come to the door — or we'll have to go out to the market for something — and I swear I *forget* how to talk like an adult.

That's how I felt at this party. I'd spent too much time in work boots and paint-stained overalls; I was out of sync. And I liked the new rhythm I was creating for myself. Easy, simple, uncomplicated.

As I floated through a pleasant-enough haze of witty small talk and clinking crystal glasses, a little voice, *a child's voice,* broke through to me.

A small boy came running up, crying. He was probably three or four. I didn't see a parent or a nanny anywhere.

"What happened?" I bent down and asked. "Are you okay, big guy?"

"I fell," he sobbed. "Look!" And when I looked down, sure enough, his knee had a nasty scrape. There was even a little blood.

"How'd he *know* you were a doctor, Suzanne?" Matt asked.

"Children know these things," I said. "I'll take him inside and clean his knee. This white dress is meant to be chic, but maybe it looked like a doctor's lab coat to him."

I put my hand out, and the little boy reached up and took it. He told me that his name was Jack Brandon. He was the son of George and Lillian Brandon, who were at the party. He explained, in a very grown-up way, how his nanny was sick and his parents had to bring him.

As he and I emerged from the screened back door, a concerned woman came up to me.

"What happened to my son?" she asked,

and actually seemed put out.

"Jack took a little fall. We were just going to find a Band-Aid," Matt said.

"It's not serious," I said. "Just a scratch. I'm Suzanne, by the way, Suzanne Bedford."

Jack's mother acknowledged my presence with a curt nod. When she tried to take Jack's hand, he turned unexpectedly and hugged my legs.

I could tell that the mother was annoyed. She turned to a friend, and I heard her say, "What the hell does she know? It's not like she's a doctor."

Nick — listen, watch closely now, this next part is magic. There is such a thing. *Believe me.*

One night after a very long day at my office, the intrepid country doctor decided to grab a bite to eat on her way home.

I was just too tired to deal with making something, or even deciding what to make. No, Harry's Hamburger would do me just fine. A burger and fries seemed perfect to end my day. I needed a little guilty pleasure.

I guess it was a little past eight when I strolled inside. I didn't notice him at first. He was sitting by the window, eating his dinner and reading a book.

In fact, I was halfway through my burger when I saw him. Picasso, my housepainter.

I'd had very little contact with him since he left me those beautiful wildflowers in the mason jar. Occasionally, I'd hear him fixing something on the roof as I was leaving for work, or catch him painting the house, but we seldom spoke more than a few words.

I got up to pay the check. I could have walked out without saying hello because his back was turned toward me, but that seemed rude, ungracious, and snobby on my part.

I stopped at his table and asked him how he was. He was surprised to see me and asked if I'd join him for a cup of coffee, dessert, anything. It was his treat.

I gave him a lame excuse, saying I had to get home to Gus, but he was already clearing a spot for me and I just sort of sat down in his booth by the window. I liked his voice — I hadn't noticed it before. I liked his eyes, too.

"What are you reading?" I asked, feeling awkward, maybe a little scared, wanting to keep the conversation going.

"Two things . . . Melville" — he held up *Moby Dick* — "and *Trout Fishing in America.* Just in case I don't catch the big one, I have a backup."

I laughed. Picasso was pretty smart, and funny. "*Moby Dick*, hmmm, is that your summer reading or a guilt hangover because you never finished it in school?"

"Both," he admitted. "It's one of those things that you have on your to-do list in life. The book just sits there looking at you, saying, 'I'm not going away till you read me.' This is the summer I'm getting all the classics out of the way so I can finally concentrate on cheap summer thrillers."

We talked for more than an hour that night, and the time just flew. Suddenly I noticed how dark it was outside.

I looked back at him. "I have to go. I start work early in the morning."

"Me, too," he said, and smiled. "My current boss is an absolute slave driver."

I laughed. "So I've heard."

I stood up at the table and for some goony reason, I shook his hand.

"Picasso," I said, "I don't even know your real name."

"It's Matthew," he said. "Matthew Harrison."

Your father.

The next time I saw Matt Harrison, he was floating high above the world, up on my roof. He was hammering shingles like a madman, definitely a good, very conscientious worker. It was a few days after we had talked at Harry's Hamburger.

"Hey, Picasso!" I yelled, this time feeling more relaxed and even happy to see him. "You want a cold drink or something?"

"Almost done here. I'll be down in a minute. I'd love something cold."

Five minutes later he entered the cottage, as brown as a burnished copper coin.

"How's it going up there where the sea-gulls play?" I asked.

He laughed. "Good and hot! Believe it or not, I'm almost done with your roof."

Damn. Just as I was starting to like having him around.

"How's it going down here?" Matt asked me, sliding into my porch rocker in his cutoff jeans and open denim shirt. The rocker went back and bumped the trellis.

"Pretty good," I said. "No tragic head-lines in the trenches today, which is always nice to report. Actually, I love my prac-tice."

Suddenly, behind Matt, the trellis broke away from its hinges and began to tumble toward us. We both leaped up simulta-neously. We managed to press the white wooden frame back into place, our heads covered with rose petals and clematis.

I began laughing as I looked over at my handyman. He looked like a bridesmaid gone wrong. He immediately responded by saying, "Oh, and you don't look like Carmen Miranda yourself?"

Matt got a hammer and nails and re-secured the trellis. My only job was to hold it steady.

I felt his strong, very solid leg brush against mine, then I could feel his chest press against my back as he hovered over me, banging in the last nail.

I shivered. *Had he done that on purpose? What was going on here?*

Our eyes met and there was a flash of something bordering on the significant be-tween us. Whatever it was, I liked it.

Impulsively, or maybe instinctively, I asked him if he'd like to stay for dinner. "Nothing fancy. I'll throw some steaks and corn on the grill . . . like that."

He hesitated, and I wondered whether there was someone else. He certainly was good-looking enough. But my insecurity evaporated when he said, "I'm kind of grubby, Suzanne. Would you mind if I took a shower? I'd love to stay for dinner."

"There are clean towels under the sink," I told him.

And so he went to wash up and I went to make dinner. It had a nice feeling to it. Regular, simple, neighborly.

That's when I realized I didn't have any steak or corn. Fortunately, Matt never knew that I ran over to Melanie's for food . . . and that she threw in wine, candles, even half a cherry pie for dessert. She also told me that she adored Matt, that everyone did, and *good for you.*

After dinner the two of us sat talking on the front porch for a long while. The time flew again, and when I looked at my watch, I saw that it was almost eleven. I couldn't believe it.

"Tomorrow's a hospital day for me," I said. "I have early rounds."

"I'd like to reciprocate," Matt said. "Take you to dinner tomorrow? May I, Suzanne?"

I couldn't take my eyes away from his.

Matt's eyes were this incredibly gentle brown. "Yes, you absolutely may take me to dinner. I can't wait," I said. It just came out.

He laughed. "You don't have to *wait*. I'm still here, Suzanne."

"I know, and I like it, but I still can't wait for tomorrow. Good night, Matt."

He leaned forward, lightly kissed my lips, and then went home.

As it always has in my life — so far, anyway — tomorrow finally arrived. It came with Gus. Every morning he goes out to the porch and fetches the *Boston Globe*. What a retriever; what a pal!

Picasso took me around the island in his beat-up Chevy truck that afternoon, and I saw it as I never had before. I felt like a tourist. Martha's Vineyard was full of picturesque nooks and crannies and stunning views that continually surprised and delighted me.

We ended up at the lovely, multicolored Gay Head Cliffs. Matt reminded me that Tashtego in *Moby Dick* was a native harpooner and a Gay Head Indian. I guess I'd forgotten.

A couple of days later, after he'd finished some work in the house, we went for another ride.

Two days later we went out to Chappaquiddick Island. There was a tiny sign on the beach: PLEASE DON'T DISTURB, NOT EVEN THE CLAMS OR SCALLOPS. Nice. We didn't disturb anything.

I know this might sound silly, or worse, but I liked just being in the car with Matt. I looked at him and thought, *Hey I'm with this guy and he's very nice. We're out looking for an adventure.* I hadn't felt like that in a long time. I missed it.

It was at that very moment Matt turned and asked me what I was thinking about.

"Nothing. Just catching the sights," I said. I felt as if I'd just been caught doing something I wasn't supposed to.

He persisted. "If I guess right, will you tell me?"

"Sure."

"If I guess right," he said, and grinned, "then we get to have another date. Maybe even tomorrow night."

"And if you guess wrong, then we never see each other again. Big stakes riding on this."

He laughed. "Remember, I'm still painting your house, Suzanne."

"You wouldn't screw up the paint job to get even?"

Matt pretended to be offended. "I'm an artist. Picasso."

He paused before winking at me, and then

nailed his guess. "You were thinking about *us*."

I couldn't even bluff, though I did blush like crazy. "Maybe I was."

"Yes!" he shouted, and raised both arms in triumph. "And so?"

"So keep your hands on the steering wheel. And so what else?"

"So what would you like to do tomorrow?"

I started to laugh, and realized I did that a lot around him. "Boy, I have no idea. I was going to give Gus a badly needed bath, do some food shopping, maybe rent a movie. I was thinking, *The Prince of Tides*."

"Sounds great, sounds perfect. I loved Pat Conroy's book, all his stuff. Never got around to the movie. Afraid they'd mess it up. If you want some company I'd love to tag along."

I had to admit, it was great fun being with Matt. He was the polar opposite of my former Boston boyfriend, Michael Bernstein, who never seemed to do anything without a logical reason, never took a day off, probably never turned down a pretty winding road just because it was there.

Matt couldn't have been more different. He seemed to take an interest in just about everything on the planet: he was a gardener, bird-watcher, avid reader, pretty good cook, basketball player, crossword-puzzle cham-

81

pion, and, of course, he was very handy around the house.

I remember looking down at my watch at one point during our ride. But I wasn't doing it because I wanted our date to be over; I was doing it because I wanted it *not* to be over. I felt so damn happy that day. Just taking a ride with him, going absolutely nowhere.

I breathed in everything around me: the sea grass, the minty blue sky, the beach, the roaring ocean. But mostly I breathed in Matthew Harrison. His freshly laundered plaid flannel shirt, his jeans, his glistening rose-brown skin, his longish brown hair.

I breathed Matt in, held him there, and never wanted to exhale. Something very nice was happening.

Now, you may be wondering about Matt Wolfe, the lawyer? Well, I called Matt several times, but all I ever got was his answering machine, and then he never called me back. It is a small island, though, so *maybe he knew.*

Nicky,

I saw Matt Harrison every day for the next two weeks. I almost couldn't believe it. I pinched myself a lot. I smiled when no one was around.

"Have you ever ridden a horse, Suzanne?" Matt asked me on Saturday morning. "This is a serious question."

"I reckon. When I was a kid," I said with a light cowgirl drawl.

"A perfect answer — because you're about to be a kid again. Right now, today. By the way, have you ever ridden a sky blue horse that has red stripes and gold hooves?"

I looked at Matt, then shook my head. "I'd remember if I had."

"I know where there's a horse like that," he said. "In fact, I know where there are lots of them."

We drove up to Oak Bluffs, and there they were. God, what a sight.

Dozens of brightly painted stallions stood in a circle beneath the most dazzling jigsaw ceiling I'd ever seen. Hand-carved horses with flared red nostrils and black glass eyes galloped in their tireless tracks in a circle of joy.

Matthew had brought me to the Flying Horses, the oldest carousel in the country. It was still open for business, for kids of all ages.

We climbed aboard as the platform tilted and rotated beneath us, and we found perfect steeds.

As the music began, I clenched the silver horse rod, rising and falling, rising and falling. I fell under the carousel's spinning spell. Matt reached out to hold my hand and even tried to catch a kiss, which he succeeded at admirably. What a horseman!

"Where did you learn to ride like that, cowboy?" I asked as we rode up and down, but also around and around.

"Oh, I've ridden for years," Matt said. "Took lessons here when I was three. You see that blue stallion up ahead? Blue the color of the sky? Wild-blue-yonder blue?"

"Reckon I do."

"He threw me a couple of times. Man, did I take a nasty spill or two. That's why I wanted to make sure you got National

Velvet first time out. She's got an even temper, lovely coat of shellac."

"She's beautiful, Matt. You know, when I was a kid I did ride some. It's all coming back to me. I used to go riding with my grandfather out in Goshen, New York. Funny I should remember that now."

Good memories are like charms, Nicky. Each is special. You collect them, one by one, until one day you look back and discover they make a long, colorful bracelet.

By the end of that day, I would have my first in a series of beautiful charms about Matthew Harrison.

Katie

Katie would never forget the very first time she saw Matt Harrison. It was in her small, comfortable office at the publishing house, and she had been looking forward to the meeting for days. She had loved *Songs of a Housepainter*, which seemed to her like the most memorable short stories, quite magical, condensed into powerful, very moving poems. He wrote about everyday life — tending a garden, painting a house, burying a beloved dog, having a child — but his *choice* of words distilled life so perfectly. She was still amazed that she had discovered his work.

And then he walked through the door of her office, and she was even more amazed. No, make that *entranced*. The most primitive parts of her brain and nervous system locked on to the image before her — the poet, *the man*. Katie felt her heart skip a beat, and she thought, *My, my. Careful, careful.*

He was taller than she was — she guessed about six foot two. He had a good nose and strong-looking chin, and everything about his face held together extremely well, like one of his poems. His hair was longish, sandy brown, clean and lustrous. He had a deep working-man's tan. He smiled at something, hopefully not her height or her gawkiness or the goofy look on her face — but she liked him, anyway. What was there not to like?

They had dinner that night, and he gallantly let her buy. He did insist on picking up the tab for a couple of glasses of expensive port a little later. Then they went to a jazz club on the Upper West Side, on a "school night" as Katie called her work nights. He finally dropped her off at her apartment at three in the morning, apologized profusely and sincerely, gave her the sweetest kiss on the cheek, and then off he went in a cab.

Katie stood on the front steps and was finally able to catch her breath, maybe for the first time since he had walked through the door of her office. She tried to remember . . . *was Matthew Harrison married?*

He was back in her office the following morning — to work — but the two of them skedaddled off to lunch at noon and didn't return for the rest of the day. They went museum hopping, and he certainly knew his art. He didn't show off, but he easily knew as much as Katie did. She kept thinking — who *is* this guy? And why am I allowing myself to feel the way I'm feeling?

And then — *why am I not trying to feel like this all the time?*

He came up to her place that night, and she continued to be astonished that any of this was happening. Katie was infamous with her friends for *not* sleeping around, for being too romantic, and way too old-fashioned about sex; but here she was with this good-looking, undeniably sexy, housepainter-poet from Martha's Vineyard, and she couldn't *not* be with him. He never, ever hustled her — in fact, he seemed almost as surprised about being in her apartment as she was that he was there.

"Hummuna, hummuna," Katie said, and they both laughed nervously.

"My sentiments exactly," Matt said.

They went to bed for the first time on that rainy night, and he made her notice the music of the raindrops as they fell on her street, the rooftop, and even the trees outside her apartment. It was beautiful, it was music; but soon they had forgotten the patter of the rain, and everything else, except for the urgent touch of each other.

He was so natural and easy and good in

bed that it scared Katie a little. It was as if he had known her for a long time. He knew how to hold her, how and where to touch her, how to wait, and then when to let everything on the inside explode. She loved the way he touched her, the gentle way he kissed her lips, her cheeks, the hollow of her throat, her back, breasts — well, everywhere.

"You're absolutely ravishing, and you don't know it, do you?" he whispered to her, then smiled. "You have the most delicate body. Your eyes are gorgeous. And I *love* your braid."

"You and my mother," Katie said. She loosened the braid and let her long hair cascade over her shoulders.

"Hmmm. I love that look, too," Matt said, and winked at her.

When he finally left her apartment the next morning, Katie had the feeling that she had never been with anyone like that, never experienced such intimacy with another person. *My God, why not?* she asked herself.

She kind of missed Matt already. It was insane, completely ridiculous, not *her;* but she *did* miss him. *My God, why not?*

When she got to her office that morning, he was already there, waiting for her. Her heart nearly stopped.

"We'd better do some work," she said.

"Seriously, Matthew."

He didn't say a word, just shut her office door and kissed her until Katie felt as if she were melting into the hardwood floor.

He finally pulled away, looked into her eyes again, and said, "As soon as I left your place, *I missed you*."

The Diary

Nicholas,

I remember all of this as if it happened yesterday. It's still vibrant and alive. Matt and I were riding on the Edgartown–Vineyard Haven road in my Jeep. Gus went along for the ride. He sat on the backseat and looked like one of the lions that guard the front of the New York Public Library.

"Can't you drive any faster?" Matt asked, tapping his fingers on the dashboard. "I walk faster than this."

I am by my own admission a slow and careful driver. Matt had found my first flaw.

"Hey I got the safety-first award in my driver's ed class in Cornwall on Hudson. I hung the diploma under my medical degree."

Matt laughed and rolled his brown eyes. He got all of my dumb little jokes.

We were driving to his mother's house. Matt thought it would be interesting for me to meet her.

Interesting? What did that mean?

"Oops, there's my mom!" Matt said just then. "Oh, man. There she is."

She was up on the roof of the house when we got there. She was fixing an ancient TV antenna. We got out of my old blue Jeep, and Matt called up to her.

"Mom, this is Suzanne. And Gus the Wonder Dog. Suzanne . . . my mother, Jean. She taught me how to fix things around the house."

His mother was tall, lanky, silver-haired. She called down to us, "Very nice to meet you, Suzanne. You, too, Gus. You three go have a seat on the porch. I'll only be a minute up here."

"If you don't fall off the roof and break both your legs," Matt said. "Fortunately, we have a good doctor in the house."

"I won't fall off the roof." Jean laughed, and went back to her work. "I only fall off extension ladders."

Matt and I took our seats at a wrought-iron table on the porch. Gus preferred the front yard. The house was an old saltbox with a northern view of the harbor. To the south lay cornfields, and then deep woods that gave you the impression you were in Maine.

"It's gorgeous here. Is this where you grew up?" I asked.

98

"No, I was born in Edgartown. This house was bought a few years after my father died."

"I'm sorry, Matt."

He shrugged. "It's another thing we have in common, I guess."

"So why didn't you tell me?" I asked him.

He smiled. "You know, I guess I just don't like to talk a lot about sad things. Now you know *my* flaw. What good does it do to talk about sad things in the past?"

Jean suddenly appeared with iced tea and a plate heaped with chocolate-chip cookies.

"Well, I promise I won't give you the once-over, Suzanne. We're too mature for that sort of thing," she said with a quick wink. "I would love to hear about your practice, though. Matthew's father was a doctor, you know."

I looked over at him. Matt hadn't told me that, either. "My dad died when I was eight years old. I don't remember too much."

"He's private about some things, Suzanne. Matthew was hurt badly when his dad died. Don't listen to him on that. I think he believes it might make other people uncomfortable to hear about how much he hurts."

She winked at Matt; he winked back at her. I could tell they were close. It was nice to see. Sweet.

"So, tell me about yourself, Jean. Unless

you're a private person, too."

"Hell, no!" she said with a laugh. "I'm an open book. What do you want to know?"

It turned out that Jean was a local artist — a painter. She walked me through the cottage and showed me some of her work. She was good, too. I knew enough to be fairly sure that her paintings could have sold at a lot of galleries in Back Bay, or even New York. Jean had framed a quote from the primitive artist Grandma Moses. It said, "I paint from the top down. From the sky, then the mountains, then the hills, then the cattle, and then the people."

Jean laughed at my praise of her work and said, "I once saw a cartoon with a couple standing before a Jackson Pollock painting. The painting had a price tag of a million dollars under it, and the man turned to the woman and said, 'Well, he comes through clear enough on the price.' " She had a good sense of humor about her work, about anything really. I saw a lot of her in Matt.

The afternoon turned into evening, and Matt and I ended up staying for dinner. There was even time to see a priceless old album of some of Matt's baby pictures.

He was a *cutie,* Nick. He had your blond hair as a boy and that spunky look you have sometimes.

"No naked bottoms on bear rugs?" I asked Jean as I went through the pictures.

She laughed. "Look hard enough, and I'm sure you'll find one. He has a nice butt. If you haven't seen it, you should ask for a look."

I laughed. Jean was a hoot.

"All right," Matt said, "show's over. Time to hit the highway."

"We were just getting into the good stuff," Jean said, and made a pouty face. "You are a party pooper."

It was about eleven when we finally got up to leave. Jean grabbed me in a hug.

She whispered against my cheek, "He never ever brings anybody home. So whatever you think of him, he must like you a lot. Please don't hurt him. He *is* sensitive, Suzanne. And he's a pretty good guy."

"Hey!" Matt finally called from the car. "Knock it off, you two."

"Too late," his mother said. "The damage is already done. I had to spill the beans. Suzanne knows enough to drop you like a bad habit."

The damage was probably already done — to me. I was falling for Matthew Harrison. I couldn't quite believe it myself, but it was happening, if it hadn't already happened.

The Hot Tin Roof is a fun nightclub at the Martha's Vineyard Airport in Edgartown. Matt and I went there to eat oysters and listen to the blues on Friday night. At that point, I would have gone anywhere with him.

A host of local celebrities floated in and out of the bar: funky, laid-back Carly Simon, Tom Paxton, William Styron and his wife, Rose. Matt thought it would be fun to sit at the raw bar and just people watch. It was, too.

"Want to slow dance?" Matt asked me after we'd had our fill of oysters and cold beer.

"Dance? No one is dancing, Matt. I don't think this is a dancing-type place."

"This is my favorite song, and I'd love to

dance with you. Will you dance with me, Suzanne?"

I did something I do infrequently. I blushed.

"Come on," Matt whispered against my cheek. "No one will tell the other doctors at the hospital."

"All right. One dance."

"Done well, one dance will always lead to another," he said.

We began to slow dance in our little corner of the bar. Eyes started to turn our way. What was I doing? What had happened to me? Whatever it was, it felt so good to be doing it.

"Is this okay?" Matt checked.

"You know, actually it's great. What is this song, anyway? You said it was your favorite."

"Oh, I have no idea, Suzanne. I just wanted an excuse to hold you close."

With that, Matt held me a little tighter. I loved being in his arms. I loved, loved, loved it. Corny maybe, but absolutely true. What can I say? I felt a little dizzy as we spun around in rhythm with the music.

"I have a question to ask you," he whispered against the side of my ear.

"Okay," I whispered back.

"How do you feel about us? So far?"

I kissed him. "Like that."

He smiled. "That's how I feel, too."

"Good."

"I lived with somebody for three years," Matt said. "We met while we were at Brown. The Vineyard wasn't right for her, but it was for me."

"Four years. Another doctor," I confessed.

Matt leaned in and lightly kissed me on the lips again. "Would you come home with me tonight, Suzanne?" he asked. "I want to do some more dancing."

I told him I would love to.

I have this wink that Matt calls "Suzanne's famous wink." I did it for the first time to Matt that night. He loved it.

Matt's house was a small Victorian covered in gingerbread lace that draped itself over the eaves and softened all the corners. The trellises, railings, and overhangs looked as if they'd been lifted off some elaborately trimmed wedding cake and carefully placed around the rim of the roof.

It was the first time I had been invited, and I was suddenly nervous. My mouth was cottony and dry. I hadn't been with anyone like this since Michael, and that was still a bad memory for me.

We went inside and I immediately noticed a library. The room had been remodeled to be made up of nothing but shelves. There were thousands of books in there. My eyes traveled up and down the bookshelves: Scott Fitzgerald, John Cheever, Virginia Woolf, Anaïs Nin, Thomas Merton, Doris Lessing. An entire wall was devoted to collections of poetry. W. H. Auden, Wallace Stevens, Hart Crane, Sylvia Plath, James

Wright, Elizabeth Bishop, Robert Hayden, and many, many more. There was an antique globe; an old English pond boat, its sails stained and listing; some nautical brass fittings; a big pine table covered in writing pads and miscellaneous papers.

"I love this room. Can I look around?" I asked.

"I love it, too. Of course you can look."

I was totally surprised by the cover page on top of a stack of pages. It read, *Songs of a Housepainter*, Poems by Matthew Harrison.

Matt was a poet? He hadn't told me about it. He really didn't like to talk about himself, did he? What other secrets did he have?

"Okay, yes," he admitted quietly. "I do some scribbling. That's all it is. I've had the bug since I was sixteen, and I've been trying to work it out since I left Brown. I majored in English and Housepainting. Just kidding. You ever write, Suzanne?"

"No, not really," I said. "But I've been thinking about starting a diary."

In the south of France there is supposedly a special time known as the Night of the Falling Stars. On this night, everything is just so. Perfect and magical. According to the French, the stars seem to pour out of the sky, like cream from a pitcher.

It was like that for us; there were so many stars, I could imagine I was up in heaven.

Matt said, "Let's take a walk down to the beach. Okay? I have an idea."

"I've noticed that you have a lot of ideas."

"Maybe it's the poet in me."

He grabbed an old blanket, his CD player, and a bottle of champagne. We walked on a winding path through high sea grass, finally finding a patch of sand to spread the blanket.

Matt popped open the champagne, and it sparkled and blinked in the midnight air. Then he pushed PLAY and the strains of Debussy whirled up into the starry night sky.

Matt and I danced again, and we were in another time and place. Around and around we went, in sync with the rhythm of the sea, turning up fountains of sand, leaving improbable footprint patterns in our wake. I let my fingers play on his back, his neck. I let my hands comb through his hair.

"I didn't know you could waltz," I said.

He laughed. "I didn't know, either."

It was late when we made our way back up from the beach, but I wasn't tired. If anything, I was more awake than ever. I was still dancing, flying, singing inside. I hadn't expected any of this to happen. Not now, maybe not ever. It seemed a thousand years from my heart attack in the Public Garden in Boston.

Nicky, I felt so lucky — so blessed.

Matt gently took my hand and led me up the stairs to his room. I wanted to go with him, but still I was afraid. I hadn't done this in a while.

Neither of us spoke, but suddenly my mouth opened wide. He had converted the top floor to one big, beautiful space, complete with skylights that seemed to absorb the evening sky. I loved what he had done to the room. He turned on the CD player in the bedroom.

Sarah Vaughan. Perfect.

Matt told me that he could count falling stars from his bed. "One night I counted sixteen. A personal record."

He came to me, slowly and deliberately, drawing me toward him like a magnet. I could feel the buttons in the back of my blouse coming undone. The little hairs on the back of my neck stood on end. His fingers traveled down to the base of my spine, playing so very gently. He slipped off my blouse, and I watched it float to the floor, milkweed in a breeze.

I stood so close to him, felt so close to Matt, barely breathing, feeling light, dizzy, magical, and very special.

He slipped his hands down onto my hips. Matt then leaned me back, gently laid me on his bed. I watched him in the moon shadows. I found him to be beautiful. *How had this happened? Why was I suddenly so lucky?*

He stretched over me like a quilt on a cold night. That's all I will say of it, all I will write.

Dear Nicky,

I hope when you grow up that everything you want comes your way, but especially love. When it's true, when it's right, love can give you the kind of joy that you can't get from any other experience. I have been in love; I *am* in love, so I speak from experience. I have also lived long periods without love in my life, and there is no way to describe the difference between the two.

We is always so much better than *I*.

Please don't listen to anyone who tells you otherwise. And don't ever become a cynic, Nicky. Anything but that!

I look at your little hands and feet. I count your toes over and over, moving them gently as if they were beads on an abacus. I kiss your belly till you laugh. You are so innocent. Stay that way when it comes to love.

Just look at you. How is it that I got so lucky? I got the perfect one. Your nose and mouth are just right. Your eyes and your smile are your very best features. Already I see your personality blossoming. It's in your eyes. What are you thinking about right now? The mobile over your head? Your music box? Daddy says you're probably thinking about girls and tools and flashy cars. He jokes that your favorite things are flashy cars, pretty girls, and birthday cake. "He's a real boy, Suzanne."

That's true, and it's probably a good thing. But do you know what you like the best? Teddy bears. You're so gentle and sweet with your little bears.

Daddy and I laugh about all the good things that wait for you. But what we want most for you is love and that it will always surround you. It is a gift. If I can, I will try to teach you how to receive such a gift. Because to be without love is to be without grace, what matters most in life.

We is so much better than *I*.

If you need proof, just look at *us*.

"It's Matt. Hi. Hello? Anybody home? Suzanne? You here?"

The banging at my kitchen door was persistent and annoying, like an unexpected visit from an out-of-town relative. I went to the door, opened it, and then stopped, my mouth open in a little circle of surprise.

It was Matt, all right, but not Matt Harrison.

My visitor was Matt Wolfe.

Behind him in my driveway, I could see his glistening green Jag convertible.

Where had he been? He *still* hadn't returned any of my calls.

"Hi," he said. "God, you look good, Suzanne. You look great, actually." He leaned in and I let him give me a kiss on the cheek.

I had no reason to feel guilty — but I did, anyway. "Matt. How are you? I just made some sun tea. Come on in."

And he did, finding a comfy, sunny place in the kitchen, leaning into what looked like

112

a catch-up mode. We definitely had some catching up to do, didn't we?

"I've been out of town for most of the month, Suzanne. I kept meaning to call, but I was in the middle of a legal fiasco. Unfortunately, it was in Thailand."

Suddenly he smiled. "And you know — blah, blah, blah, yadda, yadda, yah. So how have you been? Obviously, you got some sun. You look fantastic."

"Well, thanks . . . so do you."

I had to tell him. I even decided to give Matt Wolfe the long version of what had been happening in my life.

He listened, smiling at some parts, fidgeting nervously at others. I could tell his acceptance was somewhat bittersweet. But he kept listening intently, and when I was done, he got off the kitchen stool, put his arms around me, and gave me a hug.

"Suzanne," he said, "I'm happy for you."

He smiled bravely. "I knew in my gut I shouldn't have gone away. Now the best thing that could have happened to me has slipped through my fingers again."

I found myself laughing. I was starting to notice that Matt Wolfe was a little bit of a con man. "Oh, Matt, your flattery is so sweet. Thanks for being a friend. Thanks for being *you*."

"Hey, if I'm gonna lose the big prize, I'm going down with a little dignity. But I'm telling you, Suzie, if this guy flinches, or if I sense a crack in the dam, I'm coming back."

We both laughed, and I walked him out to the Jag. Somehow, I just knew that Matt would be all right. I doubted he'd been all by his lonesome in Thailand. And let's face it, he hadn't called in nearly a month.

I watched Matt get into his car, his pride and joy.

"I actually think you two will get along. In fact, I think the two Matts will like each other a lot," I called from the porch.

"Oh, great! Now I have to like the guy, too?" he called back.

The last thing I heard him say before he fired the convertible's powerful engine was, "He does know how to duel, doesn't he?"

"Okay What's going on? Spill the beans, Suzanne. I want the scoop. I know there's something going on with you," said my neighbor and friend Melanie Bone. "I feel it in my *bones*."

She was right. I hadn't told her how Matt and I were progressing, but she could read my face and maybe even tell by the spring in my step.

We were walking along the beach near our houses, the kids and Gus romping in front of us.

"You're smart," I told her. "And nosy."

"I know that already. So tell me what I don't know. *Spill*."

I couldn't resist any longer. It had to come out sooner or later. "I'm in love, Mel. This has never happened before. I'm head over heels in love with Matt Harrison. I have no idea what's to become of us!"

She screeched. Then Melanie jumped up and down a few times in the sand. She was so cute, and such a good friend. She screeched again.

"That is so perfect, Suzanne. I knew he was a good painter, but I had no idea about his other talents."

"Did you know he's a poet? A very good poet."

"No, you're kidding," she said.

"A beautiful dancer?"

"That doesn't surprise me. He moves pretty well on rooftops. So, how did this happen? I mean, how did it go from adding a touch of Cape Cod white to your house to *this?*"

I started to giggle and felt like a schoolgirl. After all, things like this didn't happen to grown women.

"I talked to him one night at the hamburger place."

Melanie arched an eyebrow. "*Okay.* You talked to him at the hamburger place?"

"I can talk to Matt about anything, Melanie. I've never had that happen with any man before. He even writes poems the way he talks. It's very down-to-earth and at the same time, sometimes over your head. He's passionate, exciting. He's humble, too. Maybe more than he should be sometimes."

Melanie suddenly gave me hug. "God, Suzanne, this is it! As IT as it can get. Congratulations, you're *gone for good*."

We laughed like a couple of giddy fifteen-year-olds, and headed back with Melanie's kids and Gus. That morning at her house, we talked nonstop about everything from first dates to first pregnancies. Melanie confessed that she was thinking of having a fifth baby, which blew me away. For her it was as easy as organizing a cabinet. She had her life as under control as a grocery shelf lined neatly with canned goods. Orderly, alphabetized, well stocked.

I also fantasized about having kids that morning, Nicholas. I knew I would have a high-risk pregnancy because of my heart condition, but I didn't care. Maybe there was something in me that knew you'd be here one day. A flutter of hope. A deep desire. Or just the sheer inevitability of what love between two people can bring.

You — it brought you.

Bad stuff happens, Nicholas. Sometimes it makes no sense at all. Sometimes it's unfair. Sometimes it just plain sucks.

The red pickup came tearing around the corner, going close to sixty, but the whole thing seemed to happen in slow motion.

Gus was crossing the street, heading toward the beach, where he likes to race the surf and bark at seagulls. Bad timing.

I saw the whole thing. I opened my mouth to stop him, but it was probably already too late.

The pickup swung around the blind curve like a blur. I could almost smell the rubber of the tires as they skidded along the hot tar, then I watched as the left front fender caught Gus.

A second more, and he would have cleared that unforgiving metal.

Five miles an hour slower, and the pickup would have missed him.

Or maybe if Gus had been a couple of

years younger, closer to his prime, it wouldn't have happened.

The timing was nightmarish, irrevocable, like a rock falling on the windshield of a passing car.

It was over, done, and Gus lay like a rag discarded by the side of the road. It was so sad. He'd been so defenseless, so carefree just seconds before as he romped toward the water.

"No!" I yelled. The truck had stopped, and two stubble-faced men in their twenties got out. They both wore colorful bandannas on their heads. They stared at what their speeding vehicle had done.

"Gee, I'm sorry, I didn't see him," the driver stammered, and hitched at his blue jeans as he looked at poor Gus.

I didn't have time to think, to argue, to yell at him. The only thing I needed to do was to get Gus help.

I threw the driver my keys. "Open the back of my Jeep," I snapped as I gently lifted Gus up into my arms. He was limp and heavy, but still breathing, still *Gus*.

I laid him in the back of the Jeep, bloody and tenuous. His sweet, familiar eyes were as far away as the clouds. Then he focused on me. Gus whimpered pitifully, and my heart broke into a hundred pieces.

"Don't die, Gus," I whispered. "Hold on, boy," I said as I pulled out of my driveway. "Please don't leave me."

I called Matt on my cell phone, and he met me at the vet's. Dr. Pugatch took Gus in at once, maybe because she saw the look of desperation on my face.

"The truck was going way too fast, Matt," I told him. I was reliving the scene again, and I could see every detail. Matt was even angrier than me.

"It's that damn curve. Every time you pull out, I worry about it. I need to lay you a new driveway on the other side of the house. That way you'll be able to see the road."

"This is so horrible. Gus was right there when I —" I stopped myself. I still hadn't told Matt about my heart attack. Gus knew, but Matt didn't.

I had to tell him soon.

"Shhhh, it's okay, Suzanne. It's going to be okay." Matt held me, and though it wasn't okay, it was as good as it could possibly be. I burrowed into his chest and stayed there. Then I could feel Matt shaking a little. He and Gus had become close, too. Matt had unofficially taken over most of the ball playing with Gus.

Two hours later the vet came out. It seemed like an eternity before she spoke.

Now I knew how my own patients must feel when I hesitate or am at a loss for words. Their faces seem calm but their bodies convey something else. They beg to be relieved of their anxiety with good news, *only good news.*

"Suzanne, Matt . . . ," Dr. Pugatch finally said. "I'm sorry. I'm so, so sorry. Gus didn't make it."

I began to cry, and my whole body was shaking uncontrollably. Gus had always been there with me, for me. He was my good buddy, my roommate, my jogging partner, my confidant. We had been together for fourteen years.

Bad stuff does happen sometimes, Nicholas.

Always remember that, but remember that you have to move on, somehow.

You just pick your head up and stare at something beautiful like the sky, or the ocean, and you move the hell on.

Nicholas,

An unexpected letter arrived in the mail for me the next day.

I don't know why I didn't rip it open and read it. I just stood there wondering why Matt Harrison had written me a letter when he could easily have picked up the phone or come over.

I stood at the end of the driveway in front of the weatherbeaten, off-white mailbox. I opened the letter carefully and held it tight so it wouldn't be blown away by the ocean wind.

Rather than try to paraphrase what the letter said, Nicky, I'm enclosing it in the diary.

Dear Suzanne,

You are the explosion of carnations
   in a dark room.
Or the unexpected scent of pine
   miles from Maine.

You are a full moon
that gives midnight its meaning.
And the explanation of water
For all living things.

You are a compass,
a sapphire,
a bookmark.
A rare coin,
a smooth stone,
a blue marble.

You are an old lore,
a small shell,
a saved silver dollar.
You are a fine quartz,
a feathered quill,
and a fob from a favorite watch.

You are a valentine
tattered and loved and reread a
    hundred times.
You are a medal found in the drawer
    of a once sung hero.
You are honey
and cinnamon
and West Indies spices,
lost from the boat
that was once Marco Polo's.

You are a pressed rose,
a pearl ring,
and a red perfume bottle found near
the Nile.

You are an old soul from an ancient
place,
a thousand years, and centuries and
millenniums ago.

And you have traveled all this way
just so I could love you.
I do.

Matt

What can I say, Nicholas, that your good, sweet father cannot say better? He is a stunningly good writer, and I'm not even sure he knows it.
I love him so much.
Who wouldn't?

Nicky,

I called Matt very early the next morning, as soon as I dared, about seven. I had been up since a little past four, thinking that I had to call him, even rehearsing what I should say and how I should say it. I don't really know how to be dishonest or manipulate people very well. It puts me at a great disadvantage sometimes.

This was hard.

This was impossible.

"Matt, hi. It's Suzanne. Hope I'm not calling too early. Can you come by tonight?" was all I could manage.

"Of course I can. In fact, I was going to call you and ask for a date."

Matt arrived at the house a little past seven that night. He was wearing a yellow plaid shirt and navy blue trousers — kind of formal for him.

"You want to take a walk on the beach, Suzanne? Take in the sunset with me?"

It was exactly what I wanted to do. He'd read my mind.

As soon as we crossed the beach road and had our bare feet in the still-warm sand, I said, "Can I talk? There's something I have to tell you."

He smiled. "Sure. I always like the sound of your voice."

Poor Matt. I doubted that he was going to like the sound of what was coming next.

"There's something I've wanted to tell you for a while. I keep putting it off. I'm not even sure how to broach the subject now."

He took my hand, swung it gently in rhythm with our strides. "Consider it broached. Go ahead, Suzanne."

"Why are you so dressed up tonight?" I thought to ask him.

"I'm dressed up because I have a date with the most special woman on this entire island. Is that the subject you had trouble broaching?"

I squeezed his hand a little. "Not exactly. No, it isn't. Okay, here goes."

Matt finally said, "You are scaring me a little now."

"Sorry," I whispered. "*Sorry.* Matt, right before I came to the Vineyard —"

"You had a heart attack," he said in the softest voice. "You almost died in the Public

126

Garden, but you didn't, thank God. And now, here we are, and I think we're two of the luckiest people. I know that *I* am. I'm here holding your hand, looking into your beautiful blue eyes."

I stopped walking and stared at Matt in disbelief. The setting sun was just over his shoulder, and it looked like a nimbus. Was Matthew an angel?

"How long have you known? *How* did you know?" I stammered.

"I heard before I came to work for you. This is a small island, Suzanne. I was half expecting some old biddy with a walker."

"I *did* use a walker for a couple of days in Boston. I had surgery. So you knew, but you never told me you knew."

"I didn't think it was my place. I knew that you'd tell me when you were ready. I guess you're ready, Suzanne. That's good news. I've been thinking about what happened to you a lot in the past few weeks. I even arrived at a point of view. Would you like to hear it?"

I held on to Matt's arm. "Of course I would."

"Well, I can't help thinking of this whenever we're together. I think, isn't it lucky that Suzanne didn't die in Boston and we have today to be together. Now we get to

watch this sunset. Or isn't it lucky Suzanne
didn't die and we're sitting out on her front
porch playing hearts or watching a stupid
Red Sox game. Or listening to Mozart or
even that smarmy love song you like by
Savage Garden. I keep thinking, isn't this
day, this moment, incredibly special, because you're *here*, Suzanne."

I started to cry, and that's when Matt took
me into his arms. We cuddled on the beach
for a long time, and I never wanted him to
let me go. Never ever. We fit together so
well. I kept thinking, *Isn't this moment incredibly special? Aren't I the lucky one?*

"Suzanne?" I heard him whisper, and I
felt Matt's warm breath on my cheek.

"I'm here. Hard to miss. I'm right here in
your arms. I'm not going anywhere."

"That's good. I want you to always be
there. I love having you in my arms. Now
there's something I have to say. Suzanne, I
love you so much. I treasure everything
about you. I miss you when we're apart for
just a couple of hours. Every day while I'm
working, I can't wait to see you that night.
I've been looking for you for a long time, I
just didn't know it. But now I do. Suzanne,
will you marry me?"

I pulled back and looked into the beautiful eyes of this precious man I had found

somehow, or maybe who had found me. I couldn't stop smiling, and the warm glow spreading inside me was the most incredible feeling.

"I love you, Matt. I've been looking for you for a long time, too. Yes, I'll marry you."

Katie

Katie closed the diary again.

She slammed it shut this time. It hurt her so much to read these pages. She could take them only in small doses. Matt had warned her in his letter that it might happen, and it had. *There will be parts that may be hard for you to read.* What an incredible understatement that was.

The diary continued to put her in a place of unexpected surprises. Now it was making her jealous, something she didn't think she was capable of. She *was* jealous of Suzanne. She kind of felt like a jerk, a small and petty person. Not herself. Maybe it was hormones. Or maybe it was just a normal reaction to everything abnormal that had happened to her recently.

She shut her eyes tight, and felt incredibly alone. She hugged herself with both arms. She needed to talk to someone besides Guinevere and Merlin. Ironically, the person she wanted to reach out to was on Martha's

Vineyard. As much as she wanted to, she wouldn't call him. She would call her friends Laurie or Gilda or Susan, but not Matt.

Her eyes moved over to the bookshelves she had built into her walls. Her apartment was like a small bookstore. Very independent. *Orlando, The Age of Innocence, Bella Tuscany, Harry Potter and the Goblet of Fire, The God of Small Things.* She'd been reading voraciously ever since she was seven or eight. She read everything, anything.

She was feeling a little queasy again. Cold, too. She wrapped herself in a blanket and watched *Ally McBeal* on TV. Ally turned thirty in the episode, and Katie cried. She wasn't nearly as crazy as Ally and her friends, but the show still hit a nerve.

She lay on her living-room couch and couldn't stop thinking about the baby growing inside her. "It's all right, little baby," she whispered. *I hope so, anyway.*

Katie remembered the night when she had gotten pregnant. She'd had a fantasy in bed that night, but she dismissed it, thinking, *I've never gotten pregnant before.* She hadn't ever missed a period — except one time in college, when she'd been playing varsity basketball at North Carolina and learned that her body fat was too low.

That last night with Matt, Katie had felt

that it had never been like this before. Something had changed between them.

She could feel it in the way he held her and looked at her with his luminous brown eyes. She felt some of his walls come down, felt, *This is it.* He was ready to tell her things that he hadn't been able to talk about.

Had that scared Matt? Had he felt it, too, that last night we were together? Was that what had happened?

She had never felt as close to Matt as she had that night. She always loved being with him, but that night it was urgent; they were both so needy.

Katie recalled that it had started so simply: all he had done was wrap his fingers around hers. He slid his free arm beneath her and stared into her eyes. Next their legs touched, then their entire bodies reached toward each other. She and Matt never lost eye contact, and it seemed as if they were really one in a way that they hadn't been before that night.

His eyes said, *I love you, Katie.* She couldn't have been wrong about that.

She had always wanted it to be like that, *just like that.* She'd had that thought, that dream, a thousand times before it actually happened. His strong arms were around her back, and her long legs were wrapped around his. She knew she could never forget any of those images or sensations.

He was so light when he was on top of her, supporting himself on his elbows, his knees. He was athletic, graceful, giving, dominating. He whispered her name over and over: *Katie, sweet Katie, my Katie, Katie, Katie.*

This was it, she knew — he was completely aware and attuned to her, and she had never experienced such love with anyone before. She loved it, loved Matt, and she pulled him deep inside, where they made a baby.

Katie knew what she had to do the next morning. *Seven A.M.* — but it wasn't too early for this. This was it.

She called home — Asheboro, nestled between the Blue Ridge and Great Smoky Mountains in North Carolina — where life had always been simpler. Kinder, too. Much, much kinder.

So why had she left Asheboro? she wondered as the phone began to ring. To follow her love of books? It was her passion, something she truly loved. Or had she just needed to see a world larger than the one she knew in the heart of North Carolina?

"Hey, Katie," her mother answered on the third ring. "You're up with the city birds this morning. How are you doing, sweetie?"

They had Caller ID in Asheboro now. Everything was changing, wasn't it? For better or for worse, or maybe somewhere in between.

"Hey, Mom. What's the latest?"

"You doing a little better today?" her mother asked. She knew that Katie had found someone in New York. She knew all about Matt and had loved it when Katie called to talk about him, especially when she said they would probably be getting married. Now he had left her, and Katie was suffering. She didn't deserve that. Her mother had tried to get her to come home, but Katie wouldn't do it. She was too tough — right. A big-city girl. Well, her mother knew better.

"Some. Yeah, sure. Well, actually, no. I'm still a *mess*. I'm *pitiful*. I'm *hopeless*. I swore I'd never let a man get me into a state like this — and here I am."

Katie began to tell her mother about the diary and what she had read so far. The lesson of the five balls. Suzanne's daily routine on Martha's Vineyard. How she met Matt Wolfe again.

"You know what's so strange, Mom? I actually like Suzanne. Damn it. I'm such a sap. I ought to hate her, but I can't do it."

"Of course you can't. Well, at least this dumb bunny Matt has good taste in women," her mother said, and cackled as she always did. She could be wicked-funny when she wanted to be. Katie was always grateful that she'd inherited her mom's sense of humor. But she didn't feel like joking.

Tell her, Katie was thinking to herself. *Tell her everything.*

But she couldn't. She had told her two best friends in New York — Laurie Raleigh and Susan Kingsolver — but couldn't tell her mother she was pregnant. The words just wouldn't come out of her mouth.

Why not? Katie wondered. But she knew the answer. She didn't want to hurt her mother and father. They meant too much to her.

Her mom was quiet for a moment. Holly Wilkinson was still a first-grade teacher in Asheboro, Katie's mentor for thirty years. She was always, *always* there for her, supportive, even when Katie had gone to dreaded, hated New York and her father didn't talk to her for a month.

Tell her, Katie. She'll understand. She can help you.

But Katie just couldn't get the words out. She choked on them and felt bile rising from her stomach.

Katie and her mother talked for almost an hour, and then she spoke to her father. She was almost as close to him as she was to her mom. He was a minister, much beloved in the area because he taught "God-loving" instead of "God-fearing." The only time he'd ever been really mad at Katie was when she had packed up and moved to New York. But he got over it, and he never threw it up in her face anymore.

Her mother and father were like that. Good people. And so was she, Katie thought, and knew it was true. *Good people.*

So why had Matt left her? How could he just walk out of her life? And what was the diary supposed to tell her that would somehow make her understand?

What was the deep, dark secret of the diary? That Matt had a smart, wonderful wife and a beautiful, darling child and that he had slipped up with her? Had an affair with a New York woman? Strayed for the first time in his picture-perfect marriage? *Damn him! Damn him!*

When she had finished talking to her dad, Katie sat in her study with her good buddies Guinevere and Merlin; they curled up on the couch with her and looked out the bay window at the Hudson. She loved the river, the way it changed every day or even several times in the same day. The river was a lesson, just like the lesson of the five balls.

"What should I do?" she whispered to Guinevere and Merlin. Tears welled up in her eyes, then spilled down her cheeks.

Katie picked up the phone again. She sat there nervously tapping the receiver with her fingernail. It took all the courage that she had, but she finally dialed the number.

Katie almost hung up — but she waited through ring after ring. Finally, she got the answering machine.

She choked up when she actually heard a voice. "This is Matt. Your message is important to me. Please leave it at the beep. Thanks."

Katie left a message. She hoped it was important to Matt. "I'm reading the diary," she said. That was all.

The Diary

Come to our wedding, Nicky. This is your invitation. I want you to know exactly what it was like on the day your mother and father pledged their love.

Snow was falling gently on the island. The bells were ringing in the clear, cold, crisp December air as dozens of frosty well-wishers crossed the threshold into the Gay Head Community Church, which happens to be the oldest Indian Baptist church in the country. It's also one of the loveliest.

There is only one word that can describe our wedding day . . . *joy*. Matt and I were both giddy. I was just about flying among the angels carved in the four corners of the chapel ceiling.

I really did feel like an angel, in an antique white dress strung with a hundred luminescent pearls. My grandfather came to Martha's Vineyard for the first time in fifteen years, just to walk me down the aisle. All my doctor friends from Boston made the trip in the dead of winter. Some of my septuagenarian patients came, too. The church was full, standing room only for the ecumenical service. As you might have guessed, just about everybody on the island is a friend of Matt's.

He was incredibly handsome in a jazzy black tux, with his hair trimmed for the occasion, but not too short, his eyes bright and shining, his beautiful smile more radiant than it had ever been.

Can you see it, Nicky — with the snow lightly blowing in from the ocean? It was glorious.

"Are you as happy as I am?" Matt leaned toward me and whispered as we stood before the altar. "You look incredibly beautiful."

I felt myself blush, which was unlike me. Dr. Control, Dr. Self-Confidence, Dr. Hold It Together. But a feeling of unguarded vulnerability washed over me as I looked into Matt's eyes. This was so right.

"I've never been happier, never surer of anything in my life," I said.

We made our pledge on December 31, just before the New Year arrived. There was something almost magical about becoming husband and wife on New Year's Eve. It felt to me as if the whole world were celebrating with us.

146

Seconds after Matt and I pledged our vows, everyone in the church stood and yelled, "Happy New Year, Matt and Suzanne!"

Silvery white feathers were released from dozens of satin pouches that had been carefully strung from the ceiling. Matt and I were in a blizzard of angels and clouds and doves. We kissed and held each other tightly.

"How do you like the first moment of marriage, Mrs. Harrison?" he asked me. I think he liked saying, "Mrs. Harrison," and I liked hearing it for the first time.

"If I had known how wonderful it was going to be, I'd have insisted we marry twenty years ago," I said.

Matt grinned and went along with me.

"How could we? We didn't know each other."

"Oh, Matt," I said, "we've known each other all our lives. We must have."

I couldn't help remembering what Matt had said the night he proposed on the beach in front of my house. "Isn't it lucky," he'd said, "Suzanne didn't die in Boston and we have today to be together." I was *incredibly* lucky, and it gave me a chill as I stood there with Matt on our wedding night.

That's what it felt like — that was the exact feeling — and I'm so happy that now you were there.

Nicholas,

Matt and I went on a whirlwind, three-week honeymoon that started on New Year's Day.

The first week we were on Lanai in Hawaii. It is a glorious spot, the best, with only two hotels on the entire island. No wonder Bill Gates chose it for his honeymoon, too. I soon discovered that I loved Matt even more than I had before he proposed. We never wanted to leave Lanai. He would paint houses and finish his first collection of poems. I would be an island doctor.

The second week we went to Hana on Maui, and it was almost as special as Lanai. We had our mantra: *Isn't it lucky?* We must have said it a hundred times.

Matt and I spent the third week back home on the Vineyard, but we didn't see much of anyone, not even Jean or Melanie Bone and her kids. We were luxuriating in the newness and specialness of being together for the rest of our lives.

I suppose that not all honeymoons work out so well, but ours did. Nick, here's something your father did, something so thoughtful and special that I will always hold it close to my heart.

Every single day of our honeymoon, Matt woke me in bed — with a honeymoon present. Some of them were small, some were funny jokes, and some were extravagant, but every present came straight from Matt's heart.

Isn't it lucky?

I'll never forget this. It hit me like a wave of seasickness. Unfortunately, Matt had already gone to work and I was alone in the house. I sat down on the edge of the tub, feeling as if my life were draining away.

A cold sweat broke out on the back of my neck, and for the first time in over a year, I wanted to call a doctor. It seemed odd to want a second opinion. I was always diagnosing myself.

But today I felt just bad enough to want to ask someone else, "Hey what do you think?" Instead, I threw cold water on my face and told myself it was probably a touch of the flu, which was making the rounds. I hadn't been feeling well lately.

I took something to settle my stomach, dressed, and went to work. By noon I was feeling much better, and by dinner I had forgotten about it.

It wasn't until the next morning that I found myself sitting on the edge of the tub once more

— spent, tired, and feeling nauseated.

That's when I knew.

I called Matt on the cell phone, and he was surprised to hear from me so soon after he'd left the house.

"Are you okay? Is everything all right, Suzanne?"

"I think . . . that everything just got perfect," I told him. "If you can, I'd like you to come home right now. On your way could you stop at the drugstore? Would you pick up an EPT kit? I want to be absolutely sure, but, Matt, we're pregnant."

Nicholas,

You were growing inside me.

What can I tell you, Nicky — happiness flooded our hearts and every room of the beach cottage. It came like high tide on a full moon.

After the wedding, Matt had moved into my house. It was his idea. He said it was best to rent his place out since I was so established with my patients, and my proximity to the hospital was ideal. It was considerate and sweet of him, which is his way. For a big, tough guy, he's awfully nice. Your daddy *is* the best.

I would have missed the ocean, our sweet and salty garden, and the summer shutters that clack all night against the house when it's windy. But now I don't have to.

We decided to make the sunroom of the house yours. We thought you'd love the way the morning light comes pouring over the sills to fill every nook and cranny. Daddy and I began converting it into a perfect nursery, gathering things that we thought you might love.

We hung wallpaper that danced with Mother Goose stories. There were your bears, your first books, and colorful wall quilts that hung over your crib, the same crib Daddy had when he was a baby. Grandma Jean had saved it all these years. *Just for you, pumpkin.*

We jammed the shelves with far too many variously colored stuffed animals, and every variety of ball known to sportsmen.

Daddy made an oak rocking horse that boasted a beautiful one-of-a-kind crimson and gold mane. Daddy also made you delicately balanced mobiles filled with moons-and-stars galaxies. And a music box to hang in your crib.

Every time you pull the cord, it plays "Whistle a Happy Tune." Whenever I hear that song, I think of you.

We can't wait to meet you.

Nick,

Matt is at it again. A present was on the kitchen table when I got home from work. Gold paper covered in hearts and tied with blue ribbon concealed the contents. I couldn't possibly love him any more than I do.

I shook the small package, and a tiny note dropped out from under the bow.

It read, "Working late tonight, Suze, but thinking about you as always. Open this when you get in and get relaxed. I'll be back by ten. Matt."

I wondered where Matt was working until ten, but I let it go. I unwrapped the box carefully and lifted the tiny lid.

Inside was the most beautiful antique necklace. A sapphire locket in the shape of a heart hung from a silver chain. It was probably a hundred and fifty years old.

I pressed the clasp, and the heart opened

to reveal a message that had been engraved inside.

Nicholas, Suzanne, and Matt — Forever One.

Nick —

A few years back there was a book called *The Bridges of Madison County*. Its huge success was partly due to the fact that so many people seem to be missing romance and emotion in their lives. But an underlying premise of the novel was that romance can last for only a short time; in this particular book, only a couple of days for the main characters, Robert and Francesca. Romeo and Juliet were also star-crossed lovers whose love for each other ended tragically.

Nicky, please don't believe it. Love between two people can last a long time if the people love themselves some and are ready to give love to another person.

I was ready, and so was Matt.

Your father is starting to embarrass me. He is *too* good to me and makes me so happy. Like today. He did it to me again.

The house was filled with friends and

family when I came downstairs this morning, in floppy pink pajamas no less, with a sleepy expression on my face.

I had almost forgotten that today was my birthday. My thirty-sixth.

Matt hadn't. He had made a surprise breakfast . . . and I was surprised, all right. Unbelievably surprised.

"Matt?" I said, laughing, embarrassed, wrapping my arms around my wrinkled pajamas. "I'm going to murder you."

He weaved through the people crowded into the kitchen. He was holding a glass of orange juice for me and wearing a silly grin. "You're all witnesses. You heard my wife. She looks kind of harmless and sweet, but she's a killer. Happy birthday, Suzanne."

Grandma Jean handed me her present, and she insisted I open it then and there. Inside was a beautiful blue silk robe, which I put on to hide my flannels. I gave Jean a big hug for bringing the perfect gift.

"The grub is hot, pretty good, and it's *ready!*" Matt yelled, and everyone moved toward the groaning table, which was filled with eggs, several varieties of breakfast meats, sweet rolls, Jean's homemade babka, plenty of hot coffee.

After everyone had their fill of the sumptuous breakfast — and, yes, *birthday cake* —

they filed from the house and left us alone. Matt and I collapsed onto the big, comfy couch in the living room.

"So, how does it feel, Suzie? Another birthday?"

I couldn't help smiling. "You know how most people dread a birthday. They think, *Oh God, people will start looking at me like I'm old.* Well, I feel the exact opposite. I feel that every day is an extraordinary gift. Just to be here, and especially to be with you. Thanks for the birthday party. I love you."

Then Matt knew just the right thing to do. First, he leaned in and gave me the sweetest kiss on the lips. Then he carried me upstairs to our room, where we spent the rest of my birthday morning and, I must admit, most of my birthday afternoon.

Dear Nicky,

I am still a little shaky as I write about what happened a few weeks ago.

A local construction worker was rushed into the ER about eleven in the morning. Matt knew him and his family. The worker had fallen eighteen feet from a ladder and had suffered trauma to his head. Since I had previously been the attending physician on out-of-control nights at Mass. General, I had seen my share of trauma. I had the emergency room functioning on all cylinders, full tilt, snapping orders and directives.

The man's name was John Macdowell, thirty years old, married, with four kids. The MRI showed an epidermal hematoma. The pressure on his brain had to be alleviated immediately. Here was a young man, so close to dying, I thought. I didn't want to lose this young father.

I worked as hard as I have since I was in Boston.

It took nearly three hours to stabilize his condition. We almost lost him. He went into cardiac arrest. Finally, I knew we had him back. I wanted to kiss John Macdowell, just for being alive.

His wife came in with their children. She was weak with fear and couldn't stop tearing up every time she tried to speak. Her name was Meg, and she was carrying an infant boy. The poor young woman looked as if she were carrying the weight of the world on her shoulders. She probably felt that she *was* on this particular day.

I ordered a mild sedative for Mrs. Macdowell and sat with her until she could gather herself. The kids were obviously scared, too.

I took the second smallest, two years old, into my lap and gently stroked her hair. "Daddy is going to be okay," I said to the little girl.

The mother looked on, letting my words seep in. This was meant for her even more than for the children.

"He just fell down. Like you do sometimes. So we gave him medicine and a big bandage. He's going to be fine now. I'm his doctor, and I promise."

The little girl — all of the Macdowell kids — fastened on to every word I had to say. So did their mother.

"Thank you, Doctor," she finally whispered. "We love John so much. He's one of the good guys."

"I know he is. I could tell by the concern everybody showed. His entire crew came to the ER. We're going to keep John here for a few days. When it's time for him to leave, I'll tell you exactly what you'll need to do at home. He's stable now. Why don't I watch the kids. You can go in and see him."

The little girl climbed down from my lap. Mrs. Macdowell unraveled the baby from her arms and lowered him into mine. He was so tiny, probably only two or three months old. I doubted that his mother was more than twenty-five.

"Are you sure, Dr. Bedford? You can spare the time?" she asked me.

"I have all the time in the world for you, John, and the kids."

I sat there, holding the baby boy, and I couldn't help thinking about the little boy growing inside me. And also about mortality, and how we face it every day of our lives.

I already knew I was a pretty good doctor. But it was only at that moment, when I held the little Macdowell baby, that I knew I was going to be a good mother.

No, Nick, I knew I was going to be a *great* mom.

"What was *that?*" I said. "Matt? Honey?"

I spoke with difficulty. "Matt . . . something's going on. I'm in . . . some pain. *Whew.* There's more than a little pain, actually."

I dropped my fork on the floor of the Black Dog Tavern, where we were having dinner. *This couldn't be happening. Not yet.* I was still weeks away from my delivery date. There was no way I could be having a contraction.

Matt jumped into action. He was more prepared for the moment than I was. He tossed cash onto the table and escorted me out of the Black Dog.

Part of me knew what was happening. Or so I believed. *Braxton Hicks.* Contractions that don't represent true labor. Women sometimes have these pains, occasionally even in their first trimester, but when they come in the third, they can be mistaken for actual labor.

However, my pain seemed to be *above* my uterus, spreading up and under my left lung. It came like a sharp knife. Literally took my breath away.

We got into the Jeep and headed directly to the hospital.

"I'm sure it's nothing," I said. "Nicky's just giving a heads-up, letting us know he's physically fit."

"Good," Matt said, but he kept driving.

I had been getting weekly monitoring because this was considered a high-risk pregnancy. But everything had been fine, even a joy, up until now. If I were in trouble, I would have known it. Wouldn't I? I was always on the lookout for the least little problem. The fact that I'm a doctor made me even more prepared.

I was wrong. I was in trouble. The kind of trouble you're not quite sure you want to know about before it happens.

This is the story of how we both almost died.

Nicholas,

We had the best doctor on Martha's Vineyard, and one of the best in all of New England. Dr. Constance Cotter arrived at the hospital about ten minutes after I got there with Matt.

I felt fine by then, but Connie monitored me herself for the next two hours. I could see her urgency; I could read it in the tightness of her jaw. She was worrying about my heart. Was it strong enough? She was worrying about you, Nicky.

"This is potentially dangerous," Connie said, sparing me no illusions. "Suzanne, your pressure is so high that part of me wants to start labor right now. I know it's not time, but you've got me worried. What I *am* going to do is keep you here tonight. And as many nights as I feel are warranted. No, you have no say in this."

I looked at Connie like, You must be kidding. I was a doctor. I lived right down the road from the hospital. I would come in immediately if necessary.

"Don't even think about it. You're staying. Check in, and I'll be up to see you before I go. This isn't negotiable, Suzanne."

It was strange to be checking in to the hospital where I worked. An hour or so later, Matt and I sat in my room waiting for Connie to return. I was telling him what I knew so far, in particular about a condition called preeclampsia.

"What exactly is preeclampsia, Suzanne?" he asked. He wanted every detail explained in clear layman's terms. He was asking all the right questions. So I told him, and he shifted uncomfortably in his chair.

"You wanted to know," I said.

Connie finally came in. She took my blood pressure again. "Suzanne," she said, "it's higher than it was. If it doesn't go down in the next few hours, I'm inducing labor."

I had never seen Matt look, or act, so nervous. "I'm going to stay here with you tonight, Suzanne," he said.

"Don't be silly," I told him. "Sit in an uncomfortable chair and watch me sleep? That's crazy."

But Connie looked at me and, in the clinical tone that she uses only for patients, said,

"I think that's a very good idea. Matt should stay with you, Suzanne."

Then Connie checked my pressure once more before leaving for the night.

I studied her face, looking for any kind of sign of trouble. *What kind of look was that?*

Connie stared at me oddly and I couldn't quite figure it out.

Then she said, "Suzanne, I'm not getting a strong reading from the baby's heart. The baby has to come out *now*."

Dear Nicholas,

All my life I had wanted a baby. I wanted to experience natural childbirth, just as my mother and grandmother had. Connie knew how important a natural delivery was to me. Matt and I had attended Lamaze classes together. She'd heard me go on and on about it in her office, and even over lunch.

I could see the sadness and pain in her face when she leaned over to me. She grasped my hand tightly in both of hers.

"Suzanne," she whispered, "I wanted to bring this baby into the world the way you hoped it would happen. But you know I'm not going to put either you or this baby at risk. We have to do a c-section."

Tears welled up in my eyes, but I nodded. "I know, Connie. I trust you."

Everything began to move too fast after that.

Connie inserted an IV in my arm and administered magnesium sulfate. I immediately felt sicker than ever. A blinding headache overcame me.

Matt was right there as they prepped me for the c-section. He was told by a new attending doctor that this was an emergency. He couldn't stay with me.

Thank God, Connie came back in just then and overrode the decision.

Connie then told me what was happening.

My liver was swollen. The blood platelet count was alarming, and my blood pressure was 190/130.

Worse, Nicky, *your* heartbeat was weakening.

"You're going to be okay, Suzanne," I kept hearing Connie say. Her voice was like an echo from a distant canyon. The room lights above my head appeared to be spinning out of control.

"What about Nicky?" I whispered through parched lips.

I waited for her to say, "And Nicky will be fine, too."

But Connie didn't say it, and tears came to my eyes again.

I was rolled into the operating room, where they were not only ready to deliver a baby but also to transfuse me with eight units of blood. My platelet count had dropped. I knew what was going on here. If I started bleeding internally I would die.

As I was being given the epidural anesthesia, I saw Dr. Leon, my cardiologist, standing right by the anesthetist. *Why was Leon here? Oh God, no. Please don't do this. Oh please, please, please. I beg you.* An oxygen mask was placed over my face. I tried to resist.

Connie raised her voice. "No, Suzanne. Take the oxygen."

I felt as if I were on fire. I wasn't able to logically attribute it to the magnesium sulfate. I didn't know that my kidneys were shutting down, my platelets were dangerously low, my blood pressure had risen even more, to an alarming 200/115. I didn't know that steroid injections were being administered to optimize the baby's lung function and his prospect of survival.

The next few minutes were a blur. I saw a retractor come out. There were concerned looks from Connie, and then evasive eye contact.

I heard staccato orders and cold, unfeeling machine beeps and Matt chanting only positive things. I heard a loud sucking noise as amniotic fluid and blood were cleaned out of me.

There was numbness, some dizziness, and the oddest feeling of not being there, of not being anywhere, actually.

What brought me out of my surreal feeling of having entered another world was *a cry*. A distinct and mighty cry. You had announced your arrival like a strong warrior.

I began to cry, and so did Matt and Connie. You were such a little thing, just over six pounds. But so strong. And alert. Especially considering the stress you had been through.

You looked right at Daddy and me. I'll never forget it. *The first time ever I saw your face.*

I got to hold you in my arms before you were whisked away to the NICU. I got to look into the beautiful eyes that you struggled to keep open, and I got to whisper for the first time, "I love you."

Nicholas the Warrior!

Katie

Fear and confusion swept over Katie again that night. While she read a few more diary pages, she forced herself to eat pasta primavera and drink tea. It didn't help.

Everything was moving way too fast in her head, and especially inside her sore, bloated body.

A baby boy had been born. Nicholas the Warrior.

Another child was growing inside her.

Katie had to think and be logical about this. What were all the possibilities? What could really be happening now?

Matt had been cheating on Suzanne all of these months?

Matt had been cheating and I wasn't the first?

Matt had left Suzanne and Nicholas for some reason that was yet to be revealed in the diary? They were divorced?

Suzanne had left Matt for somebody else?

Suzanne had died — her heart had finally given out?

Suzanne was alive, but very ill?

Where was Suzanne right now? Maybe she should try to call her on Martha's Vineyard. Maybe they should talk. Katie wasn't sure if that was a good idea or if it would be one of her worst blunders ever.

She tried to work it through. What did she have to lose? A little pride, but not much else. *But what about Suzanne?* What if she had no idea about Matt? Was that even faintly possible? Of course it was. Wasn't that pretty much what had happened to Katie? Anything seemed possible to her right now. Anything *was* possible. So what had really happened?

This was so overwhelming — unbearable. The man she had loved, and trusted, and thought she completely understood, had left her. Wasn't that just typical these days? Wasn't it sad?

She remembered a particular moment with Matt that kept her going. He had woken up beside her one night and was crying. She had held Matt in her arms for a long time. She stroked his cheek. Finally Matt had whispered, "I'm trying hard to get everything behind me. I will. I promise, Katie."

God, this was crazy!

Katie pounded her thigh with a closed

fist. Her pulse was racing too fast. Her breasts really hurt.

She pushed herself out of her sofa, hurried into the bathroom, and threw up the pasta she'd just eaten.

A little while later, Katie went into the kitchen and fixed herself more tea. She and Guinevere sat staring at the four walls. She had hung the kitchen cabinets herself. The guys at Chinatown Lumber knew her all too well. She had her own toolbox and prided herself on never having to call the super to fix anything. *So fix what's wrong with your heart,* Katie thought. *Fix that!*

Finally, she reached for the phone.

Merlin opened one sleepy eye as she nervously punched some numbers and heard a pickup on the other end of the line.

"Hi, Mom. It's me," she said in a voice that came out much smaller than she intended.

"I know, Katie. What's the matter, sweetheart? Couldn't you just come home for a couple of days? I think it would do us all a world of good."

This was so hard, so bad.

"Could you get Daddy to pick up, too?"

she asked. "Get Daddy, please."

"I'm here, Katie," her father said. "I'm on in the den. I picked up when the phone rang. How are you?"

She sighed loudly. "Well . . . I'm pregnant," Katie finally said.

Then all three of them were crying over the phone — because that's the way they were. But Katie's mother and father were already comforting her, saying, "It's all right, Katie, we love you, we're with you, we understand."

Because that's the way they were, too.

The Diary

Nicholas,

Just for the record. You started sleeping through the night early on. Not every night, but most, starting when you were about two weeks old, to the envy of all the other moms!

When you go through your little growth spurts, you wake up hungry. And what a little eater you are! You will eat *anything* — whether you're breast-fed, or bottle-fed formula, or water, you chow down and aren't picky.

On your first visit to the pediatrician after the initial hospital checkups, the doctor couldn't believe how you were already focusing on the toys she had laid out. She exclaimed, "He's extraordinary — sensational, Suzanne." And she said you're "so smart and so strong" because when she turned you on your tummy you lifted your head.

That's a great feat for a two-week-old. Nicholas the Warrior!

You were baptized at the Church of Mary Magdalene. It was a beautiful day. You wore my christening gown — a handmade heirloom of my aunt Romelle's family in Newburgh, New York. It was also worn by my cousins and various other relatives over the past fifty years, and it was in perfect condition. You looked sweet and were such a charmer.

Monsignor Dwyer was completely taken with you. During the baptism, you kept reaching for the service book and touching his hand. You were looking right at him, attentive as could be.

Toward the end of the service, after you hadn't missed a trick, Monsignor Dwyer said to you, "I don't know *what* you're going to be when you grow up, Nicholas. On second thought — you *are* grown up."

It's my first day back at work today. Not surprisingly, I miss you already. No, let me make that a little stronger: *I'm bereft without you.*

I wrote something as I sat thinking about you — even between patients.

> Nickels and dimes
> I love you in rhythms
> I love you in rhymes
> I love you in laughter
> Here and ever after
> Then I love you a million
> Gazillion more times!

I think I could come up with dozens of Nicky nursery rhymes if I tried. They just come to me when you do something silly, or smile, or even when you sleep. What can I say? You inspire poetry.

Matt loves them, too. And coming from him, it's a real compliment. Make no mistake about it, your daddy is definitely the writer in this family. But I still love writing these little love poems to you.

Yikes, here comes one now!

You're my little Nicky Knack
I love you so, you love me back.
I love your toes, your knees, your nose,
And everywhere a big kiss goes.
I kiss you tons, and know what then?
I have to kiss you once again.

Okay, little man, I have to go now. My next patient is here already. If she knew what I was doing behind closed doors in my office, the poor woman would flee to the free clinic in Edgartown.

I thought I'd ease into work with a half day just to get used to the routine again. But ever since I arrived this morning, all I wanted to do was look at your pictures and write silly poems.

Anyone peeking in at me would think I was in love.

I am.

Nicky, it's me again —

I heard you crying tonight and got up to see what was the matter. You looked up at me with such sad little eyes. Your eyes are so blue, and always so expressive.

I looked to see if you needed changing — but it wasn't that. Then I checked to see if you were hungry — but it wasn't that, either.

So I lifted you up and sat with you in the rocker next to your crib.

Back and forth we went, back and forth, in a rhythm about double the rush of the ocean surf.

Your eyes started slowly closing, and your tears dissolved into sweet dreams. I placed you back in your crib and watched your heart-shaped bottom rising in the air. Then I turned you over on your back and watched your little tummy rise and fall.

I think all you wanted was a little company. Could you really just have wanted to be rocked and held and talked to?

I'm here, sweetie. I'm right here, and I'm not going anywhere. I'll always be right here.

"What are you doing, Suzie?" Matt whispered. I hadn't heard him come into the nursery. Daddy can be as quiet as a cat.

"Nick couldn't sleep."

Matt looked into the crib and saw your tiny hand clenched to your mouth like a teething ring.

"God, he's beautiful," Matt whispered. "I mean it — he is gorgeous."

I looked down at you. There wasn't an inch of you that didn't make my heart leap.

Matt put his arms around my waist. "Want to dance, Mrs. Harrison?" He hadn't called me that since our wedding day. My heart fluttered like a sparrow in a birdbath.

"I think they're playing our song."

And to the high, plucky notes that came squeaking out of your music box, Matt and I danced round and round in your nursery that night. Past the stuffed animals, past Mother Goose and your homemade rocking horse, past the stars and the moons that float from your homemade mobile. We danced slowly and lovingly in the low light of your tiny cocoon.

When the music finally wound down to its

final note, Matt kissed me and said, "Thank you, Suzanne. Thank you for this night, this dance, and most of all for this little boy. My whole world is right here, in this room. If I never had another thing, I would have everything."

And then strangely — magically — as if your music box were just taking a rest, it played one more sweet refrain.

Nick,

Melanie Bone came over to baby-sit while I went to work. Full day, full load. Melanie's kids were in Maine with her mother for a week, so she gave Grandma Jean a breather. It feels strange to leave you for this long, and I can't stop thinking about what you're doing now.

And *now*.

And *now*.

The last time I felt this tired, I was working my butt off at Mass. General in Boston. Maybe it's because I'm juggling so many things again these days. Having a job and a baby is even harder than I thought. My respect for all mothers has never been higher, and it was high to begin with. Working mothers, mothers who stay at home, single mothers — they are all so amazing.

Something happened at the hospital today that made me think of your delivery.

A forty-one-year-old woman who was on vacation from New York was brought in. She was in her seventh month, and not doing well. Then all hell broke loose in the emergency room. She began to hemorrhage. It was so terrible. The poor woman ended up losing her baby, and I had to try to console her.

You probably wonder why I'm writing about this. Even I thought twice before sharing this sad story with you.

But it has made me realize more than ever how vulnerable we are, how life can be like walking on a high wire. Falling seems a tiny misstep away. Just seeing that poor woman today, and remembering how lucky we were, made me catch my breath.

Oh, Nicky, sometimes I wish I could hide you like a precious heirloom. But what is life if you don't live it? I think I know that as well as anyone.

There's a saying I remember from my grandmother: One today is worth two tomorrows.

Dear Show-off,

You are starting to hold your own bottle. No one can believe it. This little guy feeding himself at two months. Every new experience that you have, I take as a gift to me and Daddy.

Sometimes I can be such a goofball. Reduced to gauzy visions of station wagons, suburbia, and bronzed baby shoes. So I had to do it. I had to have your picture professionally taken.

Every mother has to do it once. Right?

Today is the perfect day. Daddy is off on a trip to New York, where someone has taken a liking to his poems. He's very low-key about it, but it's the greatest news. So the two of us are home alone. I have a plan.

I got you dressed in washed-out blue overalls (so cool), your little work boots (just like Daddy's), and a Red Sox baseball cap (with the peak bent just so).

The cap had to go! You freaked out over it; I guess you thought I was trying to attach antlers to your head.

Here's the whole scene, just in case you don't remember it.

When we got to the You Oughta Be in Pictures photography studio, you looked at me as if to say, *Surely you have made a grotesque mistake.*

Maybe I had.

The photographer was a fifty-year-old man who had no kidside manner at all. It wasn't that he was mean, he was just clueless. I got the idea that his real specialty might be still life, because he tried to warm you up with a variety of fruits and vegetables.

Well, one thing is certain. We now have a unique set of pictures. You begin with the surprised look, which quickly dissolves into a slightly more annoyed attitude. After that you enter the cantankerous phase, which swiftly disintegrates into the angry portion of our program. And last but not least, irreconcilable meltdown.

There is a small consolation. At least you can't tell Daddy. He'd get too much mileage out of his *I told you so*s.

Forgive me this one. I promise I will never show these pictures to new girlfriends, old fraternity brothers, or Grandma Jean. She'd have them in every shop window on the Vineyard before dusk.

Nicky,

It was a little cool out, but I bundled you up and we took a picnic basket down to Bend in the Road Beach — to celebrate Daddy's thirty-seventh birthday. *God, he's old!*

We made castles and sand angels and wrote your name in big bold letters until the surf came and washed it away.

Then we *wrote it again,* high enough up so the water couldn't reach it.

It was such a total blast to watch you and Daddy play together. You are very much a chip off the old block, two peas in a pod, Laurel and Hardy! Your mannerisms, your ways, your gestures, are Matt's. And vice versa. Sometimes when I look at you, I can imagine Daddy when he was a boy. You are both joyful, graceful, and athletic, beautiful to watch.

So there you are, just back to our blanket from fighting sand monsters and friendly sea urchins, when Matt reaches into his pocket and pulls out a letter. He hands it to me.

"The publisher in New York didn't want my collection — *yet* — but here's a consolation prize."

He had sent a poem off to a magazine called the *Atlantic Monthly.* They accepted it. He didn't even tell me he was doing it. Said he didn't want it to be out there just in case it didn't happen. But it did, Nicky, and he got the letter on his birthday.

I asked if I could read it, and Matt unfolded a separate sheet of paper. It was the poem, and he had it with him all this time.

My eyes teared up when I saw the title, "Nicholas and Suzanne."

Matt told me that he had been writing down all the things I say and sing to you, that he'd strain to overhear my little poems and rock-a-bye rhymes.

He said that this wasn't just his poem but mine, too. He told me that it was *my voice* he heard in these lines; so we had created it together.

Daddy read part of it out loud, above the crashing surf and screeching gulls.

Nicholas and Suzanne

Who makes the treetops wave their hands?

And draws home ships from foreign lands,
And spins plain straw back into gold
And has a love too large to hold . . .

Who chases the rain from the sky?
And sings the moon a lullaby,
And grants the wishes from a well
And hears whole songs sung from a shell . . .

Who has the gift of making much?
From everything they hold or touch,
Who turns pure joy back into life?
For this I thank my son, my wife.

What could be better than this?
Absolutely nothing.
Daddy said this was his best birthday ever.

Nicholas,

Something unexpected has happened, and I'm afraid it's not so good.

It was time again for your dreaded baby shots. I hated to have to put you through it. Your pediatrician on the Vineyard was on vacation, so I decided to call a doctor friend in Boston. It was time for a visit to Beantown, anyway.

While I was in Boston, I would get my own physical. It was also a chance to catch up with friends, maybe do a little window shopping on Newbury Street, eat at Harvard Gardens, and, best of all, show you off, Nicky Mouse.

We took the ferry over to Woods Hole and hit Route 3 by nine in the morning. This was our first adventure off the island. *Nicholas's Trip to the Big City!*

Your appointment was first. The children's office looked exactly as it always had.

Highlights, crayons, and blocks lay everywhere. A black clock cat moved its tail and eyes back and forth to the time. You were fixated on it.

Other babies were crying and fidgety, but you sat there as quiet as a little mouse, checking out these new surroundings.

"Nicholas Harrison," the receptionist finally called.

It was funny to hear your name announced so officially by a complete stranger. I almost expected you to answer, "Present."

It was good to see my old buddy Dan Anderson, and he couldn't believe how big you were already. He said he saw a lot of me in you, and of course that thrilled me. But in fairness I had to show him pictures of Daddy, too.

"You seem so happy, Suzanne," Dan said as he measured, tapped, and tuned you up, Nicky.

"I am, Dan. Never been happier. It's great."

"Leaving the big city did you a world of good. And just look at this future quarterback you've got here."

I beamed. "He is the best little boy on this earth. Like you've never heard that before. Right?"

"Not from you, Suzanne." He handed you

back over to me. "It's wonderful seeing you again, Mother Bedford. And as far as this one goes, he's the poster child for good health."

Of course, I already knew that.

Now it was my turn.

I sat at the edge of the examining-room table, already dressed, waiting for my doctor, Dr. "Philadelphia" Phil Berman, to come back in. Phil had been my doctor in Boston and had kept in touch with the specialist on Martha's Vineyard. They complemented each other nicely.

The physical had taken a little longer than usual. One of the nurses outside was watching over you, but I was anxious for a hug and also to hit the road back to the Vineyard. That's when Phil came in and asked me to step into his office.

We were old friends, so we exchanged small talk for a minute or two. Then Phil got down to business.

"Your stress test doesn't look too good to me, Suzanne. I noticed a few irregularities on your EKG. I took the liberty of calling downstairs to Dr. Davis. I know Gail was your cardiologist when you were here as a patient. She has your records from the island. She's going to squeeze you in today."

"Wait a minute, Phil," I said. I was stunned. This had to be wrong. I was feeling fine — *great*, actually. I was in the best shape of my life. "That can't be right. Are you sure?"

"I know your history, and I would be remiss in not insisting that Gail Davis take a look. Hey, Suzanne, you're here already. Martha's Vineyard is a long way off. Just do it. It won't take long. We'll keep Nicholas here until you're done. Our pleasure."

And then Phil continued, his tone changing ever so slightly, "Suzanne, you and I have known each other for a long time. I just want you to take care of whatever this might be. It could be absolutely nothing, but I want a second opinion. You'd give the same advice to any of your own patients."

It felt like déjà vu, walking through the halls, heading to Gail Davis's office. *Dear God, please don't let this happen again. Not now. Oh please, God. Everything in my life is so good.*

I entered the waiting room as if I were walking in a misty fog in a bad dream. I couldn't focus or think.

The ominous mantra that kept repeating loudly in my brain was *Tell me this isn't happening.*

A nurse walked right up to me. Actually, I knew her from the hospital visits after my heart attack. "Suzanne, you can come with me now."

I followed her like a prisoner about to be executed.

Tell me this isn't happening.

I was in there for nearly two hours. I think I was given every cardiology test known. I was worried about you, even though I knew you were in good hands at Dr. Berman's office.

When it was finally over, Gail Davis came in. She looked grave but Gail usually does, even at parties where I've seen her socially. I reminded myself of that, but it didn't really help.

"You have *not* had another heart attack, Suzanne. Let me put your mind at ease about that. But what I detect is some weakness in *two* of your valves. I suspect it was caused by the last cardiac infarction. Or possibly the pregnancy.

"Because the valves are damaged, your heart is having some difficulty pumping blood. You know where I'm going, Suzanne, but I feel compelled to alert you. This is a warning, a very lucky warning."

"I don't feel very lucky," I said.

"Some people never get a warning, and so

they don't get a chance to fix what could be about to break. When you get back to Martha's Vineyard, there'll be more tests, then we can talk about your options. Valves may have to be replaced, or possibly not."

Now I was having trouble catching my breath. I absolutely refused to cry in front of Gail. "It's so strange," I said. "Everything can be going along just great, and then one day, *whack*, you're blind-sided — a lousy, crummy blow you didn't see coming."

Gail Davis didn't say anything; she just put her hand gently on my back.

Nicky,

In the words of a feisty little Italian girl, Michele Lentini, who used to be my best friend back in Cornwall, New York, *oh, marone.*

Or, in the words of the Blues Brothers, *They're not going to catch us, we're on a mission from God!*

I watched you in the rearview mirror, your little feet kicking up and down, your arms reaching toward me. The world swept past us on both sides, and it felt to me that we were falling home instead of just going there.

I talked to you, Nicky, really talked.

"My life feels so connected to you. It seems impossible that something bad could happen to me now. But I guess that's just the false sense of security that love gives."

I thought about that for a second. Falling in love with Matt, and being so much in love with him now, *had* given me a feeling of security.

How could anything harm us? How could anything really bad happen?

And you give me this same sense, Nick. How could anything happen to break us apart? How could I *not* see you grow up? That would be too cruel for God to let happen.

The tears I had held back in Dr. Davis's office suddenly flooded my eyes. I quickly wiped them away. I concentrated on the road home and kept our journey at my usual slow and steady pace.

I talked to you in the little rearview mirror I have that looks directly at your car seat. "So let's make a plan. All right, baby boy? Every time I can make you smile means that we have one more year together, a whole year for every smile. Magical thinking, Nicky, that's what this is. Already we have a dozen more years together, because you've smiled at least that many times on this car ride. At this rate, I'll be a hundred and thirty-six, you a spry eighty-two."

I started to laugh at my own crazy humor.

Suddenly you broke into the biggest smile I have ever seen you make. You made me laugh so hard, I just looked back and whispered, "Nicholas, Suzanne, and Matt — Forever One."

That is my prayer.

Nicholas,

Four long, nervous weeks have passed since I received the troubling news in Boston. Matt is out with you riding in the Jeep, and I'm sitting in the kitchen with the sun falling through the window like yellow streamers in a parade. It's so beautiful.

The medical opinions are all in. I have heart-valve disease, but it is treatable. For the moment, we won't be replacing the valves, and we definitely won't be considering a heart transplant. Everything will be treated with radiation for now.

I have been warned, though: *Life doesn't go on forever. Enjoy every moment of it.*

I can smell the morning unfolding, carrying with it the song and salt and the grassy perfume of the marshes.

My eyes are closed, and the wind chimes are being tickled by the ocean breeze outside the window.

"Isn't it lucky?" I finally say out loud.

"That I'm sitting here, looking out on this beautiful day. . . .

"That I live on Martha's Vineyard, so close to the ocean that I could throw a stone into the surf — if I were the kind of person who could throw stones far. . . .

"That I am a doctor and love what I do. . . .

"That somehow, however improbable, I found Matthew Harrison and we fell wildly in love. . . .

"That we have a little boy with the most beautiful blue eyes, and the most wonderful smile, and the nicest disposition, and a baby smell I just love.

"Isn't it lucky, Nicky? Isn't it just so lucky?"

That's what I think, anyway.

That's another of my prayers.

Nicholas,

You are growing up before our eyes, and it is such a glorious thing to watch. I *savor* each moment. I hope all the other mommies and daddies are remembering to savor these moments and have the time to do so.

You love to ride bikes with Mommy. You have your own little Boston Bruins helmet and a seat that holds you snuggly and safely on the back of my bike. I tie a water bottle with a ribbon and attach it to your seat for you to enjoy on the ride — and we're off.

You love singing, and looking at all the people and sights on the Vineyard. Fun for Mama, too.

You have a lot of the blondest of blond curls. I know that if I cut them, they'll be gone forever. You'll really be a little boy then, no longer a baby.

I love watching you grow, but at the same time I don't like seeing this time fly by so fast. It's hard to explain; I don't really know how. But there's something so precious about watching your child day after day after day. I want to hold on to every moment, every smile, every single hug and kiss. I suppose it has to do with *loving* to be needed and *needing* to give love.

I want to relive this all over again.

Every single moment since you were born.

I told you I would be a great mom.

Each day lately has felt so complete for me.

Every morning, without fail, Matt turns to me before we get up. He kisses me, and then whispers in my ear, "We have today, Suzanne. Let's get up and see our boy."

But today feels a little different to me. I'm not exactly sure why, but my intuition tells me there's something going on. I don't know if I like it. I'm not quite sure yet.

After Daddy goes off to work and I have you fed and dressed, I still don't feel right.

It is an odd feeling. Not too bad, but definitely not too good. I am lightheaded, and more tired than usual.

So tired, in fact, I have to lie down.

I must have fallen asleep after I tucked you into your crib, because when I opened my eyes again, the church bells from the town were striking.

It was noon already. Half the day was gone.

That's when I decided to find out what was going on.

And now, I know.

Nicholas,

After Daddy put you to bed tonight, the two of us sat out on the porch and watched the sun set on the ocean in a blaze of streaking oranges and reds. He has the most amazing touch and was patiently stroking my arms and legs, which I love more than almost anything on the planet. I could let him do this for hours, and sometimes I do.

He is very excited about his poetry lately. His great dream is to have a collection published, and suddenly people are interested. I love the excitement in his voice, and I let him talk.

"Matthew, something happened today," I finally said, once he had told me all his news.

He turned on the couch and sat up straight. His eyes were full of worry, his brow creased.

"I'm sorry, I'm sorry." I soothed him. "Something good happened today."

I could feel Matt relax in my arms and also saw it on his face. "So what happened, Suzanne? Tell me all about your day."

The nice thing is that your daddy really wants to hear about these things. He listens, and even asks questions. Some men don't.

"Well, on Wednesdays I don't go to work unless there's an emergency. There wasn't any today, thank God. So I stayed home with Nick."

Matt put his head in my lap and let me stroke his thick, sandy brown hair. He likes this finger combing almost as much as I like his tickling. "That sounds pretty nice. Maybe I'll start taking Wednesdays off, too," he teased.

"Isn't it lucky," I said, "that I get to spend Wednesdays with Nicky?"

Matt pulled my face to his and we kissed. I don't know how long this incredible honeymoon of ours is going to last, but I love it and don't want it to end. Matthew is the best friend I could have ever wished for. Just about any woman would be lucky to have him. And if it ever, ever came to that — another mommy for you — I'm sure Matt would choose wisely.

"Is that what happened? You and Nick had a great day together?" he asked.

I looked deeply into Matt's eyes. "I'm

pregnant," I told him.

And then Matt did just the right thing: He kissed me gently. "I love you," he whispered. "Let's be careful, Suzanne."

"Okay," I whispered back. "I'll be very careful."

Nicholas,

I don't know why, but life is usually more complicated than the plans that we make. I visited my cardiologist on the Vineyard, told him about the pregnancy, had a few tests. Then, on his recommendation, I went to Boston to see Dr. Davis again.

I hadn't mentioned the checkup to Matt, thinking it might worry him. So I went to work for a few hours, then I drove to Boston in the afternoon. I promised myself that I would talk to Matt as soon as I got home.

The porch light of the house was on when I pulled into the driveway at about seven that night. I was late. Matt was already home. He had relieved Grandma Jean of her baby-sitting duties.

I could smell the delicious aroma of home cooking: chicken, pan potatoes, and gravy warming the whole house. *Oh, my God, he made dinner,* I thought.

"Where's Nicky?" I asked as I entered the kitchen.

"I put him to bed. He was exhausted. Long day for you, sweets. You're being careful?"

"Yeah," I said, kissing him on the cheek. "I actually only saw a couple of patients this morning. I had to go to Boston and see Dr. Davis."

Matt stopped stirring the gravy. He stared at me and didn't say another word. He looked so hurt that I couldn't stand it.

"I should have told you, Matthew. I didn't want to worry you. I *knew* you would, and I didn't want you to; I knew you'd want to come to Boston with me."

It was a nervous, run-on thought, my attempt to explain what I had done. It wasn't right, but it wasn't wrong, either. Matt decided to leave my decision at that.

"Well?" he said. "What did Dr. Davis have to say?"

My mind traced back to Gail Davis's office, back to the edge of the examining table, where I had sat so tenuously in a blur of emotions: *What did she say? What did she say?*

"Well, I told her about the baby."

"Right."

"And she was . . . she was very concerned. Gail wasn't pleased."

218

The next few words locked in my throat, nearly closed off my breathing. I almost couldn't speak. Tears flooded to my eyes, and I started to shake.

"She said it was too risky for me to be pregnant. She said I shouldn't have this baby."

Matt's eyes filled with tears now, too. He took a breath. Then he spoke, splitting the silence between us.

"Suzanne, I agree with her. I couldn't bear to risk losing you."

I was crying, sobbing terribly, still shaking badly. "Don't give up on this baby, Matt."

I looked at him, waiting for some comforting words. But he was too quiet. He finally shook his head slowly. "I'm sorry, Suzanne."

Suddenly I needed to breathe some fresh air, to escape, to be by myself. I left the house in a spin. I ran through the tall sea grass until I reached the beach. Shaken, winded, fatigued. There was a loud roaring noise in the space between my ears. It wasn't the sound of the ocean.

I lay down in the sand and wept. I felt awful, so inconsolably sad for the baby inside me. I thought about Matt and you waiting for me back at the house. Was I being selfish, headstrong, foolish? I was a doctor. I knew the risks.

This baby was a precious and unexpected gift. I couldn't give it up. I held myself and rocked with that feeling for what seemed like hours. I talked to the little baby growing inside me. Then I looked up at the full moon, and I knew it was time to go back to the house.

Matt was waiting for me in the kitchen. I saw him in the mellow, yellow light as I trudged up from the beach. I started to cry again.

I did a strange thing, then, and I'm not exactly sure why I knocked on the door, then knelt on the first step. Maybe I was tired and drained from the long, stressful day. Maybe it was something else, something more important, something I still can't explain.

Maybe I was remembering the English king who had knelt in the snow hoping not to be excommunicated, to be forgiven by Pope Gregory.

I had been hurting badly out on the beach, but I also knew I had acted selfishly. I shouldn't have run away and left you and Matt alone at the house.

"Forgive me for running off like that," I said as Matt opened the screen door. "For running away from you. I should have stayed and talked it out."

"You know better," he whispered, and gently stroked my hair. "There's nothing to forgive, Suzanne."

Matt pulled me to my feet and into his

arms. A feeling of relief swept through me. I listened to the strong beating of his heart. I let him snuggle the top of my head with his chin. I let his warmth seep into me.

"It's just that I want to keep this baby Matt. Is that so terrible?"

"No, Suzanne. That isn't terrible. It's losing you that I couldn't bear. If I lost you, I don't think I could live. I love you so much. I love you and Nicky."

Oh, Nicky,

Life can be unforgiving sometimes. Learn that lesson, sweet boy. I had just gotten home from a couple of hours at the office. Routine really, nothing unusual, nothing stressful. Actually, I was feeling pretty chipper.

I drove back to the cottage to take a catnap before seeing one more patient in the afternoon. You were at Grandma's house for the day. Matt had a job over in East Chop.

I was going to take it easy, catch a nice, healthy, restful snooze. I had an appointment to see Connie in town the next day — *about the baby.*

I fell onto the bed, feeling dizzy suddenly. My heart began to pound a little. Strange. I felt a headache coming on, out of nowhere.

It was about to rain buckets, and the barometric pressure had dropped. I sometimes get headaches when that happens.

My appointment with Connie was the next

day, but I was deliberating over whether I should wait until then. Maybe I would feel better in an hour, or when the rain finally came.

I was so nervous about staying healthy that I was driving myself into neurotic symptoms, for God's sake.

Easy, Suzanne, I told myself. *Lie down and close your eyes and tell every part of your body to relax.*

Your eyes, your mouth, your chest, your belly, your arms, your legs, your feet, your toes.

Relax them all and slip under the blanket, the Golden Fleece.

All you need is an hour, a break, and when you wake up, it will all feel better.

Just fall asleep, fall asleep now, fall . . .

"Suzanne, what's the matter?"

I turned over on the daybed at the sound of Matt's gentle whisper. I still didn't feel too good. He leaned in closer, and he looked concerned. "Suzanne? Can you talk, sweetheart?"

"Seeing Connie tomorrow," I finally said. This was strange. It took all my strength just to get those few words out.

"You're seeing Connie right now," Matt said.

When we arrived at Connie's office, she took one look at me and said, "No offense, but you look less than stellar, Suzanne."

She took my blood pressure, then blood and urine samples, and finally an EKG. All through the tests, I was in a daze. I felt hollow inside, and more than a little worried.

Following my examination, she sat down with Matt and me. Connie didn't look happy. "Your blood pressure is up, but it will be a day or so before we get your blood work back. I'll put a rush on it. In some ways things are steady, but I don't like how you were feeling today. Or how you *look*. I'm inches away from admitting you. I agree with Dr. Davis about the abortion. It's your decision, of course, but you're putting yourself at grave risk."

"God, Connie," I said, "short of stopping my practice altogether, I'm doing everything else right. I'm being so careful, so good."

"Then stop working altogether," she said without missing a beat. "I'm not kidding, Suzanne. I don't like what's going on with you. If you go home and make your *number one priority* absolute rest, then we have a chance. Otherwise, I'm checking you in."

I knew Connie meant what she said. She always did. "I'm going home now," I mumbled. "I can't give up on this baby."

Dear Nicholas,

I am so sorry, sweetie. A month has passed and you have kept me busy. I am also tired, and I haven't had a chance to write. I'll try to make it up to you.

At eleven months, your favorite words are *Dada, Mama, wow, watch, boat, ball, water* (*wa*), *car,* and your very favorite is *LIGHT.* You are crazy about lights. You say, "Yight."

You are like a windup toy these days. You just keep going and going and going and going and going.

I was in the middle of giving you my "be a good boy" rap when the phone rang. It was Connie Cotter's nurse, who put me on hold for the doctor.

It seemed to take forever before Connie got on the line. You came over and wanted to take the phone away from me. "Sure. Why don't you talk to Dr. Cotter," I said.

"Suzanne?"

"Yeah, I'm here, Connie. Taking it easy at home."

"Listen . . . we got your most recent bloods back. . . ."

Oh, that awful doctor's pause, that search for just the right wording. I know it only too well.

"And . . . I'm not happy. You're heading into the danger zone. I want to check you in right away. Start you on fluids. I'll show you the results on your bloods when you get here. How soon can that be?"

The words roared through my head with the force of a gale, taking all my strength with it. I was devastated. I had to sit down immediately. With the phone still to my ear, I lowered my head between my legs.

"I don't know, Connie. I'm here with Nicky. Matt's at work."

"Unacceptable, Suzanne. You could be in trouble, sweetie. I'll call Jean if you won't."

"No, no. I'll call her. I'll do it right now."

I hung up, and you held on to my hand like a strong little soldier. You knew just what to do — you must have learned it from your daddy.

I remember tucking you into your crib and pulling the cord on your music box. "Whistle a Happy Tune" begins to play. *It's so beautiful* — even in my nervous state of mind.

I remember turning on your night-light and closing the curtains.

I remember that I was on my way downstairs to call Grandma Jean, then Matt.

That's all I remember.

Matt found me lying as limp as a rag doll at the bottom of the stairs. I had a deep gash alongside my nose. Had I fallen down the entire flight? He called Grandma Jean and rushed me to the ER.

From there, I was transferred to the Critical Care Unit. I awoke to a whir of frantic activity around my bed. *Matt wasn't there anymore.*

I cried out for Matt, and both he and Connie were at my side in seconds. "You took a bad fall, Suzanne." Matt was the first to speak. "You passed out at the house."

"Is the baby okay? Connie, my baby?"

"We have a heart rate, Suzanne, but the situation isn't good. *Your* pressure is off the charts, your proteins are skyrocketing and . . ."

She paused long enough for me to know there was another big *and*.

"And what?" I asked.

"And you have toxemia. That could be

why you passed out at the house."

I knew what this abnormal condition meant, of course. My blood was poisoning both the baby and me. I had never heard of it occurring this early in a pregnancy, but Connie couldn't be wrong.

I was hearing what Connie was telling me in disjointed sound bites. I wasn't able to form whole sentences in my head. I felt as if I were being lobotomized. I thought I could actually feel the toxic blood swelling up inside me as if I were a dam about to break.

Then I heard Matt being ordered out of the room, and an emergency team rushing in. Doctors and nurses were swarming all around me. I could feel the oxygen mask covering my nose and my mouth.

I knew what was happening to me. In layman's terms:

My kidneys were shutting down.

My blood pressure was dropping.

My liver was barely functioning as guardian against the poisons.

My body was beginning to convulse.

Fluids and medications were given through an IV to stop the convulsions, but then I started hemorrhaging.

I knew I was shutting down. I knew so much more than I wanted to. I was scared. I was floating out of my body and then falling into a dark tunnel. The passing black walls were narrowing and squeezing the breath out of me.

I was dying.

Matt sits vigil by my bedside, day and night. Daddy never leaves me alone, and I worry about him. I have never loved him more than I do now. He is the best husband, the best friend, a girl ever had.

Connie visits constantly three or four times a day. I never knew what a great doctor she is, and what a great friend.

I hear her, and I hear Daddy. I just can't respond to either of them. I'm not sure why.

From what I can tell listening to them, I know that I've lost the baby. If I could cry, I would weep for all eternity. If I could scream, I would. I can do neither, so I mourn in the most awful silence imaginable. The sadness is bottled up inside and I ache to let it out.

Grandma Jean comes and sits with me for long stretches at a time, too. So do friends of mine from around the Vineyard, doctors from the hospital and even from Boston.

Melanie Bone and her husband, Bill, visit every day. Even Matt Wolfe, my lawyer friend, came by and whispered kind words to me.

I hear bits and pieces of what people are saying around me.

"If it's okay I'm going to bring Nicky in this afternoon," Daddy says to Connie. "He misses his mother. I think it's important he sees her." And then Matt says, "Even if it's for the last time. I think I should call Monsignor Dwyer."

Matt brings you to my hospital room, Nicholas. And then you and Daddy sit by my bedside all afternoon, telling me stories, holding my hand, saying good-bye.

I hear Matt's voice cracking, and I'm worried about him. A long time ago, his father died. He was only eight, and he never got over it. He won't even talk about his father. He's so afraid of losing someone again. And now it's me he's going to lose.

I just hold on. At least I think I'm still here. What other explanation can there be?

How could I possibly hear your laughter, Nicky? Or you calling out, "Mama," to me, in the black hole of my sleep?

But I do.

Your sweet little voice reaches down into my abyss and finds me in this deep dark place where I'm trapped. It is as if you and Daddy were calling me out of a strange dream, your voices like a beacon guiding me.

I struggle upward, reaching toward the sound of your voices — up, up, up.

I need to see you and Daddy one more time. . . .

I need to talk to you one more time. . . .

I feel a dark tunnel closing behind me, and I think that maybe I've found my way out of this lonely place. Everything is getting brighter. There is no more darkness surrounding me, just rays of warmth, and maybe the welcoming light of Martha's Vineyard.

Was I in heaven? Am I in heaven now? What is the explanation for what I'm feeling?

That's when the unexpected happens.

I *open* my eyes.

"Hello, Suzanne," Matt whispers. "Thank God, you came back to us."

Katie

There was only so much of the diary that Katie could take at any given time. Matt had warned her in his note: *there will be parts that may be hard for you to read.* Not just hard, Katie knew now, but overwhelming.

It was difficult for her to imagine right now, but there *were* happy endings in life.

There were normal, semisane couples like Lynn and Phil Brown, who lived in Westport, Connecticut, on a really cool little farm with their four kids, two dogs, and one rabbit and who were still in love as far as she or any of their other friends could tell.

The next day Katie called Lynn Brown and volunteered to sit for the kids that night, a one-night-only offer. She *needed* to be with the Browns. She needed the warmth and comfort of a family around her.

Lynn was immediately suspicious. "Katie, what's this all about? What's going on?"

"Nothing, I just miss you guys. Consider

it a pre-anniversary present for you and Phil. Don't look a gift horse in the mouth. I'm *in* Grand Central Station right now. I'm on my way."

She took the train to Westport and was at Lynn and Phil's by seven. At least she hadn't stayed late working at the office.

The Brown kids — Ashby, Tory, Kelsey, and Roscoe — were eight, five, three, and one. They loved Katie, thought she was so neat. They loved her long braid. And they loved that she was so *tall*.

So off went Lynn and Phil on their hot "date," and Katie took the kids. Actually, she was incredibly grateful to Lynn and Phil for "taking her in." They had met and liked Matt Harrison, and basically they knew what had happened between him and Katie. They didn't understand any of it, either. Lynn had predicted that Katie and Matt would be married within the year.

What a great night it turned out to be. The Browns had a small guest house that Phil was always threatening to fix up and make respectable. That was where Katie always went to hang out with the four kids.

They loved to play tricks on her, like hiding her suitcase and clothes or taking her makeup and putting it on (Roscoe included). She took the kids' pictures with her

Canon camera. They washed Lynn's Lexus SUV. Went on a group bike ride. Watched the movie *Chicken Run.* Ate an "everything" pizza.

When Lynn and Phil got home about eleven, they found Katie and the kids asleep on pillows and quilts thrown all over the guest-house floor.

She was actually awake and heard Lynn whisper to Phil, "She's so cool. She'll be a great mom." It brought tears to Katie's eyes, and she had to choke back a sob as she pretended to be asleep.

She stayed at the Brown house through Saturday afternoon. She finally took the six o'clock train back to New York. Before she left, she told Lynn that she was pregnant. She was exhausted, but she also felt alive again, rejuvenated — better, anyway. She believed in small miracles. She had hope. She knew there were some happy endings in life. She believed in families.

About halfway into the trip, Katie reached down into her bag and pulled out the diary.

She got off the train from Westport at the gorgeously renovated and restored Grand Central Station, and she needed to walk some. It was a little past seven-thirty and Manhattan was filled with traffic, most of it honking taxis or cars returning from weekend and vacation homes, the drivers *already* on edge.

She was on edge, too. The diary was doing that to her more and more.

She still didn't have the answer she needed to move on with her life. She wasn't over Matt — and she wasn't over Suzanne and Nicholas.

She was thinking about something she'd read earlier in the diary, the lesson of the five balls: work, family, health, friends, and integrity.

Work was a rubber ball, right?

Suzanne had figured that out, and her life had suddenly become peaceful and manageable. She had gotten away from all of *this:*

work, stress, pressure, deadlines, crowds pushing and shoving, road rage, life rage.

Immersing herself in someone else's reality had made Katie reexamine things that she had been doing on autopilot for the past nine years. She'd gotten her job at twenty-two, fresh out of the honors program at the University of North Carolina at Chapel Hill. She had been lucky enough to intern for two summers at Algonquin Press in Chapel Hill, which had opened important doors for her in Manhattan. So she had settled into New York City with the best of intentions, and loved so many things about it; yet she never felt that she truly fit, that New York City was where she was meant to be.

She still felt like a visitor here at times — a tall, gawky tourist.

Now she thought that maybe she knew why. Her life had been out of balance for such a long time. She had spent so many late nights at work or at home, reading and editing manuscripts, trying to make them as good as they could be. Rewarding work, but *work was a rubber ball, right?*

Family health, friends, and integrity were the precious glass ones.

The baby she was carrying was a glass ball for sure.

The following morning at about eleven, she was in a yellow cab with two of her best friends, Susan Kingsolver and Laurie Raleigh. She was going to see her gynecologist, Dr. Albert K. Sassoon, in the East Seventies.

Susan and Laurie were there for moral support. They knew about the pregnancy and had insisted on coming along. Each of them held one of Katie's hands.

"You feel okay, sweetie?" Susan asked. She was a grade-school teacher on the Lower East Side. They had met the one summer Katie had gone in on a summer house in the Hamptons, and had been best buddies ever since. Katie had been maid of honor at Susan's wedding, then a bridesmaid at Laurie's.

"I'm okay. Sure. I just can't make myself believe what's happened in the past few days. I can't believe I'm going to see Sassoon right now." *Oh, God, please help me. Please give me strength.*

As she got out of the taxi, Katie found that

she was blankly staring at pedestrians and familiar storefronts on East Seventy-eighth Street. What was she going to say to Dr. Sassoon? When Katie had been there for her yearly checkup, Albert was so incredibly excited to hear that she'd *found* someone — and now *this.*

Everything was a blur, even though Susan and Laurie were chatting amiably, keeping her *up,* doing a great job, really.

"Whatever you decide," Laurie whispered as Katie was called into Dr. Sassoon's examination room, "it will work out great. *You're* great."

Whatever she decided.

God, she just couldn't believe this was happening.

Albert Sassoon was smiling, and that made Katie think of Suzanne and her kindly way with patients.

"So," Dr. Sassoon said as Katie lay down and fitted her feet into the stirrups. Usually, Albert asked Katie not to hit him in the head with her knees. A little joke to lighten the moment. Not today, though.

"So. I was so much in love I stopped using my birth control. I guess I got knocked up," Katie said, and laughed. Then she was crying, and Albert came to her and tenderly held her head against his chest. "It's all right, Katie. It's all right. It's all right."

"I think I know what I'm going to do," Katie finally managed to say between sobs. "I think . . . I'm going . . . to keep . . . my baby."

"That's great, Katie," Dr. Sassoon said, and patted her back gently. "You'll be a wonderful mother. You'll have a beautiful child."

The Diary

Nicholas,

Today I came home from the hospital, and it's so unbelievably good to be here. Oh, I'm the luckiest girl in the world.

The familiarity of the rooms, your perfect nursery, the way the morning light comes spilling over the windowsills and lights all the things in its path. What a thrill to be here again. To be anywhere, actually.

Life is such a miracle, a series of small miracles. It really is, if you learn how to look at it with the right perspective.

I love our little cottage on Beach Road. More than ever, Nicky. I appreciate it more, every little crevice and crease.

Matt made a beautiful lunch for us. He's a pretty good cook — as handy with a spatula and skillet as he is with a hammer and nail. He laid out a picnic in the sunroom on a red-and-white-checkered blanket. A salad niçoise, fresh, twelve-grain bread, sun tea. Fabulous. After lunch the three of us sat there, and he held my hand and I held yours.

Nicholas, Suzanne, and Matt.

Happiness is this simple.

Nick, you little scamp,

Every moment with you fills me with such incredible wonder and happiness.

I took you into the Atlantic Ocean for the first time yesterday. It was the first day of July. You absolutely loved it.

The water was beautiful, with very small waves. Just your size. Even better was all the sand, your own private sandbox.

Big smiles from you.

And from me, of course.

Mommy see, Mommy do!

When we got home, I happened to show you a picture of two-year-old Bailey Mae Bone, our neighbor just down Beach Road. You started to smile, and then you *puckered* your lips. You're going to be a killer with the ladies. Be gentle, though, like your daddy.

You have good taste — for a guy. You love to look at pretty things — trees, the ocean, light sources, of course.

You also like to tickle the ivories on our piano, which is so cute.

And you love to *clean*. You push around a toy vacuum cleaner and wipe up messes with paper towels. Maybe I can take advantage of that when you're a little older.

Anyway, you are such a joy.

I treasure and hold close to my heart every giggle, every laugh, every needy cry.

"Wake up, beautiful. I love you even more today than I did yesterday."

Matt wakes me this same way every morning since I got home from the hospital. Even if I'm still half asleep, I don't mind being awakened by his soothing voice and those words.

The weeks passed, and I was getting my strength back. I began taking long walks on the beach in front of the cottage. I even saw a few patients. I exercised more than I ever had in my whole life.

A few more weeks passed, and I was even stronger. I was proud of myself, actually.

Matt was hovering over my bed again one morning. He was holding you, and smiling down on me. You both were grinning. I smelled a conspiracy.

"It's official! The three-day-long Harrison family weekend has begun. Wake up, beautiful. I love you! We're already late for today, though!"

253

"What?" I said, looking out the bedroom window. It was still dark outside.

You finally looked at your father as if he had gone completely bonkers.

"Down, *pup*," Matt said, putting you on my bed, beside me.

"Pack your bags. We're going away. Take whatever you need for three glorious days, Suzanne."

I was leaning on one elbow, staring curiously at Matt. "Three glorious days where?"

"I booked us into the Hob Knob Inn in Edgartown. King-size beds; full country breakfast, and afternoon tea. You won't have to lift a finger, wash a dish, or answer a telephone, Suzanne. Sound good?"

It sounded wonderful, exactly what I needed.

This is a love story, Nicholas. *Mine, yours, Daddy's!* It's about how good it can be if you find the right person. It's about treasuring every moment with that special one. *Every single millisecond.*

Our three-day adventure began at the Flying Horses Carousel, where we mounted the enchanted horses and circled the high hills of Oak Bluffs. There we were, riding the painted ponies under the bright umbrella, just like old times. What a rush!

We visited the beaches that we had been away from for so long. Lucy Vincent Beach off South Road, Quansoo and Hancock Beaches . . . private beaches that Matt, somehow, was able to get a key to gain entry.

We walked hand in hand in hand along Lighthouse Beach and Lobsterville Beach — and my very favorite, Bend in the Road Beach.

How invigorating it was to see those beaches again with Daddy and you. I can still see them now, and I can even see the three of us.

We took a carriage ride at Scrubby Neck Farm, and you couldn't stop laughing. You fed carrots to the horses, and you laughed so hard that I was afraid you might get sick. You glowed under the manes of the magnificent Belgian giants.

We ate at all the nicest restaurants, too. The Red Cat, the Sweet life Café, L'Etoile.

You looked like such a big boy in your high chair, sitting with us, so grown up, smiling in the candlelight.

We saw *Rumpelstiltskin* at the Tisbury Amphitheater and went to storytelling night at the Vineyard Playhouse. You were such a good boy at the *theater*.

Not far from where we were staying, there was a craft store called Splatter. We made our own cups and saucers.

You painted your plate, Nickels, drawing little splotches we took to be me and Daddy and yourself, in bright blues and soft yellows.

And then it was time to go home.

Nicky,

Do you remember any of this?

I noticed cars parked helter-skelter all along the side of Beach Road as we turned the last curve to our house. Several more cars, SUVs, and trucks were leading up to the driveway, but the strange thing was that *the driveway was no longer there.*

Instead, a new addition covered its place, and a new driveway lay on the far side of the addition, just as your daddy had promised.

"What," I asked Matt, shocked, "is all this?"

"A little extension, Suzanne. At least the humble beginnings of one. It's your new home office, and it has everything your old office didn't have. Now you can make less house calls, or *no* house calls. It's all right here in our backyard. Your office even has an ocean view."

Dozens of our friends and Matt's worker pals were on the lawn, applauding as we climbed out of the car. You started to clap your hands, too, Nicky. I think you were clapping for yourself, though.

"Suzanne! Matt!" our friends were chanting in sync with the clapping. I was in awe, speechless, struck dumb. For three days Matt's coworkers and friends must have hammered day and night to create this unbelievable space.

"I still have to do the electrical work and plumbing," Matt said in an apologetic tone.

"This is too much," I said as I hugged him tight.

"No," he whispered back, "it isn't nearly enough, Suzanne. I'm just so happy to have you home."

Nicholas, sweet Nicholas,

Everything seems to be moving in the right direction again. The time is really flying. Tomorrow, you will be one! Isn't that something? *Dang!*

What can I say, except that it is a godsend to watch you grow up, to see your first tooth, watch you take your first step, say your first word, make a half sentence, develop your little personality day by day.

This morning you were playing with Daddy's big, bad work boots that he keeps at the bottom of the closet; when you came out, you were standing in them. You started to laugh; you must have thought this was the funniest joke anyone has ever played. Then I was laughing, and Daddy came in, and he started laughing, too.

Nicholas, Suzanne, and Matt! What a trio.

We're going to celebrate your first twelve months tomorrow. I have your gifts all picked out. One of them is the pictures from our vacation. I selected the best couple of shots, and I'm having them framed. I won't tell you which picture I like best; that'll be a surprise.

But I will tell you that it will be in a silver frame with carved moons and stars and angels all around it. Just your style.

It's almost time to sing "Happy Birthday!"

Nicholas,

It's late, and Daddy and I are being silly geese. It's a little past midnight, so it's *officially your birthday. Hoo-ray! Congratulations, you!*

We couldn't resist, so we sneaked into your room and watched over you for several moments. We held hands and blew you kisses. You know how to blow kisses, too. You're so smart.

Daddy brought along one of your birthday presents, a bright red Corvette convertible. He placed it carefully at the foot of your crib. You and your dad are both caraholics: you boys live for cars; you feel the need for speed.

Matthew and I hugged each other as we watched you sleep — which is one of the greatest pleasures in the world — *don't miss watching your child sleep.*

Then I got a little playful, and I pulled the cord on your music box. It played that simple, beautiful song "Whistle a Happy Tune," which I know I will always associate with you sleeping in your crib.

Matt and I held each other and swayed to the music. I think we could have stayed there all night. Holding each other, watching you sleep, dancing to your music-box tune.

You didn't wake up, but a little smile crossed your face.

"Isn't it lucky?" I whispered to Matt. "Isn't this the best thing that could ever happen to anyone?"

"It is, Suzanne. It's so simple, but it's so right."

Finally, Daddy and I went to bed, and experienced the second best thing. Matt eventually fell asleep in my arms — guys do that if they really like you; and I got up to write this little note to you.

Love you, sweetie. See you in the morning. I can't wait.

Matthew

Hello, my sweet Nicholas, it's Dada.

Have I told you how much I love you? Have I told you how precious you are to me? There — *now I have*. You are the best little boy, the best anyone could ever hope for. I love you so much.

Yesterday morning something happened. And that's why I'm writing to you today instead of Mommy.

I am compelled to write this. I don't know anything for sure right now, except that I have to get this out. I have to talk to you.

Fathers and sons need to talk more than they do. A lot of us are so afraid to show our emotions, but I never want us to be like that. I always want to be able to tell you what I'm feeling.

Like right now.

But this is so hard, Nicky.

It's the hardest thing I have ever had to say to anybody.

Mommy was going to the store to pick up your birthday present, your beautiful framed pictures. She was incredibly happy. She looked so pretty, deeply tanned and toned from all her walks on the beach. I remember seeing her leave, and I can't get that image out of my mind.

Suzanne had such a beautiful smile on her face. She was dressed in a yellow jumper and gauzy white blouse. Her blond hair was full of curls and swung with her body as she walked. She was humming *your* song, "Whistle a Happy Tune."

I should have gone to her, should have kissed Suzanne good-bye, should have hugged her in my arms. But I just called, *"Love you,"* and since her hands were full, she blew me a kiss.

I keep seeing Suzanne blowing me that kiss. I see her walking away, looking back, giving me her famous wink. Imagining that playful wink of hers makes me tear up as I try to write this.

Oh, Nicky, Nicky, Nicky. How can I say this? How can I write these words?

Mommy had a heart attack on the way into town, sweet baby. Her heart, which was so big, so special in so many ways, could no longer hold out.

I can't imagine that it really happened; I

can't get it into my head. I was told that Suzanne was unconscious before she crashed into the guardrail on Old Pond Bridge Road. Her Jeep dropped into the water, landing on its side. I haven't gone to look at the actual scene of the accident. That is an image I don't need inside my head. What I can see already is too much.

Dr. Cotter says that Suzanne died instantly after the massive coronary, but who really knows about those final seconds? I hope she didn't feel any pain. I hate to think that she did. It would be too cruel.

She was unimaginably happy the last time I saw her. She looked so pretty, Nick. Oh God, I just want to see Suzanne one more time. Is that too much to ask? Is it unreasonable? It doesn't seem so to me.

It's important to me that you know it wasn't Mommy's fault. She was such a safe driver; she would never have taken any chances. I always teased her about her driving.

I loved Suzanne so much, and I can't begin to explain how lucky it is to find someone you can love that much and who, miracle of miracles, loves you that much back.

She was the most generous-hearted person I have ever known, the most caring and compassionate. Maybe what I loved best about her was that she was a great, great listener. And she was funny. She would make a joke, right now. I know she would. And maybe she is. *Are you smiling now, Suzanne?* I'd like to think that you are. I believe you must be.

I went today to the cemetery on Abel's Hill, to choose Mommy's special place. She was just thirty-seven when she died. How sad, how completely unthinkable to me, and everyone else who knew her. What a shame; what a waste. Sometimes it makes me so angry — and I get this strange, irrational urge to *break glass*. I don't know where it comes from, but I want to break glass!

Tonight I sit in your nursery and watch your clown lamp throw happy shadows against the walls in the half-light. The oak rocking horse I made for you reminds me of the Flying Horses Carousel. Remember when we all went there on our vacation and rode the colorful horses? *Nicholas, Suzanne, and Matt.*

I held you in front of me, and you loved to stroke the real horsehair mane. I can see Mommy riding ahead of us on National Velvet. She turns — and there's that famous wink of hers.

Oh, Nick, I wish I could turn back time to last week, or last month, or last year. I almost can't bear to face tomorrow.

I wish this had a happy ending.

I wish I could say just one more time: *Isn't it lucky?*

Dear sweet Nick,

There is one image that keeps coming back to me about Suzanne. It captures who she was, and what was so special and unique about her.

She is kneeling on our front porch one night. She wants my forgiveness, even though there is nothing to forgive. If anything, I should have been seeking her forgiveness. She had gotten some sad news that day but, in the end, could only think about how she might have hurt me. Suzanne always thought about other people first, but especially about the two of us. God, did she spoil us, Nicholas.

I was startled out of my thoughts and reveries this afternoon by an unexpected phone call.

It was for Mommy.

Obviously someone had no idea what had happened, and for the first time, those strange and awful words passed through my lips like heavy weights: "Suzanne has passed away."

There was a long silence on the other end, followed by quiet apologies, and then nervous condolences. It was the man from the frame shop on the other side of the island, in Chilmark Center. Mommy had never made it there, and the pictures she had framed for you were still at the store.

I told the shop owner that I would come around for the photos. Somehow, I would manage to do it. I feel so out of it all the time. I have a hollow feeling inside me, and it seems I could crumble like old tissue paper and blow away. At other times, there is a stone column inside my chest.

I never used to be able to cry, but now I cry all the time. I keep thinking that I'll run out of tears, but I don't. I used to think it wasn't manly to cry, but now I know that isn't so.

I walk aimlessly from room to room, trying desperately to find a place where I can feel at peace with myself. Somehow, I always end up back in your room, sitting in the same rocker Mommy so often did when she talked to you and read to you and recited her goofy rhymes.

And so I sit here now, looking at the pictures of us I finally picked up this afternoon in Chilmark.

We are all sitting in front of the Flying Horses Carousel on a perfect, blue-skied afternoon.

You are wedged between us, Nick. Mommy has her arm around you and her legs crossed on mine. You're kissing Mommy, and I'm tickling you, and everyone is laughing, and it's just so beautiful.

Nicholas, Suzanne, and Matt — Forever One.

It's time to tell you a story, Nick. It's a story that I will share only with you. It's just between the two of us.

Man to man, my little buddy.

Actually, it is the saddest story that I've ever heard, certainly the saddest one I've ever told.

I'm finding it hard to breathe right now. I'm shaking like a leaf. I have goose bumps all over my skin.

Years ago, when I was just eight, my father died very suddenly while he was at work. We didn't expect it, so we never got to say good-bye. For years, my father's death has haunted me. I've been so afraid of losing someone like that again. I think it's why I didn't get married earlier, before I met Suzanne. I was afraid, Nicky. Big, strong Daddy was so terribly afraid he might lose someone he loved. That's a secret I never told anyone before I met your mother. And now, I've told it to you.

I pull the cord on your music box in your crib, and it begins to play "Whistle a Happy Tune." I love this song, Nicky. It makes me cry, but I don't care. I love your music and I want to hear it again.

I reach into the crib and I touch your sweet cheek.

I tousle your golden blond hair, always so soft and fragrant. I wish I had listened to Mommy and never cut it.

I do a nose to nose, gently touching my nose to yours. I do another nose to nose and you smile gloriously. One of your smiles is worth the universe to me. That's the truth.

I place an index finger in each of your small hands and let you squeeze. You're so strong, buddy.

I listen to your beautiful laugh, and it almost makes me laugh.

"Whistle a Happy Tune" continues to play.

Oh my dear, darling little boy. Oh, my darling baby.

The music plays, but you *aren't* in your crib.

I remember Mommy leaving on her errand that morning. I called out "I love you," and she blew me a kiss. Then she crinkled up her nose the way she does. You know what I mean. You know that look of hers. Then she gave me her "famous wink," and I can see it right now. I can see Suzanne.

Her arms were full, because she was carrying *you*, sweet baby. She wanted you to be the first to see the beautiful framed photographs. That's why she took you with her to town on your birthday morning.

Suzanne carried you outside and carefully strapped you in your car seat. You were in the Jeep with Mommy when she crashed on Old Pond Bridge Road. The two of you were together. I still can't bear to think about it.

I should've been there, Nicholas. I should've been there with you and Mommy! Maybe I could have helped; maybe I could have saved you somehow. At least I could have tried, and that would have meant everything to me.

Oh sweetness, I need to hear your laugh one more time. I ache to look into your bright blue eyes. To nuzzle your soft cheek next to mine.

Oh my dear little boy, my innocent little sweetheart, my baby son forever. I miss you so much, and it destroys me that you will never know how I feel, that you will never hear how much your daddy loves you. I miss you so much, *I miss you so much*, sweet baby. I always will.

But isn't it lucky that I knew you, held you

and loved you, for the twelve months before God took you away?

Isn't it lucky that I got to know you, sweet little boy, my darling, darling son?

Katie

Katie slowly raised her face toward the bathroom ceiling and shut her eyes as tight as she could. A soft moan rose from her throat. Tears squeezed under her eyelids and rolled down both cheeks. Her chest was heaving. She wrapped both arms around herself.

Merlin was in the doorway, whining, and Katie whispered, "It's okay, boy."

A column of pain rose inside her like a hot poker cutting into her lungs. *Oh God, why would you let something like that happen?*

Finally, Katie opened her eyes again. She could barely see through her tears. There was an envelope taped inside the diary, on the very last page.

It said, simply, *Katie.*

She wiped away her tears with both hands. She took a deep, calming breath. And another. The breaths didn't help much. She opened the plain white envelope that was addressed to her.

The letter inside was in Matt's handwriting. Her fingers trembled as she unfolded it. The tears started again as she began to read.

Katie, dear Katie,

Now you know what I haven't been able to tell you all these months. You know my secrets. I wanted to tell you, almost since the day that we met. I have been grieving for such a long time, and I couldn't be comforted. So I kept my past from you. You, of all people. There are words from a poem about the local fishing boats and their crews that have been carved into the bar of Docks Tavern on the Vineyard. The longed-for ships / Come empty home or founder on the deep / And eyes first lose their tears and then their sleep. *I saw the words one night at Docks, when I couldn't cry anymore, and couldn't sleep, and I was almost crushed by the awful truth in them.*

Matt

That was all that he wrote, but Katie needed more. She had to find Matt.

She had always been a fighter. She'd conquered her fears to come to New York by herself. She'd always had the courage to do what she had to do.

Katie took the shuttle to Boston first thing in the morning. At Logan Airport, she was met by a car service that would take her to Woods Hole and the ferry to Martha's Vineyard.

She entered the Steamship Authority terminal in Woods Hole, bought her ticket, and got on a two-decker ferry called the *Islander*.

She had to talk to Matt. It was wrong not to let him know everything. It was just plain wrong, and she couldn't live that way. Matt needed to know about the baby.

During the seven-mile, forty-five-minute ride, she thought of Suzanne, and *her* arrival on the Vineyard after she left Boston. She wondered if Suzanne had been on board the *Islander*, too. She remembered the last words Suzanne had written to Nicholas: *I can't wait to see you in the morning.*

Katie realized she hadn't brought a manu-script to read on the plane or the ferry. *Work is a rubber ball,* she thought. *Yes, it is.*

God, look at what she would have missed if she had brought along paperwork: the rhythmic chop of the waves against the ancient ferry's bow, the picturesque island of Martha's Vineyard getting closer and closer, the queasiness in her stomach every time a big wave splashed into the ship.

Matt was a glass ball. He had been scuffed, marked, damaged, but maybe he hadn't been shattered. Or maybe he had been.

The mystery would never be solved unless she found him.

As the *Islander* got closer and closer to the Vineyard, Katie couldn't take her eyes off the old Oak Bluffs ferry terminal. It was a gray clapboard building, a one-story structure that looked a hundred years old if it was a day. She could see a beach on one side of the terminal, and the small town of Oak Bluffs on the other.

Her eyes searched the terminal building, the beach, the town — looking for Matt.

She didn't see him anywhere.

The town buildings of Oak Bluffs were across the street from the ferry terminal. There were several odd-colored taxis parked out front. And, of course, Matt wasn't waiting there for her to appear. He didn't know she was coming, and even if he had, he might not have come.

Katie spotted Docks Tavern as she started toward the taxi stand. Her heart skipped a beat. This had to be a sign, no? Had to be something. She walked toward the bar instead of searching for a cab.

Was Matt in there? Probably not, but Docks was where he had read the lines carved into the bar, which he had included in his note in the diary.

It was dark inside, a little smoky, pleasant enough, though. A Bruce Springsteen song played from an old Wellington jukebox. About a dozen patrons were at the bar, and several people were seated in the weathered wooden booths on either side. Most of them looked up at her as she entered. She knew she was having a bad hair day, bad clothes day, bad life day.

"I come in peace," Katie said, and smiled.

She was incredibly nervous, though. She had decided she was coming to Martha's Vineyard about three in the morning. She had to see Matt again. She wanted to be in his arms and to hold him, even if that might not happen. Katie needed a hug badly.

Her eyes roamed slowly over the faces, which seemed right out of *The Perfect Storm*. Her heart sped up some. She didn't see Matt. Well, thank God, he wasn't a regular at least.

She went looking for the poem carved into the bar. It took her a few minutes to find it at the far end, near a dartboard and a public phone. She read the words again:

The longed-for ships.
Come empty home or founder on the deep.
And eyes first lose their tears and then their sleep.

"Help you with something? Or is your interest wholly literary?"

She looked up at the sound of the male voice. She saw a bartender, mid-thirties, red-bearded, ruggedly good-looking. Maybe a sailor himself.

"I'm just looking for someone. A friend. I think he comes in here," she said.

"He has good taste in taverns, anyway. Does he have a name?"

She took in a breath and tried to keep the tremor out of her voice. "Matt Harrison," Katie said.

The bartender nodded, but his dark brown eyes narrowed. "Matt comes in here for dinner sometimes. He paints houses on the island. You say you're a friend of his?"

"He also writes books," Katie said, feeling a little defensive now. "Poetry."

The bartender shrugged, and continued to look at her suspiciously. "Not that I know of. At any rate, Matt's not here today. As you can see for yourself." The red-bearded man finally smiled at her. "So what will it be? You look like a Diet Coke to me."

"No, nothing, thanks. Could you tell me how to get to his place? I'm a friend of his. I'm his editor. I have the address."

The bartender thought about it, and then he tore a sheet off his order pad. "You driving?" he asked as he began to write down a few directions.

"I'll probably take a cab."

"They'll know the place," the man said, but didn't elaborate. "Everybody knows Matt Harrison."

Katie slowly climbed into a rusted sky blue Dodge Polaris cab at the ferry terminal. Suddenly she was feeling tired. She said to the driver, "I'd like to go to the Abel's Hill Cemetery. Do you know it?"

By way of an answer, the cabdriver simply pulled away from the curb. She guessed he knew where everything was on the island. She certainly hadn't meant to offend him.

Abel's Hill was a good twenty minutes away, a small, picturesque place that looked at least as old and historic as any of the houses they had passed on the way there.

"I won't be too long," she said to the driver as she struggled out of the backseat. "Please wait for me."

"I'll wait, but I have to keep the meter running."

"That's fine. I understand," she told him, and shrugged. "I'm from New York City. I'm used to it."

The cab waited while she slowly and reverently walked from row to row in Abel's Hill, checking all the headstones, but especially the newer ones. During the ride over, the cabdriver had told her that John Belushi and the writer Lillian Hellman were buried here.

Her chest felt tight, and there was a lump in her throat as she searched for the grave. She felt as if she were intruding.

Finally she found it. She saw the carved lettering on a stone set on a hill, *Suzanne Bedford Harrison.*

Her heart clutched again, and she felt dizzy. She bent and went down on one knee.

"I had to come, Suzanne," she whispered. "I feel as if I know you so well by now. I'm Katie Wilkinson."

Her eyes traveled across the inscription. *Country doctor, much loved wife of Matthew, perfect mother of Nicholas.*

Katie offered up a prayer, one that her father had taught her when she was only three or four.

She turned to the smaller stone right beside Suzanne's. She sucked in a breath.

Nicholas Harrison, a real boy, cherished son of Suzanne and Matthew.

"Hello, sweet baby boy. Hello, Nicholas. My name is Katie."

She began to sob uncontrollably then.

She clutched her chest with both arms, and her whole body shook like a weeping willow in a storm. She mourned for poor baby Nicholas. She couldn't begin to understand how Matt had survived this.

She imagined him in Nicholas's room, playing the music box on the crib over and over, trying to remember how it had been with his baby son, trying to bring Nicholas back.

There were flowers, daisy poms, carnations, and gladiolas at both of the graves. *Someone has been here recently, maybe even today.* Matt had always given her roses. He was a good man, sweet and kind. She'd been right about that. She hadn't made a bad choice, just an unlucky one.

And then Katie noticed something else, the date that was carved into the two headstones.

July 18, 1999.

She felt a shiver vibrate through her, and her knees were weak again. July 18 was two years to the day of the party she'd had planned for Matt on her terrace in New York, the night she'd given him the copy of his book of poems. No wonder he ran away. And now, where was Matt?

Katie had to see him — one more time.

It took another twenty minutes for the creaky island cab to bump its way from the cemetery to the old boathouse that she immediately recognized as Suzanne's.

It was painted white now. The barnlike doors and the trim were gray. There was a flower garden full of hydrangeas, azaleas, and daylilies.

She could see why Suzanne had loved it so much. Katie did, too. It was a real home.

She slowly got out of the cab. An ocean breeze played with her hair. She felt the wind gently pat her face and her bare legs. Her heart was back into its pounding routine again.

"Should I wait?" the driver asked.

Katie nibbled her upper lip, crossed and uncrossed her long arms. She looked at her watch: 3:28. "No. Thanks. You can go this time. I'll be here for a while."

She paid the driver, and he sped off.

Her heart was stuck in her throat as she

walked up the gravel path to the house. Her eyes did a once-over of the property. She saw no sign of Matt. No car. Maybe it was in back.

She knocked on the front door, waited, fidgeted, then used the old wooden knocker.

No one answered.

God, it was so weird to be here.

Her heart just wouldn't stop pounding.

She didn't see a sign of anyone at the house, but she was determined to wait for Matt. She could almost imagine him showing up now: old jeans, a khaki shirt, work boots, welcoming smile.

Would Matt smile if he saw her here? She needed to talk to him, to get some things off her chest. It was her turn to talk. She deserved that much. She had secrets she needed to share.

So she waited and waited. Then Katie sat on the front lawn for a while, massaging her stomach gently, listening to the waves. Eventually she crossed Beach Road . . . where Suzanne's dog, Gus, had been struck by a speeding red truck.

She sat on the beach where Matt and Suzanne had danced in the moonlight. She could *see* them. And then she imagined dancing with Matt again. He wasn't a great dancer, but she had loved being in his strong arms. She didn't like admitting it now, but it was the truth. It would always be the truth.

She thought that she probably had most of the mystery solved: Matt couldn't get Suzanne and Nicholas out of his mind, couldn't stop grieving. He probably didn't think that he ever could. Maybe he couldn't bear the thought of losing someone again. He had lost his wife and year-old child, and even his father when he was just a boy.

She couldn't blame him; she really couldn't. Not since she'd read the diary and understood what he had been through. If anything, and this really hurt her, she loved Matt more now than she ever had.

Katie picked her head up and saw a small, dark-haired woman in a pale blue dress, but barefoot. She was walking toward her across Beach Road. Katie didn't take her eyes off her.

When the woman was close, she said, "You're Melanie Bone, aren't you?"

Melanie had the nicest, friendliest smile, just what she would have imagined. "And you're Katie. You're Matthew's editor from New York. He told me about you. He said you were willowy and pretty; that you usually wore your dark hair in a braid but sometimes loose strands fell across your cheeks."

Katie wanted so much to ask Melanie what else Matt had said, but she didn't, couldn't. "Do you know where he is?" she asked.

Melanie grimaced and shook her head. "He's not here. I'm sorry, Katie. I don't know where Matt is. We're all worried about him, actually. I was hoping that he was with you in New York."

"He's not," Katie said. "I haven't seen him, either."

Late in the afternoon Melanie gave Katie a ride back to the ferry terminal in Oak Bluffs. The kids rode in the back of the station wagon. They were just about as good-natured as their mother. They liked Katie right away and she liked them.

"Don't give up on him," Melanie said as Katie was about to walk away to board the *Islander.* "He's worth it. Matt's had the worst experience of anyone I know. But I think he'll recover. He's a really good person. Handy around the house, too. And Katie, I know he loves you."

Katie nodded, and she waved good-bye to the Bone family. Then she left Martha's Vineyard the way she had come there, alone.

Another long, bad week passed for her. Katie fell deeper and deeper into her work, but she thought a lot about going home to North Carolina. For good. She would have the baby there, among the people she loved and who loved her.

Katie hadn't been in the office very long that Monday morning when she heard her name being called.

She had just transferred her tea from the blue Le Croissant paper cup to the antique china one she kept on her desk. Her stomach didn't feel too bad that morning. Or maybe she was just getting used to it.

"Katie? Come over here right now. Katie! Now."

She was slightly annoyed. "What, *what?* I'm coming. Hold your horses."

Her assistant, Mary Jordan, was poised behind a floor-to-ceiling window that looked down on East Fifty-third Street. She motioned for Katie to come to the window.

293

"Come *here!*"

Curious, she walked to the window and looked down on the street. She spilled hot tea on herself, nearly dropping her antique cup, until Mary reached out and deftly snatched it from her.

Katie then walked past Mary, down the short hallway of the publishing-house offices, to the single elevator. Her knees were weak, her head spinning. She was self-consciously brushing strands of hair away from her face. She didn't know what to do with her hands.

She passed the publisher and owner, who was getting out of the elevator. "Katie, I need to talk —" He started to say something, but she cut him off with a raised hand and a shake of the head. "I'll be right back, Larry," she said, then rushed into the elevator, which was just starting back down. The publishing-house offices were on the top floor.

Time to compose yourself, she thought.

No, not enough time. Not even close.

The elevator descended to the first floor without making any stops.

Katie stood in the lobby and forced herself to be very still inside. Her thoughts were amazingly concise, actually. Suddenly everything seemed so clear and simple to her.

She thought about Suzanne, about Nicholas, and about Matt.

She thought about the lesson of the five balls.

Then Katie walked outside the building and onto the streets of New York. She took a deep breath as the warmth of the sunshine struck her face.

Dear God, make me strong enough for whatever is going to happen now.

She saw Matthew on Fifty-third Street.

He was kneeling on the sidewalk, less than a dozen feet away from where Katie stood, right in front of her office building. His head was bowed slightly. He was courteous and considerate enough to have placed himself out of the main pedestrian flow. She couldn't take her eyes off him.

Of course, *everyone* looked at him as they passed. How could they resist? Rubbernecking was an art in New York City.

He looked good: tan, trim, his hair a little longer than usual; jeans, a clean but frayed chambray shirt, dusty work boots. He looked like the Matt she knew, the Matt she had loved, and realized now that she still did.

Kneeling in front of her building. Right there in front of her.

Just as Suzanne had knelt that one night on their porch — to ask forgiveness, even though there was nothing to forgive.

Katie believed she knew what she had to

do. She followed her instincts on this, followed her heart.

She took a breath, then she got down on one knee beside Matt, facing him, very close to him, as close as she could get. Her heart was thundering. *Thump-thump, thump-thump.*

She had wanted to see Matt one more time, and here he was. Now what?

Pedestrians were starting to clog up the sidewalk. A few of them made unkind remarks, complaining about the loss of a few precious seconds on their journeys to work, or wherever it was that they rushed off to every morning.

Matt reached out his hand. Katie hesitated, but then she let him take her long, thin hands in his.

She had missed his touch. Oh God, she had missed this.

She had missed a lot about him, but especially the way she felt at peace when he was with her.

Strangely, she was starting to feel calm now. What did that mean? What was supposed to happen next?

Why was he here? To apologize or explain in person? What?

Finally Matt raised his head and looked at her. She had missed those soft brown eyes, even more than she thought. She'd missed his strong cheekbones, the furrowed brow, his perfect lips.

Matt spoke, and, God, she had missed the sound of his voice. "I love looking into your eyes, Katie, the honesty I see there. I love your country drawl. You're so unique, and I treasure that. I love being with you. I never tire of it. Not for one minute since I've known you. You are a great editor. You're a great carpenter, too. You *are* tall, but you *are* ravishing."

Katie found that she was smiling. She couldn't help it. Here they were, the two of them, on their knees in midtown. Nobody could possibly understand what they were doing and why. Maybe not even they themselves understood.

"Hello, stranger," she said. "I went looking for you, Matt. I traveled to the Vineyard. I finally got up there."

Matt smiled now. "So I heard. From Melanie and the kids. They thought you were ravishing, too."

"What else?" Katie asked. She needed to know more, to learn more, anything that he would tell her. God, she was so glad to see him again. She couldn't have imagined how glad she would be, how this would feel.

"What else? Well, the reason I'm here, on my knees, is I want to give myself over to you,

Katie. I'm sure of it. I'm finally ready. I'm yours, if you'll have me. I want to be with you. I want to have children with you. I love you. I'll never leave you again. I promise, Katie. I promise with all my heart."

And then, they finally kissed.

That October on the gorgeous Outer Banks of North Carolina, Katie Wilkinson and Matt Harrison were married at the Kitty Hawk Chapel.

The Wilkinson and Harrison families hit it off famously right from the start. The two families immediately became one. Katie's friends from New York all came down, spent a few extra days at the beach, and got lobster pink, of course. Her North Carolina friends preferred the cover of porches and shade trees. Both groups of friends reached agreement on the mint juleps.

Katie was thin, but she wasn't showing too much. Only a few of the wedding guests knew that she was going to have a baby. When she had told Matt, he hugged and kissed her and said he was the happiest, luckiest person in the world.

"Me, too," said Katie. "Actually, me three."

It was a simple but beautiful wedding and reception, held under cloudless blue skies with temperatures hovering in the low seventies. Katie looked like an angel, white, with wings. Tall. Ravishing. The wedding was completely unpretentious from beginning to end. The tables were decorated with family photographs. The bridesmaids carried pale pink hydrangeas.

While they were exchanging vows, Katie couldn't help thinking to herself, *Family, health, friends, integrity — the precious glass balls.*

She understood it now.

And that was how she would live the rest of her life, with Matt and their beautiful baby.

Isn't it lucky.

We hope you have enjoyed this Large Print book. Other Thorndike Press or Chivers Press Large Print books are available at your library or directly from the publishers.

For more information about current and upcoming titles, please call or write, without obligation, to:

Thorndike Press
295 Kennedy Memorial Drive
Waterville, Maine 04901 USA
Tel. (800) 223-1244
Tel. (800) 223-6121

OR

Chivers Press Limited
Windsor Bridge Road
Bath BA2 3AX
England
Tel. (0225) 335336

All our Large Print titles are designed for easy reading, and all our books are made to last.

Methods in Enzymology

Volume 415
GLYCOBIOLOGY

METHODS IN ENZYMOLOGY

EDITORS-IN-CHIEF

John N. Abelson   Melvin I. Simon

DIVISION OF BIOLOGY
CALIFORNIA INSTITUTE OF TECHNOLOGY
PASADENA, CALIFORNIA

FOUNDING EDITORS

Sidney P. Colowick and Nathan O. Kaplan

Methods in Enzymology

Volume 415

Glycobiology

EDITED BY

Minoru Fukuda

GLYCOBIOLOGY PROGRAM
CANCER RESEARCH CENTER
THE BURNHAM INSTITUTE FOR MEDICAL RESEARCH
LA JOLLA, CALIFORNIA

AMSTERDAM • BOSTON • HEIDELBERG • LONDON
NEW YORK • OXFORD • PARIS • SAN DIEGO
SAN FRANCISCO • SINGAPORE • SYDNEY • TOKYO

Academic Press is an imprint of Elsevier

ELSEVIER

Academic Press is an imprint of Elsevier
525 B Street, Suite 1900, San Diego, California 92101-4495, USA
84 Theobald's Road, London WC1X 8RR, UK

This book is printed on acid-free paper. ∞

For information on all Elsevier Academic Press publications
visit our Web site at www.books.elsevier.com

ISBN-13: 978-0-12-182820-2
ISBN-10: 0-12-182820-4

PRINTED IN THE UNITED STATES OF AMERICA
06 07 08 09 9 8 7 6 5 4 3 2 1

Table of Contents

Section I. N-Glycan Processing

Section II. Structural Analysis

Section III. Carbohydrate Synthesis and Antibiotics

Section IV. Carbohydrate Ligand Specificity

Contributors to Volume 415

Article numbers are in parentheses following the names of contributors.
Affiliations listed are current.

NICHOLAS J. AGARD (15), *Departments of Chemistry and Molecular and Cell Biology, Howard Hughes Medical Institute, University of California, Berkeley, California*

RICHARD A. ALVAREZ (18), *Department of Biochemistry & Molecular Biology, University of Oklahoma Health Sciences Center, Oklahoma City, Oklahoma*

JEREMY M. BASKIN (15), *Departments of Chemistry and Molecular and Cell Biology, Howard Hughes Medical Institute, University of California, Berkeley, California*

VICENTE VEREZ BENCOMO (10), *Faculty of Chemistry, University of Havana, Habana, Cuba*

CAROLYN BERTOZZI (15), *Departments of Chemistry and Molecular and Cell Biology, Howard Hughes Medical Institute, University of California, Berkeley, California*

MICHEL BEURRET (10), *Netherlands Vaccine Institute, Bilthoven, The Netherlands*

OLA BLIXT (9, 18), *Department of Molecular Biology, The Scripps Research Institute, La Jolla, California*

ISAAC S. CARRICO (15), *Departments of Chemistry and Molecular and Cell Biology, Howard Hughes Medical Institute, University of California, Berkeley, California*

WENGANG CHAI (20), *The Glycosciences Laboratory, Imperial College London, Harrow Middlesex, United Kingdom*

PAMELA V. CHANG (15), *Departments of Chemistry and Molecular and Cell Biology, Howard Hughes Medical Institute, University of California, Berkeley, California*

ROBERT A. CHILDS (20), *The Glycosciences Laboratory, Imperial College London, Harrow Middlesex, United Kingdom*

LOURDES COSTA (10), *Center for Genetic Engineering and Biotechnology, University of Havana, Habana, Cuba*

CHRISTOPHER L. DE GRAFFENRIED (14), *Department of Cell Biology, Yale School of Medicine, New Haven, Connecticut*

JOSE L. DE PAZ (17), *Laboratory for Organic Chemistry, Swiss Federal Institute of Technology (ETH), Zürich, Zürich, Switzerland*

KISABURO DEGUCHI (6), *Biological Sciences Laboratory of Glyco-Finechemistry, Hokkaido University, Sapporo, Japan*

ANNE DELL (5), *Department of Molecular Biosciences, Imperial College London, London, United Kingdom*

DANIELLE H. DUBE (14), *Department of Chemistry, Stanford University, Stanford, California*

YOUJI EBE (21), *Yokohama Technical Center, Moritex Corporation, Nano/bio Science Research Lab, Kanagawa, Japan*

TEN FEIZI (20), *The Glycosciences Laboratory, Imperial College London, Harrow Middlesex, United Kingdom*

MINORU FUKUDA (11), *Glycobiology Program, Cancer Research Center, The Burnham Institute for Medical Research, La Jolla, California*

ANJALI S. GANGULI (15), *Departments of Chemistry and Molecular and Cell Biology, Howard Hughes Medical Institute, University of California, Berkeley, California*

NINGGUO GAO (1), *Department of Pharmacology, UT-Southwestern Medical Center, Dallas, Texas*

DAVID GOLDBERG (5), *The Scripps-PARC Institute, Advanced Biomedical Sciences, Palo Alto, California*

MATTHEW J. HANGAUER (15), *Departments of Chemistry and Molecular and Cell Biology, Howard Hughes Medical Institute, University of California, Berkeley, California*

GERALD HART (8), *Department of Biological Chemistry, Johns Hopkins University School of Medicine, Baltimore, Maryland*

STUAR HASLAM (5), *Department of Molecular Biosciences, Imperial College London, London, United Kingdom*

HIROSHI HINOU (13), *Hokkaido University, Graduate School of Advanced Life Science, Sapporo, Japan*

JUN HIRABAYASHI (19, 21), *AIST, Research Center for Glycoscience, Ibaraki, Japan*

KOJI HORIO (21), *Yokohama Technical Center, Moritex Corporation, Nano/bio Science ResearchLab, Kanagawa, Japan*

TIM HORLACHER (17), *Laboratory for Organic Chemistry, Swiss Federal Institute of Technology (ETH), Zürich, Zürich, Switzerland*

LUIS PEÑA ICART (10), *Faculty of Chemistry, University of Havana, Habana, Cuba*

YUKI ITO (11), *Department of Pathology, Shinshu University School of Medicine, Matsumoto, Japan*

JIHYE JANG-LEE (5), *Department of Molecular Biosciences, Imperial College London, London, United Kingdom*

MASATOMO KAWAKUBO (11), *Department of Pathology, Shinshu University School of Medicine, Matsumoto, Japan*

MOTOHIRO KOBAYASHI (11), *Department of Pathology, Shinshu University School of Medicine, Matsumoto, Japan*

JENNIFER J. KOHLER (14), *Department of Chemistry, Stanford University, Stanford, California*

SHIORI KOSEKI-KUNO (21), *AIST, Research Center for Glycoscience, Ibaraki, Japan*

ATSUSHI KUNO (21), *AIST, Research Center for Glycoscience, Ibaraki, Japan*

MASAKI KUROGOCHI (13), *Graduate School of Advanced Life Science, Hokkaido University, Sapporo, Japan*

SCOTT T. LAUGHLIN (15), *Departments of Chemistry and Molecular and Cell Biology, Howard Hughes Medical Institute, University of California, Berkeley, California*

HEESEOB LEE (11), *Glycobiology Program, Cancer Research Center, The Burnham Institute for Medical Research, La Jolla, California*

MARK A. LEHRMAN (1), *Department of Pharmacology, UT-Southwestern Medical Center, Dallas, Texas*

WILLIAM J. LENNARZ (4), *Department of Biochemistry and Cell Biology, Stony Brook University, Stony Brook, New York*

YAN LIU (20), *The Glycosciences Laboratory, Imperial College London, Harrow Middlesex, United Kingdom*

ANDERSON LO (15), *Departments of Chemistry and Molecular and Cell Biology, Howard Hughes Medical Institute, University of California, Berkeley, California*

BRUCE A. MACHER (7), *Department of Chemistry and Biochemistry, San Francisco State University, San Francisco, California*

STEVEN W. MAST (3), *Department of Biochemistry & Molecular Biology, University of Georgia Complex Carbohydrate Research Center, Athens, Georgia*

KELLEY W. MOREMEN (3), *Department of Biochemistry & Molecular Biology, University of Georgia Complex Carbohydrate Research Center, Athens, Georgia*

HOWARD MORRIS (5), *Department of Molecular Biosciences, Imperial College London, London, United Kingdom*

HIROAKI NAKAGAWA (6), *Biological Sciences Laboratory of Glyco-Finechemistry, Hokkaido University, Sapporo, Japan*

SACHIKO NAKAMURA-TSURUTA (19), *AIST, Research Center for Glycoscience, Ibaraki, Japan*

JUN NAKAYAMA (11), *Department of Pathology, Shinshu University School of Medicine, Matsumoto, Japan*

SHIN-ICHIRO NISHIMURA (13), *Graduate School of Advanced Life Science, Hokkaido University, Sapporo, Japan*

SIMON J. NORTH (5), *Department of Molecular Biosciences, Imperial College London, London, United Kingdom*

MARIA PANICO (5), *Department of Molecular Biosciences, Imperial College London, London, United Kingdom*

JENNIFER A. PRESCHER (15), *Departments of Chemistry and Molecular and Cell Biology, Howard Hughes Medical Institute, University of California, Berkeley, California*

NAHID RAZI (9), *Department of Molecular Biology, The Scripps Research Institute, La Jolla, California*

VIOLETA FERNANDEZ SANTANA (10), *Faculty of Chemistry, University of Havana, Habana, Cuba*

PETER H. SEEBERGER (17), *Laboratory for Organic Chemistry, Swiss Federal Institute of Technology (ETH), Zürich, Zürich, Switzerland*

KEITH A. STUBBS (16), *Department of Chemistry, Simon Frasier University, Burnaby, British Columbia, Canada*

MARK SUTTON-SMITH (5), *Department of Molecular Biosciences, Imperial College London, London, United Kingdom*

TADASHI SUZUKI (4), *Department of Biochemistry, Osaka University Graduate School of Medicine, Osaka, Japan*

TADASHI TAI (2), *Seikagaku Corporation, Central Research Laboratories, Tokyo, Japan*

KAORI TANABE (4), *Department of Biochemistry, Osaka University, Graduate School of Medicine, Osaka, Japan*

NOBORU UCHIYAMA (19, 21), *AIST, Research Center for Glycoscience, Ibaraki, Japan*

DAVID J. VOCADLO (16), *Department of Chemistry, Simon Frasier University, Burnaby, British Columbia, Canada*

ERIC WESTHOF (12), *Biologie Moléculaire et Cellulaire du CNRS, Université Louis Pasteur, Strasbourg Cedex, France*

STEPHEN A. WHELAN (8), *Department of Biological Chemistry, Johns Hopkins University School of Medicine, Baltimore, Maryland*

JULA WIRMER (12), *Biologie Moléculaire et Cellulaire du CNRS, Université Louis Pasteur, Strasbourg Cedex, France*

MASAO YAMADA (21), *Yokohama Technical Center, Moritex Corporation, Nano/bio Science Research Lab, Kanagawa, Japan*

TEN-YANG YEN (7), *Department of Chemistry and Biochemistry, San Francisco State University, San Francisco, California*

Preface

In the past decade, we have seen an explosion of progress in understanding the roles of carbohydrates in biological systems. This explosive progress was made with the efforts in determining the roles of carbohydrates in immunology, neurobiology, and many other disciplines, examining each unique system and employing new technology. Thanks to Academic Press Editorial Management, particularly to Ms. Cindy Minor, three books, namely Glycobiology (vol. 415), Glycomics (vol. 416), and Functional Glycomics (vol. 417), in the series of *Methods in Enzymology*, have been dedicated to disseminate information on methods in determining the biological roles of carbohydrates. These books are designed to provide an introduction of new methods to a large variety of readers who would like to participate in and contribute to the advancement of glycobiology. The methods covered include structural analysis of carbohydrates, biological and chemical synthesis of carbohydrates, expression and determination of ligands for carbohydrate-binding proteins, gene expression profiling including micro array, and generation of gene knockout mice and their phenotype analyses. The book also covers recent advances in special topics such as chaperones for glycosyltransferase, the roles of glycosylation in signal transduction, chemokine and cytokine binding, muscle development, glycolipids in development, and cell-cell interaction. I believe that we have a collection of outstanding contributors who represent their respective expertise and field.

The current Glycobiology vol. 415 covers methods on N-glycan processing, structural analysis, carbohydrate synthesis and antibiotics, and carbohydrate ligand specificity including glycan array analysis. I believe that this book will be useful to a wide variety of readers from graduate students, researchers in academics and industry, to those who would like to teach glycobiology at various levels. We hope this book will contribute to explosive progress in Glycobiology.

MINORU FUKUDA

METHODS IN ENZYMOLOGY

Section I

N-Glycan Processing

[1]   Non-Radioactive Analysis of Lipid-Linked Oligosaccharide Compositions by Fluorophore-Assisted Carbohydrate Electrophoresis

By NINGGUO GAO and MARK A. LEHRMAN

Abstract

Lipid-linked oligosaccharides (LLOs) are the donors of glycans that modify newly synthesized proteins in the endoplasmic reticulum (ER) of eukaryotes, resulting in formation of N-linked glycoproteins. The vast majority of LLO analyses have relied on metabolic labeling with radioactive sugar precursors, but these approaches have technical limitations resulting in many important questions about LLO synthesis being left unanswered. Here we describe the application of a facile non-radioactive technique, fluorophore-assisted carbohydrate electrophoresis (FACE), which circumvents these limitations. With FACE, steady-state LLO compositions can be determined quantitatively from cell cultures and animal tissues. We also present FACE methods for analysis of phosphosugars and nucleotide sugars, which are metabolic precursors of LLOs.

Overview

Analysis of LLO Biosynthesis by Incorporation of Radiolabeled Precursors

The lipid-linked oligosaccharide (LLO) glucose$_3$mannose$_9$N-acetylglucosamine$_2$-P-P-dolichol (G$_3$M$_9$Gn$_2$-P-P-Dol), or some variant of it, can be found in all eukaryotes. It is the donor substrate of oligosaccharyltransferase (OT) for N-linked glycosylation of newly synthesized polypeptides in the lumen of the endoplasmic reticulum (ER). By far, the most widely used approach for characterization of LLOs has been labeling of cells with [2-^{3}H]mannose, which is converted successively to [^{3}H]mannose-6-P (M6P), [^{3}H]mannose-1-P (M1P), GDP-[^{3}H]mannose, and [^{3}H]mannose-P-dolichol (MPD). GDP-mannose is the donor for the mannosyl residues of M$_5$Gn$_2$-P-P-Dol, whereas MPD is the donor for the subsequent mannosyl residues forming M$_9$Gn$_2$-P-P-Dol. This tried-and-true technique therefore permits efficient labeling of LLOs with multiple mannosyl residues. It is also quite sensitive: in our hands, a 100-mm culture dish with approximately 5×10^6 CHO-K1 cells routinely provides enough ^{3}H-LLO for several HPLC analyses using a standard in-line

METHODS IN ENZYMOLOGY, VOL. 415
0076-6879/06 $35.00
DOI: 10.1016/S0076-6879(06)15001-6

tritium detector. By pulse-labeling, it can be used to assess the rates of synthesis of LLO intermediates. It is also advantageous because labeling is effectively limited to mannosyl conjugates, as well as fucosyl conjugates after conversion of GDP-mannose to GDP-fucose. Although M6P is readily converted to F6P by phosphomannose isomerase, leading to G6P, labeling does not extend into other major classes of glycoconjugates because isomerization of M6P to F6P results in loss of tritium with formation of $[^3H]H_2O$. Although used less frequently, LLOs can also be labeled at GlcNAc residues with radiolabeled glucosamine, at glucosyl residues with radiolabeled galactose or glucose, and in the dolichol moiety with radiolabeled acetate or mevalonate. When labeled with sugar precursors, it is most common to cleave the glycan from the lipid with mild acid treatment and characterize the free glycan.

Yet, metabolic labeling techniques present some problems for LLO analysis. Depending on the source, LLO pools generally turn over after 5 to 20 min, so labeling must be done at least that long if one attempts to determine steady-state compositions. Systems must be amenable to labeling with high concentrations of radioactive precursor, a challenge for analysis of LLOs in animal tissues, or cells that may not actively incorporate precursor due to inhibitory culture conditions or genetic defects. Although identities and relative amounts of individual LLO intermediates can be determined, extra steps must be taken to estimate the actual chemical yield of each LLO intermediate. This is because the specific activities of the mannosyl residues in LLOs will be much lower than that of the starting material due to isotopic dilution with intracellular mannosyl compounds. One approach has been to determine the exact specific activity of the cell-associated GDP-$[^3H]$mannose, and to apply that value to the radiochemical yield of LLO to calculate the chemical amount of LLO (Rush and Waechter, 1995). The requirement for radiolabeling makes LLO analysis of clinical specimens laborious and time consuming. For example, after collection of a skin biopsy specimen, diagnosis of LLO defects in type I Congenital Disorders Of Glycosylation requires growth of fibroblast cultures amenable to labeling with $[^3H]$mannose rather than immediate analysis of the specimen. Finally, effects of certain interesting metabolic alterations (such as inhibition of protein synthesis) on the LLO pathway are very difficult to study this way precisely because they block labeling of LLOs (Gao and Lehrman, 2002b).

Approaches for Analysis of Non-Radioactive LLOs

In principle, many of the problems associated with radiolabeling can be circumvented by nonradioactive methods. Several approaches have been published (Table I), although not all have the simplicity required for a

TABLE I

METHODS FOR DETERMINING COMPOSITIONS OF NON-RADIOACTIVE LLO PREPARATIONS

References	Separation mode	Chemical modification	Separation medium	Detection	LLO sources
Badet and Jeanloz, 1988	endo-H cleaved glycans	reduction with $NaBH_4$	amino HPLC	absorbance at 190 nm	calf pancreas
Badet and Jeanloz, 1988	acid cleaved glycans	reduction with $NaBT_4$	amino HPLC	scintillation counting	calf pancreas
Hall et al., 1989	acid cleaved glycans	reduction with $NaBT_4$	TLC	autoradiography	ovine NCL tissues
Hall et al., 1989	acid cleaved glycans	reduction with $NaBT_4$ and peracetylation	Spherisorb HPLC	scintillation counting	ovine NCL tissues
Faust et al., 1994	acid cleaved glycans	perbenzoylation	reverse-phase HPLC	absorbance at 230 nm	normal and NCL mouse brain
Gibbs and Coward, 1999	acid cleaved glycans	none	Biogel P-4	sugar analysis	bovine pancreas
Kelleher et al., 2001	intact LLO	none	amino HPLC	HPLC after transfer to [^{125}I]-peptide	porcine pancreas
Gao and Lehrman, 2002a	acid cleaved glycans	reduction with ANTS or ANDS	FACE polyacrylamide gel	fluorescence	dermal fibroblasts, CHO cells, and various mouse tissues
Grubenmann et al., 2004	acid cleaved glycans	reduction with 2-aminobenzamide	GlycoSep-N HPLC	fluorescence	dermal fibroblasts

routine laboratory technique able to handle many samples, or a modest requirement for equipment and operator expertise. One method (Kelleher *et al.*, 2001) is unique in that it separates intact LLO intermediates. The identity and quantity of each intermediate are determined by using it as a glycan donor in an OT assay with [^{125}I]-acceptor peptide, followed by chromatographic characterization of the resulting [^{125}I]-glycopeptide. This approach was used to analyze LLOs in various pathogens (Samuelson *et al.*, 2005). In contrast, most of the methods listed in Table I require release of the glycan; chemical modification with a chromophore, fluorophore, or isotope; chromatographic or electrophoretic separation; and detection based on absorbance, fluorescence, or liquid scintillation spectroscopy.

LLO Analysis by Fluorophore-Assisted Carbohydrate Electrophoresis

The general principles of fluorophore-assisted carbohydrate electrophoresis (FACE) have been well described (Jackson, 1996; Starr *et al.*, 1996). FACE fluorophores, all of which bear an amino group, are linked to reducing termini of saccharides by reductive amination. For oligosaccharides with three or more sugars, fluorophores such as 8-amino-naphthalene-1,3, 6-trisulfonate (ANTS) or 7-amino-1,3-naphthalenedisulfonic acid (ANDS) with anionic sulfate modifications are used (Fig. 1). For monosaccharides and disaccharides, 2-aminoacridone (AMAC), which lacks sulfates, is used. The AMAC system should thus be used for $GlcNAc_{1-2}$-P-P-Dol, and the ANTS or ANDS systems should be used with $Glc_{0-3}Man_{1-9}GlcNAc_2$-P-P-Dol. The modified samples are then applied to high-percentage polyacrylamide gels (containing borate for AMAC-saccharides) in a "mini-gel" format and separated by high-voltage electrophoresis with cooling. Multiple saccharide samples can be run on a single gel, facilitating direct comparison. The positions of the fluorescent compounds are documented by illuminating the gel sandwich under ultraviolet light. Sensitivity using a fluorescence detector with charge-coupled device (CCD) camera is in the range of 1 to 2 pmol, with baseline resolution of oligosaccharides in the 14-sugar range differing by single sugars. As discussed later, all supplies and equipment are commercially available, either as kits (through ProZyme, which carries the Glyko line of FACE supplies) or as independent components (Gao and Lehrman, 2002a).

In some regards, FACE is less advantageous than metabolic labeling for LLO analysis. With FACE, it is more difficult to follow the rate of synthesis or consumption of LLOs, which must be "chased" by adding tunicamycin (Gao and Lehrman, 2002a) rather than by removal of radiolabeled precursor. In our hands, FACE is about 5- to 10-fold less sensitive than metabolic labeling, typically requiring at least 10^7 cells per analysis to achieve the sensitivity needed to detect minor intermediates. However, for

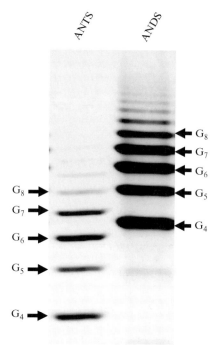

Fig. 1. ANTS- and ANDS-labeled glucose oligosaccharide standards. Glucose oligomers (4 to 8 residues) are shown on an oligosaccharide profiling gel. ANTS-oligomers (which fluoresce green) migrate faster, but ANDS-oligomers (which fluoresce blue) are detected with about 5-fold greater sensitivity. In each case, the G6 oligomer is 50 pmol.

LLO analysis, FACE offers many advantages over metabolic labeling techniques: (i) It is equally suitable for cell cultures and animal tissues (Figs. 2 and 3). (ii) Results reflect true steady-state compositions. (iii) Quantitative yields can be determined for each LLO intermediate. (iv) Detection is independent of the number of sugar residues, eliminating detection bias toward larger LLO intermediates. (v) Unlike metabolic labeling, FACE can be done with cells cultured in conventional media, eliminating potential artifacts from use of media depleted for components that compete with radiolabeled precursors. The advantages of FACE were instrumental in recent work from our laboratory concerning the mechanism of loss of LLO labeling following translation arrest (Gao and Lehrman, 2002b), the steady-state compositions of LLO pools in mouse tissues (Gao and Lehrman, 2002a), the association of LLO accumulation with neuronal ceroid lipofuscinosis in various mouse models of the disease (Cho *et al.*,

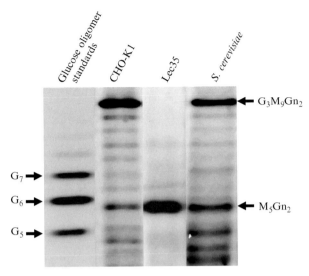

FIG. 2. LLOs detected in cultured cells with FACE. Shown are G_5 to G_7 glucose oligomer standards, and LLO glycans from normal CHO-K1 cells, CHO cells with Lec35 gene defects, and *Saccharomyces cerevisiae*. In CHO-K1 and *S. cerevisiae*, the major LLO is $G_3M_9Gn_2$-P-P-Dol, whereas in Lec35 mutants the major LLO is M_5Gn_2-P-P-Dol. All samples were labeled with ANDS separated with an oligosaccharide profiling gel.

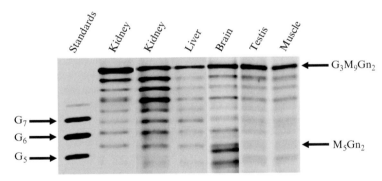

FIG. 3. LLOs detected in mouse tissues detected by FACE. Shown are G_5 to G_7 glucose oligomer standards and LLO glycans from kidney (two lanes), liver, brain, testis, and muscle. The major LLO in mouse tissues is $G_3M_9Gn_2$-P-P-Dol, but as shown, we noticed unusual variability with kidney LLOs. All samples were labeled with ANDS separated with an oligosaccharide profiling gel.

2005), and the link between use of low-glucose medium and the apparent detection of LLO biosynthetic abnormalities in CDG-Ia fibroblasts (Gao *et al.*, 2005).

In this article, we present detailed methods for FACE analysis of LLOs. We also present methods for FACE analysis of relevant phosphosugar (Fig. 4) and nucleotide sugar precursors of LLOs (Table II).

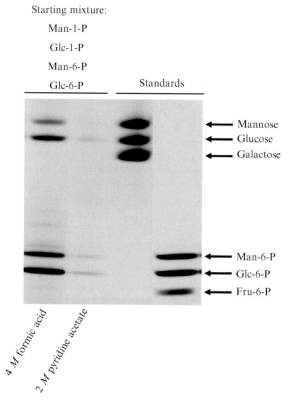

FIG. 4. Detection of 1-phosphosugars and 6-phosphosugars by FACE. A mixture of Man-1-P, Glc-1-P, Man-6-P and Glc-6-P was loaded onto a Dowex AG1–2X column, which was eluted with 4 *M* formic acid and then 2 *M* pyridine acetate. The eluates were treated with acid, derivatized with AMAC, and characterized on a monosaccharide profiling gel. Phosphosugars are efficiently eluted with 4 *M* formic acid. Note that 1-phosphosugars are sensitive to acid and are detected as free sugars, whereas 6-phosphosugars are resistant to acid.

TABLE II

FRACTIONATION METHOD AND FACE DETECTION BY SACCHARIDE CLASS

Saccharide class	Extract	DEAE-cellulose elution	Dowex AG1-2X elution	Acid pretreatment	FACE product	FACE gel
free sugar	water	—	water	no	sugar-AMAC	mono
1-phosphosugar	water	—	4 M formic acid	yes	sugar-AMAC	mono
6-phosphosugar	water	—	4 M formic acid	yes	6-phosphosugar-AMAC	mono
nucleotide sugar	water	—	2 M pyridine acetate	yes	sugar-AMAC	mono
Man-P-Dol, Glc-P-Dol	CM	3 mM ammonium acetate	—	yes	Man/Glc-AMAC	mono
free oligosaccharide	water	—	water	no	oligosaccharide-ANDS/ANTS	oligo
Gn_{1-2}-P-P-Dol	CM	100 mM ammonium acetate	—	yes	Gn_{1-2}-AMAC	mono
$G_{0-3}M_{1-9}Gn_2$-P-P-Dol	CMW	100 mM ammonium acetate	—	yes	$G_{0-3}M_{1-9}Gn_2$-ANDS/ANTS	oligo

Materials

Unless specified otherwise, reagents are stored at room temperature.

Phosphosugar, Neutral Sugar, Nucleotide-Sugar, and Neutral Free Glycan Extraction

1. 70% ethanol.
2. AG1–2X anion exchange resin, 200 to 400 mesh, formate form (Bio-Rad), stored at 4°.
3. 4 M formic acid.
4. 2 M pyridine acetate, pH 5.0, stored in the dark.
5. Poly-Prep Columns (Bio-Rad).

Neutral Sugar, Phosphosugar, Nucleotide Sugar, and $Gn_{1-2}P$-P-Dol Analysis: Derivatization with 2-aminoacridone

1. 0.1 M AMAC (Molecular Probes) in dimethyl sulfoxide (DMSO) containing 15% acetic acid. Stable at −80° for at least 3 months. Exposure of stock solutions to room light should be minimized.
2. 1 M NaBCNH3 (Aldrich) in DMSO. Stable at −80° for at least 3 months.

Monosaccharide Profiling by FACE

1. Resolving buffer (4×): 0.75 M Tris-0.5 M boric acid, adjusted to pH 7.0 with concentrated HCl, stored at 4°.
2. Stacking buffer (4×): 0.5 M Tris-0.5 M boric acid, adjusted to pH 6.8 with concentrated HCl, stored at 4°.
3. Running buffer (5×): 0.5 M glycine, 0.6 M Tris base, and 0.5 M boric acid, final pH 8.3, store at 4°.
4. Resolving gel stock solution: 38% (w/v) acrylamide and 2% (w/v) N, N'-methylenebisacrylamide, stored at 4° in the dark.
5. Stacking gel stock solution: 10% (w/v) acrylamide and 2.5% (w/v) N, N'-methylenebisacrylamide, stored at 4° in the dark.
6. 10% ammonium persulfate (prepared daily).
7. 100% N,N,N'N'-tetramethylenediamine (TEMED), stored at 4°.
8. Loading buffer (2×): 0.01% thorin I (Sigma) in 20% glycerol.
9. Gel box with cooling system. For FACE, a specialized gel box is offered by ProZyme. This apparatus consists of two chambers: a lower chamber maintained at −4° by circulating coolant, and an upper chamber contacting the bottom chamber and containing electrolyte, in which gel sandwiches were

mounted. The upper chamber reaches a running temperature of 4° during electrophoresis. This apparatus is convenient because gels can easily be removed and replaced for periodic inspection with ultraviolet light. A temperature-controlled gel box with a matching gel-casting apparatus (Owl Separation Systems, Model P8DS, Portsmouth, NH) may also be used, although removal of the gel sandwich from the apparatus for periodic inspection with ultraviolet light is more cumbersome.

10. Gel-forming components: Gels are sandwiched between 1.0-mm-thick glass plates (10 cm wide and 10 cm high) and formed with 0.5-mm-thick spacers. Combs with eight 8-mm-wide teeth (2 mm between teeth) are typically used to form loading wells, but 12-tooth combs can also be used. Plates are made of 1-mm-thick EVR glass with very low ultraviolet absorption and are components of preassembled FACE gels from ProZyme. For most investigators who need small quantities of plates and spacers, it will be adequate to recycle them from used gels. We also obtained EVR plates in bulk by special order from Erie Scientific Company (Portsmouth, NH)—order #SMC-2101 for front notched plate, and #SMC-2102 for back plate. The combs and spacers can be prepared by any competent machine shop.

11. Imager: Many of the available fluorescence imagers using CCD cameras are suitable for image acquisition and analysis. We have used the Bio-Rad Fluor-S MultiImager with a 530DF60 filter and Quantity-One software supplied with the imager.

LLO and Monosaccharide-P-Dolichol Extraction

1. CM: chloroform:methanol 2:1, stored in a dark bottle.
2. CMW: chloroform:methanol:water 10:10:3, stored in a dark bottle.
3. DEAE-cellulose (Sigma) converted to acetate form by washing with 10 bed volumes of 1 M acetic acid in CMW, stored in methanol.
4. 3 M NH$_4$OAc stock solution: Prepared by dissolving 3 moles NH4OAc into a total volume of 1000 ml methanol with 3% acetic acid.
5. 0.1 N HCl in 50% isopropanol.
6. Water-saturated butanol: Shake equal volumes of water and butanol in a glass bottle and allow to separate into upper (butanol) and lower (water) phases.
7. AG50–8X cation exchanger resin (Bio-Rad), hydrogen form, store at 4°.
8. AG1–8X anion exchanger resin (Bio-Rad), formate form, store at 4°.
9. 15 ml conical disposable centrifuge glass tubes with caps (Kimble Glass).
10. Water bath sonicator. We have used one from Fisher Scientific.

$G_{0-3}M_{1-9}Gn_2$-P-P-Dol and Free Glycan Analysis: Derivatization with ANTS or ANDS

1. 0.15 M 8-aminonophthalene-1,3,6-trisulfonate (ANTS, from Molecular Probes) in 15% acetic acid, stable at $-80°$ for at least 3 months. Exposure of stock solutions to room light should be minimized.

2. 0.15 M 7-amino-1,3-naphthalenedisulfonic acid (ANDS, from Aldrich) in 15% acetic acid, stable at $-80°$ for at least 3 months. Exposure of stock solutions to room light should be minimized.

3. 1.0 M sodium cyanoborohydride (Aldrich) in DMSO. Stable at $-80°$ for at least 3 months.

4. Fluorescent oligosaccharide standards. Suitable standards can be obtained from ProZyme. We use a maltooligosaccharide mixture, formerly offered by Pfanstiehl Laboratories (#M-138), from which we determined the glucose-equivalent mobility for each LLO glycan (Fig. 5). Fragments of glycogen or other glucose polymers can also be used. These are labeled with ANTS or ANDS as described later. Stable at $-80°$ for at least 12 months.

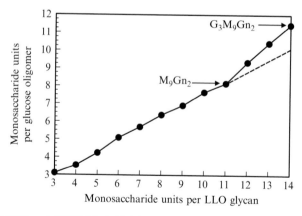

FIG. 5. Mobilities of ANDS-labeled LLO glycans compared with ANDS-labeled glucose oligomers. "Glucose-equivalent" mobilities (plotted on Y-axis) of ANDS-labeled LLO glycans were determined graphically with respect to ANDS-labeled glucose oligomers. This graph is the result of averaging four FACE gels. For reference, data points for M_9Gn_2-ANDS and $G_3M_9Gn_2$-ANDS are labeled. The slope of the line segment representing addition of glucosyl residues to LLO glycans is greater than the slope representing addition of mannosyl residues as indicated by the dashed line.

Oligosaccharide Profiling by FACE

All the reagents and equipment are the same as for monosaccharide profiling (previously) except the following:

1. Resolving gel buffer (8×): 1.5 *M* Tris-Cl, pH 8.9.
2. Stacking gel buffer (8×): 1.0 *M* Tris-Cl, pH 6.8.
3. Running buffer (10×): 1.92 *M* glycine, 0.25 *M* Tris base, pH 8.3.

Procedures

Drying of Samples

To dry samples, our laboratory uses a SpeedVac centrifugal concentration system from Thermo Electron Corporation, consisting of a SC210A concentrator, a RVT4104 vapor trap ($-109°$ operating temperature), and an OFP400 vacuum pump. The system vacuum is typically about 10 torr during drying, and it falls below 1 torr when drying is complete. The concentrator is set to low or medium heating rate; use of high heating rates can cause boiling of samples. Alternatively, drying can be done by warming samples and using a stream of nitrogen gas, such as with an N-EVAP device.

Partial Purification and Separation of Phosphosugars and
   Nucleotide Sugars

1. Grow cells in 15-cm dishes until there is an 80% to 90% confluence.
2. Wash twice with ice-cold phosphate-buffered saline (PBS).
3. Harvest the cells in 10-ml 70% ethanol with a cell lifter. Transfer suspension to a glass tube.
4. Sonicate in a water bath sonicater for 10 to 15 min.
5. Put the tube on ice for 15 min.
6. Centrifuge, 3000 × *g* at 4° for 15 min.
7. Collect the supernatant and dry.
8. Dissolve the dried residue in 1 ml water, and load onto a Poly-Prep Column packed with 1 ml of AG1-X2 resin. During loading, washing, and eluting, recovery is optimal if the flow rate is kept below 20 ml/hr. This can usually be achieved by restricting the flow with a 200-μl flat-end gel loading tip attached to the bottom of the column.
9. Wash the column with 40 ml water to remove all neutral or positively charged contaminants. This wash also contains neutral sugars and soluble neutral glycans. If these are to be characterized, the wash fraction should be treated sequentially with AG50W-X8 and AG1-X8

(see *"LLO Extraction and Partial Purification"*), dried, then reacted with AMAC, ANTS, or ANDS as discussed later.

10. Elute the column with 20 ml 4 *M* formic acid, collect the eluate, and dry. This contains all monophosphoryl sugars. Dissolve samples in 1 ml 0.1 N HCl, and transfer to a 1.5-ml plastic centrifuge tube. Heat at 100° for 15 min to selectively convert 1-phosphosugars to free sugars, then dry. 1-Phosphosugars do not contain aldehydes and thus cannot be labeled by AMAC, whereas the free sugars are detected as neutral conjugates with AMAC. In contrast, 6-phosphosugars have available aldehydes, survive the acid treatment, and react directly with AMAC. Conjugates of neutral sugars with AMAC are well separated from conjugates of 6-phosphosugars on FACE monosaccharide profiling gels (see Fig. 4).

11. Elute the column with 20 ml 2 *M* pyridine acetate, collect the eluate, and dry. This contains nucleotide sugars (Spiro, 1984). Dissolve in 1 ml 0.1 N HCl and transfer to a 1.5-ml plastic centrifuge tube. React at 100° for 15 min to hydrolyze nucleotide sugars to free sugars (exposing the reactive aldehyde), then dry.

AMAC Derivatization

1. Add 5 μl AMAC to acid-treated, dried phosphosugar and nucleotide sugar samples in 1.5-ml tubes, then mix with a vortex mixer. If the sample is less than 200 pmol, use 0.5-ml tubes and 1 μl AMAC. GlcNAc from Gn_1-P-P-Dol, chitobiose from Gn_2-P-P-Dol, mannose from MPD, and glucose from GPD (obtained from CW extracts, treated sequentially with AG50W-X8 and AG1-X8, and dried as described later under *"LLO Extraction and Partial Purification"*) should also be reacted with AMAC.
2. Add 5 μl $NaBCNH_3$ to the tube and vortex.
3. Incubate at 37° for 18 h in the dark.
4. Dry.
5. Dissolve in 30% DMSO and store at −80°.

Monosaccharide Profiling FACE

1. Assemble the gel-casting apparatus, and prepare a resolving gel solution by mixing the following: resolving gel stock solution (4 ml); stock resolving gel buffer, 4× (2 ml); water (2 ml); and 10% ammonium persulfate (20 μl).
2. Polymerization is initiated by addition of TEMED (20 μl), and the solution is poured into the casting apparatus, leaving 0.5 cm below the bottom of the teeth of the comb. Immediately overlay the gel solution with 1 cm of water. Polymerization occurs after 15 min.

3. Prepare the stacking gel solution by mixing the following: stacking gel stock solution (1 ml); stacking gel buffer, 4× (0.5 ml); water (0.5 ml); and 10% ammonium persulfate (6 μl). Polymerization is initiated by adding TEMED (5 μl).

4. Pour off the water overlay, and fill the remaining space in the mold with stacking gel solution. Insert a comb to form sample wells. The stacker should polymerize within 15 min.

5. Dilute the 5× mono running buffer 5-fold, and cool to 4°. Pour into the electrophoresis apparatus (anode compartment) connected to a circulating cooler set to the appropriate temperature to maintain the gel at 4°.

6. Insert the gel sandwich into the apparatus. Add 1× running buffer to the cathode compartment.

7. Mix the AMAC labeled sample (dissolved in 30% DMSO) with an equal volume of 2× loading buffer. Apply 1 to 2 μl to the wells with flat end tips (USA Scientific #1022–2610).

8. Connect to a power supply set at a constant current of 15 mA (a voltage in the range of 200 to 1200 V will result). Run until the thorin I marker dye exits the bottom of the gel, usually in 1 h. Turn off the current, remove the gel, and place in the gel imager with both glass plates still attached. A delay of more than 20 min may result in some diffusion of the fluorescent bands. Although this will result in good separation of AMAC-labeled 6-phosphosugars, nonphosphorylated AMAC-labeled sugars usually require an additional 2 h.

LLO Extraction and Partial Purification

1a. (For cell cultures) Culture dishes (typically 90% confluent, $>10^7$ cells) are washed twice with ice-cold PBS. Monolayers are then covered with room temperature methanol and harvested by scraping with a cell lifter. The suspension is transferred to a 15-ml glass tube, sonicated for 5 to 10 min in a waterbath sonicator, and dried. To avoid overdrying the samples (causing the residue to harden), stop when samples are slightly damp with no liquid visible. Then add 5 to 10 ml CM, resuspend the pellet by sonication, and dry completely to drive out remaining water and leave the residue as a loose, powdery substance.

1b. (For animal tissues) Animal tissues should be used immediately after harvest or snap-frozen in liquid nitrogen and can be disrupted by either (i) dispersing fresh or frozen tissues into 10 volumes of methanol with a Bronson homogenizer (Polytron) for 30 seconds at a setting of 6, or (ii) breaking frozen tissues into small pieces with a mortar and pestle chilled with liquid nitrogen, then dispersing into 10 volumes of methanol with a motorized Teflon-coated homogenizer. For both disruption methods, the methanolic

suspensions are dried, then dispersal and drying are repeated two or three times to fully drive out residual water. This should yield a loose powder, from which LLO recovery is high.

2. Add 10 ml CM to the tube, sonicate for 5 to 10 min with occasional vortexing, centrifuge (3000 × g for 10 min at room temperature), and set aside the liquid phase. Repeat once with the pellet. Combine the liquid phases (which contain Gn_{1-2}P-P-Dol, MPD, and GPD), backwash twice with water, dry, and dissolve in CMW. Set aside for purification with DEAE-cellulose (see later).

3. Resuspend the residue from the previous step in 2 ml CM by sonication, and dry completely so that the residue will be loose and powdery.

4. Add 10 ml water to the tube, sonicate for 5 to 10 min with occasional vortexing, centrifuge, and discard the liquid phase. Repeat once.

5. Resuspend the residue in 2 ml methanol by sonication, and dry completely to drive out residual water.

6. Add 10 ml CMW to the tube, and sonicate 5 to 10 min with occasional vortexing. Centrifuge and collect the liquid phase, which contains $G_{0-3}M_{1-9}$ Gn_2-P-P-Dol. Repeat two or three times, and combine the liquid phases. The remaining residue contains the glycoprotein fraction, which can be kept for glycan studies if needed.

7. Load the CMW extracts onto separate Poly-Prep Columns packed with DEAE-cellulose pre-equilibrated with CMW, using 1 ml bed volume per 5 × 10^7 cells or 500 mg wet tissue. Use of the DEAE-cellulose column is optimal when all steps are kept below 0.5 ml/min, such as by attaching a 200-μl flat-end gel loading tip to the bottom.

8. Wash with 10 bed volumes of CMW.

9. Elute with 10 bed volumes of 3 mM NH_4OAc in CMW. When starting with a CM extract, this fraction contains MPD and GPD.

10. Elute with 10 bed volumes of 300 mM NH_4OAc in CMW. This fraction contains Gn_{1-2}P-P-Dol when starting with a CM extract, and $G_{0-3}M_{1-9}Gn_2$-P-P-Dol when starting with a CMW extract.

11. Combine 10 ml of each eluate with 4.3 ml of chloroform and 1.2 ml of water, and mix thoroughly. Centrifuge (3000 × g, 10 min). Remove the upper aqueous phase (containing salts) carefully, without disturbing the middle layer. Dry the combined middle layer and lower phase.

12. Add 2 ml 0.1 N HCl (in 50% isopropanol), vortex, incubate at 50° for 60 min, and dry.

13. Add 1 ml butanol-saturated water to the tube, vortex, then add 1 ml water-saturated butanol. Vortex, centrifuge, and collect the lower phase containing the released oligosaccharide. Dry.

14. Resuspend the dried material in 1 ml water, and add 200 μl (packed volume) of AG50W-X8 (hydrogen form) cation exchange resin. Mix for 5 min, centrifuge, and collect the liquid.

15. Add 200 μl (packed volume) of AG1-X8 (formate form) anion exchange resin to the tube, mix for 5 min, centrifuge, and collect the liquid.

Derivatization of Oligosaccharides with ANTS and ANDS

Samples containing mono- or disaccharides should be derivatized with AMAC (see previous text). Samples of oligosaccharides (e.g., released from LLO) should be derivatized with ANTS or ANDS (discussed here). In our hands, ANDS is approximately five times more sensitive than ANTS (see Fig. 1). However, ANTS carries a greater negative charge (-3) than ANDS (-2), so ANTS-labeled oligosaccharides have a greater electrophoretic rate.

1. Dry oligosaccharide samples released from $G_{0-3}M_{1-9}Gn_2$-P-P-Dol, or samples containing deionized neutral glycans in a 1.5-ml microcentrifuge tube (use a 0.5-ml tube if the sample is below 200 pmol).

2. Add 5 μl (1 μl if the sample is below 200 pmol) of ANTS or ANDS solution. Vortex and centrifuge briefly.

3. Add 5 μl (or 1 μl if the sample is below 200 pmol) of sodium cyanoborohydride solution. Vortex, centrifuge briefly, and react at 37° for 18 h.

4. Dry the reaction mixture. Dissolve in the desired volume of water.

Oligosaccharide Profiling by FACE

1. Assemble the gel-casting apparatus provided with the gel box, and prepare the resolving gel solution by mixing the following: resolving gel stock solution (4 ml); stock-resolving gel buffer, 8× (1 ml); water (3 ml); and 10% ammonium persulfate (20 μl).

2. Polymerization is initiated by addition of TEMED (20 μl), and the solution is poured into the casting apparatus to a height of 0.5 cm below the bottom of the teeth of the comb. Immediately overlay the gel solution with 1 cm of water. Polymerization occurs after 15 min.

3. Prepare the stacking gel solution by mixing the following: stacking gel stock solution (1 ml); stacking resolving gel buffer, 8× (0.25 ml); water (0.75 ml); and 10% ammonium persulfate (6 μl). Polymerization is initiated by adding TEMED (5 μl).

4. Pour off the water overlay, and fill the remaining space in the mold with stacking gel solution. Insert a comb to form sample wells. The stacker should polymerize within 15 min.

5. Dilute the stock electrode buffer 10-fold, and cool to 4°. Pour into the electrophoresis apparatus (anode compartment), connected to a circulating cooler set to the appropriate temperature to maintain the gel at 4°.

6. Insert the gel sandwich into the apparatus. Add electrode buffer to the cathode compartment. Mix the sample (dissolved in water) with an equal volume of 2× loading buffer. Apply 2 to 4 μl of mixture to the wells with flat-end tips (USA Scientific #1022–2610).

7. Connect to a power supply, and set at a constant current of 15 mA (a voltage in the range of 200 to 1200 V will result). Run until the thorin I marker dye exits the bottom of the gel, usually in 1 h. Turn off the current, remove the gel, and place in the gel imager with both glass plates still attached. A delay of more than 20 min may result in some diffusion of the fluorescent bands. Examples of LLO glycans detected with ANDS are presented in Figs. 2 and 3.

Acknowledgments

The authors are generously supported by NIH grants GM38545 and DK70954, and Welch grant I-1168. Our thanks go to Jeffrey Rush for valuable comments on the manuscript.

References

Badet, J., and Jeanloz, R. W. (1988). Isolation and partial purification of lipid-linked oligosaccharides from calf pancreas. *Carbohydr. Res.* **178**, 49–65.

Cho, S. K., Gao, N., Pearce, D. A., Lehrman, M. A., and Hofmann, S. L. (2005). Characterization of lipid-linked oligosaccharide accumulation in mouse models of Batten disease. *Glycobiology* **15**, 637–648.

Faust, J. R., Rodman, J. S., Daniel, P. F., Dice, J. F., and Bronson, R. T. (1994). Two related proteolipids and dolichol-linked oligosaccharides accumulate in motor neuron degeneration mice (*mnd/mnd*), a model for neuronal ceroid lipofuscinosis. *J. Biol. Chem.* **269**, 10150–10155.

Gao, N., and Lehrman, M. A. (2002a). Analyses of dolichol pyrophosphate-linked oligosaccharides in cell cultures and tissues by fluorophore-assisted carbohydrate electrophoresis. *Glycobiology* **12**, 353–360.

Gao, N., and Lehrman, M. A. (2002b). Coupling of the dolichol-P-P-oligosaccharide pathway to translation by perturbation-sensitive regulation of the initiating enzyme, GlcNAc-1-P transferase. *J. Biol. Chem.* **277**, 39425–39435.

Gao, N., Shang, J., and Lehrman, M. A. (2005). Analysis of glycosylation in CDG-Ia fibroblasts by fluorophore-assisted carbohydrate electrophoresis: Implications or extracellular glucose and intracellular mannose-6-phosphate. *J. Biol. Chem.* **280**, 17901–17909.

Gibbs, B. S., and Coward, J. K. (1999). Dolichylpyrophosphate oligosaccharides: Large-scale isolation and evaluation as oligosaccharyltransferase substrates. *Bioorganic Medicinal Chem.* **7**, 441–447.

Grubenmann, C. E., Frank, C. G., Hulsmeier, A. J., Schollen, E., Matthijs, G., Mayatepek, E., Berger, E. G., Aebi, M., and Hennet, T. (2004). Deficiency of the first mannosylation step in the N-glycosylation pathway causes congenital disorder of glycosylation type Ik. *Hum. Mol. Genet.* **13**, 535–542.

Hall, N. A., Jolly, R. D., Palmer, D. N., Lake, B. D., and Patrick, A. D. (1989). Analysis of dolichyl pyrophosphoryl oligosaccharides in purified storage cytosomes from ovine ceroid-lipofuscinosis. *Biochim. Biophys. Acta* **993**, 245–251.

Jackson, P. (1996). The analysis of fluorophore-labeled carbohydrates by polyacrylamide gel electrophoresis. *Molecular Biotechnol.* **5**, 101–123.

Kelleher, D. J., Karaoglu, D., and Gilmore, R. (2001). Large-scale isolation of dolichol-linked oligosaccharides with homogeneous oligosaccharide structures: Determination of steady-state dolichol-linked oligosaccharide compositions. *Glycobiology* **11**, 321–333.

Rush, J. S., and Waechter, C. J. (1995). Method for the determination of cellular levels of guanosine-5′-diphosphate mannose based on a weak interaction with concanavalin A at low pH. *Anal. Biochem.* **224**, 494–501.

Samuelson, J., Banerjee, S., Magnelli, P., Cui, J., Kelleher, D. J., Gilmore, R., and Robbins, P. W. (2005). The diversity of dolichol-linked precursors to Asn-linked glycans likely results from secondary loss of sets of glycosyltransferases. *Proc. Natl. Acad. Sci.* **102**, 1548–1553.

Spiro, M. J. (1984). Effect of diabetes on the sugar nucleotides in several tissues of rat. *Diabetologia* **26**, 70–75.

Starr, C. M., Masada, R. I., Hague, C., Skop, E., and Klock, J. C. (1996). Fluorophore-assisted carbohydrate electrophoresis in the separation, analysis, and sequencing of carbohydrates. *J. Chromatograph. A* **720**, 295–321.

[2] Identification of *N*-Glycan–Binding Proteins for E3 Ubiquitin Ligases

By Tadashi Tai

Abstract

N-glycans serve as a degradation signal by the SCF^{Fbx2} ubiquitin ligase complex in the cytosol. Fbx2, an F-box protein, binds specifically to proteins attached with *N*-linked high-mannose type oligosaccharides, and subsequently contributes to ubiquitination of glycoproteins. Pre-integrin $\beta1$ is identified as one of the Fbx2 targets. These two proteins bind in the cytosol after inhibition of the proteasome. These results indicate that SCF^{Fbx2} ubiquitinates *N*-linked glycoproteins, which are translocated from the endoplasmic reticulum to the cytosol by the quality control mechanism. This chapter describes methods, including a binding protein assay for *N*-glycans,

METHODS IN ENZYMOLOGY, VOL. 415   0076-6879/06 $35.00
Copyright 2006, Elsevier Inc. All rights reserved.   DOI: 10.1016/S0076-6879(06)15002-8

a ubiquitination assay for *N*-linked glycoproteins with SCF[Fbx2] ubiquitin ligase complex, an overlay assay for the detection of Fbx2 binding proteins, and a pull-down assay for the interaction between Fbx2 and *N*-linked glycoproteins, used to identify *N*-glycan–binding proteins for E3 ubiquitin ligases.

Overview

N-glycosylation of the proteins occurs when newly synthesized proteins enter the secretory pathway through the translocation channel at the membrane of the endoplasmic reticulum (ER). *N*-glycans have recently been shown to play an important role in glycoprotein transport and sorting, in particular, at the initial step of secretion that occurs in the ER compartment (Ellgaard and Helenius, 2003; Helenius and Aebi, 2001; Parodi, 2000). *N*-linked glycoproteins are subjected to the quality control of glycoproteins through which aberrant proteins are distinguished from properly folded proteins and retained in the ER. The quality control system includes the calnexin-calreticulin cycle, a unique chaperone system that recognizes $Glc_1Man_{9-6}GlcNAc_2$ and assists refolding of misfolded or unfolded proteins. When the improperly folded or incompletely assembled proteins fail to restore their functional states, they are degraded by the ER-associated degradation (ERAD) system, which involves retrograde transfer of proteins from the ER to the cytosol followed by degradation by the proteasome (Lilley and Ploegh, 2004; Plemper and Wolf, 1999; Tsai *et al.*, 2002).

Together with ubiquitin-activating enzyme E1 and ubiquitin-conjugating enzyme E2, E3 ubiquitin ligases catalyze the ubiquitination of a variety of protein substrates for targeted degradation via the 26S proteasome. E3 ubiquitin ligases constitute a large protein family including SCF (Deshaies, 1999; Glickman and Ciechanover, 2002; Hershko and Ciechanover, 1998). F-box proteins are also diverse and are pivotal in selecting target proteins for ubiquitination (Kipreos and Pagano, 2000; Winston *et al.*, 1999). Modification of target proteins is a prerequisite for the recognition by certain SCF-type E3 ubiquitin ligases and subsequent destruction. These include phosphorylation and proline hydroxylation (Ivan *et al.*, 2001; Jaakkola *et al.*, 2001). We have found that *N*-glycosylation also acts as a target signal to eliminate intracellular glycoproteins by Fbx2-dependent ubiquitination and proteasome degradation (Yoshida *et al.*, 2002). It has recently been reported that Fbx2 belongs to a subfamily consisting of at least five homologous F-box proteins (Ilyn *et al.*, 2002). Fbx6b/FBG2 protein exhibits 75% identity with Fbx2/NFB42/FBG1. We have determined that Fbx6b/FBG2 binds several *N*-linked glycoproteins, but other F-box proteins fail to bind any of the glycoproteins tested, and Fbx2/FBG1 is expressed only on

mouse brain and testis, whereas Fbx6b/FBG2 is widely distributed in a variety of mouse tissues (Yoshida et al., 2003). This chapter deals with biochemical protocols used to identify sugar-binding proteins for ubiquitin ligases.

Binding Protein Assay for N-Glycans

Reagents

1. Asialofetuin type II and di-N-acetylchitobiose (chitobiose) (Sigma-Aldrich, St. Louis, MO).
2. β-galactosidase (*Streptococcus* 6646K) and richius communis agglutinin (RCA) lectin agarose (Seikagaku Corporation, Tokyo, Japan).
3. Affi-Gel 15 (Bio-Rad, Richmond, CA).
4. Protease inhibitor mixture (complete ethylenediamine tetraacetic acid [EDTA] free; Roche Applied Science, Mannheim, Germany).

Methods

1. GlcNAc-terminated fetuin (GTF) is prepared by incubating asialo-fetuin (10 mg) with β-galactosidase (0.5 units) in 100 mM citrate/phosphate buffer (pH, 6.5) at 37° for 24 h and passed through RCA lectin agarose column. GTF is immobilized to Affi-Gel 15 (0.5 ml). The preparation of the beads is followed by a manufacture procedure provided.

2. Eight-week-old ICR mouse brains are homogenized in 10 volumes of TBS (20 mM Tris-HCl; pH, 7.5; 150 mM NaCl) containing 5 mM CaCl$_2$ and protease inhibitor mixture. After centrifugation of the homogenate at 30,000g for 30 min at 4°, the supernatant is incubated with GTF-immobilized beads for 18 h at 4°.

3. The beads are washed with TBS containing 0.1% TritonX-100 (TBS-T). The adsorbed proteins are eluted successively by 0.5 M NaCl, 20 mM EDTA/ethyleneglycol-bis-(β-aminoethylether)-N,N,N',N'-tetra-acetic acid (EGTA), 0.2 M GlcNAc or 0.1 M chitobiose in TBS-T. Two proteins, 45 and 23 kDa, are specifically eluted with chitobiose, but not with NaCl, EDTA/EGTA, or GlcNAc (Fig. 1).

4. Eluted proteins are separated by sodium dodecyl sulphate (SDS)-polyacrylamide gel electrophoresis (SDS-PAGE), excised and digested with lysyl endopeptidase, and their sequences are determined by using a protein sequencer. These two proteins have been identified as Skp1 and Fbx2/NFB42 (Erhardt et al., 1998).

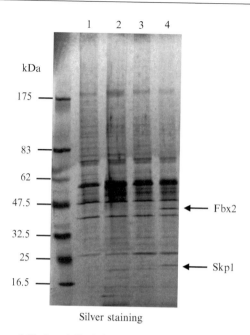

FIG. 1. Detection of Fbx2 and Skp1 for GlcNAc-binding proteins. GlcNAc-terminated fetuin (GTF)-bound proteins are eluted with the indicated compounds: 1, NaCl; 2, ethylenediamine tetraacetic acid (EDTA)/ethyleneglycol-bis-(β-aminoethylether)-*N,N,N′*, *N′*-tetraacetic acid (EGTA); 3, GlcNAc; 4, chitobiose. Revised from Yoshida *et al.* (2002).

In Vitro Ubiquitination Assay for *N*-linked Glycoproteins with SCF^{Fbx2}

Reagents

1. HiTrap HP column and Chemiluminescence detection system (GE Healthcare BioSciences Corporation, Piscataway, NJ)
2. Bovine ubiquitin (Sigma-Aldrich)
3. Immobilon-P transfer membrane (Millipore, Bedford, MA)
4. Rat polyclonal antibody to fetuin (Chemicon, Temecula, CA).

Methods

1. Recombinant His-Ubc4 (E2) and GST-ubiquitin are produced in *Escherichia coli*. Recombinant His-E1 and SCF^{Fbx2} (Flag-Skp1/Cul1-HA/His-Fbx2/T7-Roc1) are produced by baculovirus-infected HiFive insect cells. The SCF^{Fbx2} complex is obtained by simultaneously infecting four

baculoviruses. These proteins are affinity purified by a HiTrap HP column. SCF$^{\beta TrCP1}$ complex used for a standard ubiquitin ligase E3 and other recombinant proteins used for *in vitro* ubiquitination assay are prepared as described previously (Kawakami *et al.*, 2001).

2. GTF (1 μg) is incubated in the reaction mixture (50 μl) containing ATP-regenerating system, E1 (0.5 μg), E2 (1 μg), SCF^{Fbx2} (2 μg), and recombinant GST-ubiquitin (6.5 μg) at 30° for 2 h. After terminating the reaction by the addition of SDS-PAGE sample buffer (25 μl), the proteins in the boiled supernatants (8 μl) are separated with 4% to 20% SDS-PAGE and transferred into a membrane.

3. After the blotted membranes are blocked with phosphate buffered saline (PBS) containing 3% bovine serum albumin, immunoblotting is performed with an antibody against fetuin and visualized with horseradish peroxidase-conjugated secondary antibody using the enhanced chemiluminescence detection system. Ubiquitinated GTF is detected as the high-molecular-mass proteins (Fig. 2).

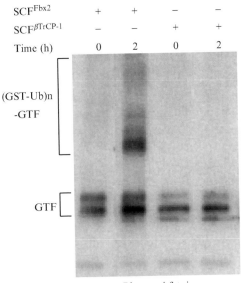

Blot: anti-fetuin

FIG. 2. *In vitro* ubiquitination of GTF by the SCF^{Fbx2} ubiquitin ligase system. Revised from Yoshida *et al.* (2002).

Overlay Assay for the Detection of Fbx2 Binding Proteins

Reagents

1. pTracer-EF vector (Invitrogen, Carlsbad, CA)
2. [^{35}S]-methionine (GE Healthcare BioSciences Corporation)
3. TNT Coupled Reticulocyte Lysate System (Promega, Madison, WI)
4. endo-β-N-acetylglucosaminidase H (endo H) enzyme (Seikagaku Corporation)
5. PNGase F (*Flavobacterium meningosepticum*; Roche Applied Science).

Methods

1. Mouse neuroblastoma (Neuro2a) cells and brains are homogenized in 10 volumes of TBS containing 0.5% Nonidet P40 and protease inhibitor mixture. After freeze-thawing, the homogenate is centrifuged at 24,000g for 20 min at 4°. The cytoplasmic fraction (supernatant) and membrane fraction (pellet), which is solubilized with 1% Triton X-100, are separated by SDS-PAGE and blotted onto a transfer membrane.

2. To prepare [^{35}S]-labeled Fbx2 probes, Fbx2–Metx3 (full and ΔN-2) plasmids expressing amino-terminally Flag-tagged full-length Fbx2 and ΔN-2 with an additional three methionines at the C-terminus are prepared by subcloning into pTracer-EF vector. [^{35}S]-labeled Fbx2 and ΔN-2 are generated with [^{35}S]-methionine and pTracer-EF-Fbx2 and -ΔN-2 in TNT Coupled Reticulocyte Lysate System. The coding residues of the Fbx2 deletion ΔN-2 are from amino acids 95 to 296.

3. The blotted membranes are blocked by 5% skim milk in TBS and are washed and incubated with [^{35}S]-labeled probes in TBS containing 1% skim milk at 4° for 18 h with gentle shaking. Membranes are washed with TBS containing 0.05% Tween20, air-dried, and analyzed with a Molecular Imager FX.

4. For treatments with endo H or PNGase F, the blotted membranes are blocked with 3% bovine serum albumin in PBS. Each membrane is treated with recombinant endo H (1 unit) in 50 mM citrate/phosphate buffer (pH, 5.5) or PNGase F (1 unit) in 50 mM sodium phosphate buffer (pH, 8.2) at 37° for 18 h. The enzyme-treated membranes are washed with TBS and incubated in TBS containing 1% skim milk before the overlay.

5. Several proteins are detected with full-length Fbx2 and ΔN-2 probes, mainly in the membrane fractions of mouse brains and neuroblastoma cells. These interactions are specifically inhibited by chitobiose. Treatment with PNGase completely inhibits Fbx2 binding. Furthermore, endo H treatment also interferes with the binding.

6. The two proteins p120 and p75 are significant for Fbx2 binding, and p120 is identified as integrin $\beta1$ from sequence analysis.

Pull-Down Assay for the Interaction Between Fbx2 and *N*-Linked Glycoproteins

Reagents

1. Fetuin and ribonuclease B (RNase B) (Sigma-Aldrich)
2. β-galactosidase, β-*N*-acetylhexosaminidase (jack bean), α-mannosidase (jack bean) and WGA lectin-agarose (Seikagaku Corporation)
3. Lipofectamine Plus (Invitrogen)
4. Mouse monoclonal antibody against Flag (M2-horseradish peroxidase conjugate; Sigma-Aldrich).

Methods

1. Man-terminated fetuin (MTF) is prepared by incubating GTF (2 mg) with β-*N*-acetylhexosaminidase (0.2 units) in 50 mM sodium citrate buffer (pH, 5.0) at 37° for 18 h. Deglycosylated fetuin (DGF) is prepared by incubating asialofetuin (2 mg) with recombinant PNGase F (0.2 units) in 50 mM sodium phosphate buffer (pH, 8.2) at 37° for 24 h. The enzyme-treated fetuins are loaded onto successive WGA and RCA-lectin agarose columns. The flow-through fraction from both columns is used as DGF. RNase B having ManβGlcNAc$_2$ and GlcNAc is prepared by incubating RNase B (2 mg) with α-mannosidase (0.2 units) in 50 mM sodium acetate buffer (pH, 4.5) and endo H (0.2 units) in 50 mM citrate/phosphate buffer (pH, 5.5), respectively, at 37° for 24 h. Each glycoprotein is immobilized to Affi-Gel 15 (0.5 ml).

2. HEK293T cells are transfected with pTracer-EF-ΔN-2 plasmid by lipofection. Cell extracts (25 μg) are prepared with TBS containing 0.5% Nonidet P40 and incubated with various glycoprotein-immobilized beads (15 μl) for 18 h at 4°. After washing the beads with TBS containing 0.5% Nonidet P40, bound proteins are eluted by boiling with SDS sample buffer and are separated by spin filtration.

3. The bound proteins are separated by SDS-PAGE and transferred into a membrane. After blocking with PBS containing 3% bovine serum albumin, immunoblotting is performed with an antibody against Flag and visualized, using the enhanced chemiluminescence detection system.

4. Fbx2 binds to MTF more efficiently than to asialofetuin and GTF, but not to DGF or intact fetuin. Moreover, Fbx2 binds to intact RNase B,

but its binding is reduced when RNase B is treated with α-mannosidase and endo H.

Interaction of Fbx2 with Integrin β1 in Neuro2a Cells

Reagents

1. Lipofectamine Plus (Invitrogen)
2. Tunicamycin (Wako Pure Chemicals Industries, Osaka, Japan)
3. Mouse monoclonal antibody against Flag (M2-agarose; Sigma-Aldrich)
4. Rat monoclonal antibody to integrin β1 (MAb1997; Chemicon).

Methods

1. Neuro2a cells are transfected with pTracer-EF-ΔN-2 plasmid by lipofection. To inhibit biosynthesis of N-glycans, cells are treated with tunicamycin (5 μg/ml) for 24 h. Cell extracts are prepared in TBS containing 0.5% Nonidet P40 and are solubilized with 1% Triton X-100.

2. After immunoprecipitation of lysates with mouse monoclonal antibody against Flag, the bound proteins are separated by SDS-PAGE and immunoblotted with rat monoclonal antibody against integrin β1. Integrin β1 proteins with different oligosaccharide chains migrate as two separate bands on SDS-PAGE (Akiyama *et al.*, 1989; Koivisto *et al.*, 1994).

3. Fbx2 is coimmunoprecipitated with the precursor of integrin β1, but not with the mature form. Furthermore, Fbx2 fails to interact with deglycosylated fetuin in tunicamycin-treated cells but binds to pre-integrin β1 (Fig. 3).

In an attempt to determine the biological roles of N-glycans in the central nervous system, mouse brain extracts were screened for proteins that bind to various sugar probes. When GTF-immobilized beads were used, two proteins (i.e., 45 and 23 kDa) were detected (see Fig. 1). These two proteins were identified as the molecules that associated with the ubiquitin-proteasome system. Critical points in the finding are considered as follows: (i) brain was used for materials; (ii) glycoproteins, but not glycopeptides or oligosaccharides were used for sugar probes; (iii) the cytosol fraction was surveyed, and (iv) chitobiose was used for elution.

A number of approaches have previously been used for elucidating the biological function of glycol conjugates in mammalian cells. Among them, to isolate sugar-binding proteins is one of the most simple and effective techniques. In fact, many sugar-binding proteins have been isolated and

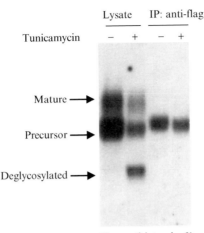

Blot: anti-integrin β1

FIG. 3. *In vivo* interaction of Fbx2 with integrin β1 containing high-mannose oligosaccharides. Neuro2a cells transfected with Flag-tagged ΔN-2 are untreated ($-$) or treated with tunicamycin ($+$). After immunoprecipitation of ΔN-2 from cell extracts, integrin β1 proteins are analyzed by immunoblotting. Cited from Yoshida *et al.* (2002).

characterized in the past 3 decades. For example, a galactose-binding protein on hepatocytes (Ashwell and Morell, 1974), a mannan–binding protein (Ikeda *et al.*, 1987), galectins, beta-galactoside-binding proteins, (Barondes *et al.*, 1994), Siglecs, sialic acid–binding immunoglobulin-like lectins (Crocker, 2002), and selectins, which belong to C-type lectins and are involved in cell–cell adhesion (Lowe, 2002), have been reported. These sugar-binding proteins described previously function mainly on the outside of plasma membranes.

On the other hand, binding proteins occur in the inside of cells, including mannose-6-phosphate receptor, which is involved in the sorting of lysosomal enzymes (Kornfeld, 1992; Nolan and Sly, 1987; Von Figura and Hasilik, 1986); ERGIC-53 and VIP36, which recognize *N*-linked-high mannose oligosaccharides and are involved in protein sorting in cells (Fiedler and Simons, 1995; Hara-Kuge *et al.*, 2002); and calnexin/calreticulin, a unique chaperone system that recognizes $Glc_1Man_{9-6}GlcNAc_2$ and assists refolding of misfolded or unfolded proteins (Helenius and Aebi, 2001; Parodi, 2000). Interestingly, glycans that are recognized by these binding proteins belong to *N*-linked high-mannose type oligosaccharides. The procedures for these studies have been reviewed in the volume 363 of *Methods in Enzymology* (2003). These studies have led to the

demonstration of essential functions for a number of glycoconjugates in basic life science as well as in clinical medicine. Further studies on the isolation and characterization of sugar-binding proteins will be expected to reveal unknown and important functions of glycans in cells and tissues in the near future.

Acknowledgments

I wish to thank my colleagues and collaborators, especially those at the Department of Tumor Immunology, Tokyo Metropolitan Institute of Medical Science, for their contributions in this work.

References

Akiyama, S., Yamada, S. S., and Yamada, K. M. (1989). Analysis of the role of glycosylation of the human fibronectin receptor. *J. Biol. Chem.* **264**, 18011–18018.

Ashwell, G., and Morell, A. (1974). The dual role of sialic acid in the hepatic recognition and catabolism of serum glycoproteins. *Biochem. Soc. Symp.* **40**, 117–124.

Barondes, S. H., Cooper, D. N., Gitt, M. A., and Leffler, H. (1994). Galectins: Structure and function of a large family of animal lectins. *J. Biol. Chem.* **269**, 20807–20810.

Crocker, P. R. (2002). Siglecs: Sialic-acid–binding immunoglobulin-like lectins in cell-cell interactions and signaling. *Curr. Opin. Struct. Biol.* **12**, 609–615.

Deshaies, R. J. (1999). SCF and Cullin/Ring H2-based ubiquitin ligases. *Annu. Rev. Cell Dev. Biol.* **15**, 435–467.

Ellgaard, L., and Helenius, A. (2003). Quality control in the endoplasmic reticulum. *Nat. Rev. Mol. Cell. Biol.* **4**, 181–191.

Erhardt, J.A, Hynicka, W., DiBenedetto, A., Shen, N., Stone, N., Paulson, H., and Pittman, R. N. (1998). A novel protein, NFB42, is highly enriched in neurons and induces growth arrest. *J. Biol. Chem.* **273**, 35222–35227.

Fiedler, K., and Simons, K. (1995). The role of *N*-glycans in the secretory pathway. *Cell* **81**, 309–312.

Glickman, M. H., and Ciechanover, A. (2002). The ubiquitin-proteasome proteolytic pathway: Destruction for the sake of construction. *Physiol. Rev.* **82**, 373–428.

Hara-Kuge, S., Ohkura, T., Ideo, H., Shimada, O., Atsumi, S., and Yamashita, K. (2002). Involvement of VIP36 in intracellular transport and secretion of glycoproteins in polarized Madin-Darby canine kidney (MDCK) cells. *J. Biol. Chem.* **277**, 16332–16339.

Helenius, A., and Aebi, M. (2001). Intracellular functions of *N*-linked glycans. *Science* **291**, 2364–2369.

Hershko, A., and Ciechanover, A. (1998). The ubiquitin system. *Annu. Rev. Biochem.* **67**, 425–479.

Ikeda, K., Sannoh, T., Kawasaki, N., Kawasaki, T., and Yamashina, I. (1987). Serum lectin with known structure activates complement through the classical pathway. *J. Biol. Chem.* **262**, 7451–7454.

Ilyn, G. P., Serandour, A. L., Pigeon, C., Rialland, M., Glaise, D., and Guguen-Guillouzo, C. (2002). A new subfamily of structurally related human F-box proteins. *Gene* **296**, 11–20.

Ivan, M., Kondo, K., Yang, H., Kim, W., Valiando, J., Ohh, M., Salic, A., Asara, J. M., Lane, W. S., and Kaelin, W. G., Jr. (2001). HIFα targeted for VHL-mediated destruction by proline hydroxylation: Implication for O$_2$ sensing. *Science* **292,** 464–468.

Jaakkola, P., Mole, D. R., Tian, Y. M., Wilson, M. I., Gielbert, J., Gaskell, S. J., Kriegsheim, A., Hebestreit, H. F., Mukherji, M., Schofield, C. J., Maxwell, P. H., Pugh, C. W., and Ratcliffe, P. J. (2001). Targeting of HIF-α to the von Hippel-Lindau ubiquitylation complex by O$_2$-regulated prolyl hydroxylation. *Science* **292,** 468–472.

Kawakami, T., Chiba, T., Suzuki, T., Iwai, K., Yamanaka, K., Minato, N., Suzuki, H., Shimbara, N., Hidaka, Y., Osaka, F., Omata, M., and Tanaka, K. (2001). NEDD8 recruits E2-ubiquitin to SCF E3 ligase. *EMBO J.* **20,** 4003–4012.

Kipreos, E. T., and Pagano, M. (2000). The F-box protein family. *Genome Biol.* **1,** 3002.1–3002.7.

Koivisto, L., Heino, J., Hakkinen, L., and Larjava, H. (1994). The size of the intracellular β1-integrin precursor pool regulates maturation of β1-integrin subunit and associated α-subunits. *Biochem. J.* **300,** 771–779.

Kornfeld, S. (1992). Structure and function of the mannose 6-phosphate/insulinlike growth factor II receptors. *Annu. Rev. Biochem.* **61,** 307–330.

Lilley, B. N., and Ploegh, H. L. (2004). A membrane protein required for dislocation of misfolded proteins from the ER. *Nature* **429,** 834–840.

Lowe, J. B. (2002). Glycosylation in the control of selectin counter-receptor structure and function. *Immunol. Rev.* **186,** 19–36.

Nolan, C. M., and Sly, W. S. (1987). Intracellular traffic of the mannose 6-phosphate receptor and its ligands. *Adv. Exp. Med. Biol.* **225,** 199–212.

Parodi, A. J. (2000). Protein glycosylation and its role in protein folding. *Annu. Rev. Biochem.* **69,** 69–93.

Plemper, R. K., and Wolf, D. H. (1999). Retrograde protein translocation: ERADication of secretory proteins in health and disease. *Trends Biochem. Sci.* **24,** 266–270.

Tsai, B., Ye, Y., and Rapoport, T. A. (2002). Retro-translocation of proteins from the endoplasmic reticulum into the cytosol. *Nature Rev. Mol. Cell. Biol.* **3,** 246–255.

Von Figura, K., and Hasilik, A. (1986). Lysosomal enzymes and their receptors. *Annu. Rev. Biochem.* **55,** 167–193.

Winston, J. T., Koepp, D. M., Shu, C., Elledge, S. J., and Harper, J. W. (1999). A family of mammalian F-box proteins. *Curr. Biol.* **9,** 1180–1182.

Yoshida, Y., Chiba, T., Tokunaga, F., Kawasaki, H., Iwai, K., Suzuki, T., Ito, Y., Matsuoka, K., Yoshida, M., Tanaka, K., and Tai, T. (2002). E3 ubiquitin ligase that recognizes sugar chains. *Nature* **418,** 438–442.

Yoshida, Y., Tokunaga, F., Chiba, T., Iwai, K., Tanaka, K., and Tai, T. (2003). Fbs2 is a new member of the E3 ubiquitin ligase family that recognizes sugar chains. *J. Biol. Chem.* **278,** 43877–43884.

[3] Family 47 α-Mannosidases in N-Glycan Processing

By STEVEN W. MAST and KELLEY W. MOREMEN

Abstract

α-Mannosidases in eukaryotic cells are involved in both glycan biosynthetic reactions and glycan catabolism. Two broad families of enzymes have been identified that cleave terminal mannose linkages from Asn-linked oligosaccharides (Moremen, 2000), including the Class 1 mannosidases (CAZy GH family 47 (Henrissat and Bairoch, 1996)) of the early secretory pathway involved in the processing of N-glycans and quality control and the Class 2 mannosidases (CAZy family GH38 [Henrissat and Bairoch, 1996]) involved in glycoprotein biosynthesis or catabolism. Within the Class 1 family of α-mannosidases, three subfamilies of enzymes have been identified (Moremen, 2000). The endoplasmic reticulum (ER) α1,2-mannosidase I (ERManI) subfamily acts to cleave a single residue from Asn-linked glycans in the ER. The Golgi α-mannosidase I (GolgiManI) subfamily has at least three members in mammalian systems (Herscovics *et al.*, 1994; Lal *et al.*, 1994; Tremblay and Herscovics, 2000) involved in glycan maturation in the Golgi complex to form the $Man_5GlcNAc_2$ processing intermediate. The third subfamily of GH47 proteins comprises the ER degradation, enhancing α-mannosidase-like proteins (EDEM proteins) (Helenius and Aebi, 2004; Hirao *et al.*, 2006; Mast *et al.*, 2005). These proteins have been proposed to accelerate the degradation of misfolded proteins in the lumen of the ER by a lectin function that leads to retrotranslocation to the cytosol and proteasomal degradation. Recent studies have also indicated that ERManI acts as a timer for initiation of glycoprotein degradation via the ubiquitin-proteasome pathway (Hosokawa *et al.*, 2003; Wu *et al.*, 2003). This article discusses methods for analysis of the GH47 α-mannosidases, including expression, purification, activity assays, generation of point mutants, and binding studies by surface plasmon resonance.

Overview

The biosynthesis of glycoproteins containing Asn-linked oligosaccharides is initiated by the transfer of a $Glc_3Man_9GlcNAc_2$ oligosaccharide from a lipid-linked precursor to nascent polypeptides in the lumen of the endoplasmic reticulum (ER) (Kornfeld and Kornfeld, 1985). After

METHODS IN ENZYMOLOGY, VOL. 415
0076-6879/06 $35.00
DOI: 10.1016/S0076-6879(06)15003-X

the glycan transfer, a series of processing steps occurs, wherein hybrid or complex oligosaccharides can be formed from the initial high-mannose precursor. The initial steps of this process involve the trimming of glucose residues by glucosidase I and II, enzymes that are localized in the ER. These enzymes are responsible for the excision of two of the three glucose residues to form the $GlcMan_9GlcNAc_2$ structure necessary for recognition by the lectin chaperones, calnexin and calreticulin. Further trimming by glucosidase II removes the final Glc residue to result in the release from calnexin/calreticulin interactions. Subsequent trimming of mannose residues from the precursor oligosaccharide is a prerequisite for the maturation of N-glycans to hybrid and complex-type structures, and in mammalian cells these cleavage reactions are accomplished by the action of ERManI, members of the GolgiManI subfamily of enzymes, and an $\alpha1,3/\alpha1,6$mannosidase in the Golgi complex, termed Golgi α-mannosidase II (Moremen, 2000). ERManI and the GolgiManI isozymes are Class 1 (GH47) α-mannosidases, whereas Golgi α-mannosidase II is a Class 2 (GH38) mannosidase and has been reviewed elsewhere (Moremen, 2000, 2002). Alternative pathways for high-mannose glycan maturation have also been proposed involving enzymes termed ER α-mannosidase II and endo α-mannosidase, which can trim $Man_9GlcNAc_2$ to $Man_8GlcNAc_2$ and $Glc_{3-1}Man_9GlcNAc_2$ to Man_8-$GlcNAc_2$, respectively. These latter enzymes have also been previously reviewed (Spiro and Spiro, 2000; Weng and Spiro, 1996).

Class 1 α-mannosidases recognize and cleave only α-1,2–linked mannose residues from the $Man_9GlcNAc_2$ high-mannose precursor, but each of the subfamilies has unique specificities for recognition of terminal α-1,2-mannose residues in the high-mannose structure. The ERManI subfamily contains a single member in all eukaryotes that have been examined, and the enzyme primarily cleaves a single mannose residue from $Man_9GlcNAc_2$ to produce a unique $Man_8GlcNAc_2$ isomer (Man8 isomer B; Fig. 1). ERManI is the first mannosidase to act in this pathway, and it has been suggested that it mediates quality control in the ER by creating a specific glycoform that is recognized by members of the putative degradation family of lectins, termed the EDEM proteins. The GolgiManI subfamily of enzymes comprises three isoforms in mammalian organisms: Golgi-ManIA (Lal et al., 1994), GolgiManIB (Herscovics et al., 1994), Golgi-ManIC (Tremblay and Herscovics, 2000). The substrate specificities of each of the isoforms are similar, and they are all complementary to the action of ERManI (Gonzalez et al., 1999). Thus, the GolgiManI isoforms can efficiently trim the $Man_8GlcNAc_2$ isomer B down to $Man_5GlcNAc_2$, but they poorly cleave the central branch mannose residue that is the target of ERManI action (see Fig. 1) (Lal et al., 1998). Subsequent transfer of a terminal GlcNAc residue by GlcNAc transferase I and mannose trimming

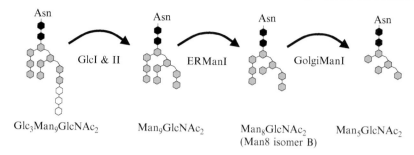

FIG. 1. Diagram of the early *in vivo* N-glycan processing steps of $Glc_3Man_9GlcNAc_2$-Asn to $Man_5GlcNAc_2$-Asn. The processing steps catalyzed by glucosidases I and II remove the three glucose residues. Trimming by ERManI generates the Man8 isomer B, and trimming by the GolgiManI subfamily of enzymes generates $Man_5GlcNAc_2$. The GlcNAc residues are indicated by black shaded hexagons. Mannose residues are gray hexagons. Glc residues are white hexagons.

by Golgi α-mannosidase II result in the formation of the $GlcNAcMan_3GlcNAc_2$ intermediate that acts as the committed step in the synthesis of complex-type *N*-glycans.

The EDEM subfamily of proteins is composed of three isoforms, and two members (EDEM1 and EDEM2) do not appear to have catalytic activity. Recent data suggest that the third isoform (EDEM3) may have glycosidase activity toward high-mannose oligosaccharides (Hirao *et al.*, 2006). All three of these mannosidase homologs have been shown to participate in the targeting of misfolded glycoproteins from the ER to the cytosol for degradation by the ubiquitin-proteasomal system. The mechanism for this protein disposal function has not been determined, but it is thought that EDEM proteins act as lectins, using the same oligosaccharide binding site employed by the catalytic Class I α-mannosidases (Mast *et al.*, 2005).

ERManI was first isolated, characterized, and cloned from *Saccharomyces cerevisiae* (Herscovics, 1999), but subsequent cDNA cloning, expression, and characterization of the human ERManI homolog (Gonzalez *et al.*, 1999) have revealed similarities in structure and specificity between the fungal and human enzymes. Recombinant forms of human (Karaveg *et al.*, 2005; Vallee *et al.*, 2000a) and yeast ERManI (Vallee *et al.*, 2000b) have been expressed in large quantities in *Pichia pastoris,* and the structures of the catalytic domains of each enzyme have been determined by X-ray diffraction in the presence and absence of bound enzymatic products, small molecule inhibitors, and uncleavable pseudosubstrates.

Members of the GolgiManI subfamily of enzymes have been isolated from a variety of species sources (e.g., rat, rabbit, pig, and calf, reviewed by

Moremen *et al.* [1994]), and recombinant forms of the human (Bause *et al.*, 1993) and mouse enzymes (Herscovics *et al.*, 1994; Lal *et al.*, 1994) have been expressed and characterized. Each of the members of the GolgiManI subfamily has been expressed in *P. pastoris*, and the structure of mouse GolgiManIA has been solved as a co-complex with a bound glycan enzymatic product (Tempel *et al.*, 2004). In addition, structures have been determined for the secreted GH47 α-mannosidases from *Penicillium citrinum* (Lobsanov *et al.*, 2002) and *Trichoderma reesei* (Van Petegem *et al.*, 2001), fungal enzymes that have substrate specificities similar to the GolgiManI enzymes.

The membrane topologies of the ERManI and GolgiManI subfamily members are similar, each containing a short cytoplasmic tail, transmembrane domain, linker region, and luminal catalytic domain (Moremen, 2000). The catalytic domains of both ERManI and GolgiManIA are both barrel-shaped $(\alpha\alpha)_7$ structures composed of two concentric layers of α helices containing a broad open cleft at one end of the barrel structure and a β-hairpin plugging the opposite end of the barrel. The oligosaccharide substrates bind to the open cleft, and extended interactions across the cleft provide specificity for insertion of glycan branches into the barrel core containing the enzyme active site. Structural data indicate that the cores of the catalytic pockets (+1 and −1 substrate-binding subsites) are conserved between the ERManI and GolgiManI subfamily members; however, outside of this area (>+1 residues), including the broad open clefts adjacent to the catalytic sites, the enzymes have distinct oligosaccharide binding sites that provide differences in branch specificities for these enzyme subfamilies (Tempel *et al.*, 2004). The energetic contributions of oligosaccharide substrate binding to the catalytic pocket and the broad open cleft were examined in a series of surface plasmon resonance binding studies between a catalytic mutant of human ERManI (E330Q) and a series of high-mannose *N*-glycan structures (Karaveg *et al.*, 2005). These studies revealed that the majority of the binding energy was contributed through interactions with the substrate residue in the +1 subsite (∼52% of the substrate binding energy). Considerably less energetic contribution was provided by interactions with the −1 subsite residue (∼21% of the binding energy). The remainder of the binding energy (∼27% of the binding energy) was provided by substrate residues in the >+1 substrate binding subsites. Despite the relatively minor contributions of the >+1 subsite residues to the binding energy, the stereochemistry of the interactions in the broad substrate-binding cleft clearly provides the branch specificity for the selective cleavage of terminal α1,2-mannose residues. Swapping of a key substrate-binding residue from GolgiManIA to ERManI led to broader specificity for glycosidase cleavage but poor catalytic efficiency and a significantly reduced binding affinity.

Thus, the catalytic pockets and adjacent substrate-binding clefts for the Class 1 (GH47) α-mannosidases provide for the high-affinity and high-specificity substrate interactions that lead to the unique substrate recognition characteristics for the individual subfamilies. This chapter describes the methods for enzyme expression and characterization for wild-type and mutant forms of the Class 1 mannosidases that can be used as paradigms for similar studies in other enzyme systems.

Overview of Class 1 (GH47) Expression and Purification

Recombinant expression of wild-type and mutant Class 1 mannosidases in *P. pastoris* has proven to be an effective strategy for large-scale production of enzymes for biochemical and structural studies (Karaveg and Moremen, 2005; Karaveg *et al.*, 2005; Tempel *et al.*, 2004). The advantages of this expression system include the ease in generating recombinant constructs and transformants, the short-time course of induction, the high levels of protein expression, the ability to generate recombinant proteins that are secreted into the culture media for easy purification, and the ability to generate recombinant products that contain eukaryotic post-translational modifications (i.e., disulfide bond formation, N-glycosylation). We have generally used commercial expression vectors (Invitrogen pPICZα series or pHIL-S1) that encode an NH_2-terminal *S. cerevisiae* α-factor signal sequence or *PHO1* secretion signal followed by a series of restriction sites. cDNA sequences encoding the COOH-terminal catalytic domains of the Class 1 mannosidases (with deletions of the NH_2-terminal cytoplasmic tails, transmembrane domains, and stem regions) have been inserted in-frame into the vectors, and transformants were generated in either Mut^+ (X-33, GS115) or Mut^S (KM71H) strains. Following growth of cell mass on glycerol as a carbon source, enzyme expression was initiated by shifting to methanol as a carbon source and inducing agent for the AOX1 promoter. These transformants generate translation products that are targeted to the yeast secretory pathway and released into the culture media as potentially glycosylated products. Both *S. cerevisiae* ERManI (Vallee *et al.*, 2000b) and mouse Golgi-ManIA (Tempel *et al.*, 2004) coding regions contain acceptor sequences for N-glycosylation (Table I) while human ERManI (Vallee *et al.*, 2000a) contains no N-glycosylation sites and is secreted as a non-glycosylated protein. In general, affinity tags have not been used for the generation of α-mannosidase recombinant constructs, since purification of the enzyme from the culture media is readily accomplished with two or three chromatography steps (see later). In recent studies, the mannosidase homology domain of human EDEM2 has also been expressed in *P. pastoris* using pPICZα constructs. These latter constructs have been generated as epitope and/or His-tagged

TABLE I

CONSTRUCTS FOR EXPRESSION OF CLASS 1 α-MANNOSIDASES IN *PICHIA PASTORIS*

Class 1 α-mannosidase[a]	*Pichia* host strain	Expression vector	Residues of coding region expressed	Amino acids in full-length coding region	Predicted *N*-glycan acceptor sites
Yeast ERManI	GS115, KM71	pHILS-1	34–549	549	3
Human ERManI	X33	pPICZ-α	226–699	699	0
Mouse GolgiManIA	GS115	pHILS-1	66–655	655	1
Human EDEM2	KM71H	pPICZ-α	18–492	578	4

[a] Information on the constructs and strains for the expression data of yeast ERManI (Lipari and Herscovics, 1996), human ERManI (Karaveg *et al.*, 2005), mouse GolgiManIA (Lal *et al.*, 1998), and human EDEM2 (unpublished data) were obtained from the respective sources.

fusions at the NH$_2$-terminus of the α-mannosidase–homology domain, because previous data have indicated that COOH-terminal fusions yield in inactive ERManI constructs. A summary of α-mannosidase constructs generated for expression in *P. pastoris* is shown in Table I.

Generation of Class 1 α-Mannosidase Site-Directed Mutants

A number of site-directed mutants of ERManI have been generated to analyze the effects of altering selected residues on glycan binding and catalytic efficiency (Karaveg and Moremen, 2005; Karaveg *et al.*, 2005). Residues were selected based on their positions in the α-mannosidase structure, and mutations were generally either conservative substitutions or replacements with Ala residues. Mutagenesis was accomplished in the pPICZα vector using the QuikChange mutagenesis kit (Stratagene) using protocols provided by the manufacturer. After the site-directed mutagenesis, the coding regions were fully sequenced to confirm that only the desired mutation was introduced into the recombinant protein product. All mutant proteins were expressed and purified under identical conditions as the wild-type enzyme, and in most instances the protein yields were similar to the wild-type enzyme. For catalytically inactive mutant enzymes, protein purification profiles were followed by immunoblots using antibodies raised to the recombinant wild-type proteins.

Enzyme Expression and Purification

The Class 1 α-mannosidases can be readily expressed in *P. pastoris*, either in shake flasks (Karaveg *et al.*, 2005) or small fermentor cultures (Tempel *et al.*, 2004). For shake flask cultures, yeast transformants were

grown for 2 days in 1L BMGY (400 mg of biotin, 0.5% methanol, 1% glycerol, 1% yeast extract, 2% peptone, 0.1 M potassium phosphate [pH 6], and 1.34% yeast nitrogen base), at 30°, with agitation at 300 rpm to generate cell mass; recombinant protein production was initiated by the addition of 0.5% methanol to the culture. Additional methanol (0.5% final concentration) was then added at 12-h intervals for 3 to 5 days. Alternatively, protein production can be generated in a New Brunswick BioFlow 3000 fermentor (1- or 2.5-l vessels) in BMGY medium supplemented by a limiting feeding strategy using glycerol as a carbon source until a cell density of approximately 200 g/l is attained. The glycerol feed was then stopped, and the culture was maintained at 30° with agitation until oxygen consumption fell, indicating that the carbon source was consumed. Methanol was then pumped into the fermentor vessel and maintained at 0.5% by the use of a methanol probe (Raven Biotech, Vancouver, BC) controlling the methanol feed pump. The culture was then maintained at this methanol concentration for an additional 3 days of culture before harvesting.

Initial purification of the secreted α-mannosidase catalytic domains followed a clarification of the culture supernatant by centrifugation (6000 × g, 20 min) and filtration through a 0.4-μm filter. For GolgiManIA (Tempel *et al.*, 2004), the resulting supernatant was dialyzed against four volumes of 10 mM sodium succinate (pH, 6.5) with two changes of buffer, followed by adjustment of the pH to 5.5 with HCl immediately before chromatography. The human ERManI (Karaveg *et al.*, 2005) culture supernatant was adjusted to pH 6.0 and applied to the cation exchange column after 1:2 dilution with H_2O. Both enzyme preparations were loaded on SP Sepharose columns (Amersham-Pharmacia Biotech, 5 × 13 cm column), washed with column buffer, and elution was performed with a linear NaCl gradient (0 to 0.5 M) in the same buffer. For GolgiManIA purification, the column buffer was 10 mM sodium succinate (pH, 5.5); and for the ERManI purification, the column buffer was 10 mM sodium succinate (pH, 6.0), 1 mM CaCl$_2$. ERManI preparations were sufficiently enriched after the SP Sepharose step to proceed directly to gel filtration. GolgiManIA preparations were further purified by adjustment of the solution to 1.0 M (NH$_4$)$_2$SO$_4$, 0.1 M Na-HEPES (pH, 6.5), followed by loading on a Phenyl Sepharose column (Amersham-Pharmacia Biotech; 1.5 × 20 cm). The column was washed with 1.0 M (NH$_4$)$_2$SO$_4$ and 10 mM Na-HEPES (pH, 6.5) and eluted using a linear gradient of 1.0 to 0.5 M (NH$_4$)$_2$SO$_4$ in 10 mM Na-HEPES (pH, 6.5). For both enzyme preparations, fractions containing enzyme activity were pooled and concentrated using an Amicon YM10 membrane after the addition of NDSB-201 (Calbiochem) to 0.75 M. Aliquots of 5 ml of concentrated enzyme sample were applied to a Superdex 75 column (Amersham-Pharmacia Biotech, 1 × 100 cm) pre-equilibrated

with 20 mM Na-MES (pH, 6.5 for mouse GolgiManIA; pH, 7.0 for human ERManI), 150 mM NaCl, 5 mM CaCl$_2$, and 0.25 M NDSB-201. The final purified enzyme preparations were concentrated to 30 mg/ml in the same buffer and stored at 4°. The inclusion of NDSB-201 in all of the buffers after the Phenyl Sepharose step was an important element in the purification strategies. Significant losses in the recovery of the recombinant enzymes by aggregation and precipitation were avoided by inclusion of NDSB-201 in all of the buffers.

Oligosaccharide Isolation for Enzyme Assays and Binding Studies

Man$_9$GlcNAc$_2$ glycan structures were preparatively isolated from crude soybean agglutinin (SBA) generated from soybean meal by acid extraction and (NH$_4$)$_2$SO$_4$ precipitation (Karaveg and Moremen, 2005; Karaveg et al., 2005). Crude SBA was subsequently denatured, reduced, and carboxyamidomethylated in 8 M guanidine HCl containing 50 mM dithiothreitol and 100 mM iodoacetamide before proceeding to trypsin/elastase digestion. Glycopeptides were recovered by affinity chromatography using concanavalin A-Sepharose (ConA-Sepharose, Amersham-Pharmacia Biotech) and further purified by high performance liquid chromatography (HPLC) on a Cosmosil C18 column for surface plasmon resonance (SPR) studies (Karaveg et al., 2005). Free oligosaccharides were liberated from the peptide by peptide:N-glycosidase F digestion and derivatized with pyridylamine. Man$_8$GlcNAc$_2$-PA isomers, Man$_6$GlcNAc$_2$-PA, and Man$_5$GlcNAc$_2$-PA were generated by digestion with either ERManI or GolgiManIA and isolation by reverse phase HPLC, as described (Karaveg and Moremen, 2005; Karaveg et al., 2005).

α-Mannosidase Assays and Kinetic Analyses

Class I mannosidase enzyme activity can be measured by monitoring the cleavage of Man-α1,2-Man disaccharides or pyridylamine-tagged high-mannose oligosaccharides (Man$_9$GlcNAc$_2$-PA). Assays using tagged high-mannose oligosaccharides as substrates are more sensitive than the disaccharide assays, but they are more labor intensive and not as readily adapted to a high-throughput assay format. An alternative assay using a [^{3}H]mannose-labeled Man$_9$GlcNAc$_2$ substrate followed by detection of cleavage by substrate binding to ConA has also been described (Herscovics and Jelinek-Kelly, 1987), but this requires the generation and handling of hazardous radiolabeled substrates.

Disaccharide assays for α-mannosidase activity make use of a modified glucose oxidase/horseradish peroxidase assay (Lal et al., 1998). Assays were assembled in flat-bottomed 96-well microliter plates containing

10 mM potassium phosphate (pH, 6.0), 2 to 10 mM Manα1,2Manα-O-CH$_3$ (Sigma), and 5 to 15 μl of enzyme solution (25 μl total volume) followed by incubation at 37° for 30 min. Reactions were terminated by adding 25 μl of 1.25 M Tris-Cl (pH, 7.6), and the amount of mannose released was detected by incubation with 250 μl of developing solution containing glucose oxidase (55 U/ml), horseradish peroxidase (1 purpurogallin unit/ml), and o-dianisidine dihydrochloride (70 μg/ml) for 3 h at 37°. The final color intensity was determined by measuring the absorbance at 450 nm. Free mannose was used as a standard. One unit of mannosidase activity is defined as the amount of enzyme that releases 1 nmole of mannose in 1 minute at 37°.

Assays using oligosaccharide substrates (Lal *et al.*, 1998) contained 100 pmol Man$_9$GlcNAc$_2$-PA, 50 mM MES (pH, 7), 150 mM NaCl, and 5 mM CaCl$_2$ (20 μl total volume), and reactions were performed at 37° for varied periods of time, depending on the concentration of the enzyme. Reactions were stopped by the addition of 20 μl 1.25 M Tris-HCl (pH, 7.6), followed by separation of the digested products on a Hypersil APS-2 NH$_2$ column (4.6 × 250 mm, Alltech), using a gradient from 90% buffer A (80% acetonitrile : 20% 50 mM Na phosphate; pH, 4) to 35% Buffer B (100 mM Na phosphate; pH, 4) run over 50 min at flow rate of 1 ml/min. Elution of the PA-tagged oligosaccharide was monitored by an in-line fluorescence detector (excitation wavelength, 320 nm; emission wavelength, 400 nm), with identification of the cleavage products based on a comparison to the elution positions of Man$_{9-5}$GlcNAc$_2$-PA standards. One unit of enzyme activity generated 1 μmol/min of Man$_8$GlcNAc$_2$ from Man$_9$GlcNAc$_2$ at 37°. An alternative solvent protocol provided a shorter HPLC separation time. Using 40% acetonitrile in 60 mM Na phosphate buffer (pH, 4) with an isocratic flow of 1.0 ml/min, Man$_{9-5}$GlcNAc$_2$-PA could be resolved in less than 15 min. Variations in the flow rate (0.7 to 1.4 ml/min) or an increase in the percentage of acetonitrile by 5% to 15% can be used to shift the elution positions of the high-mannose glycans for optimal resolution.

For kinetic analysis (Karaveg and Moremen, 2005; Karaveg *et al.*, 2005), initial rates (*v*) for the enzyme reactions were determined at various substrate concentrations ranging from 10 to 300 μM. Catalytic coefficient (k_{cat}), and Michaelis constant (K_m) values were determined by fitting initial rates to a Michaelis-Menten function by nonlinear regression analysis using SigmaPlot (Jandel Scientific, San Rafael, CA). k_{cat}/K_m values could also be derived from reciprocal plots of *v* and [S]. For kinetic pH rate analysis (Karaveg *et al.*, 2005), values of k_{cat} and K_m were determined from initial rates of enzyme reactions in the pH range from 4 to 10 using a 4× universal buffer (80 mM succinic acid, 80 mM MES, 80 mM HEPBS, 80 mM HEPES,

and 80 mM CHES adjusted to pH with 5 M NaOH) in place of the MES assay buffer described previously. Plots of log(k_{cat}/K_m) versus pH could then be used to estimate the pK_a values for the acidic and basic limbs of the pH curves for the wild-type and mutant enzymes. In studies on the temperature dependence of catalysis (Karaveg and Moremen, 2005), values for k_{cat} were obtained between 5° and 40° at 5° intervals and were used to calculate activation energies (E_a) from the slopes ($-E_a/R$) of Arrhenius plots (ln(k_{cat}) as a function of $1/T$). The thermodynamic activation parameters can then be calculated from the plots as previously described (Karaveg and Moremen, 2005).

Glycan Binding Studies by SPR

The combination of mutagenesis and substrate binding studies by SPR provided a useful set of tools for investigating the determinants of binding kinetics and glycan affinity for the recombinant α-mannosidases (Karaveg and Moremen, 2005; Karaveg et al., 2005). Among the various strategies for binding studies by SPR, we have chosen to immobilize the enzymes to the sensor chip surface and screen for glycan binding as the mobile phase analyte (Fig. 2).

SPR analyses were conducted using a Biacore 3000 apparatus (Biacore AB), with recombinant enzymes immobilized on the chip surfaces at 25° by amine coupling. The flow cells were activated by injecting a mixture of 50 mM N-hydroxysuccinimide and 200 mM 1-ethyl-3-(dimethylaminopropyl) carbodiimide over the CM5 sensor chip surface for 7 min at 5 μl/min. The recombinant proteins prepared in 10 mM sodium succinic acid (pH, 6.0) were passed through a 0.2-μm filter and diluted in the same buffer to obtain a concentration of 5 μg/ml before injection onto the activated surface. The desired immobilization level can be achieved by specific contact time with the chip surface. The immobilization efficiency for ERManI was about 2500 RU/min at a flow rate of 5 μl/min in HPB-EP buffer (10 mM HEPES; pH, 7.4; 150 mM NaCl; 3.4 mM EDTA; 0.01% polysorbate P20) at 25°. The remaining reactive groups were blocked by injection of 1 M ethanolamine-HCl at pH 8.5 for 7 min at 5 μl/min. Mock derivatized flow cells served as reference surfaces. Binding analyses were generally performed at 10° with a continuous flow (30 μl/min) of running buffer (10 mM MES; pH, 7.0, 300 mM NaCl; and 5 mM CaCl$_2$) except in temperature-dependent interaction studies, which have been performed between 5° and 35° at 5° intervals, consecutively, in an automated method. Analytes were prepared in running buffer by twofold serial dilution to obtain appropriate concentration ranges. A concentration series of Man$_9$GlcNAc$_2$-glycopeptide (0.4 to 400 μM) was analyzed over a low-density immobilization surface of

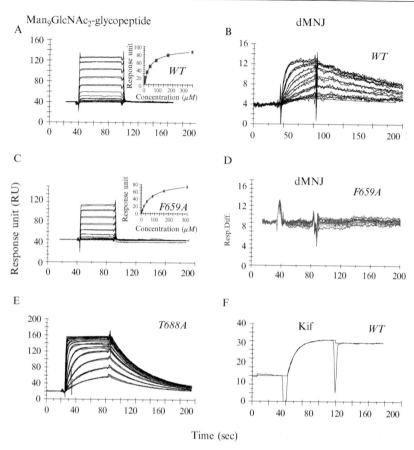

FIG. 2. Examples of SPR binding of Man$_9$GlcNAc$_2$, dMNJ, or Kif ligands to wild-type and mutant ERManI. Wild-type (A, B, and F) or mutant (C, D, and E) ERManI forms were immobilized on the SPR chip surface and various concentrations of Man$_9$GlcNAc$_2$ (left panels), dMNJ (B and D), or Kif (F) ligands were tested for binding. Representative SPR sensorgrams in the ligand concentration series are shown. If the on- and off-rates (k_a and k_d, respectively) were sufficiently slow (B and E), curve fitting of the sensorgrams was performed to determine values for the equilibrium dissociation constants ($K_D = k_d/k_a$). Where the kinetics for ligand binding were too rapid, equilibrium sensorgram values were used to plot a saturation curve (insets in A and C) and calculate values for K_D. In the F659A mutant, binding of Man$_9$GlcNAc$_2$ was minimally affected (C), whereas dMNJ binding was completely abolished, as indicated by the absence of a deflection in the SPR sensorgram trace (D). For panel F, the tight-binding kinetics of Kif resulted in no appreciable dissociation from the chip surface. Reproduced in part from Karaveg and Moremen (2005), with permission of American Society for Biochemistry and Molecular Biology via Copyright Clearance Center.

recombinant protein (3000 RU), whereas disaccharides (16 to 1000 μM) and small molecule inhibitors, such as 1-deoxymannojirimycin (dMNJ) (2 to 1000 μM) and kifunensine (Kif), were analyzed over a high-density immobilization surface of recombinant protein (10,000 RU). Samples were injected for 1 min, followed by at least 3 min of dissociation phase. In most cases, the baseline returned to the original response level in 5 min without a further regeneration of the chip surface, except when Kif was used as the analyte (see Fig. 2). This latter inhibitor exhibits tight-binding kinetics and does not dissociate from the chip surface even with extensive washing. SPR data for each concentration of analyte were collected in duplicate and globally fit to a 1:1 Langmuir binding algorithm model to calculate the on-rate (k_a), the off-rate (k_d), and the equilibrium dissociation constant $(k_d/k_a = K_D)$ using BiaEvaluation version 3.1 software supplied with the instrument. Alternatively, the maximal equilibrium sensorgram values were used to plot a saturation binding curve and calculate values for the equilibrium dissociation constant (K_D) directly. K_D values measured at different temperatures can also be used to calculate thermodynamic parameters (ΔG, ΔH, $T\Delta S$) of binding (Karaveg and Moremen, 2005) using the van't Hoff equation (Day et al., 2002) to plot the linear relationship of $y = \ln(K_D)$ versus $x = 1/T$, which gives a slope of $\Delta H/R$ and an intercept of $-\Delta S/R$. The effects of temperature on the individual association rates (k_a) and dissociation rates (k_d) can also be independently determined using the Eyring equation (Day et al., 2002), which allows thermodynamic parameters to be determined from measured k_a and k_d values at different temperatures by a linear relationship of $y = R\ln(hk_a/k_BT)$ or $y = R\ln(hk_d/k_BT)$ versus $x = 1/T$, where the slope and the intercept of the Eyring plots are $\Delta H\dagger$ and $\Delta S\dagger$, respectively (Karaveg and Moremen, 2005).

Summary

The biochemical characterization of Class 1 α-mannosidases has made use of a combination of approaches including recombinant expression of wild-type and mutant enzymes, purification by classical chromatography approaches, analyses of enzyme kinetics, and binding affinity and kinetic measurements using SPR (Karaveg and Moremen, 2005; Karaveg et al., 2005). In the case of human ERManI, these approaches have been complemented by structural determination of co-complexes between the enzyme and small molecule glycone mimics, uncleavable disaccharide pseudo-substrates, and large glycan enzymatic products (Karaveg et al., 2005; Tempel et al., 2004; Vallee et al., 2000a,b). The result of these studies was a detailed map of the energetic contributions to substrate recognition, binding, and catalysis (Fig. 3) (Karaveg and Moremen, 2005) that will

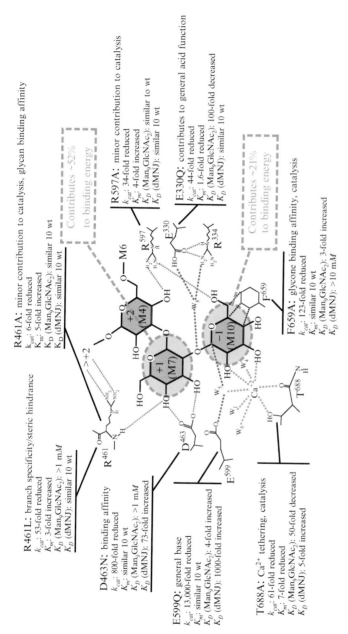

FIG. 3. Summary of the enzyme kinetics and binding studies for key protein and glycan residues in the ERManI active site. The catalytic and binding components in the ERManI active site are shown, highlighting the protein residues interacting with the -1, $+1$, and $+2$ subsite glycan residues. Fold differences for kinetic parameters (values for k_{cat} and K_m) and SPR binding parameters (K_D values for binding to $Man_9GlcNAc_2$ and dMNJ) for the mutant enzymes relative to values for wild-type ERManI are indicated adjacent to each respective amino acid. In addition, the respective proposed functions for each residue based on the binding and kinetic analyses are indicated adjacent to each amino acid. Energetic contributions to binding within the -1 and $+1$ subsites are also indicated by the dotted lines with shading. Reproduced from Karaveg and Moremen (2005), with permission from American Society for Biochemistry and Molecular Biology via Copyright Clearance Center.

establish a framework for studying the differences in recognition between the different Class 1 α-mannosidase subfamilies. Since these proteins play distinct roles in glycoprotein maturation and ER quality control, further study should lead to a greater understanding of how these common sequences and structural folds can accomplish their unique biological functions. The strategies used for the biochemical characterization of the α-mannosidases can also be applied to the study of many other families of enzymes that play critical cellular functions.

Acknowledgment

This work was supported by National Institutes of Health Research Grants GM047533 and RR05351 (to K.W.M.).

References

Bause, E., Bieberich, E., Rolfs, A., Volker, C., and Schmidt, B. (1993). Molecular cloning and primary structure of Man9-mannosidase from human kidney. *Eur. J. Biochem.* **217,** 535–540.

Day, Y. S., Baird, C. L., Rich, R. L., and Myszka, D. G. (2002). Direct comparison of binding equilibrium, thermodynamic, and rate constants determined by surface- and solution-based biophysical methods. *Protein Sci.* **11,** 1017–1025.

Gonzalez, D. S., Karaveg, K., Vandersall-Nairn, A. S., Lal, A., and Moremen, K. W. (1999). Identification, expression, and characterization of a cDNA encoding human endoplasmic reticulum mannosidase I, the enzyme that catalyzes the first mannose trimming step in mammalian Asn-linked oligosaccharide biosynthesis. *J. Biol. Chem.* **274,** 21375–21386.

Helenius, A., and Aebi, M. (2004). Roles of N-linked glycans in the endoplasmic reticulum. *Annu. Rev. Biochem.* **73,** 1019–1049.

Henrissat, B., and Bairoch, A. (1996). Updating the sequence-based classification of glycosyl hydrolases. *Biochem. J.* **316**(Pt. 2), 695–696.

Herscovics, A. (1999). Processing glycosidases of *Saccharomyces cerevisiae. Biochim. Biophys. Acta* **1426,** 275–285.

Herscovics, A., and Jelinek-Kelly, S. (1987). A rapid method for assay of glycosidases involved in glycoprotein biosynthesis. *Anal. Biochem.* **166,** 85–89.

Herscovics, A., Schneikert, J., Athanassiadis, A., and Moremen, K. W. (1994). Isolation of a mouse Golgi mannosidase cDNA, a member of a gene family conserved from yeast to mammals. *J. Biol. Chem.* **269,** 9864–9871.

Hirao, K., Natsuka, Y., Tamura, T., Wada, I., Morito, D., Natsuka, S., Romero, P., Sleno, B., Tremblay, L. O., Herscovics, A., Nagata, K., and Hosokawa, N. (2006). EDEM3, a soluble EDEM homolog, enhances glycoprotein ERAD and mannose trimming. *J. Biol. Chem.* **281,** 9650–9658.

Hosokawa, N., Tremblay, L. O., You, Z., Herscovics, A., Wada, I., and Nagata, K. (2003). Enhancement of endoplasmic reticulum (ER) degradation of misfolded Null Hong Kong alpha1-antitrypsin by human ER mannosidase I. *J. Biol. Chem.* **278,** 26287–26294.

Karaveg, K., and Moremen, K. W. (2005). Energetics of substrate binding and catalysis by class 1 (glycosylhydrolase family 47) alpha-mannosidases involved in N-glycan processing and endoplasmic reticulum quality control. *J. Biol. Chem.* **280,** 29837–29848.

Karaveg, K., Siriwardena, A., Tempel, W., Liu, Z. J., Glushka, J., Wang, B. C., and Moremen, K. W. (2005). Mechanism of class 1 (glycosylhydrolase family 47) α-mannosidases involved

in N-glycan processing and endoplasmic reticulum quality control. *J. Biol. Chem.* **280,** 16197–16207.

Kornfeld, R., and Kornfeld, S. (1985). Assembly of asparagine-linked oligosaccharides. *Ann. Rev. Biochem.* **54,** 631–664.

Lal, A., Pang, P., Kalelkar, S., Romero, P. A., Herscovics, A., and Moremen, K. W. (1998). Substrate specificities of recombinant murine Golgi alpha1, 2-mannosidases IA and IB and comparison with endoplasmic reticulum and Golgi processing alpha1,2-mannosidases. *Glycobiology* **8,** 981–995.

Lal, A., Schutzbach, J. S., Forsee, W. T., Neame, P. J., and Moremen, K. W. (1994). Isolation and expression of murine and rabbit cDNAs encoding an alpha 1,2-mannosidase involved in the processing of asparagine-linked oligosaccharides. *J. Biol. Chem.* **269,** 9872–9881.

Lipari, F., and Herscovics, A. (1996). Role of the cysteine residues in the alpha1,2-mannosidase involved in N-glycan biosynthesis in *Saccharomyces cerevisiae:* The conserved Cys340 and Cys385 residues form an essential disulfide bond. *J. Biol. Chem.* **271,** 27615–27622.

Lobsanov, Y. D., Vallee, F., Imberty, A., Yoshida, T., Yip, P., Herscovics, A., and Howell, P. L. (2002). Structure of *Penicillium citrinum* alpha 1,2-mannosidase reveals the basis for differences in specificity of the endoplasmic reticulum and Golgi class I enzymes. *J. Biol. Chem.* **277,** 5620–5630.

Mast, S. W., Diekman, K., Davis, A. W., Karaveg, K., Sifers, R. N., and Moremen, K. W. (2005). Human EDEM2, a novel homolog of family 47 glycosidases, is involved in ER-associated degradation of glycoproteins. *Glycobiology* **15,** 421–436.

Moremen, K. (2000). α-mannosidases in Asparagine-linked oligosaccharide processing and catabolism. *In* Oligosaccharides in Chemistry and Biology: A Comprehensive Handbook, (B. Ernst, G. Hart, and P. Sinay, eds.), Vol. II, pp. 81–117. John Wiley & Sons Inc., New York.

Moremen, K. W. (2002). Golgi alpha-mannosidase II deficiency in vertebrate systems: Implications for asparagine-linked oligosaccharide processing in mammals. *Biochim. Biophys. Acta* **1573,** 225–235.

Moremen, K. W., Trimble, R. B., and Herscovics, A. (1994). Glycosidases of the asparagine-linked oligosaccharide processing pathway. *Glycobiology* **4,** 113–125.

Spiro, M. J., and Spiro, R. G. (2000). Use of recombinant endomannosidase for evaluation of the processing of N-linked oligosaccharides of glycoproteins and their oligosaccharide-lipid precursors. *Glycobiology* **10,** 521–529.

Tempel, W., Karaveg, K., Liu, Z. J., Rose, J., Wang, B. C., and Moremen, K. W. (2004). Structure of mouse Golgi alpha-mannosidase IA reveals the molecular basis for substrate specificity among class 1 (family 47 glycosylhydrolase) alpha1,2-mannosidases. *J. Biol. Chem.* **279,** 29774–29786.

Tremblay, L. O., and Herscovics, A. (2000). Characterization of a cDNA encoding a novel human Golgi alpha 1, 2-mannosidase (IC) involved in N-glycan biosynthesis. *J. Biol. Chem.* **275,** 31655–31660.

Vallee, F., Karaveg, K., Herscovics, A., Moremen, K. W., and Howell, P. L. (2000a). Structural basis for catalysis and inhibition of N-glycan processing class I alpha 1,2-mannosidases. *J. Biol. Chem.* **275,** 41287–41298.

Vallee, F., Lipari, F., Yip, P., Sleno, B., Herscovics, A., and Howell, P. L. (2000b). Crystal structure of a class I alpha1,2-mannosidase involved in N-glycan processing and endoplasmic reticulum quality control. *EMBO J.* **19,** 581–588.

Van Petegem, F., Contreras, H., Contreras, R., and Van Beeumen, J. (2001). *Trichoderma reesei* alpha-1,2-mannosidase: Structural basis for the cleavage of four consecutive mannose residues. *J. Mol. Biol.* **312,** 157–165.

Weng, S., and Spiro, R. (1996). Endoplasmic reticulum kifunensine-resistant a-mannosidase is enzymatically and immunologically related to the cytosolic a-mannosidase. *Arch. Biochem. Biophys.* **325,** 113–123.

Wu, Y., Swulius, M. T., Moremen, K. W., and Sifers, R. N. (2003). Elucidation of the molecular logic by which misfolded alpha 1-antitrypsin is preferentially selected for degradation. *Proc. Natl. Acad. Sci. USA* **100,** 8229–8234.

[4] A Cytoplasmic Peptide: N-Glycanase

By KAORI TANABE, WILLIAM J. LENNARZ, and TADASHI SUZUKI

Abstract

A cytoplasmic peptide:N-glycanase (PNGase) has been implicated in the proteasomal degradation of aberrant glycoproteins synthesized in the endoplasmic reticulum. The reaction is believed to be important for subsequent proteolysis by the proteasome since bulky N-glycan chains on misfolded glycoproteins may impair their efficient entry into the interior of the cylinder-shaped 20S proteasome, where the active sites of the proteases reside. The deglycosylation reaction by PNGase brings about two major changes on substrate proteins; one is a removal of N-glycan chains, and the other is the introduction of negative charge(s) into the core peptide by converting glycosylated asparagine residue(s) into aspartic acid residue(s). Therefore, PNGase action can be accurately monitored by detecting both changes using two different methods; that is, sodium dodecyl sulfate-polyacrylamide gel electrophoresis (SDS-PAGE) for deglycosylation and isoelectric focusing for detection of introduction of negative charge(s) into core proteins. This chapter will describe the simple *in vivo* as well as *in vitro* assay method to detect PNGase activity.

Introduction

Peptide:N-glycanase (PNGase; also called N-glycanase or glycoamidase; peptide-N^4-[N-acetyl-β-D-glucosaminyl]asparagine amidase; EC 3.5.1.52) cleaves the amide bond in Asn residue of N-linked glycopeptides or glycoproteins (Suzuki *et al.*, 2002). This cleavage releases an intact oligosaccharide and generates at the site of hydrolysis an aspartic acid residue on the protein/peptide chain (Suzuki *et al.*, 2002) (Scheme 1, line 1). When the pH is below 8, the 1-amino-oligosaccharide is unstable and is spontaneously converted to the reducing oligosaccharide (Scheme 1, line 2). This enzyme activity was first found in plants (Takahashi, 1977) and subsequently in bacteria (Plummer *et al.*, 1984). Since those discoveries, this enzyme has been widely used as a powerful tool reagent for analyzing

METHODS IN ENZYMOLOGY, VOL. 415   0076-6879/06 $35.00
Copyright 2006, Elsevier Inc. All rights reserved.   DOI: 10.1016/S0076-6879(06)15004-1

R-GlcNAcβ1-Asn-peptide + H$_2$O $\rightarrow$ R-GlcNAc-NH$_2$ + Asp-peptide(line1)
R-GlcNAc-NH$_2$ + H$_2$O $\rightarrow$ R-GlcNAc + NH$_3$(line2)
R : oligosaccharide

SCHEME 1. Reaction of PNGase.

the structure and functions of N-linked glycan chains on glycoproteins. The occurrence of cytoplasmic PNGase activity has been reported in a wide variety of animal origins as well as in yeast (Kitajima *et al.*, 1995; Seko *et al.*, 1999; Suzuki *et al.*, 1993, 1997, 1998). The cytoplasmic PNGase was found to be quite distinct in terms of enzymatic properties from the PNGases of plant and bacterial origin (Suzuki *et al.*, 1995). A gene encoding the cytoplasmic enzyme, *PNG1*, was first identified in *Saccharomyces cerevisiae* (Suzuki *et al.*, 2000). Structural elements of the gene are highly conserved throughout eukaryotes (Suzuki *et al.*, 2000, 2003).

Extensive studies during the last decade have clearly established that eukaryotic cells have a so-called endoplasmic reticulum–associated degradation (ERAD) system for elimination of newly synthesized misfolded/unassembled proteins (McCracken and Brodsky, 2003; Meusser *et al.*, 2005; Nishikawa *et al.*, 2005; Sayeed and Ng, 2005). In this system, misfolded/unassembled (glyco)proteins are first "dislocated" or "retrotranslocated" from the ER into the cytosol. The 26S proteasome plays a central role in degrading them in the cytosol. Now, a number of ERAD substrates are known to be deglycosylated (most likely by PNGase) during the degradation process (Table I). The most recent study showed that a model protein, RTAΔ (ricin nontoxic allele of ricin A chain), could serve as an ERAD substrate, which is deglycosylated by the cytoplasmic PNGase, in yeast (Kim *et al.*, 2006). In this chapter, we will describe the simple assay methods for *in vivo* enzyme activity using this protein and yeast cells, as well as an *in vitro* assay method using a commercially available glycoprotein substrate. Essentially, this technique can be applied for any ERAD substrates assumed to be deglycosylated by PNGase (see Table I).

In Vitro PNGase Assay

PNGase-catalyzed deglycosylation results not only in the removal of glycan(s) from the core protein/peptide, but also in introducing negative charge(s) into the core protein/peptide by converting glycosylated Asn residue(s) into Asp residue(s). Therefore, to conclusively prove the PNGase activity, these two drastic changes in the substrates should be monitored. For glycopeptide substrates, paper chromatography is normally used to detect the release of glycan from peptide, whereas paper electrophoresis is used to detect the introduction of negative charges into the core peptide (Suzuki,

TABLE I
LIST OF ERAD SUBSTRATES REPORTED TO BE DEGLYCOSYLATED *IN VIVO*

Substrates	References
β_1 Integrin	Lee *et al.*, 2004
Cystic fibrosis transmembrane conductance regulator	Bebök *et al.*, 1998
	Johnston *et al.*, 1998
	Xiong *et al.*, 1999
Ether-a-go-go–related gene (hERG) protein mutant (α subunit of the K+ channel (I_{Kr})	Gong *et al.*, 2005
γ Opioid receptor	Petäjä-Repo *et al.*, 2001
Human cytomegalovirus US2 product	Wiertz *et al.*, 1996b
Immunoglobulin M, J subunit	Mancini *et al.*, 2000
Major histocompatibility complex class I heavy chain	Wiertz *et al.*, 1996a
	Hughes *et al.*, 1997
Prion protein (PrP)	Yedidia *et al.*, 2001
	Ma *et al.*, 2002
Ribophorin I mutant	de Virgilio *et al.*, 1998
	Kitzmüller *et al.*, 2003
Ricin A chain mutant in yeast	Simpson *et al.*, 1999
	Kim *et al.*, 2006
Ricin A chain in plant cells	Di Cola *et al.*, 2001
T-cell receptor α subunit	Yu *et al.*, 1997
	Huppa and Ploegh, 1997
Tyrosinase	Halaban *et al.*, 1997
	Mosse *et al.*, 1998

2005; Suzuki *et al.*, 1994). The substrate glycopeptide can be radiolabeled at the *N*-terminal α-amino group by reductive methylation using [^{14}C] formaldehyde (Seko *et al.*, 1991) to establish the very sensitive PNGase assay. The detailed method for preparation of glycopeptide, as well as two analytical assays (paper chromatography and paper electrophoresis), has been reported (Suzuki, 2005).

The principle (i.e., that two changes should be monitored to detect the PNGase activity) can also be applied for PNGase assay when glycoprotein substrates are used. This can be achieved by using SDS-PAGE for the detection of release of glycans from glycoproteins and isoelectric focusing (IEF) for the detection of charge changes (Di Cola *et al.*, 2001; Huppa and Ploegh, 1997; Kitzmuller *et al.*, 2003; Wiertz *et al.*, 1996a).

For the detection of cytoplasmic PNGase activity using glycoprotein substrates, it has been shown that although native glycoproteins are highly

resistant to the cytoplasmic PNGase, extensively misfolded glycoprotein becomes susceptible to this enzyme (Hirsch *et al.*, 2004; Joshi *et al.*, 2005). For *in vitro* deglycosylation analysis, *S*-alkylated bovine pancreatic ribonuclease B (RNase B; Sigma) turned out to be a good substrate (Hirsch *et al.*, 2004; Suzuki, 2005). Additionally, the heat-denatured carboxypeptidase Y from Sigma (100° for 10 min) can also be used (Joshi *et al.*, 2005).

Figure 1A and B shows the elution pattern of a glycoprotein substrate (human asialotransferrin) and the deglycosylated product by bacterial PNGase (PNGase F; Roche) on SDS-PAGE and IEF, respectively. Deglycosylated protein migrates faster on SDS-PAGE (1A, lane 2) and elutes more toward the anode on IEF (1B, lane 2).

Glycoprotein Substrate for *In Vivo* PNGase Assay

Deglycosylation by the cytoplasmic PNGase seems a rate-limiting step in the ERAD reaction *in vivo* since, in most cases, the deglycosylated substrates can only be seen under conditions in which the proteasome activity was compromised. Therefore, the inhibition of the proteasome activity, either by addition of a proteasome inhibitor or by using a mutant that has a defect in proteasomal degradation, is crucial for detection of deglycosylated ERAD substrates *in vivo*. In most cases, SDS-PAGE was carried out for the detection of PNGase activity; but to distinguish the "deglycosylated" (Asp) form from the "unglycosylated" (Asn) form, the confirmation of the negative charge introduction by IEF is always desirable to demonstrate the PNGase activity.

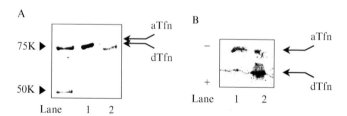

FIG. 1. Detection of PNGase activity by (A) SDS-PAGE and (B) isoelectric focusing (IEF). Human transferrin (0.5 mg/ml; Nacalai Tesque Co., Tokyo) was treated with 25 mU of *Arthrobacter ureafaciens* sialidase (Glyko) or 5 U PNGase F (Roche) in 100 μl reaction, and 1 μl of sample was applied for SDS-PAGE (7.5%) or isoelectric focusing (IEF). SDS-PAGE was stained with Coomassie-Brilliant Blue (CBB) R250 (Wako Chemical Co.). For IEF, the 1 μl of sample was applied onto a PAGplate, with a pH range of 3.5 to 9.5 (Amersham) with 1 *M* phosphoric acid as anode solution and 1.0 *M* NaOH as cathode solution. After 1.5 h of focusing at 1500 V (50 mA), the sample was stained with CBB as in the case with SDS-PAGE. Lane 1: asialotransferrin (aTfn). Lane 2: deglycosylated asialotransferrin (dTfn).

In yeast, so far there is only one substrate, RTAΔ (a ricin mutant that lacks the critical residues for its toxicity), reported to be deglycosylated by PNGase *in vivo* (Kim *et al.*, 2006; Simpson *et al.*, 1999). We have used (his)$_6$-tagged RTAΔ protein for detection, but different tags such as FLAG-tag can also be used (Kim *et al.*, 2006).

Plasmids

A nontoxic allele of RTA (pRS315-$_{GPD}$*Kar2-RTA*Δ) was originally provided by Dr. J. Michael Lord (University of Warwick). The allele was mutagenized by QuikChange mutagenesis kit (Stratagene) to add (his)$_6$-tag at its C-terminus. YEp352-$_{GPD}$*PNG1* was constructed by inserting *Eco*RI/*Xho*I fragment of *PNG1* from pRD53-*PNG1* (Suzuki *et al.*, 2001) at *Eco*RI/*Sal*I site of YEp352-GPD. The sequence was confirmed by direct sequencing.

Cell Extracts and Western Blotting

Standard yeast media and genetic techniques were used in this study (Elble, 1992; Rose *et al.*, 1990; Sherman, 2002). The yeast strains used were *cim5-1* strain (proteasome mutant; CMY 763), and isogenic *cim5-1 png1*Δ double mutant (TSY216; -ura cells were selected using 5-fluoroorotic acid (5-FOA)) (Suzuki and Lennarz, 2002). About 10^8 cells of *S. cerevisiae* carrying the RTAΔ-(his)$_6$ plasmid were harvested. Pellets were washed with dH$_2$O, diluted in 100 μl of dH$_2$O and boiled at 95° for 5 min. 120 μl of 2 × Laemmli buffer (4% SDS, 20% glycerol, 0.6 M β-mercaptoethanol, 0.12 M Tris-HCl [pH 6.8]) with 8 M urea, and 0.02% bromophenol blue was then added to the samples, which were vigorously vortexed with an equal volume of zirconium-silica beads (Biospec Products Inc.) for 3 min and heated again at 95° for 5 min. The zirconium-silica beads and cell debris were removed by centrifugation at 15,000 × g for 15 min at 4°. Samples were analyzed by SDS-PAGE with 15% polyacrylamide gel and then transferred to Immobilon transfer membranes, BioTrace (Pall Corporation). For detection of RTAΔ-(his)$_6$ proteins, membranes were incubated with a 1:1000 dilution of anti-(his)$_6$ antibody (SC-803; Santa Cruz Biotechnology, Inc) in 2% dry milk in PBS-T (PBS-0.1% Tween-20), followed by incubation with 1:2000 dilution of anti-rabbit immunoglobulin (Ig) G horseradish peroxidase-conjugated secondary antibody and 2% dry milk in PBS-T. The secondary antibodies were detected with the ECL plus system (Amersham) according to the manufacturer's protocol. Figure 2 shows the Western blotting analysis of RTAΔ-(his)$_6$ in the absence (lane 1) and the presence (lane 2) of PNGase. In *png1*Δ cells, accumulation of only g0 form was observed (lane 2), while in the absence of PNGase both g1 and g0 (most likely the unglycosylated form) were observed.

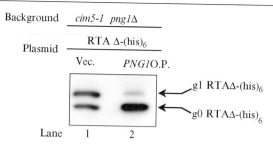

Fig. 2. Western Blotting of RTAΔ-(his$_6$) in the presence and absence of Png1. g1: singly-glycosylated RTAΔ-(his$_6$) g0: deglycosylated (or unglycosylated) RTAΔ-(his$_6$). Lane 1: *cim5-1 png1Δ* cells; lane 2: *cim5-1 png1Δ* cells overproducing Png1 (YEp352-$_{GPD}$PNG1).

Cycloheximide-Decay Experiment

To examine the stability of the RTAΔ-(his)$_6$ in proteasome mutants, a cycloheximide (CHX)-decay experiment (Taxis *et al.*, 2003) was carried out. Two hundred fifty milliliters of isogenic *cim5-1* single mutant (CMY765) or *cim5-1 png1Δ* double mutant (TSY216) harboring pRS315-$_{GPD}$Kar2-RTAΔ-*(his)$_6$* were grown in SC-leucine medium to an OD$_{600}$ of 0.5, and the culture was divided into two. In one sample, CHX in dimethyl sulfoxide (DMSO) (250 μg/ml final) was added, and in the other an equal amount of DMSO was added as a control and set to be the 0 time point. At the indicated time points, a 10-ml culture specimen was taken from the sample, and protein extracts of the cells were prepared as described previously. Immunoblotting was performed using anti-(his)$_6$ antibody as described previously.

As shown in Fig. 3, rapid conversion of g1 (glycosylated RTAΔ-(his)$_6$) to g0 (un- or deglycosylated RTAΔ-(his)$_6$) was observed only in *cim5-1* cells, indicating the PNGase action on this protein. On the other hand, RTAΔ-(his)$_6$ in *cim5-1 png1Δ* double mutant remained as g1 form even after a 2-h chase.

Theoretically, it is possible to assay the occurrence of deglycosylation reaction with Western blotting at steady state (see Fig. 2); however, pulse-chase experiments or the equivalent, such as cycloheximide-decay experiments, are always favored to confirm the deglycosylation reaction because they will allow one to detect the conversion of the glycosylated form to the deglycosylated form (see Fig. 3).

For IEF of yeast extract, yeast pellet can be prepared by suspending the yeast pellet, obtained as described previously, with sample buffer (10 mM Tris-HCl [pH, 7.5], 5 M urea, 1 M thio-urea, 1% CHAPS, 1% Triton X-100) together with 1× complete protease inhibitor cocktail (Roche), and protein

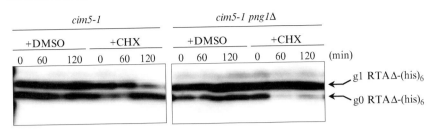

FIG. 3. Cycloheximide-decay experiment confirms the g1 to g0 conversion of RTAΔ-(his₆) by Png1. g1: singly-glycosylated RTAΔ-(his)₆; g0: deglycosylated (or unglycosylated) RTAΔ-(his)₆. CHX, cycloheximide; DMSO, dimethyl sulfoxide (vehicle control). For details, see text.

extract can be achieved by vigorous vortex with an equal volume of zirconium-silica beads for 4 min with wet-ice cooling. IEF analysis of the extract thus obtained can also be carried out as described previously. After the IEF, the gel of the PAGplate can be recovered from the plastic film using Film Remover (Amersham), and the gel can be processed for immunoblotting of the substrate, as in the case of SDS-PAGE gel.

Technical Considerations

When yeast is used for the *in vivo* detection of deglycosylation in *S. cerevisiae*, it is normally difficult because proteasomal inhibitors are not taken up by this yeast (Lee and Goldberg, 1996). Therefore, we used the *cim5-1* strain to compromise the activity of proteasome. Although a certain wild-type strain can also be used for the detection of *in vivo* PNGase activity using RTAΔ (Kim *et al.*, 2006), this substrate is so unstable that it is hardly detectable in other wild-type strain backgrounds. It is also imperative to minimize the post-homogenization proteolysis, especially when preparing cell extract. In some cases, inclusion of protease inhibitor cocktail will decrease the proteolytic degradation of the substrate, though precaution has to be taken for some reagents that will change the pI of the protein.

For IEF of *in vivo* deglycosylation assay, because the ERAD substrates are normally expected to have high-mannose–type glycans, the effect of negative charges on glycan chains is not necessary to consider. However, if the PNGase activity is detected *in vitro* using a glycoprotein that bears complex-type glycan(s) (human serum transferrin, for example), sialic acids have to be removed from the substrate used before the assay to avoid any complication on charge diversity (see Fig. 1).

Acknowledgments

We thank Prof. J. Michael Lord (University of Warvick) for providing original RTAΔ plasmid and Miki Suzuki for editing the manuscript.

Research in the Suzuki laboratory is supported in part by Grants-in-Aid for Young Scientists (A) (17687009) from the Ministry of Education, Science and Culture of Japan; CREST (Core Research for Evolutionary Science and Technology), JST (Japan Science and Technology Agency); the Astellas Foundation for Research and Metabolic Disorders (Tokyo); and the Osaka Cancer Foundation (Osaka).

References

Bebok, Z., Mazzochi, C., King, S. A., Hong, J. S., and Sorscher, E. J. (1998). The mechanism underlying cystic fibrosis transmembrane conductance regulator transport from the endoplasmic reticulum to the proteasome includes Sec61beta and a cytosolic, deglycosylated intermediary. *J. Biol. Chem.* **273,** 29873–29878.

de Virgilio, M., Weninger, H., and Ivessa, N. E. (1998). Ubiquitination is required for the retro-translocation of a short-lived luminal endoplasmic reticulum glycoprotein to the cytosol for degradation by the proteasome. *J. Biol. Chem.* **273,** 9734–9743.

Di Cola, A., Frigerio, L., Lord, J. M., Ceriotti, A., and Roberts, L. M. (2001). Ricin A chain without its partner B chain is degraded after retrotranslocation from the endoplasmic reticulum to the cytosol in plant cells. *Proc. Natl. Acad. Sci. USA* **98,** 14726–14731.

Elble, R. (1992). A simple and efficient procedure for transformation of yeasts. *Biotechniques* **13,** 18–20.

Gong, Q., Keeney, D. R., Molinari, M., and Zhou, Z. (2005). Degradation of trafficking-defective long QT syndrome type II mutant channels by the ubiquitin-proteasome pathway. *J. Biol. Chem.* **280,** 19419–19425.

Halaban, R., Cheng, E., Zhang, Y., Moellmann, G., Hanlon, D., Michalak, M., Setaluri, V., and Hebert, D. N. (1997). Aberrant retention of tyrosinase in the endoplasmic reticulum mediates accelerated degradation of the enzyme and contributes to the dedifferentiated phenotype of amelanotic melanoma cells. *Proc. Natl. Acad. Sci. USA* **94,** 6210–6215.

Hirsch, C., Misaghi, S., Blom, D., Pacold, M. E., and Ploegh, H. L. (2004). Yeast N-glycanase distinguishes between native and non-native glycoproteins. *EMBO Rep.* **5,** 201–206.

Hughes, E. A., Hammond, C., and Cresswell, P. (1997). Misfolded major histocompatibility complex class I heavy chains are translocated into the cytoplasm and degraded by the proteasome. *Proc. Natl. Acad. Sci. USA* **94,** 1896–1901.

Huppa, J. B., and Ploegh, H. L. (1997). The alpha chain of the T cell antigen receptor is degraded in the cytosol. *Immunity* **7,** 113–122.

Johnston, J. A., Ward, C. L., and Kopito, R. R. (1998). Aggresomes: A cellular response to misfolded proteins. *J. Cell Biol.* **143,** 1883–1898.

Joshi, S., Katiyar, S., and Lennarz, W. J. (2005). Misfolding of glycoproteins is a prerequisite for peptide: N-glycanase mediated deglycosylation. *FEBS Lett.* **579,** 823–826.

Kim, I., Ahn, J., Liu, C., Tanabe, K., Apodaca, J., Suzuki, T., and Rao, H. (2006). The Png1-Rad23 complex regulates glycoprotein turnover. *J. Cell Biol.* **172,** 211–219.

Kitajima, K., Suzuki, T., Kouchi, Z., Inoue, S., and Inoue, Y. (1995). Identification and distribution of peptide: N-glycanase (PNGase) in mouse organs. *Arch. Biochem. Biophys.* **319,** 393–401.

Kitzmuller, C., Caprini, A., Moore, S. E., Frenoy, J. P., Schwaiger, E., Kellermann, O., Ivessa, N. E., and Ermonval, M. (2003). Processing of N-linked glycans during

endoplasmic-reticulum–associated degradation of a short-lived variant of ribophorin I. *Biochem. J.* **376,** 687–696.

Lee, D. H., and Goldberg, A. L. (1996). Selective inhibitors of the proteasome-dependent and vacuolar pathways of protein degradation in *Saccharomyces cerevisiae. J. Biol. Chem.* **271,** 27280–27284.

Lee, I., Skinner, M. A., Guo, H. B., Sujan, A., and Pierce, M. (2004). Expression of the vacuolar H+-ATPase 16-kDa subunit results in the Triton X-100–insoluble aggregation of beta1 integrin and reduction of its cell surface expression. *J. Biol. Chem.* **279,** 53007–53014.

Ma, J., Wollmann, R., and Lindquist, S. (2002). Neurotoxicity and neurodegeneration when PrP accumulates in the cytosol. *Science* **298,** 1781–1785.

Mancini, R., Fagioli, C., Fra, A. M., Maggioni, C., and Sitia, R. (2000). Degradation of unassembled soluble Ig subunits by cytosolic proteasomes: evidence that retrotranslocation and degradation are coupled events. *FASEB J.* **14,** 769–778.

McCracken, A. A., and Brodsky, J. L. (2003). Evolving questions and paradigm shifts in endoplasmic-reticulum–associated degradation (ERAD). *Bioessays* **25,** 868–877.

Meusser, B., Hirsch, C., Jarosch, E., and Sommer, T. (2005). ERAD: The long road to destruction. *Nat. Cell Biol.* **7,** 766–772.

Mosse, C. A., Meadows, L., Luckey, C. J., Kittlesen, D. J., Huczko, E. L., Slingluff, C. L., Shabanowitz, J., Hunt, D. F., and Engelhard, V. H. (1998). The class I antigen-processing pathway for the membrane protein tyrosinase involves translation in the endoplasmic reticulum and processing in the cytosol. *J. Exp. Med.* **187,** 37–48.

Nishikawa, S., Brodsky, J. L., and Nakatsukasa, K. (2005). Roles of molecular chaperones in endoplasmic reticulum (ER) quality control and ER-associated degradation (ERAD). *J. Biochem. (Tokyo)* **137,** 551–555.

Petaja-Repo, U. E., Hogue, M., Laperriere, A., Bhalla, S., Walker, P., and Bouvier, M. (2001). Newly synthesized human delta opioid receptors retained in the endoplasmic reticulum are retrotranslocated to the cytosol, deglycosylated, ubiquitinated, and degraded by the proteasome. *J. Biol. Chem.* **276,** 4416–4423.

Plummer, T. H. J., Elder, J. H., Alexander, S., Phelan, A. W., and Tarentino, A. L. (1984). Demonstration of peptide: *N*-glycosidase F activity in endo-beta-*N*-acetylglucosaminidase F preparations. *J. Biol. Chem.* **259,** 10700–10704.

Rose, M. D., Winston, F., and Hieter, P. (1990). "Methods in Yeast Genetics." Cold Spring Harbor Laboratory, Cold Spring Harbor, NY.

Sayeed, A., and Ng, D. T. (2005). Search and destroy: ER quality control and ER-associated protein degradation. *Crit. Rev. Biochem. Mol. Biol.* **40,** 75–91.

Seko, A., Kitajima, K., Inoue, Y., and Inoue, S. (1991). Peptide:N-glycosidase activity found in the early embryos of *Oryzias latipes* (Medaka fish): The first demonstration of the occurrence of peptide: *N*-glycosidase in animal cells and its implication for the presence of a de-*N*-glycosylation system in living organisms. *J. Biol. Chem.* **266,** 22110–22114.

Seko, A., Kitajima, K., Iwamatsu, T., Inoue, Y., and Inoue, S. (1999). Identification of two discrete peptide: *N*-glycanases in *Oryzias latipes* during embryogenesis. *Glycobiology* **9,** 887–895.

Sherman, F. (2002). Getting Started with Yeast. *In* "Methods in Enzymology" (Christine Guthrie and G.R. Fink, eds.), Vol. 350, pp. 3–41. Academic Press.

Simpson, J. C., Roberts, L. M., Romisch, K., Davey, J., Wolf, D. H., and Lord, J. M. (1999). Ricin A chain utilises the endoplasmic reticulum–associated protein degradation pathway to enter the cytosol of yeast. *FEBS Lett.* **459,** 80–84.

Suzuki, T. (2005). A simple, sensitive in vitro assay for cytoplasmic deglycosylation by peptide: *N*-glycanase. *Methods* **35,** 360–365.

Suzuki, T., Kitajima, K., Emori, Y., Inoue, Y., and Inoue, S. (1997). Site-specific de-*N*-glycosylation of diglycosylated ovalbumin in hen oviduct by endogenous peptide:

N-glycanase as a quality control system for newly synthesized proteins. *Proc. Natl. Acad. Sci. USA* **94**, 6244–6249.

Suzuki, T., Kitajima, K., Inoue, Y., and Inoue, S. (1995). Carbohydrate-binding property of peptide: N-glycanase from mouse fibroblast L-929 cells as evaluated by inhibition and binding experiments using various oligosaccharides. *J. Biol. Chem.* **270**, 15181–15186.

Suzuki, T., Kwofie, M. A., and Lennarz, W. J. (2003). *Ngly1*, a mouse gene encoding a deglycosylating enzyme implicated in proteasomal degradation: Expression, genomic organization, and chromosomal mapping. *Biochem. Biophys. Res. Commun.* **304**, 326–332.

Suzuki, T., and Lennarz, W. J. (2002). Glycopeptide export from the endoplasmic reticulum into cytosol is mediated by a mechanism distinct from that for export of misfolded glycoprotein. *Glycobiology* **12**, 803–811.

Suzuki, T., Park, H., Hollingsworth, N. M., Sternglanz, R., and Lennarz, W. J. (2000). *PNG1*, a yeast gene encoding a highly conserved peptide:N-glycanase. *J. Cell Biol.* **149**, 1039–1052.

Suzuki, T., Park, H., Kitajima, K., and Lennarz, W. J. (1998). Peptides glycosylated in the endoplasmic reticulum of yeast are subsequently deglycosylated by a soluble peptide: N-glycanase activity. *J. Biol. Chem.* **273**, 21526–21530.

Suzuki, T., Park, H., Kwofie, M. A., and Lennarz, W. J. (2001). Rad23 provides a link between the Png1 deglycosylating enzyme and the 26 S proteasome in yeast. *J. Biol. Chem.* **276**, 21601–21607.

Suzuki, T., Park, H., and Lennarz, W. J. (2002). Cytoplasmic peptide: N-glycanase (PNGase) in eukaryotic cells: Occurrence, primary structure, and potential functions. *FASEB J.* **16**, 635–641.

Suzuki, T., Seko, A., Kitajima, K., Inoue, Y., and Inoue, S. (1993). Identification of peptide: N-glycanase activity in mammalian-derived cultured cells. *Biochem. Biophys. Res. Commun.* **194**, 1124–1130.

Suzuki, T., Seko, A., Kitajima, K., Inoue, Y., and Inoue, S. (1994). Purification and enzymatic properties of peptide: N-glycanase from C3H mouse-derived cultured cells: Possible widespread occurrence of post-translational remodification of proteins by N-deglycosylation. *J. Biol. Chem.* **269**, 17611–17618.

Takahashi, N. (1977). Demonstration of a new amidase acting on glycopeptides. *Biochem. Biophys. Res. Commun.* **76**, 1194–1201.

Taxis, C., Hitt, R., Park, S.-H., Deak, P. M., Kostova, Z., and Wolf, D. H. (2003). Use of modular substrates demonstrates mechanistic diversity and reveals differences in chaperone requirement of ERAD. *J. Biol. Chem.* **278**, 35903–35913.

Wiertz, E. J., Jones, T. R., Sun, L., Bogyo, M., Geuze, H. J., and Ploegh, H. L. (1996a). The human cytomegalovirus US11 gene product dislocates MHC class I heavy chains from the endoplasmic reticulum to the cytosol. *Cell* **84**, 769–779.

Wiertz, E. J., Tortorella, D., Bogyo, M., Yu, J., Mothes, W., Jones, T. R., Rapoport, T. A., and Ploegh, H. L. (1996b). Sec61-mediated transfer of a membrane protein from the endoplasmic reticulum to the proteasome for destruction. *Nature* **384**, 432–438.

Xiong, X., Chong, E., and Skach, W. R. (1999). Evidence that endoplasmic reticulum (ER)-associated degradation of cystic fibrosis transmembrane conductance regulator is linked to retrograde translocation from the ER membrane. *J. Biol. Chem.* **274**, 2616–2624.

Yedidia, Y., Horonchik, L., Tzaban, S., Yanai, A., and Taraboulos, A. (2001). Proteasomes and ubiquitin are involved in the turnover of the wild-type prion protein. *EMBO J.* **20**, 5383–5391.

Yu, H., Kaung, G., Kobayashi, S., and Kopito, R. R. (1997). Cytosolic degradation of T-cell receptor alpha chains by the proteasome. *J. Biol. Chem.* **272**, 20800–20804.

Section II

Structural Analysis

[5] Glycomic Profiling of Cells and Tissues by Mass Spectrometry: Fingerprinting and Sequencing Methodologies

By JIHYE JANG-LEE, SIMON J. NORTH, MARK SUTTON-SMITH,
DAVID GOLDBERG, MARIA PANICO, HOWARD MORRIS,
STUART HASLAM, and ANNE DELL

Abstract

Over the past decade, rapid, high-sensitivity mass spectrometric strategies have been developed and optimized for screening for the types of N- and O-glycans present in a diverse range of biological material, including secretions, cell lines, tissues, and organs. These glycomic strategies are based on matrix-assisted laser desorption/ionization (MALDI) time-of-flight mass fingerprinting of permethylated derivatives, combined with electrospray (ES) or MALDI tandem mass spectrometry (MS/MS) sequencing and gas chromatography (GC)-MS linkage analysis, complemented by chemical and enzymatic degradations. Protocols for these methods are described in the first part of this chapter. Glycomic experiments yield large volumes of MS data, and interpretation of the resulting spectra remains a time-consuming bottleneck in the process. In the second part of this chapter, we describe the use and operation of a mass spectral viewer program capable of displaying and automatically labeling spectra arising from MALDI fingerprinting of N-glycans.

Overview

The implementation of highly sensitive and rapid mass spectrometric screening strategies for defining the glycosylation repertoires of cells, tissues, and organs is helping to reveal the roles that glycans play in health and disease (Dell *et al.*, 2003; Manzi *et al.*, 2000; Patankar *et al.*, 2005; Sutton-Smith, 2000; Sutton-Smith *et al.*, 2002). These procedures can be applied to very small amounts of biological samples at relatively high throughput, thereby providing a structural "fingerprint" that can be readily compared with the profiles of other samples (e.g., wild-type versus knockout mouse tissues). The Consortium for Functional Glycomics has adopted this method for screening the glycan profiles of murine and human tissues and cell populations to generate a data resource for the glycobiology community (www.functionalglycomics.org).

METHODS IN ENZYMOLOGY, VOL. 415   0076-6879/06 $35.00
                                          DOI: 10.1016/S0076-6879(06)15005-3

Modern glycomic methodologies have evolved from procedures that were introduced in the early 1980s to study complex mixtures of glycans isolated from leukocytes, erythrocytes, and cancer cells (Fukuda *et al.*, 1984a,b, 1985). These early studies used fast atom bombardment technology (Dell *et al.*, 1983), though this has now been largely superseded by matrix-assisted laser desorption/ionization (MALDI) and electrospray (ES) instrumentation, which offer higher throughput and greater sensitivity. But the fundamental strategy remains unaltered. It is based on mass spectrometry (MS) analysis of permethylated derivatives that yield molecular ions at very high sensitivity, irrespective of the type of ionization. Moreover, fragment ions obtained from permethylated derivatives are readily assignable to unambiguous sequences (Dell, 1987). MALDI is currently the ionization method of choice for molecular weight profiling. Structures are assigned to each signal in the MALDI "fingerprint" based on the usually unique glycan composition for a given mass and prior knowledge of N- and O-glycan biosynthesis. An algorithm called *Cartoonist* is now available to assist this interpretation (see later). Antennae sequences and branching patterns of each component are subsequently rigorously determined by subjecting each molecular ion to collisional activation in MS/MS experiments, which can be carried out using either MALDI or ES ionization. A limited amount of linkage information is provided in some instances by the type and abundance of fragment ions produced, but usually it is necessary to perform gas chromatography (GC)-MS linkage analysis to firmly establish this structural feature. If necessary, assignments can be confirmed by MS and MS/MS experiments on chemical and enzymatic degradations, the choice of which is guided by the sequence information provided by the earlier experiments. Protocols for these analyses are given in the following sections.

Preparation of Glycans for Mass Spectrometric Analysis

The general outline of the glycan extraction and derivatization for mass spectrometric analysis is shown in Fig. 1. To extract the glycans from glycoproteins, the sample is first homogenized, reduced, and carboxymethylated, then is digested with trypsin. This process ensures the effective release of N-glycans by peptide N-glycosidase F (PNGase F) and subsequent chemical release by reductive elimination of O-glycans. A few milligrams of tissue or about 10 million cells is normally sufficient for sequential MALDI profiling, MS/MS sequencing, and linkage analysis.

Materials for Tissue Homogenization

1. Homogenizer with a T6 or 6.1 dispersion shaft
2. Snakeskin dialysis tubing, cutoff point of 7 kDa (Pierce, Prod # 68700)

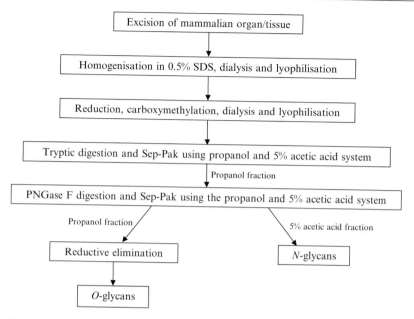

FIG. 1. Overall strategy for the preparation of glycans for mass spectrometric analysis.

3. Glass culture tubes (13 × 100 mm; Corning)
4. Culture tube caps (Bennett Scientific) with Teflon inserts (Owens Polyscience).

Solutions

1. Homogenization buffer: 0.5% (w/v) sodium dodecyl sulfate (SDS) solution in 50 mM Tris buffer; pH, 7.4
2. Dialysis buffer: 4.5 l of 50 mM ammonium hydrogen carbonate solution; pH, 7.4

Method for Homogenization

1. The organ/tissue is placed in a tube with 5 ml of detergent solution and is blasted for approximately 2 min with the homogenizer.
2. The homogenized sample is transferred into a dialysis tubing and dialyzed against the dialysis buffer at 4° with constant stirring. The dialysis process takes place over a period of 48 h; the buffer should be regularly changed, typically two to three times.

3. Once dialyzed, the samples are transferred into clean labeled glass tubes, covered with Parafilm and lyophilized.

Materials for Reduction and Carboxymethylation

1. Snakeskin dialysis tubing, cutoff point of 7 kDa (Pierce, Prod # 68700).

Solutions

1. 0.6 M Tris buffer; pH, 8.5; bubbled with nitrogen gas for 30 min
2. 2 mg/ml dithioreitol in 0.6 M Tris buffer; pH, 8.5
3. 12 mg/ml iodoacetic acid in 0.6 M Tris buffer; pH, 8.5
4. Dialysis buffer: 4.5 l of 50 mM ammonium hydrogen carbonate solution; pH, 8.5.

Method for Reduction and Carboxymethylation

1. One milliliter of 2 mg/ml of dithioreitol in 0.6 M Tris buffer (pH, 8.5; bubbled with nitrogen gas) is added to the homogenized and lyophilized material and incubated for 1 h at 37°.

2. One milliliter of 12 mg/ml of iodoacetic acid in 0.6 M Tris (pH, 8.5) is added and incubated for a further 2 h at room temperature in the dark.

3. The reaction is terminated by dialyzing against the dialyzing buffer for 48 h at 4°. The solution should regularly changed and continuously stirred.

4. After dialyzing, contents are transferred into clean glass tubes and lyophilized.

Solution for Tryptic Digestion

1. 1 mg/ml of L-1-tosylamido-2-phenylethyl chloromethyl ketone (TPCK) treated bovine pancreas trypsin (Sigma, EC 3.4.21.4) in 50 mM ammonium hydrogen carbonate solution; pH, 8.4.

Method for Tryptic Digestion

1. One milliliter of 1 mg/ml of TPCK-treated trypsin solution is added to the reduced, carboxymethylated, and lyophilized samples. Incubation takes place for 16 h at 37°. The reaction is terminated by heating at 100° for 2 min.

2. The digested sample is neutralized by adding diluted acetic acid (high pH will affect the surface pH of the Sep-Pak cartridge; pH, 7).

Neutralized samples are purified directly using propan-1-ol/5% acetic acid system as described next. If necessary, the sample should be divided to prevent cartridge overloading.

Materials for Purification of Glycopeptides

1. 20-ml glass syringe
2. Purification column: C_{18} Sep-Pak Short Body Classic Cartridges, Part # WAT051910 (Waters).

Solutions

1. Conditioning solutions: Methanol, 5% (v/v) acetic acid in water, 100% Propan-1-ol
2. Elution buffers: 20%, 40%, and 60% (v/v) Propan-1-ol in 5% (v/v) acetic acid.

Method for Purification of Glycopeptides

1. The Sep-Pak cartridge is conditioned with (i) 5 ml of methanol, (ii) 5 ml 5% acetic acid, (iii) 5 ml propanol, and (iv) 10 ml 5% acetic acid.
2. The sample is carefully loaded into the column.
3. The sample is rinsed with 30 ml of 5% (v/v) acetic acid in water. This fraction is discarded.
4. Glycopeptides are eluted with 20%, 40%, and 60% (v/v) propanol and collected in the same tube. This is dried in a vacuum centrifuge until the volume is halved and then lyophilized.

Enzyme and Digestion Buffer for the Release of N-Glycans

1. Enzyme: Peptide N-glycosidase F in glycerol (Roche, EC 3.5.1.52)
2. Digestion buffer: 50 mM ammonium hydrogen carbonate; pH, 8.4.

Method for Release of N-Glycans

1. The lyophilized sample after the tryptic digestion is dissolved in 200 μl of the digestion buffer and 3 to 4 U of PNGase F is added to this mixture.
2. Incubation takes place for 24 to 48 hrs at 37°, and a fresh aliquot of enzyme is added after 24 h.
3. Released N-glycans are purified using the propan-1-ol /5% acetic acid system.

Materials for Purification of N-*Glycans*

1. 5-ml glass syringe (Fischer/Scientific)
2. C_{18} Sep-Pak Short Body Classic Cartridges, Part # WAT051910 (Waters).

Solutions

1. Conditioning solutions: methanol, 5% (v/v) acetic acid in water, 100% propan-1-ol
2. Elution buffers: 5% (v/v) acetic acid in water, 20%, 40%, and 60% (v/v) propan-1-ol in 5% (v/v) acetic acid.

Method for Purification of N-*Glycans*

1. The Sep-Pak cartridge is conditioned with (i) 5 ml of methanol, (ii) 5 ml 5% (v/v) acetic acid, (iii) 5 ml propanol, and (iv) 10 ml 5% acetic acid.

2. The sample is carefully loaded into the cartridge and eluted with 5 ml of 5% (v/v) acetic acid and 2 ml of 20%, 40%, and 60% (v/v) propanol in 5% (v/v) acetic acid solution.

3. The 5% (v/v) acetic acid fraction is dried and lyophilized. This fraction contains the *N*-glycans; it can be derivatized (described later) or subjected to further reactions such as exo- and endo-glycosidase digestions or acid hydrolysis (described later).

4. Propanol fractions are combined and dried in a Gyrovap and lyophilized. This fraction containing the *O*-glycan–containing glycopeptides is subjected to reductive elimination.

Reagents for the Release of O-*Glycans: Reductive Elimination*

1. 1 M potassium borohydride in 0.1 M potassium hydroxide solution
2. Glacial acetic acid
3. Water.

Method for Release of O-*Glycans: Reductive Elimination*

1. To release the *O*-glycans from the peptide backbone, reductive elimination is carried out. Thus, 400 μl of the potassium borohydride solution is added to the previously obtained lyophilized peptide/glycopeptide fractions. This is incubated at 45° in a Teflon-lined screw-capped glass tube for 16 h.

2. The reaction is terminated by adding a few drops of glacial acetic acid until fizzing stops. The sample is desalted by cation exchange chromatography.

Materials for the Preparation of Dowex Beads

1. Dowex beads (50W-X8 (H) 50 to 100 mesh, Fluka).

Solutions

1. 4 *M* hydrochloric acid
2. 5% (v/v) acetic acid in water.

Method for the Preparation of Dowex Beads

1. Dowex beads (50W-X8 (H) 50 to 100 mesh) are washed three times with 4 *M* hydrochloric acid and subsequently washed with water until the pH of the supernatant is the same as that of the water.

2. The beads are washed again three times with 5% (v/v) acetic acid, and the beads are stored in 5% (v/v) acetic acid.

Materials for the Desalting of O-*Glycans*

1. Dowex beads (50W-X8 (H) 50 to 100 mesh, Fluka) (previously prepared)
2. Unplugged disposable glass Pasteur
3. Silicone tubing (length, approximately 10 cm; internal diameter, 1 to 2 mm; Fisher Scientific)
4. Glass wool
5. Adjustable clips.

Method for Desalting of O-*Glycans*

1. A Pasteur pipette is plugged at the tapered end with a small amount of glass wool; a piece of silicone tubing is inserted to the tapered end of the Pasteur pipette, and an adjustable clip is placed at the end of the silicone tubing. This is assembled onto a clamp stand.

2. The column is washed several times with approximately 5 ml of 5% (v/v) acetic acid.

3. After the wash, the column is filled again with 5% (v/v) acetic acid. The clip is slightly opened, allowing the solution to flow out slowly.

4. As the acid flows out, the Dowex beads are inserted into the column until it has reached the constriction of the column. The Dowex beads can be prevented from drying by ensuring that the 5% (v/v) acetic acid level is always above the level of the beads.

5. Beads in the column are washed by passing 20 ml of 5% (v/v) acetic acid.

6. After the wash, the clip is closed, and the sample is carefully loaded.

7. A labeled glass tube is placed under the column, and the clip is opened to allow the sample to run through the column.

8. The column is topped-up with 5% (v/v) acetic acid until approximately 5 ml of the eluent has been collected.

9. The collected sample volume is halved in a vacuum centrifuge and then lyophilized. Alternatively, if a vacuum centrifuge is not available, the sample can be dried under a stream of nitrogen gas.

Removal of Borates

To ensure that borates after the reductive elimination of *O*-glycans do not interfere with the mass spectrometric analysis, an additional step is carried out after the desalting of *O*-glycans.

Solution

1. 10% (v/v) acetic acid in methanol.

Method for Borates Removal

1. Approximately 500 μl of a 10% (v/v) acetic acid in methanol solution is added to the dried samples and dried under a stream of nitrogen gas. This process is repeated four times.

2. After this procedure, the sample is ready to be derivatized or subjected to further reactions such as glycosidase digestions or acid hydrolysis reactions.

Enzymatic and Chemical Degradations

To ascertain the anomeric configurations of glycosidic bonds and to assist in the verification of tentative sequences, purified glycans can be subjected to enzymatic or chemical degradations, the choice of which will depend on earlier results.

Buffers for Enzymatic Digestions

1. α-mannosidase digestion (jack bean, EC 3.2.1.24 [Sigma])
   Buffer: 50 mM ammonium acetate; pH, 4.5
   Condition: 0.5 U of enzyme in 100 μl of buffer
2. α-galactosidase digestion (green coffee beans, EC 3.2.1.22 [Sigma])
   Buffer: 50 mM ammonium formate; pH, 6.0
   Condition: 0.5 U of enzyme in 100 μl of buffer

3. β-galactosidase digestion (bovine testes, EC 3.2.1.23 [Sigma])
   Buffer: 50 mM ammonium acetate; pH, 4.6
   Condition: 10 mU of enzyme in 100 μl of buffer

4. Endo-β-galactosidase digestion (*Bacteroides fragilis*, EC 3.2.1.103 [Sigma])
   Buffer: 50 mM ammonium acetate; pH, 5.5
   Condition: 5 mU of enzyme in 100 μl of buffer

5. α2-3,6,8-neuraminidase digestion (*Vibrio cholerae*, EC 3.2.1.18 [Sigma])
   Buffer: 50mM ammonium acetate; pH, 5.5
   Condition: 20 U of enzyme in 100 μl of buffer

6. α2-3-neuraminidase digestion (*Streptococcus pneumoniae*, EC 3.2.1.18 [Sigma])
   Buffer: 50 mM ammonium acetate; pH, 5.5
   Condition: 0.025 U of enzyme in 100 μl of buffer

Method for Enzymatic Digestions

1. Enzymatic digestions are carried out under the conditions stated. All digestions are carried out for 24 to 48 h at 37° on non-derivatized glycans, and a fresh aliquot of enzyme is added every 12 h.

2. All reactions are terminated by lyophilization, and an appropriate aliquot is taken for subsequent derivatization and analysis by MS.

Materials and Reagents for Chemical Degradation: Acid Hydrolysis

1. Hydrofluoric acid (HF) 48% (w/v) in water
2. Beaker with ice
3. Eppendorf tube.

Method for Acid Hydrolysis

HF Treatment

Acid hydrolysis using HF acid was initially used to hydrolyze phosphodiester bonds in glycosylphosphatidylinositol-anchored proteins (Schneider, 1995). It has also been implemented to release phosphorylcholine from N-glycans (Haslam *et al.*, 1997) and can be used to analyze mannose-6-phosphate structures. Moreover, under the same conditions, this method allows us to distinguish the position of attachment of Fucose residues. Incubation with HF will readily remove α1→3-linked Fucose whereas α1→2- and 6-linked Fucose are released at a slower rate. HF is very corrosive

and "melts" glass. Therefore, it is important to handle with care and to carry out each step in a fume hood. Plastic containers and Gilson tips should be used.

1. Glycans are transferred into 1.5-ml Eppendorf tubes. This is done by dissolving them with 5% (v/v) acetic acid in water and then lyophilizing the sample.

2. 50 μl of 48% (w/v) HF is added to the dried samples and incubated at 4° for 20 h. After incubation, the reagent is removed by drying down under nitrogen gas.

3. Dried samples are transferred into screw-capped glass tubes by dissolving with 5% (v/v) acetic acid in water. Transferred samples are lyophilized and are now ready for subsequent derivatization and mass spectrometric analysis.

Derivatization and Purification of Derivatized Glycans

Glycans are permethylated before MS analysis because this process results in increased sensitivity of detection and directs fragmentation in a predictable and reliable manner during MS/MS experiments.

Materials and Reagents for NaOH Permethylation

1. Glass pestle and mortar
2. Methyl iodide
3. Dimethyl sulfoxide (DMSO)
4. Sodium hydroxide (NaOH) pellets
5. Chloroform
6. Water.

Method for Permethylation

Before starting this procedure, ensure that the pestle and mortar and the sample to be derivatized are completely dry and the cap of the culture tube is Teflon lined.

1. Approximately three to five pellets of sodium hydroxide are placed into a mortar with 3 ml of DMSO.

2. Pellets are crushed until a slurry is made. This must be done quickly to avoid excessive moisture absorption from the atmosphere.

3. A squirt of this slurry is added to the dried sample, and 0.5 ml of methyl iodide is added to this mixture.

4. The reaction mixture is mixed vigorously and shaken on an automatic shaker for 10 min at room temperature.

5. The reaction is stopped by adding water dropwise and constantly shaking between the additions. This is repeated until fizzing stops.

6. One to two milliliters of chloroform is added to the sample, and water is added to make a total volume to approximately 5 ml. This is thoroughly mixed and left to settle until two layers are formed.

7. The aqueous layer is discarded, and the chloroform (lower layer) is washed several times with water until the water removed is clear. As much water as possible is removed, and the chloroform layer is dried down under a stream of nitrogen gas.

8. Dried permethylated samples are then purified using the acetonitrile Sep-Pak purification system.

Materials for the Purification of Derivatized Glycans

1. 5-ml glass syringes (Fischer/Scientific)
2. Purification column: C_{18} Sep-Pak Short Body Classic Cartridges, Part # WAT051910 (Waters).

Solutions

1. Conditioning solutions: methanol, water, 100% acetonitrile
2. Elution buffers: 15%, 35%, 50%, and 75% (v/v) acetonitrile in water.

Method for Purification of Derivatized Glycans

1. The Sep-Pak cartridge is conditioned sequentially with (i) 5 ml methanol, (ii) 5 ml water, (iii) 5 ml acetonitrile, and (iv) 2×5 ml water.

2. The derivatized sample is dissolved in 200 μl of methanol and 200 μl water.

3. The sample is carefully loaded into the cartridge and eluted with 5 ml water and 2 ml of 15%, 35%, 50%, and 75% (v/v) acetonitrile in water. The water and 15% (v/v) acetonitrile fractions can be discarded because derivatized glycans are usually in the 35%, 50%, and 75% (v/v) fractions.

4. The collected fractions are dried in a vacuum centrifuge. These samples are now ready for mass spectrometric analyses.

MS Analyses

Figure 2 illustrates the overall MS strategy used for a rigorous structural characterization of the *N*- and *O*-glycan from a sample of interest. The principles of the MS techniques used have been well documented elsewhere

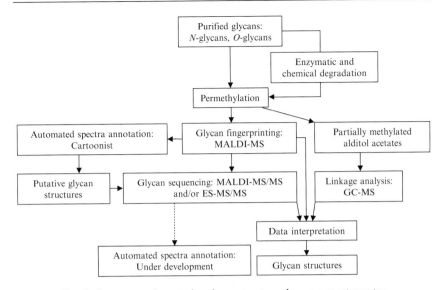

FIG. 2. Strategy to characterize glycan structures by mass spectrometry.

(Dell and Morris, 2001; Mechref and Novotny, 2002) and therefore are only briefly discussed later.

MALDI Mass Fingerprinting

MALDI is a laser-based soft ionization method in which the sample is embedded in a crystalline matrix that absorbs the laser energy, causing the target sample to vaporize, generating ions. Ions produced by MALDI are generally singly charged molecular ions such as $[M+Na]^+$, and therefore this technique serves as a useful tool to obtain an overall glycan profile— that is, a fingerprint of glycans in a sample mixture. The next section describes the methods used in our laboratory for MALDI–time-of-flight (TOF) analysis using a PerSeptive Biosystems Voyager-DE STR mass spectrometer. The methods are equally applicable to other MALDI-TOF instrumentation.

Materials for MALDI Analysis

1. Matrix for glycans: 20 mg/ml 2,5-dihydroxybenzoic acid in 80:20 (v/v) methanol:water
2. Matrix for peptide calibrants: 10 mg/ml a-cyano-4-hydroxycinnamic acid in 4:5:1 water:acetonitrile:1% (v/v) TFA
3. External peptide/protein calibrants: leucine enkephalin, bradykinin, bradykinin (fragment 1-8), angiotensin I, neurotensin, adrenocorticotropic

hormone fragment (ACTH) 1-17, ACTH fragment 1-39, ACTH fragment 7-38, ACTH 18-39, and insulin (bovine pancreas) (100 μg/ml of peptide/protein mix in 0.1% (v/v) TFA)

4. Methanol
5. 100-well MALDI metal plate.

Method for MALDI Analysis

1. The MALDI plate is cleaned with methanol and fully dried before loading the samples. If the plate is not completely dried, samples may spread on the metal surface.

2. Dried and derivatized glycan samples are dissolved in 10 μl of methanol, and an aliquot of 2 μl is mixed in a 1:1 ratio with 2,5-dihydroxybenzoic acid matrix.

3. The external calibrant (peptide/protein mixture) is dissolved in a 1:1 ratio with the matrix, a-cyanno-4-hydroxycinnamic acid.

4. A 1-μl aliquot of the samples to be analyzed is loaded onto the metal plate and dried under vacuum for approximately 20 min. It is vital that the plate be completely dried; otherwise, the pressure of the sample chamber may increase and display an error message. Once completely dried, the plate can be loaded into the mass spectrometer and analyzed.

Samples are analyzed in the positive, Reflectron mode with delayed extraction and an acceleration voltage of 20 kV. Parameters such as the grid voltage percentage and the delayed extraction time are varied according to the sample of interest. The delay time and the grid voltage are interactive parameters, and there are optimum values for each of them. An approach to optimize these settings is to keep one of these parameters constant while optimizing the other. Typical settings used in our laboratory are between 60% to 68% for the grid voltage, and the extraction delay time varies from 220 to 280 nsec.

A typical spectrum obtained from this experiment provides a MALDI fingerprint consisting of a series of singly charged molecular ions, typically [M+Na]$^+$. Occasionally, fragment ions such as A-type ions (Dell, 1987) are also observed, which are useful in determining the nature of the glycan antennae (see Sequencing Glycans section for more on fragmentation patterns). Interpretation is carried out by assigning the compositions of molecular ions in terms of the number of their monosaccharide constituents. Together with the knowledge of the glycan biosynthetic pathways, putative glycan structures can be assigned. Moreover, the algorithm Cartoonist allows an automated interpretation of MALDI spectra (see later for automated interpretation and examples of data).

Sequencing Glycans

To determine the sequence of each of the components observed in the MALDI fingerprinting experiments, and to allocate any fragment ions produced in the fingerprinting experiments to their respective molecular ions, MS/MS experiments are carried out. In this experiment, molecular ions of interest are selected and subjected to collisional activation. The result is the production of sequence informative fragment ions providing unique and important structural data. These MS/MS experiments can be carried out using either MALDI or ES ionization. The following section describes the application of the ES-MS/MS technique.

Electrospray Tandem Mass Spectrometry (ES-MS/MS)

During this technique, the sample of interest is sprayed directly into the ES source, where it is stripped of solvent to produce intact gas-phase ions that are often multiply charged. The following procedure consistently produces good-quality data from quadrupole orthogonal acceleration TOF instrumentation in our laboratory (Micromass UK Ltd). The running conditions described such as flow-rate or collision gas energy may vary slightly, depending on the instrument type.

Materials for Electrospray Ionization Mass Spectrometry (ES-MS)

1. Methanol
2. External calibrant: 1 pmol/μl solution of [Glu^1]-fibrinopeptide B in acetonitrile/5% acetic acid (1:3 (v/v))
3. NanoES spray capillaries (Protana, Denmark).

Method for Electrospray Ionization Mass Spectrometry (ES-MS)

1. The equipment is calibrated with the external calibrant of [Glu^1]-fibrinopeptide B.
2. Derivatized samples are dissolved in 10 μl of methanol, and a few microliters are loaded into NanoES spray capillary.
3. The loaded sample is analyzed by applying a potential of 1.5 kV to the nanoflow tip, ensuring a sample flow rate between 10 and 30 nanoliters/min.
4. The drying gas (N$_2$) and collision gas (Ar) are turned on, maintaining the collision gas pressure at 10^{-4} mbar or the equivalent. Depending on the size and nature of the sample, the collision gas energy is varied. For peptides, it is varied between 20 and 40 eV, and for glycans, between 30 and 80 eV.
5. Data are collected over the lifetime of the sample.

The fragmentation pathways established for permethylated derivatives using fast atom bombardment MS (Dell, 1987) are preserved in MALDI and ES MS/MS analysis. Thus, the main fragment ions arise from A-type cleavages, β-cleavages, and ring-cleavages—all of which are well documented elsewhere (Dell, 1987; Dell et al., 1994; Mechref and Novotny, 2002). Standardized nomenclature for these pathways has been introduced, and the fragment ions are designated B ions, Y ions, and A ions, respectively (Domon and Costello, 1988).

Thus, a typical MS/MS spectrum of N- and O-glycans affords a series of fragment ions from which the antennae sequences, branching patterns, and sometimes linkage can be determined. With respect to the latter, substituents at the 3' positions of an N-acetylhexosamine (HexNAc) position are readily eliminated by β-cleavage, accompanied by a water loss. This is particularly useful for determining the linkage positions of sugar residues attached to the reducing O-glycan HexNAc core (see Kui Wong et al., (2003) for examples).

When ion trap-type instruments are used (Reinhold and Sheeley, 1998; Weiskopf et al., 1998) in addition to MS/MS experiments, it is possible to perform multiple stages of fragmentation; that is, MS^n (n > 2) experiments. This type of procedure is useful when determining the structure of fragment ions and hence yields further structural information.

Linkage Analysis: Gas Chromatography Mass Spectrometry

Before GC-MS analyses, samples are converted into partially methylated alditol acetates as described previously (Albersheim, 1967). This type of analysis allows the identification of monosaccharide constituents and glycosidic bond positions, which can then be used to firmly establish glycan structures.

Solutions for Preparation of Partially Methylated Alditol Acetates

1. 2 M TFA
2. 10 mg/ml sodium borodeuteride in 2 M ammonia solution
3. Glacial acetic acid
4. Acetic anhydride
5. 10% (v/v) acetic acid in methanol
6. Hexanes
7. Water.

Method for the Preparation of Partially Methylated Alditol Acetates

1. Partially methylated alditol acetates are prepared from permethylated glycans for GC-MS linkage analysis.

2. Permethylated samples are first hydrolyzed by incubating them in 200 μl of 2 M TFA at 121° for 2 h. This results in partially methylated monosaccharides.

3. Samples are dried under a stream of nitrogen gas.

4. Once dried, samples are reduced with 10 mg/ml sodium borodeuteride in 2 M ammonium solution at room temperature for 2 h and are subsequently neutralized with few drops of glacial acetic acid.

5. After neutralization, samples are dried under nitrogen. At this stage, samples do not need to be completely dried before going into the next step.

6. Excess borates are removed by adding 4×500 μl of 10% (v/v) acetic acid in methanol.

7. Dried samples are then acetylated by incubating them with 200 μl acetic anhydride at 100° for 1 h and dried under a stream of nitrogen.

8. The resulting products are extracted with chloroform. One to two milliliters of chloroform is added, and the tube is half-filled with water. This is mixed thoroughly and left to settle.

9. The upper aqueous layer is discarded, and the chloroform layer (lower layer) is washed several times with water and dried down under a stream of nitrogen gas.

Gas Chromatography Mass Spectrometric (GC-MS) Analysis

In our laboratory, GC-MS analysis is performed on a Perkin Elmer Clarus 500 GC-MS machine fitted with an RTX-5MS column (30 m $\times$ 0.25 mm internal diameter, Restek Corp.), but similar instrumentation is available from other manufacturers.

Method for Linkage Analysis

1. Previously prepared partially methylated alditol acetates are dissolved in an appropriate amount (usually $\sim$20 μl, depending on the amount of sample) of hexanes before GC-MS linkage analysis.

2. One microliter of the dissolved sample is injected into the column at 60°, and the temperature is increased at a rate of 8°/min to a temperature of 300°.

Data interpretation from this experiment has been well documented (www.ccrc.uga.edu/specdb/ms/pmaa/pframe.html) and therefore is not included.

Automated Analysis of N-Glycan MALDI Fingerprints

Automated identification of proteins from MS and MS/MS spectra is now almost routine. A description of the most common software packages is given in Chalkley et al. (2005), and a performance comparison of several

is given in Chamrad *et al.* (2004). Although current algorithms work well, new ideas that promise improved performance continue to be made (Bern *et al.*, 2005; Frank *et al.*, 2005). The situation for glycans and glycoproteins is much more primitive. There is some literature (Cooper *et al.*, 2001; Ethier *et al.*, 2002, 2003; Gaucher *et al.*, 2000; Goldberg *et al.*, 2005; Joshi *et al.*, 2004; Lohmann and von der Lieth, 2003; Tang *et al.*, 2005) on automated identification, but there are no widely applicable programs comparable to what is available for proteins. This lack of software is due primarily to the fact that glycan identification is much more complex than protein identification. Whereas proteins are linear, glycans are branched; all amino acids (except for leucine and isoleucine) have distinct masses, whereas many of the common monosaccharide building blocks are isomers; and whereas there is only one type of peptide bond, a glycosidic bond can link to multiple positions in a sugar ring and also comes in two varieties: α and β.

As part of the Cartoonist suite of programs (Goldberg *et al.*, 2005), the PARC Mass Spectrometry Viewer (PMSV) allows interactive browsing of any mass spectrum represented in a standard ASCII format. It is also capable of viewing labeled *N*-glycan mass spectrum data in the format produced by the associated annotation program. This annotation format is still evolving and will not be documented here, though it will shortly be made available via the Consortium for Functional Glycomics (www. functionalglycomics.org). Currently, the assignment and display of labeled spectra are limited to permethylated mammalian *N*-glycan data, though with future revisions *O*-glycan assignment, alternative biosynthetic pathways and other derivatization procedures will be included.

The purpose of the Cartoonist suite is to match carbohydrate chemical structures with mass peaks observed in the spectrum, then to present the result in an interactive display. Each structure is indicated by a cartoon, which is a simple diagram for the carbohydrate structure. There may be multiple cartoons for a given mass, with the display of such structural isomers being determined by knowledge of the biosynthetic pathway associated with the sample, as well as a system of "demerits." Each known carbohydrate structure is linked directly to the Glycan Database—the carbohydrate structure database being developed by the Consortium for Functional Glycomics.

Materials

Recommended System Requirements

WINDOWS SYSTEM 2000/XP OS

1. Intel Pentium III 800 MHz or AMD Athlon 800 MHz
2. 512 MB or more of RAM
3. 10 MB available hard-drive space

4. A 56 K or better Internet connection (required for software download and Glycan Database functionality).

MAC OS X 10.3.9

1. 933-MHz or higher G4, G5, or Intel processor
2. 512 MB RAM or higher; DDR RAM recommended
3. 10 MB available hard-drive space
4. A 56 K or better Internet connection (required for software download and Glycan Database functionality).

SUN SOLARIS (SPARC OR ×86 PLATFORM) SOLARIS 8 OS OR HIGHER

1. 128 MB RAM or higher (256 MB recommended)
2. 10 MB available hard-drive space
3. A 56 K or better Internet connection (required for software download and Glycan Database functionality).

REQUIRED SOFTWARE

1. Java Runtime Environment (www.java.com/en/download/manual.jsp)
2. PARC Mass Spectrum Viewer (www.functionalglycomics.org/glycomics/publicdata/glycoprofiling.jsp)
3. Exporting Mass Spectrum Data (optional): appropriate mass spectrometer analysis software, Data Explorer, MassLynx, Analyst.

Methods

Converting and Creating Files

The mass spectrum browser is capable of viewing any spectrum in ASCII format. Converting a spectrum from the proprietary software is currently only possible in one of two ways. First, most MS software includes an "export data" function, which can be used as described later (see Exporting Mass Spectrum Data section) to produce the necessary file format. Second, the user can make use of the tools available from the Sashimi team (sashimi.sourceforge.net/software_glossolalia.html). The tools allow the user to produce mzXML files, which can then be converted to ASCII form (using the mzxml2other program) compatible with the Cartoonist algorithm and browser. Full instructions are available on their Web site.

Exporting Mass Spectrum Data

DATA EXPLORER

1. From the Data Explorer main bar, open a DAT-formatted spectrum file by selecting **File** then **Open.**
2. In the **Edit** menu on the main bar, select **Copy Data.**

3. Open a suitable text editing or notepad program, and paste the copied data.
4. Save the text file using a suitable filename, such as **Spectra1.TXT.**

MassLynx (v.4.0 and Above)

1. From the MassLynx main bar, open a RAW formatted spectrum folder by selecting **File** then **Open Data File.**
2. Navigate to the appropriate directory and select the file in the left-hand panel. Ensure the **Chromatogram** and **Spectrum** checkboxes are ticked and select **OK.**
3. Combine spectra and format as normal.
4. In the **Edit** menu on the main bar of the **Spectrum** view, select **Copy Spectrum List.**
5. Open a suitable text editing or notepad program, and paste the copied data.
6. Save the text file using a suitable filename, such as **Spectra1.TXT.**

Analyst

1. From the Analyst main bar, open a WIFF formatted spectrum folder by selecting **File** then **Open Data File.**
2. Combine spectra and format as normal.
3. In the **File** menu on the main bar, select **Save As.**
4. Save the text file using a suitable filename, such as **Spectra1.TXT.**

Other Mass Spectrometry Software

Most proprietary MS software contains a function, similar to those described previously, whereby it is possible to export the spectrum as raw text or ASCII-type information. Simply paste this information into a text editor, and save it as described previously. Alternatively, explore the Sashimi software (see Converting and Creating Files section).

Creating MSD and MSA Files

To view annotated or labeled spectra, a pair of files is required—the original exported spectrum file (the MSD file, see Exporting Mass Spectrum Data section) and an annotation file.

Creating an MSD file is a simple process, involving renaming the *.**TXT** file as a *.**MSD** file. For example, in the case of **Spectra1.TXT** on a PC system, open a file manager program such as "My Computer" and navigate to the directory containing the file. Right-click on the file **Spectra1.TXT** and select **Rename**. Delete the **TXT** extension and replace it with **MSD,** then press **Enter**.

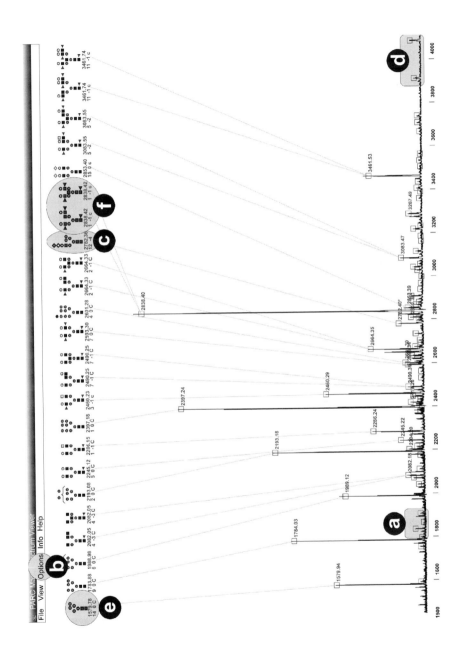

The MSA file, which contains the assignment information, is more complicated. At present, there is no simple way for the average user to create this file because the algorithm requires a UNIX environment to run under. Once a user-operable version is finished, it will be made available for download from www.functionalglycomics.org/glycomics/publicdata/gly-coprofiling.jsp. As entries are made into the Consortium for Functional Glycomics Glycan Profiling Database, the corresponding MSD and MSA files will be made available for download.

The PARC Mass Spectrum Viewer

Displaying Unlabeled or Labeled Exported Mass Spectrum Data

1. Select **File** from the main menu, then **Open** from the drop-down menu.
2. Navigate to the directory containing your files.
3. Change the **Files of Type** drop-down menu if your files have a different file extension to the standard *.**MSD.**
4. Select your file in the panel and click **Open.**
5. If the selected file has a corresponding *.**MSA** file *in the same directory,* the annotated labels will be displayed along with the spectral data.

Basic Operation

Once a spectrum has been successfully loaded (Fig. 3), it is shown within the PMSV in black. If there was an associated .MSA file, the annotated peaks will be marked with a small red square (Fig. 4). The display and scope of the spectrum can then be customized according to various options (Fig. 5). The view of a labeled spectra can be manipulated primarily by use of the quality cutoff slider, which dictates the display of annotations based on their quality measure—an amalgam of the confidence score and demerits (see Assignment Information section), with smaller numbers indicating a better fit to the data. This way, the user can not only tidy up a cluttered spectra view by only showing the best annotations through use of a low-quality cutoff, but he or she can also investigate less likely structures by setting the tolerance to a higher value (Fig. 6).

FIG. 3. An example of a labeled spectrum displayed within Cartoonist's viewer. The shaded regions are points of note and are discussed in subsequent figures. (a) Labeled peaks. (b) The options menu. (c) Unusual structures. (d) Fully expandable display. (e) Assignment information. (f) Structural isomers.

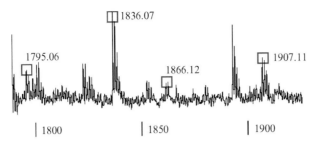

Fig. 4. *Labeled (a) in* Fig. 3. Labeled peaks. Peaks identified by Cartoonist are indicated in all views with a small box.

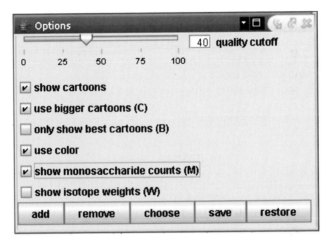

Fig. 5. *Labeled (b) in* Fig. 3. The options menu. **Quality cutoff** slider allows adjustment of the tolerance to poor confidence scores. **Show cartoons** turns the cartoon labeling of the spectrum on and off. **Use bigger cartoons** switches between large (recommended) and small cartoons. **Only show best cartoons** is used if there are multiple cartoons for a peak; this will display only those annotations with the fewest demerits. **Use color** switches between black and white and color annotations, following the nomenclature of the Consortium for Functional Glycomics. **Show monosaccharide counts** turns the monosaccharide count label beneath each cartoon on and off. When a peak is selected, **Show isotope weights** shows a series of blue circles indicating the theoretical intensity of each isotopic peak, computed using the abundances of the stable isotopes ^{13}C, ^{15}N, and ^{18}O. **Add, Remove, Choose, Save, and Restore** allow the saving and loading of option sets.

Clicking and dragging the mouse allows the user to expand any portion of the spectrum (Fig. 7), with the previous views accessible via the undo operations key (**Ctrl-Z** or **ESC**). Views can also be restored with the redo operations key (**Shift-ESC**).

2782.36
12-4

Fig. 6. *Labeled (c) in* Fig. 3. Unusual structures. By means of adjusting the tolerances in the options menu, more unusual structural possibilities can be displayed. The number of demerits and high (i.e., poor) confidence score marks this example as an unlikely assignment, but such structures can prompt further analysis.

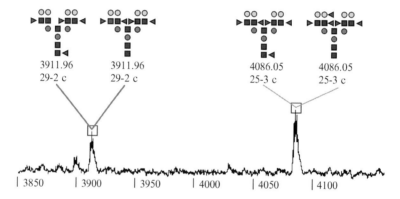

Fig. 7. *Labeled (d) in* Fig. 3. Fully expandable display. Any portion of the spectrum can be magnified, enabling the display and labeling of lower-intensity peaks or peaks with a lower signal to noise ratio that would not otherwise be displayed due to space constraints in a wider mass range window.

Notes can be individually selected by clicking a red square, or the user may choose to automatically expand the note of interest by **Ctrl**-clicking a red square (Macintosh: **Option**-click). Clicking any point on the spectrum will report the mass and intensity values at the mouse cursor. At any point, the spectrum view can be reset to the initial view by pressing the **R** key.

Assignment Information

Annotations or labels are composed of a number of parts, as shown in Fig. 8. In detail, these are as follows:

● Glycan Cartoon: For each peak identified as a match for a specific monosaccharide composition, a number of structural isomers are possible.

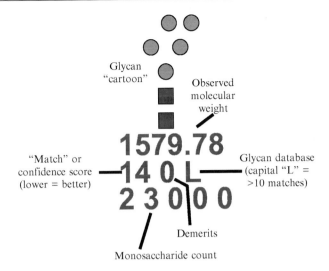

FIG. 8. *Labeled (e) in* Fig. 3. Assignment information. Each labeled peak is displayed with optional assignment information, which details the accuracy of the assignment. These are discussed in the **Basic Operation** Section.

Using information about the biosynthetic pathway involved and mass accuracy of the peak, the Cartoonist algorithm labels the peak with the most likely carbohydrate structure in cartoon form. The details of the assignment are displayed below the cartoon. The cartoons use the Consortium for Functional Glycomics nomenclature described at glycomics.scripps.edu/ CFGnomenclature.pdf.

● Observed Molecular Weight: This feature indicates the observed mass of the ^{12}C peak.

● Confidence Score: This score shows how accurate the assignment is, from a purely mass spectrometric point of view. This takes into account the practical deviation from the predicted mass of the assignment, instrument calibration, and the appearance of the ^{13}C isotope envelope. The lower the confidence score, the better the match. Higher scores indicate less confident matches. If multiple structures possess identical confidence matches and demerit scores for a given note, then the alternative structures are also shown (Fig. 9).

● Demerits: This is a system to evaluate the likelihood of an assignment, given knowledge about the species origin of the sample in question, the biosynthetic pathways involved, and any previous studies on the carbohydrate compliment. A higher number of demerits indicates the presence of structural features known to be inconsistent with or

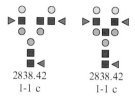

FIG. 9. *Labeled (f) in* Fig. 3. Structural isomers. Each structure has identical demerit and confidence scores, so both isomers are displayed.

uncommon in the species of interest. A low number of demerits signifies a structure compliant with the criteria listed previously. Currently, Cartoonist is designed for the analysis of mammalian *N*-glycans. However, it is possible to adjust the weighting of the demerits via the **Custom** menu under **Weights**. This will display a list of properties a glycan structure may have, such as bisecting GlcNAc. The higher the number listed for a property, the less likely it is that a cartoon featuring such a property will be chosen for display. In this example, a demerit value of 10 for bisecting GlcNAc would result in such a cartoon only being displayed if there were no other option and the quality slider was set to accept unlikely structures. Setting a value of zero would favor cartoons with bisecting GlcNAcs to be chosen above all other assignments.

● Glycan Database: The presence of an upper case **L** indicates that the structure has matching entries in the Glycan Database (www.functionalglycomics.org/glycomics/molecule/jsp/carbohydrate/carbMolecule-Home.jsp). To examine the matching records, simply mouse over and click on the cartoon of interest. This should automatically open a Web browser and bring up the appropriate page.

● Monosaccharide Count: This count shows the number of each generic monosaccharide residue present in the structure. From left to right, the fields read: HexNAc (GalNAc and GlcNAc), Hexose (Galactose, Glucose, and Mannose), Fucose, NeuAc, and NeuGc.

Menus

● File (ALT-F).
● Open (CTRL-O): Opens a directory or set of files using a platform-dependent file chooser. Opening a directory is the same as choosing all of the files in the directory.
● New (CTRL-K): Creates a new PMSV instance. The initial directory is the current directory, but no files will be loaded.

● Print (CTRL-P): Brings up a **Print** dialog window, which allows the current spectra to be printed. The printing may show additional detail for the spectra, such as additional cartoons and finer detail for the spectra. To capture the exact spectra (at screen resolution), one should use a platform-specific screen capture tool.

● Exit (CTRL-Q): Quits the current program. No confirmation is requested.

● Previous, Next ([,]): When browsing a directory of spectra, this will switch to the next (or previous) spectrum in the directory.

● View (ALT-V).

● Reset (R): Resets the view to the entire spectrum.

● Expand (X): Doubles the current view in both directions.

● Fit (F): Fits the current selected note to the entire window, attempting to get adjacent peaks viewable.

● Undo (ESC): Undoes the most recent zoom/expand/fit command.

● Redo (Shift-ESC): Re-performs the most recent command that was undone.

● Custom (ALT-U).

● Options (ALT-O): Brings up the **Options** window, which is used to edit options controlling the display.

● Weights (ALT-W): Brings up the **Cartoon Demerit Weights** window, which is used to edit the demerit weights for cartoons.

● Info (ALT-I).

● File (Shift-I): Shows information about the current file, including the title, the mass range, the intensity range, the number of pairs, and the number of notes.

● Note All (I): Shows information about all notes with the same mass as the selected note.

● Other: Shows other information for to the current running PMSV.

● Log: Displays a log of program events.

● Help (ALT-H).

● General Help (CTRL-H): Views help about the application.

● Keys (ALT-K): Views help about the single-key commands.

References

Albersheim, P., Nevins, D. J., English, P. D., and Karr, A. (1967). A method for the analysis of sugars in plant cell wall polysaccharides by gas-liquid chromatography. *Carbohydr. Res.* **5**, 340–345.

Bern, M., Goldberg, D., and Eigen, M. S. (2005). De Novo Analysis of Peptide Tandem Mass Spectra by Spectral Graph Partitioning. Ninth Research in Computational Molecular Biology (RECOMB) Cambridge, MA.

Chalkley, R. J., Hansen, K. C., and Baldwin, M. A. (2005). Bioinformatic methods to exploit mass spectrometric data for proteomic applications. *Methods Enzymol.* **402,** 289–312.

Chamrad, D. C., Korting, G., Stuhler, K., Meyer, H. E., Klose, J., and Bluggel, M. (2004). Evaluation of algorithms for protein identification from sequence databases using mass spectrometry data. *Proteomics* **4,** 619–628.

Cooper, C. A., Gasteiger, E., and Packer, N. H. (2001). GlycoMod: A software tool for determining glycosylation compositions from mass spectrometric data. *Proteomics* **1,** 340–349.

Dell, A. (1987). FAB mass spectrometry of carbohydrates. *Adv. Carbohydr. Chem. Biochem.* **45,** 19–72.

Dell, A., Chalabi, S., Easton, R. L., Haslam, S. M., Sutton-Smith, M., Patankar, M. S., Lattanzio, F., Panico, M., Morris, H. R., and Clark, G. F. (2003). Murine and human zona pellucida 3 derived from mouse eggs express identical *O*-glycans. *Proc. Natl. Acad. Sci.* **100,** 15631–15636.

Dell, A., and Morris, H. R. (2001). Glycoprotein structure determination by mass spectrometry. *Science* **291,** 2351–2356.

Dell, A., Morris, H. R., Egge, H., and von Nicolai, H. (1983). Fast-atom-bombardment mass-spectrometry for carbohydrate-structure determination. *Carbohydr. Res.* **115,** 41–52.

Dell, A., Reason, A. J., Khoo, K. H., Panico, M., McDowell, R. A., and Morris, H. R. (1994). Mass spectrometry of carbohydrate-containing biopolymers. *Methods Enzymol.* **230,** 108–132.

Domon, B., and Costello, C. E. (1988). A systematic nomenclature for carbohydrate fragmentations in FAB-MS/MS spectra of glycoconjugates. *Glycoconj.* **5,** 397–409.

Ethier, M., Saba, J. A., Ens, W., Standing, K. G., and Perreault, H. (2002). Automated structural assignment of derivatized complex N-linked oligosaccharides from tandem mass spectra. *Rapid Commun. Mass Spectrom.* **16,** 1743–1754.

Ethier, M., Saba, J. A., Spearman, M., Krokhin, O., Butler, M., Ens, W., Standing, K. G., and Perreault, H. (2003). Application of the StrOligo algorithm for the automated structure assignment of complex N-linked glycans from glycoproteins using tandem mass spectrometry. *Rapid Commun. Mass Spectrom.* **17,** 2713–2720.

Frank, A., Tanner, S., Bafna, V., and Pevzner, P. (2005). Peptide sequence tags for fast database search in mass-spectrometry. *J. Proteome Res.* **4,** 1287–1295.

Fukuda, M., Dell, A, Oates, J. E., and Fukuda, M. N. (1984a). Structure of branched lactosaminoglycan, the carbohydrate moiety of band 3 isolated from adult human erythrocytes. *J. Biol. Chem.* **259,** 8260–8273.

Fukuda, M., Spooncer, E., Oates, J. E., Dell, A., and Klock, J. C. (1984b). Structure of sialylated fucosyl lactosaminoglycan isolated from human granulocytes. *J. Biol. Chem.* **259,** 10925–10935.

Fukuda, M. N., Dell, A., Oates, J. E., and Fukuda, M. (1985). Embryonal lactosaminoglycan: The structure of branched lactosaminoglycans with novel disialosyl (sialyl alpha 2-9 sialyl) terminals isolated from PA1 human embryonal carcinoma cells. *J. Biol. Chem.* **260,** 6623–6631.

Gaucher, S. P., Morrow, J., and Leary, J. A. (2000). STAT: A saccharide topology analysis tool used in combination with tandem mass spectrometry. *Anal. Chem.* **72,** 2331–2336.

Goldberg, D., Sutton-Smith, M., Paulson, J., and Dell, A. (2005). Automatic annotation of matrix-assisted laser desorption/ionization *N*-glycan spectra. *Proteomics* **5,** 865–875.

Haslam, S. M., Khoo, K. H., Houston, K. M., Harnett, W., Morris, H. R., and Dell, A. (1997). Characterisation of the phosphorylcholine-containing N-linked oligosaccharides in the excretory-secretory 62 kDa glycoprotein of *Acanthocheilonema viteae*. *Mol. Biochem. Parasitol.* **85,** 53–66.

Joshi, H. J., Harrison, M. J., Schulz, B. L., Cooper, C. A., Packer, N. H., and Karlsson, N. G. (2004). Development of a mass fingerprinting tool for automated interpretation of oligosaccharide fragmentation data. *Proteomics* **4,** 1650–1664.

Kui Wong, N., Easton, R. L., Panico, M., Sutton-Smith, M., Morrison, J. C., Lattanzio, F. A., Morris, H. R., Clark, G. F., Dell, A., and Patankar, M. S. (2003). Characterization of the oligosaccharides associated with the human ovarian tumor marker CA125. *J. Biol. Chem.* **278,** 28619–28634.

Lohmann, K. K., and von der Lieth, C. W. (2003). GLYCO-FRAGMENT: A web tool to support the interpretation of mass spectra of complex carbohydrates. *Proteomics* **3,** 2028–2035.

Manzi, A. E., Norgard-Sumnicht, K., Argade, S., Marth, J. D., van Halbeek, H., and Varki, A. (2000). Exploring the glycan repertoire of genetically modified mice by isolation and profiling of the major glycan classes and nano-NMR analysis of glycan mixtures. *Glycobiology* **10,** 669–689.

Mechref, Y., and Novotny, M. V. (2002). Structural investigations of glycoconjugates at high sensitivity. *Chem. Rev.* **102,** 321–369.

Patankar, M. S., Jing, Y., Morrison, J. C., Belisle, J. A., Lattanzio, F. A., Deng, Y., Wong, N. K., Morris, H. R., Dell, A., and Clark, G. F. (2005). Potent suppression of natural killer cell response mediated by the ovarian tumor marker CA125. *Gynecol. Oncol.* **99,** 704–713.

Reinhold, V. N., and Sheeley, D. M. (1998). Detailed characterization of carbohydrate linkage and sequence in an ion trap mass spectrometer: Glycosphingolipids. *Anal. Biochem.* **259,** 28–33.

Schneider, P. A. F., and Ferguson, M. A. J. (1995). Microscale analysis of glycosylphosphatidylinositol structures. *Methods Enzymol.* **250,** 614–630.

Sutton-Smith, M., Morris, H. R., and Dell, A. (2000). A rapid mass spectrometric strategy suitable for the investigation of glycan alterations in knockout mice. *Tetrahedron: Asymmetry* **11,** 363–369.

Sutton-Smith, M., Morris, H. R., Grewal, P. K., Hewitt, J. E., Bittner, R. E., Goldin, E., Schiffmann, R., and Dell, A. (2002). MS screening strategies: Investigating the glycomes of knockout and myodystrophic mice and leukodystrophic human brains. *Biochem. Soc. Symp.* **69,** 105–115.

Tang, H., Mechref, Y., and Novotny, M. V. (2005). Automated Interpretation of MS/MS spectra of oligosaccharides. *13th International Conference on Intelligent Systems for Molecular Biology (ISMB)*, Detroit, MI.

Weiskopf, A. S., Vouros, P., and Harvey, D. J. (1998). Electrospray ionization-ion trap mass spectrometry for structural analysis of complex N-linked glycoprotein oligosaccharides. *Anal. Chem.* **70,** 4441–4447.

[6]   Structural Analysis of Sialyl N-Glycan Using
Pyridylamination and Chromatography Followed by
Multistage Tandem Mass Spectrometry

By HIROAKI NAKAGAWA and KISABURO DEGUCHI

Abstract

Multiple-dimensional mapping (n-DM) methods consisting of pyridylamination and high-performance liquid chromatography (HPLC) are widely
used for oligosaccharide analysis. These methods are quantitative, sensitive, and suitable for separating isomers. Oligosaccharide structures are
suggested by elution positions on two or three kinds of columns and then
confirmed by enzymatic and chemical treatments or mass spectrometry
(MS). Multiple-stage tandem MS (MS^n) analyses have been used to determine oligosaccharide structures by spectrum matching, comparing standard oligosaccharides, and offering detailed MS^n fragment analysis.
However, oligosaccharides usually exist as a mixture including isomers.
The 3-DM method provides quantitative information and well-isolated
sialyl- or neutral-oligosaccharides; on the other hand, the MS^n method
saves time by eliminating complicated enzyme degradation steps. The
combination of both methods, which support each other, represents a
reasonable and efficient strategy. This chapter describes 3-DM on HPLC
after negative-ion MS^n spectral matching for sialyl N-glycans.

Overview

Oligosaccharides are a mixture of various structures including isomers,
some of which are sialylated in mammals. Pyridylamination is a process of
fluorescence-tagging carbohydrates (Hase *et al.*, 1978), giving sugars high
sensitivity and resolution on high-performance liquid chromatography
(HPLC). Therefore, multiple-dimensional mapping (n-DM) methods using
pyridylamination and HPLC have been used extensively for quantitative
analysis of oligosaccharides since the 1980s (Hase *et al.*, 1987; Tomiya *et al.*,
1988). Such methods are sometimes combined with further analysis using
partial enzymatic degradation. On the other hand, mass spectrometry (MS)
is becoming useful for oligosaccharide analysis, and methods for detailed
analysis by multiple MS (MS^n) have been reported. MS^n should eventually
become a key technology for oligosaccharide analysis, but currently sample
purification is a very critical factor required for sensitivity. A combination

METHODS IN ENZYMOLOGY, VOL. 415   0076-6879/06 $35.00

of 3-DM methods and MS^n represents an optimal approach to analyze sialyl N-glycans. This chapter outlines structural determination of pyridylaminated (PA)-sialyl N-glycans by 3-DM methods using HPLC (Takahashi et al., 1995) and negative-ion MS^n spectral matching by nano-flow HPLC/electrospray ion trap time-of-flight mass spectrometry (ESI-IT-TOFMS) (Deguchi et al., 2006).

PA-oligosaccharides are analyzed based on elution positions using three different modes: ion-exchange (DEAE), reversed phase (octadecylsilyl silica, ODS), and amide adsorption columns using HPLC. Elution times on ODS and amide adsorption columns are converted to relative values as glucose units to obtain reproducibility. Oligosaccharide structures are suggested by matching data with databases of known PA-oligosaccharides. Suggested structures are then confirmed by partial degradation or MS^n analysis. Partial degradation has been used widely but is time consuming and requires special planning. On the other hand, MS^n analysis is growing in popularity and is a promising analytical method. Negative-ion analysis is very suitable for sialyl oligosaccharides and nano-flow realized high sensitivity. Either of the two methods described in this chapter can be used satisfactorily to determine sialyl oligosaccharide structure.

PA-Oligosaccharide Preparation

Tagging carbohydrates using reductive amination was originally developed on the basis of pyridylamination (Hase et al., 1978). This technique has become diversified with modification of labeling reagents, types of saccharides, use of analytical equipment, and varying experimental conditions. Some reports suggest that removal of sialic acid occurs during pyridylamination, but we have not experienced this outcome using a modified protocol of Yamamoto et al. (1989). Such differing outcomes can be explained by differences in heating, which depend on the equipment used. Pilot tests using the equipment at hand and standard sialyl oligosaccharide (e.g., 6'-N-acetylneuraminyl-lactose, Sigma) are recommended. (PA-lactose and PA-6'-N-acetylneuraminyl-lactose are separated on an Amide column—GU 2.0 and 2.7, respectively.)

Materials and Reagents

1. Human immunoglobulin G (IgG): I-4506 from Sigma (Sigma-Aldrich, St. Louis, MO) is recommended because it has a good N-glycan profile. The relative molar ratio of each oligosaccharide may differ among manufactured lots.
2. Trypsin and chymotrypsin: 100 μg each (Sigma) is dissolved in 200 μl 0.1 M NH_4HCO_3, 10 mM $CaCl_2$. Prepare immediately before use.

3. N-Glycosidase F: Add 1 μl water per unit of enzyme (recombinant, lyophilized) from Roche Diagnostics (Mannheim, Germany). N-Glycosidase is also called peptide:N-glycosidase (PNGase), glycoamidase, glycopeptidase, or N-glycanase. Enzymes from almond (Takahashi, 1977) and *Flavobacterium* species (Plummer *et al.*, 1984) are commercially available. The units of some commercial products are the same as *mU* on international enzyme units, so care should be taken as to the method of assay. *Flavobacterium* N-glycosidase (N-glycosidase F) cannot cleave R-(Fucα1,3-) GlcNAc-Asn(polypeptide), which is found in many plants and insects (Altmann *et al.*, 1995; Tretter *et al.*, 1991). These samples must be treated with N-glycosidase from almond (N-glycosidase A) or subjected to hydrazinolysis (Takasaki *et al.*, 1982).

4. Pronase: Calbiochem (Merck KGaA, Darmstadt, Germany).

5. PA solution: To prepare the coupling reagent, add 2 ml of HCl fuming into a fresh bottle of 5 g of 2-aminopyridine for fluorescent labeling (011–14181, Wako Chemicals USA Inc, Richmond, VA) in a draft chamber. Prepare several test tubes containing 1.8 ml water. Next, 0.2 ml of 2-aminopyridine/HCl is diluted 1:10 in a test tube, and the pH is measured with a meter (discard diluted reagent). Add HCl fuming to the bottle as required until the diluted pH is 6.8. This reagent is stored at $-20°$ under a double seal of Parafilm and should be used within 1 month.

6. Reducing reagent: Prepare immediately before use. Weigh the empty and well-dried tube with a lid before adding reagent. Quickly put a spoonful of sodium cyanoborohydride (Aldrich, Sigma-Aldrich) (Attention! Substance is a highly toxic and flammable solid.) into the tube, and close the lid immediately in a draft chamber. Weigh the tube, and calculate the amount of sodium cyanoborohydride. Add 120 μl of water per 200 mg of sodium cyanoborohydride in a draft chamber. Lightly mix contents of the tube, centrifuge, and then immediately decant the supernatant into the sample. Sodium cyanoborohydride must be Aldrich 15,615-9 (Sigma-Aldrich), since other manufactured reagents often do not work. Sodium cyanoborohydride is stored in a desiccator with well-dried silica gel; it is never vacuumed and should be used within 6 months of opening or before forming a cake.

7. Orcinol-sulfuric acid solution: 10 mg of 5-methylresorcinol (orcinol, Wako) is dissolved in 100 μl methanol, and 5 ml 2 M H_2SO_4 is added. Store it at 4° in a brown-colored atomizer bottle.

PA-Oligosaccharide Preparation from Glycoprotein

1. Five milligrams of human IgG in a 1.5-ml tube is dissolved in 500 μl 0.1 M NH_4HCO_3, heated to 90° for 15 min, and lyophilized.

2. One hundred microliters of trypsin and chymotrypsin solution is added and incubated at 37° for 16 h.

3. After heating at 90° for 15 min and then cooling, N-glycosidase F 5 U/5 μl is added and incubated at 37° for 16 h.

4. Add pronase solution (50 μg in 10 μl 0.1 M NH$_4$HCO$_3$), incubate at 37° for 16 h, and then heat to 90° for 15 min.

5. Purify by gel filtration using Bio-Gel P-4 (Bio-Rad, Hercules, CA) (1 × 38 cm, water), and fractionate at 1.0 ml each.

6. Spot 1 μl of a 10-fold concentrated fraction on a silica-gel TLC plate (cut aluminum sheet TLC [Merck] and write fraction numbers with pencil). Atomize orcinol-sulfuric acid in a draft chamber; oligosaccharides are colored by heating at 120 to 150° for several minutes with a hot plate or an oven. (Approximately 50 ng of galactose [Gal], mannose [Man], and fucose [Fuc] per spot can be detected. N-Acetylglucosamine [GlcNAc] and N-acetylneuraminic acid [NANA] are not colored.)

7. Oligosaccharide fractions, 7 to 8 ml, are lyophilized completely, and then 500 μl of coupling reagent is added and heated at 90° for 15 min with a dry block heater. Fifty microliters of reducing reagent is then added and heated for an additional 60 min. Sialic acid of a given oligosaccharide is often removed by its own acidity, so the sialyl oligosaccharide solution must be at pH 4 to 8 in buffer or a NH$_4$HCO$_3$ solution, and heating should be minimal.

8. Excess reagents are removed by gel filtration using Sephadex G-15 (Amersham Bioscience Corp, Piscataway, NJ) (1 × 38 cm, 10 mM NH$_4$HCO$_3$) and fractionated in 1.0-ml aliquots while monitoring ultraviolet absorption at 280 nm.

9. The PA-oligosaccharide fraction is confirmed by ultraviolet absorption and coloration by orcinol-sulfuric acid as described in step 6 of this section, and it is then dried by centrifugal evaporation.

3-DM Using HPLC

This section describes 3-DM methods for sialyl N-glycans, according to the HPLC analytical conditions of Takahashi et al. (1995). This method analyzes oligosaccharides quantitatively, suggests detailed structures by their elution positions, and isolates oligosaccharides. Both sulfated and sialylated oligosaccharides can be analyzed (Yagi et al., 2005). Elution position data of approximately 500 kinds of N-glycans are provided in references (Takahashi and Tomiya, 1992; Takahashi et al., 1995) and at www.glycoanalysis.info/ENG/index.html.

Materials

1. HPLC column: TSK gel DEAE-PW (7.5 × 75 mm) and Amide-80 (4.6 × 250 mm) are from Tosoh Bioscience LLC (Montgomeryville, PA). HRC-ODS (6 × 150 mm, Shimadzu Scientific Instruments, Columbia, MD) must be used. Elution positions can change depending on the type of column and analytical conditions, particularly in analysis on ODS.

2. HPLC is 7000 series (Hitachi High Technologies America, Inc., San Jose, CA) equipped with an auto-sampler, a column oven, a fluorescence detector, a 2-solvent gradient system, and an unfastened a dynamic mixer. An overly large dead volume in the line of a dynamic mixer often causes different elution positions. The column is wrapped with a paper towel to avoid direct contact with the stainless steel plate of the oven to avoid thermal gradients.

3. Acetic acid-triethylamine buffer: 50 ml acetonitrile, 13.5 ml acetic acid, and 400 ml water are mixed, and triethylamine (Sigma) is added while monitoring the pH adjusted to 7.3 in a draft chamber. Water is added to a total volume of 500 ml, and the product is stored at 4° for use within 3 months. The buffer is difficult to mix with acetonitrile when chilled, so it should be warmed up before using. This solution is 10% (v/v) acetonitrile, which must be considered in making solvents for the DEAE and the Amide columns.

4. Phosphate buffer: 0.5 M sodium phosphate buffer, with a pH of 3.8. The buffer should be stored at 4° and used within 1 month. Diluted buffer is easily contaminated with microorganisms, so solvents should be used for the ODS column within 3 days.

5. A mixture of PA-isomaltooligosaccharides (PA-glucose oligomer) and other standard PA-oligosaccharides can be purchased from Takara Mirus Bio (Madison, WI).

HPLC Analysis

Analytical conditions for each column are shown in Table I.

1. PA-oligosaccharides are separated by the number of sialic acid residues included in N-glycans on the DEAE-PW column using HPLC (Fig. 1). Separated oligosaccharide fractions are subjected to centrifugal evaporation.

2. Fractions separated on the DEAE column are dissolved in 0.1 M NH$_4$HCO$_3$ and then analyzed and isolated on the next ODS column (Fig. 2). PA-glucose oligomers are analyzed in every analytical lot as a control.

TABLE I
CONDITIONS OF HPLC ANALYSIS

	Ion exchange	Reversed phase	Amide adsorption
Column	DEAE-PW 7.5 × 75 mm	HRC-ODS 6 × 150 mm	Amide-80 4.6 × 250 mm
Solvent	A: 10% (v/v) ACN in water B: 10% (v/v) ACN in 0.5 M ATE pH, 7.3	C: 10 mM PB pH, 3.8 D: 0.5% (v/v) 1-butanol in C	E, ACN: 0.5 M TEA pH, 7.3 = 65:35 (v/v) F, ACN: 0.5 M TEA pH, 7.3 = 50:50 (v/v)
Gradient	0 min; A = 100%, B = 0% 5 min; A = 100%, B = 0% 40 min; A = 80%, B = 20%	0 min; C = 80%, D = 20% 60 min; C = 50%, D = 50%	0 min; E = 100%, F = 0% 30 min; E = 40%, F = 60%
Temperature	35°	55°	40°
Flow rate	1.0 mL/min		
Detection	Fluorescence (excitation; 320, nm; emission, 400 nm)		

ACN, acetonitrile; ATE, acetic acid-triethylamine; PB, sodium phosphate buffer.

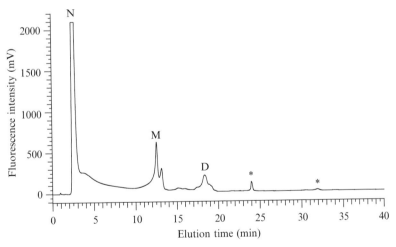

FIG. 1. Chromatogram of IgG N-glycans on the DEAE column. N, M, and D indicate elution positions of neutral, mono-sialyl and di-sialyl oligosaccharides, respectively. * indicates mechanical ghost peaks.

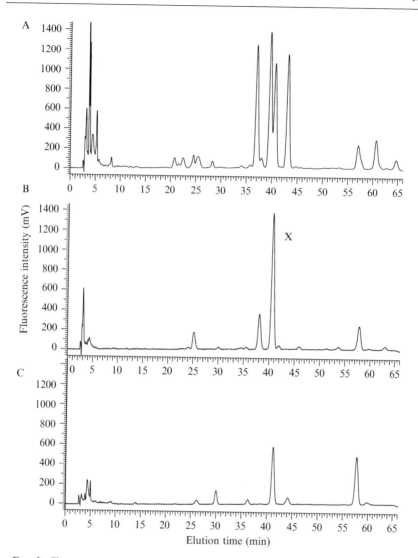

FIG. 2. Chromatograms of N-glycan ODS column fractions separated on the DEAE column according to the number of sialic acid residues. A, neutral oligosaccharide fraction; B, mono-sialyl oligosaccharide fraction; C, di-sialyl oligosaccharide fraction. Peak X, which is most abundant in mono-sialyl oligosaccharide, was further analyzed.

3. The molar ratio of PA-oligosaccharides is calculated based on areas of analysis on the ODS column, since fluorescence intensities of PA-oligosaccharides do not depend on oligosaccharide structure (Hase, 1992).

4. PA-oligosaccharide elution positions are converted to glucose units (GU), which correspond to the relative degree of polymerization (DP) of PA-isomaltooligosaccharides. For example, peak X (Fig. 3) eluted at 40.69 min on the ODS column, and it is between 39.31 min of DP13 isomaltooligosaccharide and 42.72 min of DP14. The GU of peak X is 13.4, calculated as $[(40.69 - 39.31)/(42.72 - 39.31)]+13$.

5. Fractions of the ODS column are evaporated by centrifugation. After dissolving in water, they are further analyzed and isolated on a third Amide column. PA-oligosaccharide elution positions are converted to GU in the same way as on the ODS column. The structure of an isolated oligosaccharide is suggested by comparison of its elution positions with the databases. Usually, the allowable error of GU is within 5% depending on the column manufacturing lot, column aging, solvents, and line of HPLC. Correction with standard oligosaccharides can reduce such error.

For example, peak X in a chromatogram on the ODS column is the mono-sialyl fraction on the DEAE column, 13.4 GU on the ODS, and 7.7 GU on the Amide (mono, 13.4, 7.4). This coordinate coincides with 1A1-210.4 Galβ4GlcNAcβ2Manα6(NANAα2,6Galβ4GlcNAcβ2Manα3)Manβ4GlcNAcβ4(Fucα6)GlcNAc-PA; the code number of our nomenclature is described in the references (Takahashi and Tomiya, 1992; Takahashi et al., 1995). In this case, 1A5-300.8 (mono, 13.8, 7.5) and 1A4-300.22 (mono, 14.0, 7.5) also represent other potential outcomes based on allowable error. If these structures are possible, further analysis is needed. A suggested structure is confirmed by partial degradation and further HPLC or MS^n analysis.

Determination of Structure Using Databases and Partial Degradation

An oligosaccharide is degraded according to the suggested structure shown in the 3-DM. Using HPLC section, and then analyzed by HPLC at each step. The suggested structure is confirmed by matching the elution positions of treated samples with the predicted structure. Enzyme treatment conditions are provided, either in protocols developed by manufacturers or in the references (Mizuochi, 1992; Takahashi and Tomiya, 1992).

Materials and Methods

1. Beta-galactosidase from jack bean (Associates of Cape Cod, Inc., East Falmouth, MA) is dissolved at 1 mU/μl in 0.1 M citrate-phosphate buffer (pH, 4.0) and stored in aliquots at $-20°$. Next, 100 pmol of substrate

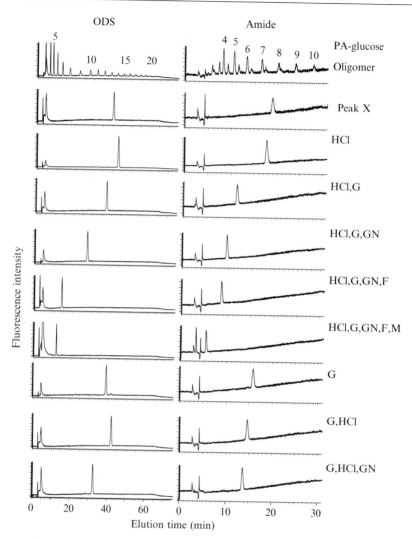

FIG. 3. Chromatograms of PA-glucose oligomers, peak X, and treated samples on ODS and Amide columns. Numbers on PA-glucose oligomers indicate DP of isomaltooligosaccharides. The following abbreviations designate peak X treatments, which were done sequentially: HCl, acid lysis; G, β-galactosidase treatment; GN, β-N-acetylhexosaminidase treatment; F, α-fucosidase treatment; M, α-mannosidase treatment.

dissolved in 10 μl is added to 5 μl of β-galactosidase solution; the pH (3.5 to 5.0) is checked using test paper, and the solution is incubated at 37° for 16 h.

2. Beta-N-acetylhexosaminidase from jack bean (Associates of Cape Cod) is dissolved at 4mU/μl in 0.1 M citrate-phosphate buffer (pH, 5.0) and stored in aliquots at −20°. Next, 100 pmol of substrate dissolved in 10 μl is added to 5 μl of β-N-acetylhexosaminidase solution, checked at pH 4.5 to 6.5 using test paper, and incubated 37° for 16 h.

3. Alpha-mannosidase from jack bean (Associates of Cape Cod) is dissolved at 10 mU/μl in 10 mM sodium-acetate buffer (pH, 6.0) and stored in aliquots at −20°. Then, 100 pmol of substrate is lyophilized, dissolved in 10 μl of 0.1 M sodium-acetate buffer (pH, 5.0) and 15 mM ZnCl$_2$, added to the 5 μl of α-mannosidase solution, checked at pH 4.0 to 5.0 using test paper, and incubated 37° for 16 h.

4. Alpha-L-fucosidase from bovine kidney (Sigma) is in suspension in 3.2 M ammonium sulfate, 10 mM sodium phosphate, and 10 mM citrate pH 6.0 (approximately 4mU/μl). Next, 100 pmol of substrate is lyophilized, dissolved in 10 μl of 0.2 M ammonium-acetate buffer (pH, 4.5), added to a 5 μl of α-fucosidase solution, checked at pH 4.0 to 5.0 by test paper, and incubated at 37° for 16 h. Alpha-L-fucosidase is stored at 4°.

5. Weak acid lysis, removing sialic acid: The sample solution is adjusted to pH 2.0 with 0.1 M HCl (check by test paper) and heated to 90° for 60 min.

This suggested structure 1A1-210.4 of peak X is confirmed by partial degradation following two pathways, one confirming neutral oligosaccharide structure and the other determining which branch binds sialic acid. First, to confirm neutral oligosaccharide structure, sialic acid is removed by weak acid lysis. The asialo peak X is then treated with β-galactosidase, β-N-acetylhexosaminidase, α-fucosidase, and α-mannosidase sequentially. Another pathway is analysis of a sialic acid position. β-Galactosidase is used to remove a galactose residue that does not bind sialic acid. Sialic acid is then removed and treated with β-N-acetylhexosaminidase to confirm the structure. Peak X is treated according to the strategy shown in Fig. 4. Each treated peak is analyzed on ODS and Amide columns at every step. Chromatograms from ODS and Amide columns of every sample are shown in Fig. 3. The analyzed GU data are compared with the predicted outcome to confirm the structure (Table II). In this way, peak X was confirmed as the structure 1A1-210.4. Linkage of sialic acid can be determined by elution positions from the Amide column. α2,6 Sialic acid increases elution time compared with the desialylated one, but the α2,3 linkage clearly reduces it. Neuraminidases that distinguish linkages are also useful (Nakagawa *et al.*, 1996), but one should be careful in interpreting their activity. Parallel treatments using standards (e.g., α2,3 and α2,6 sialyl lactose) are strongly recommended.

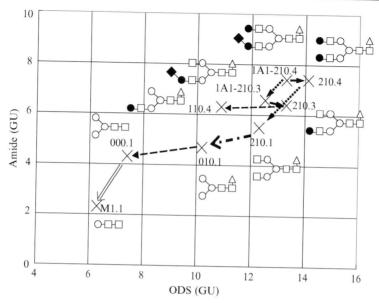

Fig. 4. Strategy of structure confirmation combining partial degradation and HPLC analysis on ODS and Amide columns. The suggested structure of peak X is 1A1-210.4, Galβ4GlcNAcβ2Manα6(NANAα2,6Galβ4GlcNAcβ2Manα3)Manβ4GlcNAcβ4(Fucα6) GlcNAc. X marks are coordinates of expected N-glycans. ◄——, acid lysis; ◄······ β-galactosidase treatment; ◄ · —, β-N-acetylhexosaminidase treatment; ◄ – –, α-fucosidase treatment; ◄═══, α-mannosidase treatment. ◆, sialic acid (N-acetylneuraminic acid); ●, galactose; □, N-acetylglucosamine; ○, mannose; △, fucose.

Structural Assignment of MS^n Spectral Matching

As shown in the Determination of Structure Using Databases and Partial Degradation section, oligosaccharide structures can be confirmed only by HPLC. However, HPLC is time consuming and requires special planning in terms of the degradation pathway, which is dependent on enzyme specificity and oligosaccharide variety. MS is another useful tool for determining oligosaccharide structures, and its use is expanding (Harvey, 2006; Mechref and Novotny, 2002; Zaia, 2004). Many MS-based methods for PA-N-glycan analysis also have been reported, such as ESI (Gu et al., 1994), LC/MS (Suzuki-Sawada et al., 1992; Takegawa et al., 2005a), and MALDI-TOF/MS-MS (Kurogochi and Nishimura, 2004; Mizuno et al., 1999; Okamoto et al., 1997). MS^n spectral matching of fragmentation pattern is useful to not only distinguish isomers, but also to determine novel oligosaccharide structures (Takegawa et al., 2005b,c). In this chapter, negative-ion

TABLE II

ELUTION POSITIONS OF SAMPLE, PEAK X, AND TREATED SAMPLE ON
ODS AND AMIDE COLUMNS

| Sample | Expected | Elution position (GU) | | | |
| | | Observed | | Reported | |
Treatment	Structure	ODS	Amide	ODS	Amide
Peak X	1A1-210.4	13.4	7.7	13.3	7.4
HCl	210.4	14.2	7.4	14.1	7.4
HCl,G	210.1	12.5	5.4	12.3	5.5
HCl,G,GN	010.1	10.0	4.6	10.2	4.7
HCl,G,GN,F	000.1	7.4	4.2	7.3	4.3
HCl,G,GN,F,M	M1.1	6.6	2.3	6.6	2.3
G	1A1-210.3	12.9	6.8	12.5	6.6
G,HCl	210.3	13.7	6.4	13.3	6.4
G,HCl,GN	110.4	11.0	6.1	10.9	6.3

HCl, acid lysis; G, β-galactosidase treatment; GN, β-N-acetylhexosaminidase treatment;
F, α-fucosidase treatment; M, α-mannosidase treatment.
Note: Treatments were done independently and sequentially.

MS^n spectral matching is introduced (Deguchi et al., 2006). This method is suitable for structural analysis of sialyl PA N-glycans.

Materials

1. Mass spectrometer: NanoFrontier system consisting of capillary HPLC based on an AT10PV nanoGR generator and ESI-IT-TOF MS (Hitachi High-Technologies).
2. Column: Develosil nano-C30 column (130 mm i.d., 5 μm long) purchased from Nomura Chemicals, Seto, Aichi, Japan.
3. Electrosprayer: Silica Tip (10 μm i.d.) from New Objective (Woburn, MA).
4. Acetonitrile and formic acid (LC/MS grade) from Wako.
5. Water is fresh Milli-Q water.

MS Analysis

1. Flow rate is set at 200 nl/min. Solvent is 50% acetonitrile in 0.1% formic acid aqueous (v/v).
2. MS conditions are as follows: ESI voltage, −1.7 kV; curtain gas, 0.7 l/min nitrogen without heating; scan range, m/z 200 to 2000.

3. Parameters of MS^2 and MS^3 used are: CID gas, 100 mTorr of He; isolation time, 1.5 msec; isolation width, 30; CID gain, 2.5 to 3.1; and CID time, 2.5 msec.
4. 1 μl of sample (isolated PA-N-glycan, approximately 10 pmol) is injected.

As an example, four isomers of mono-sialyl biantennary oligosaccharide, a sialic acid linked in an $\alpha2,3$ or $\alpha2,6$ manner on an $\alpha3$Man or $\alpha6$Man branch, were analyzed (Fig. 5). Their MS^1 spectra were almost the same, showing a predominant peak of $[M-2H]^{2-}$ ions (m/z1039.9) and small chloride-adduct ions (Fig. 6). The MS^2 spectra derived from $[M-2H]^{2-}$ ions (m/z1039.9) showed differences for the $\alpha3$Man or $\alpha6$Man branch (Fig. 7). The sialylated $\alpha3$Man branch was cleaved between $\alpha3$Man and βMan and showed $C_{4\beta}$ ions (m/z 835; nomenclature of Domon and Costello, 1988); the sialylated $\alpha6$Man branch showed relatively abundant D ions (m/z 979.3) derived from two bond cleavages, one between $\alpha3$Man and βMan and the other between βMan and GlcNAc. The $\alpha2,6$ or $\alpha2,3$ linkage of a sialic acid was determined from MS^3 spectra derived from sialyl fragment $C_{4\beta}$ ions (m/z 835) and D ions (m/z 979.3) of the MS^2spectra, which show characteristic differences in peak intensities for B_1 ions (NANA) (m/z 289.8) and B_3 ions (NANA-Gal-GlcNAc) (m/z 655.1) (Fig. 8). Specifically, B_3 ions (m/z 655.1) are more abundant than B_1 ions (m/z 289.8) in $\alpha2,6$-linked sialic acids, which is the opposite from what is seen in $\alpha2,3$-linked sialic acids.

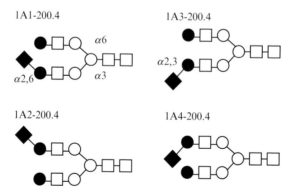

FIG. 5. Four isomers of monosialyl biantennary oligosaccharide formed by a combination of two sialic acid linkages ($\alpha2,3$ or $\beta2,6$) and two binding sites ($\alpha3$Man branch or $\alpha6$Man branch). Code numbers are nomenclature by Takahashi *et al.*, 1995. Other linkages are shown in Fig. 3. ◆, sialic acid (N-acetylneuraminic acid); ●, galactose; □, N-acetylglucosamine; ○, mannose.

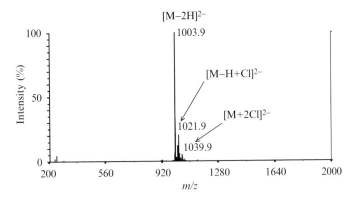

FIG. 6. MS^1 spectrum of 1A1-200.4. Other isomers also showed the same spectra. (Reproduced from ref Deguchi *et al.*, 2006 with permission from John Wiley & Sons Limited.)

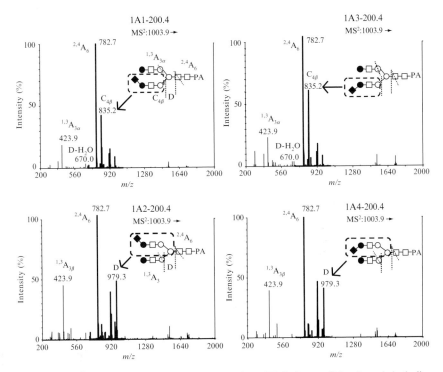

FIG. 7. MS^2 spectra of four isomers. D ions (m/z 979.3) from α6Man branch including βMan and $C_{4\beta}$ ions (m/z 835.2) from the α3Man branch are diagnostic ions used to distinguish sialic acid binding Gal. Fragmentation ions are named using the nomenclature of Domon and Costello, 1988. (Reproduced from Deguchi *et al.*, 2006, with permission from John Wiley & Sons, Linited.)

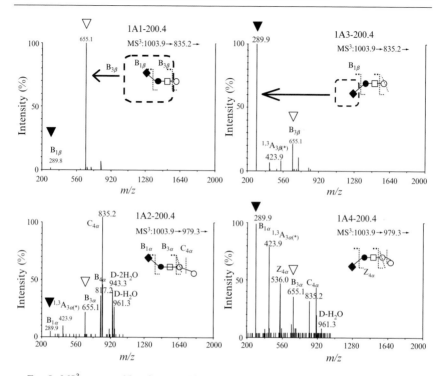

FIG. 8. MS^3 spectra of four isomers. Linkages of sialic acid are confirmed by comparison of B$_1$ ions (m/z 289.9, X ▼) and B$_3$ ions (m/z 655.1, ▼). ^{1,3}A$_3$(*) indicates a cross-ring cleavage of αMan and loss of sialic acid. (Reproduced from Deguchi *et al.*, 2006, with permission from John Wiley & Sons Limited.)

In this way, a detailed structure including sialic acid linkage and branch can be determined.

Acknowledgments

The author thanks Dr. K. Deguchi, Mr. Y. Takegawa, Dr. N. Takahashi, and Dr. S.-I. Nishimura for advice, and Ms. K. Okada and Ms. S. Kudo for technical assistance.

References

Altmann, F., Schweiszer, S., and Weber, C. (1995). Kinetic comparison of peptide: N-glycosidase F and A several differences in substrate specificity. *Glycoconjugate J.* **12,** 84–93.
Deguchi, K., Takegawa, Y., Ito, H., Miura, N., Yoshioka, S., Nagai, S., Nakagawa, H., and Nishimura, S-I. (2006). Structural assignment of isomeric 2-aminopyridine–derivativized

monosialylated biantennary N-linked oligosaccharide using negative-ion multistage tandem mass spectral matching. *Rapid Commun. Mass Spectrom.* **20**, 412–418.

Domon, B., and Costello, C. E. (1988). A systematic nomenclature for carbohydrate fragmentations in FAB-MS/MS spectra of glycoconjugates. *Glycoconjugate J.* **5**, 397–409.

Gu, J., Hiraga, T., and Wada, Y. (1994). Electrospray ionization mass spectrometry of pyridylaminated oligosaccharide derivatives: Sensitivity and in-source fragmentation. *Biol. Mass Spectrom.* **23**, 212–217.

Harvey, D. J. (2006). Proteomic analysis of glycosylation: Structural determination of N- and O-linked glycans by mass spectrometry. *Expert Rev. Proteomics* **2**, 87–101.

Hase, S. (1992). Analysis of sugar chains by pyridylamination. *In* "Glycoprotein Analysis in Biochemistry: Methods in Molecular Biology" E. F. Hounsell, ed.), volume 14, pp. 69–80. Humana Press, Totowa, NJ.

Hase, S., Ikenaka, T., and Matsushima, Y. (1978). Structure analysis of oligosaccharides by tagging of the reducing end sugar with a fluorescent compound. *Biochem. Biophys. Res. Commun.* **85**, 257–263.

Hase, S., Natsuka, S., Oku, H., and Ikenaka, T. (1987). Identification method for twelve oligomannose-type sugar chains thought to be processing intermediates of glycoproteins. *Anal. Biochem.* **167**, 321–326.

Kurogochi, M., and Nishimura, S.-I. (2004). Structural characterization of N-glycopeptides by matrix-dependent selective fragmentation of MALDI-TOF/TOF tandem mass spectrometry. *Anal. Chem.* **76**, 6097–6101.

Mechref, Y., and Novotny, M. V. (2002). Structural investigations of glycoconjugates at high sensitivity. *Chem. Rev.* **102**, 321–369.

Mizuno, Y., Sasagawa, T., Dohmae, N., and Takio, K. (1999). An automated interpretation of MALDI/TOF postsource decay spectra of oligosaccharides. 1. Automated peak assignment. *Anal. Chem.* **71**, 4764–4771.

Mizuochi, T. (1992). Microscale sequencing pf N-linked oligosaccharides of glycoproteins using hydrazinolysis, Bio-Gel P-4, and sequential exoglycosidase digestion. *In* "Glycoprotein Analysis in Biochemistry: Methods in Molecular Biology" E. F. Hounsell, ed.), volume 14, pp. 55–68. Humana Press, Totowa, NJ.

Nakagawa, H., Zheng, M., Hakomori, S.-I., Tsukamoto, Y., Kawamura, Y., and Takahashi, N. (1996). Detailed oligosaccharide structures of human integrin $\alpha 5 \beta 1$ analyzed by a three-dimensional mapping technique. *Eur. J. Biochem.* **237**, 76–85.

Okamoto, M., Takahashi, K-I., Doi, T., and Takimoto, Y. (1997). High-sensitivity detection and postsource decay of 2-aminopyridine–derivatized oligosaccharides with matrix-assisted laser desorption/ionization mass spectrometry. *Anal. Chem.* **69**, 2919–2926.

Plummer, T. H., Jr., Elder, J. H., Alexander, S., Phelan, A. W., and Tarentino, A. L. (1984). Demonstration of peptide: N-glycosidase F activity in endo-β-N-acetylglucosaminidase F preparations. *J. Biol. Chem.* **259**, 10700–10704.

Suzuki-Sawada, J., Umeda, Y., Kondo, A., and Kato, I. (1992). Analysis of oligosaccharide by on-line high-performance liquid chromatography and ion-spray mass spectrometry. *Anal. Biochem.* **207**, 203–207.

Takahashi, N. (1977). Demonstration of a new amidase acting on glycopeptides. *Biochem. Biophys. Res. Commun.* **76**, 1194–1201.

Takahashi, N., Nakagawa, H., Fujikawa, K., Kawamura, Y., and Tomiya, N. (1995). Three-dimensional elution mapping of pyridylaminated N-linked neutral and sialyl oligosaccharides. *Anal. Biochem.* **226**, 139–146.

Takahashi, N., and Tomiya, N. (1992). Analysis of *N*-linked oligosaccharides: application of glycoamidase A. *In* "CRC Handbook of Endoglycosidases and Glycoamidases" T. Muramatsu and N. Takahashi, eds.), pp. 199–332. CRC Press, Boca Raton, FL.

Takasaki, S., Mizuochi, T., and Kobata, A. (1982). Hydrazinolysis of asparagine-linked sugar chains to produce free oligosaccharides. *Methods Enzymol.* **83**, 263–268.

Takegawa, Y., Deguchi, K., Ito, S., Yoshioka, S., Nakagawa, H., and Nishimura, S.-I. (2005a). Simultaneous analysis of 2-aminopyridine–derivatized neutral and sialylated oligosaccharides from human serum in the negative-ion trap mass spectrometry. *Anal. Chem.* **77**, 2097–2106.

Takegawa, Y., Deguchi, K., Nakagawa, H., and Nishimura, S.-I. (2005b). Structural analysis of an *N*-glycan with "β1–4 bisecting branch" from human serum IgG by negative-ion MSn spectral matching and exoglycosidase digestion. *Anal. Chem.* **77**, 6062–6068.

Takegawa, Y., Deguchi, K., Ito, S., Yoshioka, S., Nakagawa, H., and Nishimura, S.-I. (2005c). Structural assignment of isomeric 2-aminopyridine–derivatized oligosaccharides using negative-ion MSn spectral matching. *Rapid Commun. Mass Spectrom.* **19**, 937–946.

Tomiya, N., Awaya, J., Kurono, M., Endo, S., Arata, Y., and Takahashi, N. (1988). Analyses of *N*-linked oligosaccharides using a two-dimensional mapping technique. *Anal. Biochem.* **171**, 73–90.

Tretter, V., Altmann, F., and März, L. (1991). Peptide-N^4-(*N*-acetyl-b-glucosaminyl) asparagine amidase F cannot release glycans with fucose attached α1-3 to the asparagine-linked *N*-acetylglucosamine residue. *Eur. J. Biochem.* **199**, 647–652.

Yagi, H., Takahashi, N., Yamaguchi, Y., Kimura, N., Uchimura, K., Kannagi, R., and Kato, K. (2005). Development of structural analysis of sulfated *N*-glycans by multidimensional high performance liquid chromatography mapping methods. *Glycobiology* **15**, 1051–1060.

Yamamoto, S., Hase, S., Fukuda, S., Sano, O., and Hase, S. (1989). Structures of the sugar chains of interferon-γ produced by human myelomonocyte cell line HBL-38. *J. Biochem.* **105**, 547–555.

Zaia, J. (2004). Mass spectrometry of oligosaccharides. *Mass Spectrom. Rev.* **23**, 161–227.

[7] Determination of Glycosylation Sites and Disulfide Bond Structures Using LC/ESI-MS/MS Analysis

By TEN-YANG YEN and BRUCE A. MACHER

Abstract

Significant progress has been made in discovering and cloning a host of proteins, including a range of glycoproteins. The availability of their predicted amino acid sequences provides useful information, including potential *N*-linked glycosylation sites. However, only a limited number of protein structures have been solved, and very little is known about the structures of membrane proteins. One of the important structural elements

METHODS IN ENZYMOLOGY, VOL. 415   0076-6879/06 $35.00
Copyright 2006, Elsevier Inc. All rights reserved.   DOI: 10.1016/S0076-6879(06)15007-7

of a protein is its disulfide bonds. These covalent bonds place conformational constraints on the overall protein structure, and thus, their identification provides important structural information.

A second important posttranslational modification found in proteins is N-linked glycosylation. Although potential sites of N-linked glycosylation can be predicted from a protein's primary sequence based on the presence of N-X-S/T sequences, not all of the predicted sites will be glycosylated. Therefore, N-linked glycosylation sites must be located by structural analysis.

We have developed a simple and sensitive method for determining the presence of free cysteine (Cys) residues and disulfide-bonded Cys residues, as well as the N-linked glycosylation sites in glycoproteins by liquid chromatography/electrospray ionization-tandem mass spectrometry (LC/ESI-MS/MS) in combination with protein database searching using the programs Sequest and Mascot. The details of our method are described in this chapter.

Overview

Although our knowledge of the three-dimensional structure of proteins has substantially increased over the past 20 years, there are still a limited set of structures available. Similarly, the structures of only a few glycosyltransferases have been solved. The availability of the known predicted primary sequences of glycosyltransferases, in conjunction with other previously obtained biochemical and chemical results, provided the opportunity to make predictions about the structural/functional importance of certain highly conserved amino acids (Drickamer, 1993; Martin *et al.*, 1997; Oriol *et al.*, 1999). For example, three of the members of the human α1,3/4-FT family (FT III, V, and VI) were shown to have highly homologous catalytic domain amino acid sequences, yet these enzymes were known to have significantly different acceptor substrate specificities. Results from studies of the determination of which of the nonhomologous amino acids among these FTs account for their differences in acceptor substrate specificity have led us to predict that the N- and C-termini of the catalytic domains of these glycosyltransferases were located in close spatial proximity. Specifically, we predicted that the highly conserved cysteine (Cys) residues in each of these FTs might be involved in one or two disulfide bonds that would facilitate the folding of the enzymes such that the N- and C-termini of the catalytic domains would be brought close together in space. To validate this hypothesis, we have developed a mass spectrometry (MS)-based methodology to determine which Cys residues in a protein are free and which are involved in disulfide bonds, as well as which potential

N-linked sites are glycosylated. Through several collaborations, we have applied this approach to a range of eukaryotic glycosyltransferases (Angata et al., 2001; de Vries et al., 2001; Holmes et al., 2000; Li et al., 2000; Shetterly et al., 2001; Yen et al., 2000, 2003). In this chapter, we present a detailed description of this methodology.

Materials

Dithiothreitol (DTT), iodoacetamide (IAM), N-ethylmaleimide (NEM), and urea were obtained from Sigma Chemical Co. (St. Louis, MO). Maleimide-biotin (M-biotin) was purchased from Pierce (Rockford, IL). Chymotrypsin and endoproteinase Glu-C sequence grade were purchased from Roche Applied Science (Indianapolis, IN). Trypsin was obtained from Promega (Madison, WI). Peptide:N-glycosidase (PNGase) F was purchased from ProZyme (San Leandro, CA). All other chemicals and solvents were either obtained from Sigma-Aldrich or Fisher Scientific (Santa Clara, CA) and were of the highest purity available.

Methods

The general strategy to determine which Cys residues in a protein are free, which participate in disulfide pairs, and which residues are involved in each disulfide pair, as well as which potential N-linked glycosylation sites, Asn-Xxx-Ser/Thr (Xxx ≠ Pro), are glycosylated is as follows: (1) alkylation of free Cys residues with an excess of M-biotin or NEM (protocol 1); (2) deglycosylation with PNGase F (protocol 2); (3) proteolytic digestion of the protein with trypsin, chymotrypsin, or endoproteinase Glu-C to obtain peptides that contain M-biotin or NEM-labeled Cys and disulfide bonds (protocol 3); (4) capillary liquid chromatography/electrospray ionization-tandem mass spectrometry (LC/ESI-MS/MS) analysis of the chromatographically separated peptides (protocol 4); (5) identification of peptides that contain the M-biotin or NEM-labeled Cys residues, and the N-linked glycosylation sites using protein database searching programs, Sequest and Mascot; (6) reduction of a fraction of the protein digest with DTT and alkylation with IAM to label the Cys residues of the peptides that are involved in disulfide bonds (protocol 5); and (7) search MS/MS spectra of the unreduced protein digest for the peptides that are involved in disulfide bonds.

The task of identifying the disulfide bonds in a protein can be very challenging because the number of possible disulfide linkages increases with the number of Cys residues that exist in the protein. For example, a protein with 5 cysteines only has 10 possible intramolecular disulfide bonds, whereas there are 28 possible intramolecular disulfide bonds for a protein with 8 cysteines. The identification of the free cysteines that are not

involved in disulfide bonds is a key component that can greatly simplify the complexity of possible disulfide matching. Once the free Cys residues are found, we need only focus on searching for those possible disulfide linkages which are produced by other cysteines.

Since free Cys residues can react with various chemicals during sample preparation for analysis, these residues should be alkylated with reagents such as IAM, iodoacetic acid, or 4-vinylpyridine before MS analysis (protocol 1). However, low-molecular-weight, Cys-containing peptides modified with these reagents are often eluted in the void volume of the reverse phase column and thus go undetected. This problem can often be overcome with the use of biotinylated alkylating reagents, such as M-biotin. Moreover, M-biotin and NEM are more reactive with free Cys residues than IAM when the pH of solution is about 7, and the M-biotin–labeled peptides yield a greater electrospray response than IAM-labeled peptides. We have also found that alkylation of the protein during denaturation with 8 M urea minimizes the occurrence of thiol-disulfide exchange (Yen et al., 2000).

After alkylation and denaturation, the labeled and denatured protein is deglycosylated with PNGase F (protocol 2) followed by proteolytic digestion with trypsin, chymotrypsin, or endoproteinase Glu-C to generate peptides (protocol 3). Peptide mixtures are analyzed by LC/ESI-MS/MS, and individual peptides are identified based on the information in the MS/MS spectrum (N-terminal b-type ions and C-terminal y-type ions) using the protein database searching programs, Mascot and Sequest (protocol 4).

The peptide mixtures are chromatographically resolved by capillary LC, and they form ions by electrospray in an ion-trap mass spectrometer. The most abundant peptide ions are detected and selected for fragmentation by collision-induced dissociation to generate an MS/MS spectrum. The peptide fragments are primarily generated by cleavage at the amide bond along the backbone of the peptide, and the charge is retained on the N-terminus (b-type ions) or C-terminus (y-type ions). The amino acid sequence is identified by correlating the fragmentation pattern of the peptide observed in the MS/MS spectrum with a predicted fragmentation pattern in the protein database. Based on the mass of the peptide and its MS/MS fragmentation pattern, the sequence of the peptide can be conclusively verified.

Identifying Cys-Containing Peptides and N-Linked Glycopeptides

NEM- or M-biotin–labeled, Cys-containing peptides can be identified from LC/ESI-MS/MS analyses of the protein digest by incorporating the differentially modified mass of 125 Da (NEM-labeled) or 525.2 Da (M-biotin–labeled) with the Cys residue into the protein database–searching programs, Mascot and Sequest. This setup allows one to simultaneously

monitor unlabeled and labeled Cys-containing peptides. When a Microcon device is used in sample preparation, we have found that the hydrated form of M-biotin is observed, and, thus, a mass of 543.2 Da for a hydrated form of the M-biotin–labeled Cys should be used when searching for alkylated, free Cys-containing peptides.

N-linked glycosylation sites in a glycoprotein can be identified using a combination of PNGase F treatment (see Note 4) and MS analysis. Before PNGase F treatment, peptides containing a glycosylated Asn-Xxx-Ser/Thr sequence are not detected because the mass of their glycan cannot be anticipated and thus cannot be included in the database search programs. However, after PNGase F treatment, peptides that contained a glycosylated Asn-Xxx-Ser/Thr sequence can be identified. The mass of an N-linked glycosylated peptide that has undergone cleavage of its glycan by PNGase F is one unit larger than expected for the peptide because the Asn (114 Da) residue at the site of oligosaccharide attachment is converted to an Asp (115 Da) (Angata et al., 2001; de Vries et al., 2001). One needs to carefully inspect the MS and MS/MS spectra to avoid possible false-positive identification of N-linked sites because the mass difference between Asn and Asp is within the range of the mass tolerance setting in the Mascot and Sequest programs.

Identifying Disulfide-Bonded Pairs of Peptides

After identifying free Cys residues in a protein, we recommend that a portion of the digested protein be reduced and alkylated (protocol 5). This produces a set of peptides in which the Cys residues that were engaged in disulfide bonds are now alkylated. Each individual peptide that is involved in a disulfide linkage is identified as an IAM-labeled, Cys-containing peptide with a mass increased by 57 Da compared with the unlabeled Cys-containing peptide.

Thus, protocol 5 also allows one to detect each Cys-containing peptide and provides information on whether the peptides produced are those that would be predicted based on the type of proteolytic enzyme used, or whether there have been any missed cleavages. This analysis is especially important when the proteolytic digestion is carried out with chymotrypsin, which is less specific than trypsin. It also is beneficial to limit the protein database for searching with Sequest to the sequence of the protein of interest. Finally, one should select the Sequest software option that does not designate a specific proteolytic enzyme constraint (i.e., no specific cleavage sites). In our experience, cleavage at Arg-C or Lys-C can occur in the chymotryptic peptides. Using the option to search with no proteolytic enzyme designated will allow Sequest to identify these nonspecific cleavages.

Once the sequences of Cys-containing peptides involved in disulfide linkages are known, the masses of the unlabeled Cys-containing peptides can be obtained from the MS-Digest program at the UCSF-Protein Prospector Web site (prospector.ucsf.edu) by inputting the corresponding sequences into the program. To identify every disulfide-bonded pair of peptides, a mass table of possible disulfide pairs is generated (Table I). For example, the masses of 6 individual Cys-peptide involved in disulfide linkages are equal to M1, M2, M3, M4, M5, and M6. Using a spreadsheet program, the possible masses of disulfide pairs can be calculated by summing the mass of each peptide, minus 2 Da (two S-H replaced by S-S) for each peptide pair as shown in Table I. The shaded part of the table designates masses for either potential intermolecular disulfide pairs or duplicates of intramolecular disulfide pairs found elsewhere in the table.

The next step is to analyze the data obtained from the LC/ESI-MS/MS analysis of an unreduced protein digest to check for multiply charged precursor ions that match the calculated mass of each possible disulfide pair. These ions are likely to be doubly or triply charged for pair of peptides between 1 and 4 kDa. When a match for one of the theoretically calculated masses of disulfide-linked peptide pairs is found, the fragments of the corresponding MS/MS spectrum are compared with the calculated fragments generated from the proposed disulfide-containing pair of peptides to confirm their sequences. The expected fragments of the potential disulfide-containing pair of peptides are calculated from fragments that can be generated from each individual peptide and fragments that would comprise parts from each of the peptides, assuming that the disulfide linkage is uncleaved (Yen $et\ al.$, 2000). Figure 1 shows an example of the MS/MS spectrum for a disulfide bonded pair of peptides (^{100}Cys-^{172}Cys) from C2-GnT-I after tryptic digestion followed by Glu-C digestion (Yen $et\ al.$, 2003). Fragments denoted B_n (n = 5–10) were

TABLE I

MASS TABLE OF POSSIBLE DISULFIDE PAIRS FOR A PROTEIN THAT CONTAINS 6 CYS RESIDUES[a]

Peptide Mass	M1	M2	M3	M4	M5	M6
M1	M1+M1−2	M1+M2−2	M1+M3−2	M1+M4−2	M1+M5−2	M1+M6−2
M2	M2+M1−2	M2+M2−2	M2+M3−2	M2+M4−2	M2+M5−2	M2+M6−2
M3	M3+M1−2	M3+M2−2	M3+M3−2	M3+M4−2	M3+M5−2	M3+M6−2
M4	M4+M1−2	M4+M2−2	M4+M3−2	M4+M4−2	M4+M5−2	M4+M6−2
M5	M5+M1−2	M5+M2−2	M5+M3−2	M5+M4−2	M5+M5−2	M5+M6−2
M6	M6+M1−2	M6+M2−2	M6+M3−2	M6+M4−2	M6+M5−2	M6+M6−2

[a] The shaded cells represent the nonredundant combinations of disulfide bonds.

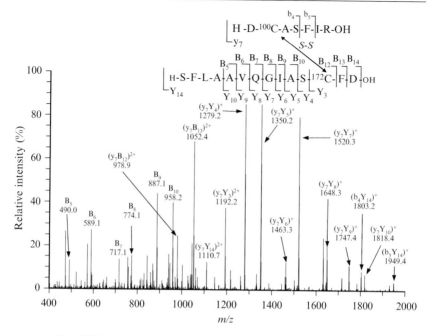

FIG. 1. The MS/MS spectrum of the disulfide-bonded pair of peptides between [100]Cys and [172]Cys of C2-GnT-I subjected to tryptic digestion followed by Glu-C digestion. The observed fragments were mostly from the N-terminal fragments of the [172]Cys-containing peptide, and the C-terminal fragments of the [172]Cys-containing peptide paired with the [100]Cys-containing peptide.

generated from the N-terminus of the [172]Cys-containing peptide, whereas fragments denoted (y_7Y_n, $n = 4$–10) were produced from the disulfide-bonded pair of peptides between the C-terminal fragments of the [172]Cys-containing peptide and the intact [100]Cys-containing peptide. We used either the MS-Digest program or the PepStat of the Sequest Browser to obtain the calculated fragments using the sequence of the peptide of interest as the input information. In some cases, the disulfide pair of peptides can be masked by other co-eluting peptides that have more intense ion signals, and thus their MS/MS spectra would not be detected. To identify these latter peptides, one needs to reinject the nonreduced protein digest for another LC/ESI-MS/MS analysis using the data-dependent scan method, in which the MS/MS scans are only performed for a set of precursor ions that match the masses of disulfide-bonded pairs of peptides. In many cases, the complete set of disulfide-bonded pairs can only be detected from a combination of tryptic and chymotryptic digests, or tryptic and Glu-C digests.

Alkylation of Free Cys Residues (Protocol 1)

1. Prepare the protein of interest (see Note 1) in Tris buffer (0.1 M Tris-HCl; pH, 7–7.5; 1 mM EDTA and 1 mM NaN$_3$).

2. Alkylate the protein with 2- to 20-fold molar excess of either M-biotin or NEM to block potential free Cys residues in the protein followed by denaturation with 8 M urea. The total volume of the mixture should be kept as low as possible.

3. Incubate the mixture in the dark for 1 h at room temperature.

4. Transfer the sample to a Microcon YM-30 (Millipore) unit and centrifuge at 8000 × g until most of the solvent has passed through the filter to reduce the amount of reagent in the protein sample (see Note 2).

5. Add 250 μl of 50 mM ammonium bicarbonate solution (AB buffer) to the Microcon unit and centrifuge at 8000 × g until most of the solvent has passed through the filter. Repeat this washing process once.

6. Redissolve the sample by adding 60 μl of AB buffer in the Microcon unit before carrying out the proteolytic digestion (see Note 3).

Peptide: N-Glycosidase (PNGase) F Digestion (Protocol 2) (See Note 4)

1. Dilute the PNGase F solution from the manufacture (40 μl of 20 mM Tris-HCl buffer; 50 mM NaCl; 1 mM EDTA; pH, 7.5) 5-fold with 160 μl of AB buffer to a concentration of 0.5 unit/ml.

2. Add 2 μl of this preparation to the Microcon unit.

3. Seal the Microcon unit with Parafilm and incubate overnight at 37° (see Note 5).

Digestion with Trypsin, Chymotrypsin, or Endoproteinase
    Glu-C (Protocol 3)

1. Prepare trypsin, chymotrypsin, or endoproteinase Glu-C in AB buffer at a concentration of 50 ng/μl.

2. Remove the Parafilm from the Microcon unit, and add trypsin, chymotrypsin, or endoproteinase Glu-C solution to the sample mixture (the PNGase F digest after protocol 2, or the protein in AB buffer after protocol 1) to obtain a ratio of 1/20 to 1/40 ratio (w/w) of the proteolytic enzyme to protein.

3. Seal the Microcon unit with Parafilm, and incubate overnight at 37°.

4. Remove the Parafilm from the Microcon unit, and use a pipette with a gel-loading pipette tip to recover the protein digest solution, and place in a 0.5 ml microcentrifuge tube (see Note 6).

5. Add 20 μl water to the Microcon unit, recover the solution, and transfer it into a microcentrifuge tube that contains the protein digest.

Identification of Free Cys-Containing and Disulfide-Bonded Peptides by LC/ESI-MS/MS (Protocol 4)

1. Use a capillary liquid chromatography system such as a MicroLC system (Micro-Tech Scientific; Vista, CA) and an LC/ESI-MS/MS system such as a Thermo Finnigan LCQ ion-trap mass spectrometer (San Jose, CA).

2. Use a capillary C18 column (100 mm × 75 μm; Nucleosil, 5-μm particle size) with a pre-etched 15-μm tip (Uncoated Fused-Silica PicoFrit Tips from New Objective; Woburn, MA). Set the LC flow rate at 0.3 μL/min and use a 5-μl sample loop to conduct the high-performance liquid chromatography (HPLC) analysis. Inject 3 to 5 μL ($\sim$0.5 to 0.8 μg) of the protein digest solution into the capillary HPLC system. To minimize the void volume, use a glass capillary (50 μm internal diameter (i.d.); 355 μm, outer diameter (o.d.)) for the LC tubing.

3. Use a three-step linear gradient of 5% to 10% B during the first 10 min, followed by a 10% to 35% B gradient during the next 40 min, and 35% to 50% B during the final 10 min to separate the enzymatically digested peptides. Mobile phase A is 0.1% HCOOH/water, and mobile phase B is 0.1% HCOOH in acetonitrile.

4. Use an automatic LC/ESI-MS/MS data-acquisition system to collect ion signals from the eluted peptides using the data-dependent scan procedure with a cyclic series of three different scan modes: full scan, zoom scan, and MS/MS scan (see Note 7).

5. A positive voltage of 1.8 to 2.2 kV is applied to the electrospray setup, a micro-tee (Upchurch; Oak Harbor, WA); one end is connected with a 50 μm (i.d.) glass capillary to the MicroLC system; the vertical end is connect to a platinum wire for high voltage, and the other end is connect to a capillary C18 column. The temperature of the stainless steel heating capillary of the LCQ is maintained at 220°. The voltages at the exit end of the heating capillary and the tube lens are held at 17 V and 3 V, respectively, to minimize source-induced dissociation and optimize the ESI signal of the peptides. Ion injection is controlled by the automatic gain control setting to avoid space charge effects. The full scan mass spectrum is acquired from $m/z = 300$ to $m/z = 2000$. MS/MS analysis is carried out with a relative collision energy of 38%.

Complete Reduction with DTT and Alkylation with IAM (Protocol 5)

1. Completely reduce a fraction of the protein digest (5 to 10 μl or 0.5 to 1.5 μg) using 35 mM of DTT in AB buffer at 55° or less for 25 min, followed by alkylation with a 2.5-fold molar excess of IAM versus DTT at room temperature in the dark (see Note 8).

Notes

1. The minimum and ideal amounts of protein required for the analysis are 5 μg and 30 μg or more, respectively. The purity of the protein needs to be 70% or higher. A capillary HPLC system and a mass spectrometer with MS/MS capability are required to perform LC/ESI-MS/MS analysis.

2. Do not use the Microcon filtration device when the amount of protein is less than 10 μg because it is difficult to recover the protein. Instead, directly dilute the alkylated and denatured protein solution to reduce the concentration to less than 2 M of urea, and skip steps 4, 5, and 6. It is preferable to keep the concentration of proteins as high as possible. Concentrate the sample using a SpeedVac as needed.

3. Keep the protein inside the Microcon tube during the proteolytic digestion to prevent sample loss. Ammonium bicarbonate (50 mM) buffer is preferred for use in the protein digest because it is compatible with LC/MS analysis.

4. PNGase F treatment can also be done after protocol 3 is completed.

5. Sealing the tube prevents the sample from drying.

6. Keep the gel-loading pipette tip away from the membrane in the bottom of Microcon unit.

7. Set the MS control program to detect the most abundant ion with the threshold ion intensity greater than the background, and include the dynamic exclusion feature to exclude the redundancy of the few dominant ions. We set a dynamic exclusion in a time window of 2 min to exclude the same m/z ions for MS/MS analysis.

8. Measure the pH of the solution using pH paper. Add a basic solution such as 5% ammonia to increase the pH of the sample if the pH of solution is less than 7; otherwise, the alkylation will be incomplete.

Acknowledgments

This work was supported in part by funding from a grant (P20 MD000262) from the Research Infrastructure in Minority Institutions Program, National Center on Minority Health and Health Disparities, NIH.

References

Angata, K., Yen, T.-Y., El-Battari, A., Macher, B. A., and Fukuda, M. (2001). Unique disulfide bond structures found in ST8Sia IV polysialyltransferase are required for its activity. *J. Biol. Chem.* **276,** 15369–15377.
de Vries, T., Yen, T.-Y., Joshi, R. K., Storm, J., van den Eijnden, D. H., Knegtel, R. M. A., Bunschoten, H., Joziasse, D. H., and Macher, B. A. (2001). Neighboring cysteine residues in human Fucosyltransferase VII are engaged in disulfide bridges, forming small loop

structures: A proposed 3D model based on location of cysteines, and threading and homology modeling. *Glycobiology* **11**, 423–432.

Drickamer, K. (1993). A conserved disulphide bond in sialyltransferases. *Glycobiology* **3**, 2–3.

Holmes, E. H., Yen, T.-Y., Thomas, S., Joshi, R., Nguyen, A., Long, T., Gallet, F., Maftah, A., Julien, R., and Macher, B. A. (2000). Human α1,3/4 Fucosyltransferases: Characterization of highly conserved cysteine residues and N-linked glycosylation sites. *J. Biol. Chem.* **275**, 24237–24245.

Li, J., Yen, T.-Y., Allende, M. L., Joshi, R. K., Cai, J., Pierce, W. M., Jaskiewicz, E., Darling, D. S., Macher, B. A., and Young, W. W., Jr. (2000). Disulfide bonds of GM2 synthase homodimers: Antiparallel orientation of the catalytic domain. *J. Biol. Chem.* **275**, 41476–41486.

Martin, S. L., Edbrooke, M. R., Hodgman, T. C., van den Eijnden, D. H., and Bird, M. I. (1997). Lewis X biosynthesis in *Helicobacter pylori:* Molecular cloning of an α (1,3)-fucosyltransferase gene. *J. Biol. Chem.* **272**, 21349–21356.

Oriol, R., Mollicone, R., Cailleay, A., Balanzino, L, and Breton, C. (1999). Divergent evolution of fucosyltransferase genes from vertebrates, invertebrates, and bacteria. *Glycobiology* **9**, 323–334.

Shetterly, S., Tom, I., Yen, T.-Y., Joshi, R., Lee, L., Wang, P. G., and Macher, B. A. (2001). α1,3 galactosyltransferase: New sequences and characterization of conserved cysteine residues. *Glycobiology* **11**, 645–653.

Yen, T.-Y., Joshi, R. K., Yan, H., Seto, N. O. L., Palcic, M. M., and Macher, B. A. (2000). Characterization of cysteine residues and disulfide bonds in proteins by liquid chromatography/electrospray ionization tandem mass spectrometry. *J. Mass Spectrom.* **35**, 990–1002.

Yen, T.-Y., Macher, B. A., Bryson, S., Chang, X., Tvaroška, I., Tse, R., Takeshita, S., Lew, A. M., and Datti, A. (2003). Highly Conserved Cysteines of Mouse Core 2 β1,6 N-Acetylglucosaminyltransferase-I Form a Network of Disulfide Bonds and Include a Thiol That Affects Enzyme Activity. *J. Biol. Chem.* **278**, 45864–45881.

[8] Identification of O-GlcNAc Sites on Proteins

By Stephen A. Whelan and Gerald W. Hart

Abstract

O-linked N-acetylglucosamine (O-GlcNAc) is a monosaccharide post-translational modification that modifies serine/threonine residues of nucleocytoplasmic proteins in metazoans. O-GlcNAc, like phosphorylation, is dynamic and responsive to numerous stimuli in diverse regulatory pathways. O-GlcNAc may also be found adjacent to or at the same sites as phosphorylation, demonstrating the potential for a reciprocal function on some of these proteins. Like most posttranslational modifications, O-GlcNAc is substoichiometric and may be found at multiple sites with other posttranslational modifications present. Additionally, there is no consensus sequence defining the addition of O-GlcNAc to the peptide

METHODS IN ENZYMOLOGY, VOL. 415   0076-6879/06 $35.00
      DOI: 10.1016/S0076-6879(06)15008-9

backbone, further complicating identification and site mapping. This chapter describes several strategies to confirm that proteins are O-GlcNAc modified and provide subsequent determination of O-GlcNAc attachment sites. We have listed the strengths and limitations of each protocol to allow readers to decide which suits their system and availability of resources. These protocols include galactosyltransferase labeling, immunoblotting, using mass spectrometry based on β-elimination followed by Michael addition with dithiothreitol, and chemoenzymatic labeling, enrichment, and detection.

Introduction

Since the discovery of O-linked N-acetylglucosamine (O-GlcNAc) more than 22 years ago (Torres and Hart, 1984), a number of techniques and tools have been developed for the detection and study of O-GlcNAc (Comer et al., 2001; Greis and Hart, 1998; Khidekel et al., 2003; Roquemore et al., 1994; Vocadlo et al., 2003; Wells et al., 2002; Zachara et al., 2004a). These tools have led to the discovery of many proteins modified by O-GlcNAc, and in some cases they have led to the site-specific mapping and the role O-GlcNAc plays in a proteins function (Khidekel et al., 2003; Whelan and Hart, 2003; Zachara and Hart, 2004). O-GlcNAc is a serine/threonine-directed single-sugar posttranslational modification, similar to phosphorylation, and appears to be as abundant as phosphorylation on nucleocytoplasmic proteins (Roquemore et al., 1994). The hexosamine biosynthetic pathway converts 2% to 5% of glucose to UDP-GlcNAc (McClain, 2002), which is used by O-GlcNAc transferase (OGT) to modify proteins with O-GlcNAc (Haltiwanger et al., 1992). OGT itself is necessary for life at the single-cell level (Shafi et al., 2000). O-GlcNAc β-N-acetylglucosaminidase (O-GlcNAcase) is the enzyme that removes O-GlcNAc from proteins similar to the removal of phosphorylation by phosphatases (Dong and Hart, 1994). O-GlcNAc has also been shown to act reciprocal to phosphorylation at adjacent and same sites (Cheng and Hart, 2001; Chou et al., 1995). O-GlcNAc is involved in a number of cellular functions including transcription, cell cycle progression, cancer, stress, diabetes, metabolic sensing, and nutrient status (Slawson and Hart, 2003; Slawson et al., 2001; Wells et al., 2003; Zachara et al., 2004b). Site mapping the location of O-GlcNAc within the protein has been addressed by several methods to help elucidate the role of the posttranslational modification. The detection and site mapping of O-GlcNAc are more difficult than mapping phosphorylation and may require a combination of methods including galactosyltransferase labeling, immunoblotting, mass spectrometry identification of β-eliminated followed by Michael addition with dithiothreitol, and chemoenzymatic

labeling, enrichment, and detection. As with most methods, however, there are limitations, and several laboratories are working on improving methods for the detection and study of *O*-GlcNAc.

Immunoblot Detection of *O*-GlcNAc

Immunoblot detection of *O*-GlcNAc may be performed with several monoclonal antibodies including C-terminal domain of pol II (CTD) 110.6 (Comer *et al.*, 2001), RL-2 (Holt *et al.*, 1987; Snow *et al.*, 1987), HGAC85 (Turner *et al.*, 1990), and MY95 (Matsuoka *et al.*, 2002). Although these monoclonal antibodies simplify the *O*-GlcNAc detection process, each is raised to a specific *O*-GlcNAc–dependent epitope and only recognizes a subset of *O*-GlcNAc–modified proteins. These antibodies also overlap in the recognition of similar proteins. Therefore, multiple immunoblot analysis with several antibodies is usually necessary and recommended to determine whether the proteins of interest are modified with *O*-GlcNAc.

In this method, we are highlighting the use of CTD 110.6 antibody raised against the *O*-GlcNAc–modified C-terminal domain of the RNA polymerase II large subunit, but we also use RL-2 raised against the epitope of an *O*-GlcNAc–modified nuclear pore (NP62) protein. CTD 110.6 antibody is the least dependent on protein structure of all of the available antibodies and recognizes the widest range of *O*-GlcNAcylated proteins. In addition, succinylated wheat-germ agglutinin (sWGA) is also used to lectin blot for *O*-GlcNAc–modified proteins. However, sWGA recognizes any terminal β-GlcNAc residue; therefore, the samples must first be treated with peptide:N-glycosidase F (PNGase F) to remove the N-linked sugars. All immunoblots are performed with specific controls that include competing the antibody signal away with 100 mM free GlcNAc and performing electrophoresis with BSA-GlcNAc as a positive control and ovalbumin as a negative control.

Materials

1. Control peptides: Ovalbumin (Sigma-Aldrich), a protein modified by GlcNAc-terminating N-linked oligosaccharide, which is used as a negative control for immunoblotting with CTD 110.6 (use 100 ng) and a positive control for sWGA (use 100 ng) and galactosyltransferase labeling (use 2 μg).

2. Bovine serum albumin (BSA, Sigma-Aldrich), a nonglycosylated protein that can be used as a negative control.

3. BSA-GlcNAc (Sigma-Aldrich), a protein chemically modified to contain GlcNAc residues, which is used as a positive control for immunoblotting with CTD 110.6, RL-2, and sWGA (use 1 to 5 ng).

4. Purified or crude protein separated by sodium dodecyl sulfate polyacrylamide gel electrophoresis (SDS-PAGE) and electroblotted to polyvinylidene difluoride (PVDF). Protein samples may be in a modified Radioimmunoprecipitation (RIPA)-type buffer of the researcher's choice but should contain protease inhibitors, including the protease inhibitor cocktails 1 and 2 (PIC1: 1 mg/ml leupeptin, 2 mg/ml antipain, 10 mg/ml benzamide dissolved in 10,000 units/ml aprotinin and PIC2: 1 mg/ml chemostatin and 2 mg/ml pepstatin dissolved in dimethyl sulfoxide) and *O*-GlcNAcase inhibitors such as *N*-acetylglucosamine or O-(2-Acetamido-2-deoxy-D-gluropyranosylidene) amino N-phenyl carbamate (PUGNAc).

5. Tris-HCl–buffered saline with tween-20 (TBST): 10 mM Tris-HCl (pH, 7.5), 150 mM NaCl, 0.1% (v/v) Tween-20.

6. Blocking buffer: TBST, 3% (w/v) bovine serum albumin (Sigma-Aldrich, St. Louis, MO).

7. CTD 110.6 (Covance, Richmond, CA) ascites diluted 1/2500, sWGA–horseradish peroxidase (HRP) diluted 0.1 μg/ml (EY Labs, SanMateo, CA), or RL-2 diluted 1/4000 (Affinity BioReagents, Golden, CO) in blocking buffer.

8. Anti-Mouse immunoglobulin (Ig) M-HRP (Sigma-Aldrich, St Louis, MO) diluted 1/5000 in blocking buffer.

9. Enhanced chemiluminescence (Amersham Biosciences, Piscataway, NJ).

10. Primary antibody competition: 10 mM GlcNAc.

Procedure

1. Proteins of interest are blotted onto PVDF. Twenty to thirty micrograms of total cell extract and 25 to 50 ng of a purified neoglycoconjugate is sufficient for immunoblotting (Comer *et al.*, 2001).

2. Block blots with blocking buffer for 60 min at room temperature.

3. Incubate blots with CTD 110.6 (RL-2 and sWGA may also be used) in blocking buffer with and without the 100 mM GlcNAc overnight at 4° while shaking. RL-2 is used under similar conditions as CTD 110.6, except at a concentration of 1/4000 and with secondary antibody anti-mouse IgG-HRP diluted 1/5000 (Amersham Biosciences, Piscataway, NJ). sWGA immunoblotting is also conducted under similar conditions as CTD 110.6, except sWGA-HRP is used at 0.1 μg/ml overnight at 4° and requires 6×10-min washes in high-salt TBST (1 M NaCl). The sWGA antibody competition assay requires 1 M GlcNAc. Milk cannot be used as the blocking agent because sWGA reacts with the many milk proteins modified by glycans.

4. Wash blots in TBST 3×10 min.

5. Incubate blots with corresponding secondary antibody in blocking buffer for 60 min.

6. Wash blots in TBST 5 × 10 min.

7. Develop the HRP reaction. The CTD 110.6 *O*-GlcNAc antibody often cross-reacts with pre-stained molecular weight markers.

Site Mapping by Galactosyltransferase Labeling and Edman Degradation

Researchers Zachara and Gooley (2000) have shown that glycosylated amino acids can be recovered and characterized during automated Edman degradation (minimum of 20 pmol of starting material, where at least 20% of the sample is glycosylated). In the past, we have used manual Edman degradation, but currently we are optimizing automated Edman degradation of *O*-GlcNAc–modified peptides. Edman degradation is used to site map reverse-phase high-performance liquid chromatography (HPLC) purified peptides that are labeled with galactosyltransferase ([³H]Gal). Edman degradation allows for the differentiation of isobaric masses of sugar modifications on peptides. Since Edman degradation requires purified peptides for amino acid sequencing, the starting material should be from single purified protein or simple mixtures so that there are no co-elution of peptides in the HPLC fractions. If the mixture is too complex for HPLC resolution, the sample may need to be pre-fractionated by another means, such as SDS-PAGE. The labeled proteins can be digested in solution or in gel according to the manufacturer's suggestions (Promega, Madison, WI). HPLC fractions that are identified to contain radioactive-labeled *O*-GlcNAc peptides via liquid scintillation counting are covalently attached to arylamine derivatized PVDF disks via the activation of peptide carboxyl groups using water-soluble *N*-ethyl-*N'*-dimethylaminopropyl carbodiimide (EDC; Kelly *et al.*, 1993). Both the C-terminus and any acidic residue will couple to the membrane, yielding blank cycles.

Autogalactosylation of Galactosyltransferase

As galactosyltransferase is modified with *N*-linked glycans, it is necessary to block these before using this enzyme to probe other proteins for terminal GlcNAc.

Materials

1. 10× galactosyltransferase buffer: 100 m*M* 4-(2-hydroxyethyl)-1-piperazineethanesulfonic acid (HEPES) (pH, 7.4), 100 m*M* Gal and 50 m*M* MnCl$_2$.

2. Galactosyltransferase storage buffer: 2.5 mM HEPES (pH, 7.4), 2.5 mM MnCl$_2$, 50% (v/v) glycerol.
3. Saturated ammonium sulfate, more than 17.4 g (NH$_4$)$_2$SO$_4$ in 25 ml Milli-Q purified water.
4. 85% ammonium sulfate 14 g (NH$_4$)$_2$SO$_4$ in 25 ml Milli-Q purified water.
5. 25 mM 5'-AMP, in Milli-Q purified water (pH, 7.0).
6. Aprotinin, β-mercaptoethanol, UDP-Gal, 30- to 50-ml centrifuge tubes.

Procedure

1. Suspend 25 U of galactosyltransferase (Sigma-Aldrich) in 1 ml of 1× galactosyltransferase buffer.
2. Transfer sample to 30- to 50-ml centrifuge tube. The centrifuge tubes should withstand a centrifugal force of 15,000 × g.
3. Remove a 5-μl aliquot for an activity assay.
4. Add 10 μl of aprotinin, 3.5 μ. of β-mercaptoethanol, and 1.5 to 3.0 mg of UDP-Gal.
5. Incubate the sample on ice for 30 to 60 min.
6. Add 5.66 ml of pre-chilled saturated ammonium sulfate in a dropwise manner, and incubate on ice for 30 min.
7. Centrifuge at more than 10,000 × g for 15 min at 4°; pour off supernatant.
8. Suspend pellet in 5 ml of cold 85% ammonium sulfate, and incubate on ice for 30 min.
9. Centrifuge at more than 10,000 × g for 15 min at 4°; pour off supernatant.
10. Suspend pellet in 1 ml of galactosyltransferase storage buffer.
11. Aliquot enzyme (50 μl).
12. Assay 5 μl of the autogalactosylated and non-galactosylated against a known substrate to determine the specific activity.
13. Store at −20° for up to 1 year.

Galactosyltransferase Labeling

Protein samples must first be denatured with 10 mM dithiothreitol (DTT) and up to 0.5% (w/v) SDS before labeling with galactosyltransferase. It is necessary to titrate out the concentration of SDS by 10× more NP-40 (v/v) or up to 5% (w/v). Galactosyltransferase is also active in solutions containing 5 mM DTT, 0.5 M NaCl, up to 2% (v/v) Triton-X 100, up to 2% (v/v) NP-40, and 1 M urea. Digitonin should be used with care because it is a substrate for galactosyltransferase. The protein sample must then be

diluted to less than 2% NP-40. In addition, galactosyltransferase requires 1 to 5 mM Mn^{2+} for activity, but it is inhibited by concentrations greater than 20 mM and is inhibited by Mg^{2+}. Once protein samples are denatured, all N-linked sugars must be removed with PNGase F. After proteins are labeled, they are separated by SDS-PAGE and undergo trypsin in gel digestion; the peptides are separated by HPLC, and pure peptide radioactive fractions are ready for site mapping by Edman degradation. Alternatively, labeled protein separated by SDS-PAGE may be detected by autoradiography. This method has been the gold standard for identifying O-GlcNAc–modified proteins for more than 20 years.

Materials

1. Protein sample(s).
2. Buffer H: 50 mM HEPES (pH, 6.8), 50 mM NaCl, 2% (v/v) Triton-X 100.
3. 10× labeling buffer: 100 mM HEPES (pH, 7.5), 100 mM Gal, 50 mM MnCl$_2$.
4. 25 mM 5′-adenosine monophosphate (5′-AMP), in Milli-Q water (pH, 7.0).
5. UDP-[^{3}H]Gal, 1.0 mCi/ml (specific activity, 17.6 Ci/mmol) in 70% (v/v) ethanol.
6. UDP-Gal (no label).
7. Stop solution: 10% (w/v) SDS, 0.1 M ethylenediamine tetraacetic acid (EDTA).
8. Desalting column, Sephadex G-50 (30 × 1cm) equilibrated in 50 mM ammonium formate, 0.1% (w/v) SDS.

Procedure

1. Remove solvent from UDP-[^{3}H]Gal label in a SpeedVac or under a stream of nitrogen, and dry at approximately 1 to 2 μCi/reaction. Ethanol can inhibit the galactosyltransferase reaction, but if less than 4 μl is required, the label can be added directly to the reaction (final reaction volume, 500 μl).

2. Suspend the label in 25 mM 5′-AMP, 50 μl per reaction. The AMP is included to inhibit possible phosphodiesterase reactions, which might compete for label during the experiment.

3. Set up reactions as follows: sample (final concentration, 0.5 to 5 mg/ml from 1–50 μl), buffer H 350 μl, 10× labeling buffer 50 μl, UDP-[^{3}H]Gal/ 5′-AMP 50 μl, galactosyltransferase 30 to 50 U/ml up to 2 to 5 μl, calf intestinal alkaline phosphatase 1 to 4 U, Milli-Q water, to a final volume of

500 μl. Free UDP is also an inhibitor, and for studies where complete labeling of the GlcNAc is preferable, such as site mapping, calf intestinal alkaline phosphatase is included in the reaction to degrade UDP.

4. Labeling is typically performed at 37° for 2 h or at 4° overnight. Gal-transferase labeling can be performed in conjunction with immunoprecipitation. However, immunoglobulin contains large amounts of GlcNAc-terminating N-linked sugars, which will be preferentially labeled over O-GlcNAc sites. So the cell extract should be labeled first, and then the protein of interest should be immunoprecipitated.

5. Add cold UDP-Gal to a final concentration of 0.5 to 1.0 mM and another 2 to 5 μl of galactosyltransferase. When site mapping is performed, it is necessary to completely label GlcNAc. Therefore, the reactions are chased with unlabeled UDP-Gal and fresh galactosyltransferase.

6. Add 50 μl of stop solution to each sample, and heat to 100° for 5 min.

7. Resolve the protein from unincorporated label using a Sephadex G-50 column (30 × 1cm, equilibrated in 50 mM ammonium formate, 0.1% (w/v) SDS), collect 1-ml fractions. We recommend the use of Sephadex G-50 to desalt samples. You may also use TCA precipitation, spin filtration/buffer exchange, or other forms of size exclusion chromatography (e.g., Pharmacia PD10 desalting column). We advise the addition of carrier proteins such as BSA ($\sim$67 kDa) and Cytochrome C ($\sim$12.5 kDa) to samples and buffers to reduce the amount of protein lost due to nonspecific protein adsorption.

8. Count an aliquot (50 μL) of each fraction using a liquid scintillation counter.

9. Approximately 2 × 10^6 DPM of ^{3}H-Gal should be incorporated into 2 μg of ovalbumin.

10. Combine the fractions containing the void volume, and lyophilize the sample to dryness.

11. Suspend the samples in 100 to 1000 μL) Milli-Q water and acetone precipitate.

12. Add 3 to 5 volumes of cold acetone ($-20°$) to the sample. Incubate for 2 to 18 h at $-20°$.

13. Pellet protein at 4° for 10 min at 15,000 × g.

14. Samples can be treated with PNGase F to remove all label incorporated into N-linked carbohydrates (see next section).

15. Samples may then be separated by SDS-PAGE or two-dimensional gel electrophoresis and detected by autoradiography, and/or they are now ready for trypsin digest and manual/automated Edman degradation and characterization of sugar to prove it is O-GlcNAc.

Control Experiments for Galactosyltransferase Labeling

Removing N-*Linked Sugars*

Digestion of sample and control proteins with PNGase F is a useful way to show that reactivity with lectins and galactosyltransferase is not due to *N*-linked glycans. PNGase F is distinct from endoglycosidase F, which cleaves only a subset of *N*-linked sugars. In addition, PNGase F will not cleave *N*-linked sugars with a core α1–3fucose or *N*-linked sugars at the N- or C-terminus of a protein or peptide. A positive control such as ovalbumin should be included for the PNGase F reaction. Ovalbumin, which contains one *N*-linked glycosylation site, will increase in mobility by several kilodaltons on a gel (10% to 12% SDS-PAGE) after treatment with PNGase F. This mobility shift is difficult to detect on a 7.5% SDS-PAGE gel.

Materials

1. PNGase F (New England BioLabs Inc.).
2. PNGase F denaturing buffer: 5% (w/v) SDS, 10% (v/v) β-mercaptoethanol in 50 mM sodium phosphate buffer (pH, 7.5).
3. PNGase F reaction buffer: 500 mM sodium phosphate buffer (pH, 7.5)
4. 10% (v/v) NP-40.

Procedure

1. Add 1/10 sample volume of 10× PNGase F denaturing buffer to each sample, and heat to 100° for 10 min. Protein samples may contain protease inhibitor cocktails 1 and 2 (PIC 1 and PIC 2).
2. Add 1/10 sample volume of 10× PNGase F reaction buffer and 10% (v/v) NP-40, and mix. PNGase F is inhibited by SDS, so it is essential to add 10% (v/v) NP-40 to the reaction mixture.
3. Add 1 μl of PNGase F, and incubate samples at 37° for 1 h to overnight.

Removing O-*GlcNAc by Hexosaminidase Digestion*

Terminal β-GlcNAc and O-GlcNAc can be removed from proteins using commercial hexosaminidases, and these enzymes will also cleave terminal β-GalNAc residues. Unlike O-GlcNAcase, commercial hexosaminidases have low pH optima—typically 4.0 to 5.0.

Materials

1. N-acetyl-β-D-glucosaminidase, from jack bean (V-Labs Inc., Covington, LA).
2. 2% (w/v) SDS.
3. 2× reaction mixture: 80 mM citrate-phosphate buffer (pH, 4.0), 1 U N-acetyl-β-D-glucosaminidase (V-labs), 8% (v/v) Triton X-100 or (v/v) NP-40, 0.01 U aprotinin, 1 μg of leupeptin, 1 μg α_2-macroglobulin.

Procedure

1. Add protein samples (include a positive control such as ovalbumin).
2. Mix sample 1:1 with 2% (w/v) SDS, and boil for 5 min.
3. Mix sample 1:1 with reaction mixture, and incubate at 37° for 4 to 24 h.

Edman Degradation

Materials

1. Sequelon-AA Reagent Kit (Millipore Cat# Gen920033, Billerica, MA) contains Mylar sheets, carbodiimide, and coupling buffer.
2. Coupling reagent: 1 mg carbodiimide, 100 μl coupling buffer (supplied).
3. Sequencing reagent: 7 ml methanol (HPLC grade), 1 ml triethylamine (sequencing grade), 1l phenylisothiocyanate (sequencing grade), 1 ml Milli-Q water.
4. Stable at 4° for 24 h: methanol (HPLC grade), CH_3CN (HPLC grade), trifluoroacetic acid (TFA; sequencing grade).
5. 100 mM Tris-HCl (pH, 7.4).
6. Sample peptide and control peptide.
7. Screw-capped polypropylene tubes.

Procedure for Coupling the Sample the PVDF Disk for Manual Edman Degradation

1. Suspend peptide in 10% to 30% (v/v) CH_3CN and up to 0.1% (v/v) TFA.
2. Place disk on a Mylar sheet on a heating block at 55°.
3. Wet disk with 10 μl of methanol; allow excess methanol to evaporate.
4. Apply sample in 10-μl aliquots, allowing the membrane to come to near dryness between each aliquot.

5. Dry the membrane after all of the sample has been applied.

6. Mix approximately 1 mg of EDC in 100 μl of coupling reagent, and carefully pipet approximately 50 μl on each disk.

7. Incubate samples at 4° for 30 min. Incubation of the sample at 4° increases the yield, and the membrane should not be incubated more than 30 min.

8. Save coupling reagent.

9. Wash the membrane alternately 3× in 1 ml of methanol and 1 ml of Milli-Q water. Combine each wash with the coupling reagent.

10. Dry the membrane. Samples are stable at −20° on disks for at least 6 months.

11. Count an aliquot of the coupling reagent/wash steps, and determine the coupling efficiency.

12. Determine the site of *O*-GlcNAc on a peptide by manual/automated Edman degradation.

13. For manual Edman degradation, place the disk in a screw-capped polypropylene microcentrifuge tube.

14. Add 0.5 ml of the sequencing reagent to the disk, and incubate at 50° for 10 min. This step derivatizes the N-terminus of the peptide, forming the phenylthiocarbamyl derivative. (Steps 14 through 22 represent a "cycle." This method is only efficient for 10 to 20 cycles.)

15. Wash disk with 5 × 1 ml of methanol.

16. Dry the disk in a SpeedVac for 5 min. Add 0.5 ml of TFA; incubate at 50° for 6 min.

17. The derivatized amino acid is released from the peptide.

18. Save the supernatant.

19. Wash the disk with 1 ml of methanol.

20. Save the supernatant, and combine it with the supernatant from step 18.

21. Wash the disk 5 × 1 ml methanol.

22. Return to step 14 and repeat the sequence.

23. Dry the combined supernatants from steps 6 and 8 on a heating block, or on the bench overnight.

24. Add 0.5 ml of 100 m*M* Tris-HCl (pH, 7.4).

25. Add 15 ml of liquid scintillation fluid, and count.

Site Mapping by β-Elimination and Michael Addition with Dithiothreitol

One of the major challenges in detection and site mapping posttranslational modifications such as *O*-GlcNAc is that they are present at substoichiometric levels. Unlike phosphorylation, current radioactive-labeling

techniques of O-GlcNAc with [³H]-Gal are far less sensitive than [³²P]-ATP labeling and require at least 10 to 20 pmol of purified peptide for site mapping by Edman degradation. Conventional mass spectrometry techniques are rarely used successfully to map O-GlcNAc sites because the glycosylation suppresses signal and β-GlcNAc is especially labile (Wells et al., 2002). As a result, we and others (Haynes and Aebersold, 2000; Vosseller et al., 2005; Wells et al., 2002) have developed several techniques taking advantage of mass spectrometry sensitivity with sequencing peptides at fmol and amol levels. First, it is necessary to chemically modify the O-GlcNAc residue since O-GlcNAc is labile and often released at lower collision energies than is required to sequence the peptide. In addition, the O-GlcNAc reduces the signal by 5-fold, and the unglycosylated peptide also further suppresses the signal.

Greis and coworkers (Greis and Hart, 1998; Greis et al., 1996) used mild base to β-eliminate O-GlcNAc from Ser (89 amu) and Thr (101 amu) residues, resulting in signature parent masses 2-aminopropenoic acid (69 amu) and α-aminobutyric acid (83 amu), respectively. However, β-elimination removes all O-linked residues including phosphorylation; therefore, all samples are treated with alkaline phosphatase before β-elimination. In an adaptation of this method, Wells and coworkers (Vosseller et al., 2005; Wells et al., 2002; Whelan and Hart, 2003) combined β-elimination with Michael addition of dithiothreitol (BEMAD). This technique has the advantage of tagging the O-GlcNAc–modified peptides with a stable chemical modification that can also be enriched via thiol affinity chromatography. The DTT tag adds 136.2 kDa to either the Ser or Thr, which is used to facilitate database searches. The BEMAD method may also be used to site map phosphorylation sites by pretreating peptides with hexosaminidase to remove all the O-GlcNAc modifications. Harsher BEMAD conditions are necessary to site map O-GlcNAc as well as phosphorylation on Ser and Thr with an adjacent carboxyl proline.

One of the problems with β-elimination is that the harsh alkali often degrades the peptide's backbone, and the extent of base-catalyzed degradation of the peptide is sequence dependent. There is also the possibility that Ser and Thr may undergo β-elimination in the absence of any modification. Sulfonation of Ser and Thr residues must also be accounted for. Although BEMAD has its uses, we continue to optimize and develop new technologies for site mapping O-GlcNAc.

Materials

1. Denaturing buffer: 50 mM Tris-HCl (pH, 8), 2 to 5 mM DTT.
2. Trypsin, sequencing grade (Promega, Madison, WI).

3. TFA.

4. Performic acid oxidation buffer (made fresh): 45% (v/v) formic acid, 5% (v/v) hydrogen peroxide, in Milli-Q water.

5. 40 mM ammonium bicarbonate (pH, 8.0), MgCl$_2$.

6. Alkaline phosphatase (Promega, Madison, WI).

7. DTT, high purity (Amersham Biosciences, Piscataway, NJ).

8. Mild BEMAD solution (made fresh): 1% (v/v) triethylamine, 0.1% (v/v) NaOH, 10 mM DTT.

9. Strong BEMAD solution 1: 55 mM Ba(OH)$_2$.

10. Strong BEMAD solution 2: 25 mM Na$_2$CO$_3$ (pH, 11) and 100 mM DTT.

11. Dry ice.

12. C$_{18}$ reversed-phase Macro Spin columns (The Nest Group, Inc., Southborough, MA; see manufacturer's instructions).

13. Buffer A: 1% (v/v) TFA.

14. Buffer B: 1% (v/v) TFA, 75% (v/v) acetonitrile.

15. Thiol column buffer (made fresh), degassed: TBS-EDTA: 20 mM Tris (pH, 7.6), 150 mM NaCl, 1 mM EDTA.

16. Thiol column elution buffer (made fresh), degassed: TBS-EDTA: 20 mM Tris-HCl (pH, 7.6), 150 mM NaCl, 1 mM EDTA, 20 mM DTT.

17. Thiopropyl Sepharose 6B (Amersham Biosciences, Piscataway, NJ).

18. Empty Macro Spin column (The Nest Group, Southborough, MA).

19. Savant SpeedVac concentrator.

20. Finnigan LCQ with nanospray source.

21. Control peptides at approximately 1 to 100 pmol of protein sample in 40 mM ammonium bicarbonate (pH, 8.0). As a control, the sample should be spiked with 10 to 20 pmol of known phosphorylated and *O*-GlcNAc–modified peptides. We commonly use a synthesized glycosylated BPP peptide (Greis *et al.*, 1996), PSVPVS(*O*-GlcNAc)GSAPGR, and commercially available phospho-AKT, KHFPQFS(P)YSAS (Upstate Biotechnology, Lake Placid, NY).

22. Seal-Rite Natural microcentrifuge tubes (USA Scientific, Inc., Ocala, FL): To avoid plastic contamination, we recommend the use of Seal-Rite Natural microcentrifuge tubes. All plastic tubes and columns should be rinsed with 50% acetonitrile before use and never autoclaved. (In addition, fresh clear pipet tips should be used whenever possible.)

Procedures for Site Mapping by BEMAD

Protein Digestion

1. Suspend the sample in denaturing buffer.

2. Heat at 95° for 15 to 20 min, then allow reaction to cool to room temperature.

3. Add sequencing-grade trypsin (Promega Corporation, Madison, WI) to a protease:protein ratio of 1:20 to 1:100, and incubate overnight at 37°.

4. Stop trypsin digest by adding TFA to a 1% (v/v) final concentration (Trypsin is inactivated below a pH of 4).

5. Clean digested peptides on a C_{18} reversed-phase column.

6. Dry peptides using a SpeedVac concentrator.

Performic Acid Oxidation

Performic acid oxidation is necessary to oxidize the cysteines rather than alkylating the cysteines with iodoacetamide because the alkylated cysteines are susceptible to β-elimination and Michael addition of DTT. If the cysteines are labeled with DTT, these peptides will compete with the thiol chromatography enrichment of peptides containing *O*-GlcNAc sites labeled with DTT and drown out the signal during mass spectrometry analysis. However, the differential labeling of cysteines from control and experimental conditions with DTT (cysteine mass increase of 120.2) and deuterated DTT (cysteine mass increase of 126.2) may be used for quantifying protein expression similar to the ICAT procedure, except at far less expense. Performic acid oxidation causes mass additions of $+16$, $+32$, or $+48$ daltons to cysteine and tryptophan and $+16$ or $+32$ daltons to methionine and histidine.

1. Spike protein sample with 10 to 20 pmol of control peptides.

2. Suspend protein sample in 300 μl performic acid oxidation buffer.

3. Incubate on ice for 1 h.

4. Dry down in a SpeedVac.

Phosphatase Treatment

(Note: This method may be substituted with treatment of hexosaminidase to remove *O*-GlcNAc in order to use BEMAD for phosphorylation site mapping.) Suspend peptides in 40 m*M* citrate-phosphate buffer pH to 4.5 and hexosaminidase at 5 U/20 μl at 37° for 16 h. Allow β-elimination of phosphorylation sites to react for 4 or more hrs.

1. Suspend peptides in 40 m*M* ammonium bicarbonate, 1 m*M* $MgCl_2$.

2. Add alkaline phosphatase at 1 U/10 μl, and incubate at 37° for 4 h.

3. Clean digested peptides on a C_{18} reversed-phase column.

4. Dry peptides using a SpeedVac concentrator.

Mild BEMAD Treatment

The mild BEMAD conditions work well for peptide sequences similar to the control peptide BPP peptide (PSVPVS(*O*-GlcNAc)GSAPGR).

Longer incubation periods may be needed for different peptide sequences and peptides with phosphorylation sites. A small percentage of Ser and Thr residues may undergo β-elimination and Michael addition of DTT. The mild BEMAD condition does not work for either O-GlcNAc or phosphorylation of Ser or Thr residues adjacent to carboxyl terminal prolines. Harsher conditions are necessary when trying to site map these types of modified peptides.

1. Suspend peptides in 200 μl BEMAD solution, and adjust pH to 12.0 to 12.5 with triethylamine, if necessary.
2. Incubate reaction at 50° for 4 h.
3. Stop reaction by adding TFA to 1% (v/v) final concentration.
4. Clean digested peptides on a C_{18} reversed-phase column.
5. Dry peptides using a SpeedVac concentrator.
6. Proceed to thiol affinity chromatography later.

Strong BEMAD Treatment

Harsher BEMAD conditions allow for about 50% DTT labeling of the most difficult peptides with carboxyl terminal prolines to Ser and Thr such as the modified c-Myc peptide KKFELLPT(O-GlcNAc)PPLSPSRR and KKFELLPT (PO_3)PPLSPSRR.

1. Suspend peptides in 200 μl barium hydroxide buffer.
2. Incubate at 37° for 12 h.
3. Precipitate barium hydroxide by adding dry ice.
4. Pellet precipitated barium hydroxide by centrifugation at 14,000 × g for 5 min.
5. Remove solution containing the peptides, and dry using a SpeedVac concentrator.
6. Suspend peptides in 25 mM sodium carbonate buffer and 100 mM DTT, place under argon gas for 48 h at 37°, and mix periodically.
7. Stop the reaction by adding TFA to 1% (v/v) final concentration.
8. Clean digested peptides on a C_{18} reversed-phase column (The Nest Group Inc, Southborough, MA; see manufacturer's instructions).
9. Dry peptides using a SpeedVac concentrator.
10. Proceed to thiol affinity chromatography later.

Thiol Affinity Chromatography

1. Swell and wash Thiopropyl Sepharose 6B resin in degassed Thiol column buffer.
2. Transfer 200 μl of 50% slurry to an empty Macro Spin column (Nest Group), and wash at least seven times with 500 μl Thiol column buffer by

centrifugation at $1500 \times g$ for 30 s. It is important to thoroughly wash the Thiopropyl Sepharose 6B to eliminate contaminants and to maximize the DTT-modified peptide binding.

3. Suspend peptides in up to $400\mu l$ Thiol column buffer.

4. Mix suspended peptides with Thiopropyl Sepharose for 4 hrs at room temperature by rotation.

5. Wash Thiopropyl Sepharose seven times with $500\mu l$ Thiol column buffer.

6. Elute peptides by mixing with Thiol elution buffer for 1 h before collecting eluent.

7. Acidify peptide eluent by adding TFA to 1% (v/v) final concentration.

8. Clean peptides over C_{18} reversed-phase column to remove free DTT and salts.

9. Dry peptides using a SpeedVac concentrator.

10. Peptides are now ready for analysis by the researcher's choice of mass spectrometry instrumentation. The amount of starting material will vary depending on the capacity of the C_{18} column, the sensitivity of the LC-MS/MS instrument, and the complexity of the sample. We use the electrospray ion traps and Thermo Finnigan LCQ Classic and LTQ, which are able to reach sensitivities in the fmol range and amol range, respectively. Therefore, researchers will have to optimize for their type of instrumentation. We have found that site mapping O-GlcNAc at a stoichiometry less then 10% is particularly difficult. Additional information on BEMAD may be found in works by Greis et al. (1996), Wells et al. (2002), Whelan and Hart (2003), and Vosseller et al. (2005).

11. Allow for a differential mass increase of 136.2 daltons (142.2 daltons for deuterated DTT experiments) to Ser and Thr residues.

Chemoenzymatic Enrichment of O-GlcNAc–Modified Proteins

One alternative to sWGA or O-GlcNAc affinity chromatography is an O-GlcNAc chemoenzymatic enrichment method developed by Hsieh-Wilson and coworkers (Khidekel et al., 2003, 2004). They use an engineered mutant galactosyltransferase (Y289L GalT) enzyme to tag O-GlcNAc with a ketone moiety UDP substrate (Khidekel et al., 2003), which is chemically reacted with an aminooxy biotin derivative (N-(aminooxyacetyl)-N'-(D-biotinoyl) hydrazine) and modified proteins isolated by biotin affinity chromatography. The engineered mutant Y289L GalT has an enlarged binding pocket that allows it to recognize the unnatural UDP analogue with a ketone at the C-2 position and enhances the catalytic activity

without compromising specificity. Similar to the galactosyltransferase labeling mentioned previously, this method is not limited to the specificity of *O*-GlcNAc antibody enrichment nor the need for multiple *O*-GlcNAc modifications required to interact with sWGA. One drawback of this technique is that the Ser and Thr *O*-GlcNAc linkage is still present, and the chemically modified *O*-GlcNAc is still labile in mass spectrometry, making BEMAD or another approach necessary to map the site. One significant advantage to this method is that chemoenzymatic-labeled proteins can be separated by SDS-PAGE and subjected to Western blot analysis with streptavidin-HRP with almost 400-fold increase in sensitivity compared with wild-type GalT tritium labeling.

Materials

1. Chemoenzymatic reaction buffer: 20 mM HEPES (pH, 7.3), 0.1 M KCl, 0.2 mM EDTA, 0.2% Triton-X, 10% glycerol.
2. 5 mM MnCl$_2$, 1.25 mM adenosine 5'-diphosphate, 0.5 mM ketone moiety UDP substrate (Hang and Bertozzi, 2001; Khidekel *et al.*, 2003), 20 μg/ml mutant Y289L GalT and PNGase F.
3. Denaturing buffer A: 5 M urea/50 mM NH$_4$HCO$_3$ (pH, 7.8), 100 mM NaCl.
4. 2.7 M NaOAc (pH, 3.9).
5. Aminooxy biotin (*N*-(aminooxyacetyl)-*N'*-(D-biotinoyl) hydrazine; Dojindo Molecular Technologies, Gaithersburg, MD).
6. 3 M NH$_4$HCO$_3$ (pH, 9.6).
7. Dialysis buffer: 6 M urea, 50 mM NH$_4$HCO$_3$ (pH, 7.8), 100 mM NaCl.
8. Denaturing buffer B: 4 M urea, 50 mM NH$_4$HCO$_3$ (pH, 7.8), 10 mM NaCl.
9. Iodoacetamide.
10. 50 mM NH$_4$HCO$_3$ (pH, 7.8).
11. Trypsin, sequencing grade (Promega, Madison, WI).
12. 20% TFA.
13. Macrotrap cartridges (Michrome Bioresources, Auburn, CA).
14. Strong cation exchange kit (Applied Biosystems, Foster City, CA).
15. 5 mM KH$_2$PO$_4$ buffers containing: 40, 100, 200, and 350 mM KCl.
16. Avidin affinity chromatography kit (Applied Biosystems, Foster City, CA).

Procedure

1. Dialyze protein isolated from tissue or cells into chemoenzymatic reaction buffer.

2. Protein extract is supplemented with 5 mM MnCl$_2$, 1.25 mM adenosine 5'-diphosphate, 0.5 mM unnatural UDP substrate, 20 μg/ml mutant Y289L GalT, and 2500 U/ml PNGase F.

3. Incubate for 12 to 14 h at 4°.

4. Dialyze into denaturing buffer three times for 2 h.

5. Adjust pH with 2.7 M NaOAc (pH, 3.9) to a final concentration of 50 mM (pH, 4.8).

6. Add aminooxy biotin (N-(aminooxyacetyl)-N'-(D-biotinoyl) hydrazine) to 5 mM final concentration for 24 h at a room temperature of 23°.

7. Dilute extracts with 3 M NH$_4$HCO$_3$ (pH, 9.6) to a final concentration of 50 mM (pH, 8).

8. Dialyze one time for 2 h in dialysis buffer and one time for 10 h in denaturing buffer B.

9. Alkylate proteins with 15 mM iodoacetamide for 45 min in the dark, dilute with 50 mM NH$_4$HCO$_3$ (pH, 7.8), and digest with trypsin 20 μg/μl overnight at 37°.

10. Peptides from proteolytic digest are desalted with macrotrap cartridges.

11. To reduce the complexity of the peptide mixture, acidify 1 to 3 mg of digested peptides with 1% TFA, and dilute in strong cation exchange buffer according to manufacturer's protocol.

12. Elute peptides with a step gradient of 40, 100, 200, and 350 mM KCl in 5 mM KH$_2$PO$_4$ containing 25% CH$_3$CN.

13. Enrich for the biotin-tagged O-GlcNAc–modified peptides in each fraction by avidin chromatography as described by manufacturer, except triple the wash volume.

14. Enriched peptides are now ready for MS/MS analysis by the researcher's choice of mass spectrometry.

15. Enriched labeled peptides may also be subjected to BEMAD to locate the site of O-GlcNAc modification.

Future Directions

These methods and tools have facilitated the identification and site-mapping of O-GlcNAc on more than 100 proteins. These data provide a solid framework from which one can obtain an understanding of the many roles of O-GlcNAc in regulating cellular function. However, the challenges are greater, and fewer O-GlcNAc methods have been developed compared with the tools available for studying phosphorylation. The key areas of focus for future methods development are as follows: (1) development of site-specific and general O-GlcNAc antibodies; (2) improvement of methods for enrichment and labeling; and (3) use of new technologies, such as

Fourier-transform mass spectrometry instruments with electron capture detection, or ion traps with electron transfer dissociation capabilities that are capable of sequencing *O*-GlcNAc–bearing peptides without the loss of the saccharide residue, allowing for direct site mapping. Another recent advance, which can be used on less expensive instrumentation, includes the LTQ mass spectrometer and software designed by Thermo Finnigan, which can be programmed to recognize the neutral loss of labile posttranslational modifications such as *O*-GlcNAc and phosphorylation and focus on that peptide for subsequent sequencing. Although the peptide that contained the modification is identified, there is no site information. Other groups have synthesized an unnatural GlcNAc analogue (peracetylated *N*-(2-azidoacetyl)-glucosamine (azido-GlcNAc)) that may be used in cell culture to metabolically label nucleocytoplasmic proteins (Vocadlo *et al.*, 2003). Proteins containing the azido-GlcNAc are isolated by the Staudinger ligation method using a biotinylated phosphine capture reagent (Sprung *et al.*, 2005; Vocadlo *et al.*, 2003). This method has led to the identification of 41 putative *O*-GlcNAc modified proteins by Yingming Zhao's group (Sprung *et al.*, 2005). Further development of these technologies and new methods will be necessary to enhance the proteomic study of *O*-GlcNAc.

Acknowledgments

Original research was supported by the National Institutes of Health, National Heart, Lung, and Blood Institute contract N01-HV-28180, and National Institute of Diabetes, Digestion, and Kidney Diseases grant R0 DK61671. Under a licensing agreement between Covance Research Products, Sigma Chemical Co., and Johns Hopkins University School of Medicine, G. W. Hart receives a percentage of royalties received by the university on sales of the CTD 110.6 antibody. The terms of this arrangement are in accordance with the conflict of interest policy at the Johns Hopkins University. G. W. Hart is on the S. A. B of Sigma Chem. Co. biotechnology division.

References

Cheng, X., and Hart, G. W. (2001). Alternative *O*-glycosylation/*O*-phosphorylation of serine-16 in murine estrogen receptor beta: post-translational regulation of turnover and transactivation activity. *J. Biol. Chem.* **276**, 10570–10575.

Chou, T. Y., Hart, G. W., and Dang, C. V. (1995). c-Myc is glycosylated at threonine 58, a known phosphorylation site and a mutational hot spot in lymphomas. *J. Biol. Chem.* **270**, 18961–18965.

Comer, F. I., Vosseller, K., Wells, L., Accavitti, M. A., and Hart, G. W. (2001). Characterization of a mouse monoclonal antibody specific for *O*-linked *N*-acetylglucosamine. *Anal. Bioche.* **293**, 169–177.

Dong, D. L., and Hart, G. W. (1994). Purification and characterization of an *O*-GlcNAc selective *N*-acetyl-beta-D-glucosaminidase from rat spleen cytosol. *J. Biol. Chem.* **269**, 19321–19330.

Greis, K. D., and Hart, G. W. (1998). Analytical methods for the study of *O*-GlcNAc glycoproteins and glycopeptides. *Methods Mol. Biol.* **76,** 19–33.

Greis, K. D., Hayes, B. K., Comer, F. I., Kirk, M., Barnes, S., Lowary, T. L., and Hart, G. W. (1996). Selective detection and site-analysis of *O*-GlcNAc–modified glycopeptides by beta-elimination and tandem electrospray mass spectrometry. *Anal. Biochem.* **234,** 38–49.

Haltiwanger, R. S., Blomberg, M. A., and Hart, G. W. (1992). Glycosylation of nuclear and cytoplasmic proteins. Purification and characterization of a uridine diphospho-*N*-acetylglucosamine:Polypeptide beta-*N*-acetylglucosaminyltransferase. *J. Biol. Chem.* **267,** 9005–9013.

Hang, H. C., and Bertozzi, C. R. (2001). Ketone isoteres of 2-*N*-acetamidosugars as substrates for metabolic cell surface engineering. *J. Am. Chem. Soc.* **123,** 1242–1243.

Haynes, P. A., and Aebersold, R. (2000). Simultaneous detection and identification of *O*-GlcNAc–modified glycoproteins using liquid chromatography-tandem mass spectrometry. *Anal. Chem.* **72,** 5402–5410.

Holt, G. D., Snow, C. M., Senior, A., Haltiwanger, R. S., Gerace, L., and Hart, G. W. (1987). Nuclear pore complex glycoproteins contain cytoplasmically disposed *O*-linked *N*-acetylglucosamine. *J. Cell Biol.* **104,** 1157–1164.

Kelly, W. G., Dahmus, M. E., and Hart, G. W. (1993). RNA polymerase II is a glycoprotein: Modification of the COOH-terminal domain by *O*-GlcNAc. *J. Biol. Chem.* **268,** 10416–10424.

Khidekel, N., Arndt, S., Lamarre-Vincent, N., Lippert, A., Poulin-Kerstien, K. G., Ramakrishnan, B., Qasba, P. K., and Hsieh-Wilson, L. C. (2003). A chemoenzymatic approach toward the rapid and sensitive detection of *O*-GlcNAc posttranslational modifications. *J. Am. Chem. Soc.* **125,** 16162–16163.

Khidekel, N., Ficarro, S. B., Peters, E. C., and Hsieh-Wilson, L. C. (2004). Exploring the *O*-GlcNAc proteome: Direct identification of *O*-GlcNAc–modified proteins from the brain. *Proc. Natl. Acad. Sci. USA* **101,** 13132–13137.

Matsuoka, Y., Matsuoka, Y., Shibata, S., Yasuhara, N., and Yoneda, Y. (2002). Identification of Ewing's sarcoma gene product as a glycoprotein using a monoclonal antibody that recognizes an immunodeterminant containing *O*-linked *N*-acetylglucosamine moiety. *Hybrid Hybridomics* **21,** 233–236.

McClain, D. A. (2002). Hexosamines as mediators of nutrient sensing and regulation in diabetes. *J. Diabetes Complications* **16,** 72–80.

Roquemore, E. P., Chou, T. Y., and Hart, G. W. (1994). Detection of *O*-linked *N*-acetylglucosamine (*O*-GlcNAc) on cytoplasmic and nuclear proteins. *Methods Enzymol.* **230,** 443–460.

Shafi, R., Iyer, S. P., Ellies, L. G., O'Donnell, N., Marek, K. W., Chui, D., Hart, G. W., and Marth, J. D. (2000). The *O*-GlcNAc transferase gene resides on the X chromosome and is essential for embryonic stem cell viability and mouse ontogeny. *Proc. Natl. Acad. Sci. USA* **97,** 5735–5739.

Slawson, C., and Hart, G. W. (2003). Dynamic interplay between *O*-GlcNAc and *O*-phosphate: The sweet side of protein regulation. *Curr. Opin. Struct. Biol.* **13,** 631–636.

Slawson, C., Pidala, J., and Potter, R. (2001). Increased *N*-acetyl-beta-glucosaminidase activity in primary breast carcinomas corresponds to a decrease in *N*-acetylglucosamine containing proteins. *Biochim. Biophys. Acta* **1537,** 147–157.

Snow, C. M., Senior, A., and Gerace, L. (1987). Monoclonal antibodies identify a group of nuclear pore complex glycoproteins. *J. Cell Biol.* **104,** 1143–1156.

Sprung, R., Nandi, A., Chen, Y., Kim, S. C., Barma, D., Falck, J. R., and Zhao, Y. (2005). Tagging-via-substrate strategy for probing *O*-GlcNAc modified proteins. *J. Proteome Res.* **4,** 950–957.

Torres, C. R., and Hart, G. W. (1984). Topography and polypeptide distribution of terminal N-acetylglucosamine residues on the surfaces of intact lymphocytes: Evidence for O-linked GlcNAc. J. Biol. Chem. **259,** 3308–3317.

Turner, J. R., Tartakoff, A. M., and Greenspan, N. S. (1990). Cytologic assessment of nuclear and cytoplasmic O-linked N-acetylglucosamine distribution by using anti-streptococcal monoclonal antibodies. Proc. Natl. Acad. Sci. USA **87,** 5608–5612.

Vocadlo, D. J., Hang, H. C., Kim, E. J., Hanover, J. A., and Bertozzi, C. R. (2003). A chemical approach for identifying O-GlcNAc–modified proteins in cells. Proc. Natl. Acad. Sci. USA **100,** 9116–9121.

Vosseller, K., Hansen, K. C., Chalkley, R. J., Trinidad, J. C., Wells, L., Hart, G. W., and Burlingame, A. L. (2005). Quantitative analysis of both protein expression and serine/threonine post-translational modifications through stable isotope labeling with dithiothreitol. Proteomics **5,** 388–398.

Wells, L., Vosseller, K., Cole, R. N., Cronshaw, J. M., Matunis, M. J., and Hart, G. W. (2002). Mapping sites of O-GlcNAc modification using affinity tags for serine and threonine post-translational modifications. Mol. Cell Proteomics **1,** 791–804.

Wells, L., Vosseller, K., and Hart, G. W. (2003). A role for N-acetylglucosamine as a nutrient sensor and mediator of insulin resistance. Cell Mol. Life Sci. **60,** 222–228.

Whelan, S. A., and Hart, G. W. (2003). Proteomic approaches to analyze the dynamic relationships between nucleocytoplasmic protein glycosylation and phosphorylation. Circ Res **93,** 1047–1058.

Zachara, N. E., Cheung, W. D., and Hart, G. W. (2004a). Nucleocytoplasmic glycosylation, O-GlcNAc: Identification and site mapping. Methods Mol. Biol. **284,** 175–194.

Zachara, N. E., and Gooley, A. A. (2000). Identification of glycosylation sites in mucin peptides by Edman degradation. Methods Mol. Biol. **125,** 121–128.

Zachara, N. E., and Hart, G. W. (2004). O-GlcNAc a sensor of cellular state: The role of nucleocytoplasmic glycosylation in modulating cellular function in response to nutrition and stress. Biochim. Biophys. Acta **1673,** 13–28.

Zachara, N. E., O'Donnell, N., Cheung, W. D., Mercer, J. J., Marth, J. D., and Hart, G. W. (2004b). Dynamic O-GlcNAc modification of nucleocytoplasmic proteins in response to stress: A survival response of mammalian cells. J. Biol. Chem. **279,** 30133–30142.

Section III

Carbohydrate Synthesis and Antibiotics

[9] Chemoenzymatic Synthesis of Glycan Libraries

By OLA BLIXT and NAHID RAZI

Abstract

The expanding interest for carbohydrates and glycoconjugates in cell communication has led to an increased demand of these structures for biological studies. Complicated chemical strategies in glycan synthesis are now more frequently replaced by regio- and stereo-specific enzymes. The exploration of microbial resources and improved production of mammalian enzymes have established glycosyltransferases as an efficient complementary tool for glycan synthesis. In this chapter, we demonstrate the feasibility of preparative enzymatic synthesis of different categories of glycans, such as blood group and tumor-associated poly-N-acetyllactosamines antigens, ganglio-oligosaccharides, N- and O-glycans. The enzymatic approach has generated over 100 novel oligosaccharides in amounts allowing milligram to gram distribution to many researchers in the field. Our diverse library has also formed the foundation for the successful developments of both the noncovalent enzyme-linked immunosorbent assay glycan array and the covalent printed glycan microarray.

Introduction

Carbohydrate groups of glycoproteins and glycolipids exhibit a great deal of structural diversity and are major components of the outer surface of animal cells (Sharon and Lis, 1993). The lack of availability and affordability of appropriate oligosaccharides has always been a major limitation in the field. To overcome these shortcomings, the Consortium for Functional Glycomics' (CFG) initiative was launched to generate an efficient and rapid production of essential oligosaccharides in sufficient quantities to build up a compound library for distribution, as well as for building glycan microarray platforms. To date, the library and the developed glycan microarrays (see Chapter 18) have been a tremendous success story and a widely used resource for advancing the research in the field of glycobiology.

One of the strategic key points in the formation of a diverse functional glycan library is to build glycan structures that can readily be adapted to various needs—for example, coupling to proteins, solid-phase or used as is. A short neutral and versatile linker (2-azidoethyl) or a protected amino acid (F-moc) was introduced chemically to the penultimate monosaccharide, followed by chemical or enzymatic elongation. The linker and the amino

METHODS IN ENZYMOLOGY, VOL. 415
0076-6879/06 $35.00
DOI: 10.1016/S0076-6879(06)15009-0

acid moiety can readily be converted to an amine functionality for further derivatization (Blixt et al., 2003, 2004). Chemical synthesis of oligosaccharides is now very sophisticated, and it is possible to synthesize essentially any glycosidic linkage, although with differing degrees of difficulty. Alternatives to traditional chemical solution-phase synthesis (Garegg, 2004; Lee et al., 2004; Nicolaou and Mitchell, 2001), such as solid-phase synthesis and one-pot reactivity-based glycosylations (Bartolozzi and Seeberger, 2001; Tanaka et al., 2002; Ye and Wong, 2000), have the advantage of avoiding intermediate isolation and several purification steps. Consequently, it dramatically shortens and simplifies chemical synthesis. Despite such improvements, chemical synthesis is still hampered by time-consuming multiple protection/de-protection building-block strategies and complex isomeric reaction mixtures (Boons and Demchenko, 2000).

A second powerful and complementary approach to chemical synthesis is to use regio- and stereo-specific glycosylating enzymes, glycosyltransferases (Auge and Crout, 1997; Koeller and Wong, 2001). Glycosyltransferases generate glycosidic linkages in one-step reactions between an unprotected acceptor and a sugar donor nucleotide; thus, multistep protection group strategies are not required (Hanson et al., 2004). With new advances in molecular biology, recombinant enzymes and accessory enzymes for sugar-nucleotide regeneration are now being readily expressed in large quantities for synthesis of complex oligosaccharides (Blixt and Razi, 2004; Endo et al., 1999; Johnson, 1999). Although enzymes have relatively strict substrate specificities, many of them offer substantial acceptor substrate flexibility for synthesis. They can be applied directly onto a simple monosaccharide acceptor, followed by subsequent elongation to a more complex structure, or they can be introduced after de-protection of chemically synthesized intermediates. For example, sialic acid, a common monosaccharide that frequently terminates oligosaccharide sequences on various glycoproteins and glycolipids, is notoriously difficult to apply in chemical glycosylation but can be easily introduced enzymatically as a final step by a sialyltransferase (Blixt et al., 2002, 2005).

In this chapter, we illustrate our enzyme production and chemoenzymatic approach for large-scale synthesis of terminal glycans commonly found on glycoproteins and glycolipids. Many are blood group related, tumor associated, and specific for the C-type lectins, galectins, and Siglecs subgroup glycan-binding protein families.

Materials

Enzyme Production Materials

Escherichia coli strain AD202 (CGSG 7297), Aspergillus niger DVK1, Sf-9 and HiFive insect cells (Invitrogen, CA), and CHO cells were used

in bacterial, fungal, baculoviral, and mammalian expression systems, respectively. Yeast Extract Trypton, 2xYT (Difco, NJ), Isopropyl-1-thio-β-D-galactopyranoside (Calbiochem-Merck, Germany), HyQ SFX (Hyclone, Utah), DMEM (Invitrogen, CA), FuGene (Roche, IN), Effectene (Qiagen, CA), uridine 5′-diphospho-N-acetylglucosamine (UDP-GlcNAc), and guanidine-5-diphospho-fucose (GDP-Fuc) were a gift from Tokyo Research Laboratories, KyowaHakko Kogyo Co. Ltd. The radioactive nucleotide sugars were diluted with unlabeled nucleotide sugars to obtain the desired specific radioactivity. GDP-[^{14}C]Fuc and CMP-[^{14}C] NeuAc came from GE Healthcare—formerly Amersham Biosciences (Little Chalfont, UK)—and the UDP-[^{3}H] sugar nucleotides from PerkinElmer Life and Analytical Sciences (Shelton, CT). All other chemicals, supplements, and resins were of highest purity and purchased from Sigma-Aldrich (St. Louis, MO). Large-scale enzyme production (100 l, Braun Fermentor), Microfluidizer (Microfluidics, Model M-110S), and Tangential Flow Filtration (Pall Corporation) were used in the extraction and concentration of the enzyme-containing media.

Methods

Enzyme Expression Methods

Recombinant enzymes were produced in different expression systems to be used in carbohydrate synthesis. All the bacterial enzymes and a rat UDP Gal/GalNAc epimerase were expressed in E. coli strain AD202. Other enzymes that require posttranslational modifications were produced in A. niger (rat ST3Gal III, human FUT V and FUT VII), insect cells (human ST6GalI, pig ST3GalI, chicken ST6GalNAcI, human FUTs II, III, IV, and VI), and the CHO cells (Core 2 β1,6GlcNAcT). The procedure for each cloning and expression has been fully described previously (Blixt et al., 2001, 2002, 2005; Fukuta et al., 1997; Schwientek et al., 2000; Uchimura et al., 1998; Vasiliu et al., 2006), as well as being incorporated into the CFG database (www.functionalglycomics.org).

General Procedure for the Bacterial Expression

E. coli strain AD202 was used in bacterial expression either in shaker flasks or in a 100-l Fermentor. Bacteria was cultured in 2×YT/ampicillin (150 μg/ml) and induced with isopropyl-thiogalactopyranoside (1 μM) at A_{600} = 0.3–0.6, depending on each enzyme. Cells were harvested by spinning at 5000 g for 30 min after completing the incubation. Pellet was weighed and resuspended in an appropriate buffer (see assay buffers in Table I) in a 1 g/ml ratio of cell weight to the buffer. The cell paste was

TABLE I

CONDITIONS FOR ENZYME PRODUCTION AND GLYCAN SYNTHESIS

Entry	Complete name (abbreviation)	Source	Expression cell	Activity (U/L)	Assay acceptor	Buffer/pH (100–50 mM)	Sugar-nucleotide	Additives 10 mM/1 mM	Assay development	Purification
a	β1-3GlcNAc T(lgtA)	*N. meningitidis*	AD202	114 U/L	LacNAcβsp[a]	Caco/7.5	UDP-^{3}HGN	MnCl$_2$	Dowex/PO4	Cell lysate
b	β1-6GlcNAc T(Core-2)	Human	CHO	2 U/L	GalNAc-αOBn[b]	Caco/7.0	UDP-^{3}HGN	MnCl$_2$	Dowex/Cl⁻	Protein C
c	α1-3GalT (BraGalT1)	Bovine	AD202	375 U/L	Lactose	Caco/7.5	UDP-^{3}HGal	MnCl$_2$	Dowex/Cl⁻	Ni-Agarose
d	α1-3GalT (GTB)	Human	Sf-9	15 U/L	LacNAcβsp	Caco/7.0	UDP-^{3}HGal	MnCl$_2$/DTT	Dowex/Cl⁻	SP-Seph.
e	β1-3GalT(GalT5)	Human	AD202	0.2 U/L	GlcNAcβsp	Mes/6.0	UDP-^{3}HGal	MnCl$_2$	Dowex/Cl⁻	Concentrate
f	β1,3GalT(cgtB)	*C. jejuni*	AD202	66 U/L	GlcNAcβOMe[c]	Mes/6.0	UDP-^{3}HGal	MnCl$_2$/DTT	Dowex/Cl⁻	Lysate
g	α1-4GalT/UDPGalE (lgtC)	*N. meningitidis*	AD202		Lactose??	Tris/7.5	UDP-^{3}HGlc	MnCl$_2$/DTT	Dowex/Cl⁻	Lysate
h	β1-4GalT/UDPGalE (lgtB)	*N. meningitidis*	AD202	72 U/L	GlcNAcβOMe	Tris/7.5	UDP-^{3}HGlc	MnCl$_2$	Dowex/Cl⁻	Lysate
i	α1-3 GalNAc (GTA)	*Human*			LacNAcβsp	Caco/7.0	UDP3HGalN	MnCl$_2$/DTT	Dowex/Cl⁻	SP-Seph.
j	β1-3GalNACT/ UDPGalNAcE (lgtD)	*P. shigelloides*	AD202	40 U/L	Galβ4Lac/Lac	Hepes/7.5	UDP-^{3}HGN	MnCl$_2$/DTT	Dowex/Cl⁻	Ni-Agarose
k	β1-4GalNAcT (cgtA)	*C. jejuni*	AD202	166 U/L	SiaLactose	Hepes/7	UDP-^{3}HGN	MnCl$_2$	Dowex/PO4	Lysate
l	β1-4GalNAcT (mutant)	Bovine	BL21(D3)	40 U/L	GlcNAc-sp	Tris/8	UDP-^{3}HGN		Dowex/Cl⁻	Concentrate
m	α1-2FucT (FUT2)	Human	Sf-9	10 U/L	Lactose	Tris/7.5	GDP-^{14}CFuc	MnCl$_2$	Dowex/Cl⁻	Concentrate
n	α1-3/4FucT (FUT3)	Human	Sf-9	15 U/L	Lac/LacNAc	Tris/7.5	GDP-^{14}CFuc	MnCl$_2$	Dowex/Cl⁻	Concentrate
o	α1-3FucT (FUT4)	Human	Sf-9	8 U/L	Lac/LacNAc	Tris/7.5	GDP-^{14}CFuc	MnCl$_2$	Dowex/Cl⁻	NH4 precip.
p	α1-3FucT (FUT5)	Human	Aspar.niger	19 U/L	Lac/LacNAc	Tris/7.5	GDP-^{14}CFuc	MnCl$_2$	Dowex/Cl⁻	Concentrate
q	α1-3FucT (FUT6)	Human	Sf-9	25 U/L	Lac/LacNAc	Tris/7.5	GDP-^{14}CFuc	MnCl$_2$	Dowex/Cl⁻	NH4 Precip.
r	α1-3FucT(FUT7)	Human	A. Niger	18 U/L	SiaLacNAc	Tris/7.5	GDP-^{14}CFuc	MnCl$_2$	Dowex/Cl⁻	PEG precip.
s	α2-3SiaT/CMP Synt (ST3)	*N. meningitidis*	AD202	124 U/L	Lactose	Caco/6.5	CMP-^{14}CSia	MnCl$_2$	Dowex/PO4	NH4 precip.
t	α2-3SiaT (rST3GalIII)	Rat	Aspar.niger	38 U/L	Lactose	Caco/6.5	CMP-^{14}CSia	MnCl$_2$	Dowex/PO4	lysate
u	α2-3SiaT/α2-8SiaT (cstII)	*C. jejuni*	AD202	50 U/L	Lactose	Hepes/7.5	CMP-^{14}CSia	MnCl$_2$	Dowex/PO4	CDP affinity
v	α2-3SiaT(pST3GalI)	Porcine	Sf-9	15 U/L	Lactose	Caco/6.5	CMP-^{14}CSia		Dowex/PO4	SP ion exch.
w	α2-6SiaT (hST6GalI)	Human	Hi Five	14 U/L	Asialofetuin	Caco/7.5	CMP-^{14}CSia	NaCl	Dowex/PO4	CDP affinity
x	α2-6SiaT(chST6GalNAcI)	Chicken	Hi Five	20 U/L	Asialofetuin	Caco/7.5	CMP-^{14}CSia	NaCl	Seph. G-50	
y	UDP-GlcNAc/GalNac-4-E	Rat	AD202	1100 U/L	UDPGNAc/LN	Hepes/7.5	UDP-^{3}HGlc	MnCl$_2$/NAD	Dowex/Cl⁻	Lysate
z	UDP Glc/GalE (GalE)	*P. aeruginosa*	AD202	205 U/L	UDPGlc/lactose	Hepes/7.5	UDP-^{3}HGlc	MnCl$_2$/NAD	Dowex/Cl⁻	Lysate

[a] sp; 2-azidoethyl.
[b] Bn; benzyl.
[c] me; methyl.

aliquoted and stored at $-20°$. The enzymes remained stable in cell paste for at least 1 year.

General Procedure for the Baculoviral Expression System

The baculoviral expression was performed in Spodoptera frugip, Sf-9, and HiFive cells at $27°$ in serum-free media, HYQ SFX-INSECT supplemented with 10% antibiotic/antimycotic. Insect cell growth and maintenances require special care (see Notes). Viral amplification was generally performed in Sf-9 cells with a multiplicity of infection, MOI $= 1$ and up to 48 hours' incubation. Enzyme expression was performed in either Sf-9 or HiFive cells with the MOI about 3 to 10, empirically determined for each enzyme. The production was regularly monitored by the enzyme assay for up to 7 to 8 days. When the activity started to decrease, the media was collected by centrifugation at $5000g$ for 30 minutes. The media was concentrated and was either directly used in synthesis or was further purified by different methods.

General Procedure for the Fungus Expression System

Enzyme expression in *A. niger* DVK1 was performed by conventional fungal protein expression in baffled Fernbach Flask (2.8 L) in 1 l growth broth of Sheftone media supplemented by a series of nutritional additives (see "protocols" in www.functionalglycomics.org). An initial pre-culture (inoculum) was prepared by adding 300 μl *A. niger* spore stock into a 100-ml media in a 500-ml shaker flask and incubated for 48 h, shaking at 180 rpm at $37°$. Nine hundred milliliters of Sheftone broth was inoculated by 100 ml of the prepared inoculum and the incubation continued at $32°$, shaking at 200 rpm, and the pH was monitored and adjusted to 6.00 twice daily. After 48 h of incubation, the expression was monitored by the enzyme assay until the activity remained constant for 12 h. At this stage, the product was collected by spinning at $5000g$ for 30 min, and the supernatant was aliquoted and stored at $-20°$. The enzyme remained stable in this condition for at least 1 year.

General Procedure for the Mammalian Expression System

Transient transfection of CHO cells was performed by Effectene according to the manufacturer's instruction. Stable transfection of CHO-K1 cells was performed by FuGene 6, according to the instruction, using the conventional mammalian single-colony selection by filter paper. Transfected colonies were selected using Geneticin (G418).

Enzyme Extraction, Concentration, and Purification

Different methods of extraction, concentration, and purification were used based on the expression systems and the enzymes specificities and stability. Except for the β-1,4 GalNAcT mutant that was extracted from the inclusion bodies (Ramakrishnan and Qasba, 2002), all the bacterial enzymes were extracted by sonication or the microfluidizer method (see later).

Bacterial Enzyme Extraction by Microfluidizer

The bacterial enzymes were extracted from the cell paste by breaking the cells using microfluidizer with 1/1 volume of the buffer to the resuspended cell paste. The released enzyme was used either as a cell lysate or was partially purified on different columns. The cell lysate remained stable for at least 1 year in 20°.

Precipitation and Purification of Fungal Expressing Enzymes

The *A. niger*–expressed recombinant enzymes were separated from the harvest supernatant by filtration and precipitation with ammonium sulfate, followed by sulfo-propyl (SP)-sepharose ion-exchange chromatography. The precipitation is performed in two steps. First, 20% (114 g/L) NH_4SO_2 was mixed for 30 min at room temperature or over night at 4°. The pellet was collected and the supernatant was further precipitated with 60% NH_4SO_2 (276 g/l). Pellets obtained from both precipitants were mixed and resuspended in 20 mM Mes (pH, 6.0). The suspension was diafiltered with the same buffer to adapt the conductivity of the buffer. The suspension needed to be passed a few times through a 5-μm filter before the final filtration on a 1.0-μm filter. The filtered product was further purified on a SP-sepharose ion-exchange column, equilibrated with 20 mM Mes, 50 mM NaCl (pH, 6.0), and eluted with 1 M NaCl.

Enzyme Concentration

The media from the baculovirus expression was concentrated by Tangential Flow Filtration. The concentrated fucosyltransferases were directly used in glycan synthesis and sialyltransferases (chicken ST6Gal-NAc-I and pig ST3Gal-I) were further purified on a cytodine-5'-diphosphate-Sepharose after concentrating the cultured media.

Enzyme Purification

Different methods of chromatography were applied for purification of the enzymes. The process was selected based on the enzyme specificity and

stability in crude mixture. The required method for each enzyme is shown in Table I.

Enzyme Assays

The enzyme assays were generally performed in 100 μl total volume of the appropriate buffer, ^{3}H or ^{14}C sugar nucleotide donor substrate with the appropriate specific activities, metal ions, acceptor substrate, and the desired enzyme (see Table I). The incubation continued at 37° for 60 min, and the reaction was stopped by adding 700 μl of cold water. The reaction mixture was loaded on a 2-ml Dowex ion exchange, a 10 × 0.5-cm Sepharose G-50 size exclusion, or C18 reverse phase, depending on the acceptor substrate and the enzyme reaction (see Table I). The radioactive incorporation to the substrate was detected by scintillation counter.

Chemical Synthesis of Functionalized Acceptor Substrates

Each synthesized compound is linked to a short neutral flexible spacer 2-azidoethyl (sp$_1$ = OCH$_2$CH$_2$N$_3$) or with the α-threonine aglycon (Blixt *et al.*, 2002), which enables broad diversification such as attachment to proteins, solid supports (e.g., affinity matrices, polystyrene), and biotin or other functional groups (Blixt and Razi, 2004; Blixt *et al.*, 2005; Eklind *et al.*, 1996). These spacered precursor glycans were subsequently elongated using recombinant glycosyltransferases as follows.

General Isolation and Purification of Synthesized Glycans

Enzymatic synthesis of carbohydrates amounts measured in milligrams is generally a straightforward procedure in terms of isolation and purification of the final structure. Ion exchange and size exclusion chromatography are the common procedures for product isolation. However, serious purification problems can be encountered when scaling up to multi-gram reactions. Additives such as nucleotide sugars, buffer salts, and crude enzymes are among the factors that interfere with efficient purification with the conventional chromatographic techniques used on milligram scale. In general, the isolation strategy used depends on the scale and the compounds to be prepared.

Dowex Resin and Size Exclusion Chromatography

Typically, the enzymatic reaction mixture (1 to 20 g solid) was centrifuged and then loaded (5 to 10 ml) onto a column of Sephadex G15 (5 to 1.8 cm × 170 cm) equilibrated and eluted with 5% nBuOH in water.

Appropriate fractions were collected, passed through a Dowex (formate, 1×3 cm) column, and lyophilized. The residue was further purified by Sephadex G15 as described, and appropriate fractions were collected to give 0.5 to 5 g of oligosaccharide in about 70% to 90% yields.

Peracetylation and Extraction

The enzymatic mixture was lyophilized, and solids were suspended and coevaporated three times in pyridine (10 g solids/200 ml), followed by addition of pyridine/acetic anhydride mixture (2:1, 10 g solid/300 ml). The mixture was agitated for 24 hrs, at which point TLC (Tol:EtOAc, 1:1 by volume) indicated complete acetylation of the oligosaccharide. The reaction was evaporated and coevaporated two times with toluene. The residues were dissolved in dichloromethane (500 ml) then washed one time with aqueous sulfuric acid (1 M), two times with aqueous sodium bicarbonate (1 M), and one time with ice-cold water. The pooled organic layers were dried over anhydrous magnesium sulphate, then filtered, and the filtrate was evaporated to dryness. The residual syrup was dissolved in methanol (200 ml) followed by the addition of sodium methoxide (20 ml, 0.5 M). The mixture was left overnight at room temperature, neutralized with a methanol-washed cat-ion exchange resin Dowex H$^+$, filtered, and evaporated to dryness. The deacetylated white-yellowish, residue was taken up into water and loaded onto a column of Sephadex G15 (5×160 cm) equilibrated and eluted with 5% BuOH. Fractions containing product were collected then lyophilized to give 3 to 6 g of oligosaccharide in typically 60% to 80% yields.

Preparative High-Performance Liquid Chromatography

High-performance liquid chromatography (HPLC) was carried out on an Alltech system equipped with an HPLC pump model 627, an Alltech automated gradient controller, and a Rheodyne injector (3725i-038). The oligosaccharide samples were run at 5 ml/min on a 5-μm Altima amino column (250×10 mm). Up to 500 mg/5 ml samples were manually injected, with an injection loop of 10 ml. Isocratic or a gradient of acetonitrile and deionized water (with or without 0.1% trifluoroacetic acid [TFA]) varying from 95:5 to 70:30 was used as mobile phase at room temperature. Standard compounds were also run in similar conditions. Chromatographic separations were monitored by an evaporative light-scattering detector (Alltech ELSD, 2000). The chromatographic system was connected to a personal computer for data acquisition and analysis using the scientific software EZChrom Elite. The elution of the separated oligosaccharides was collected using a Bio-Rad Biologic BioFrac fraction collector. Appropriate fractions were pulled and lyophilized.

General Enzymatic Glycosylation

The general approach can be taken for most enzymatic reactions: Reaction mixtures containing acceptor substrates (20 to 50 mM) and corresponding sugar-nucleotide donor substrate (1 to 3 mole equivalents) were dissolved in enzyme specific buffer (see Table I for details). Reaction was initiated by adding enzymes (typically 1 to 5 U/mmole acceptor) and slowly agitating at room temperature for 24 to 48 h. UDP-Gal(NAc)-4'-epimerase (Table I, entry y) (30 to 50 U/mmole donor) was added when UDP-Glc(NAc) donors were used. Depending on oligosaccharide product, the mixture was purified as described previously. Typical yields are 50% to 90% with a purity of 90% to 95%.

Synthesis of Poly-N-Acetyllactosamine of Type 1 and Type 2 Series

Elongated type 1 (Galβ1–3GlcNAc-) and type 2 (Galβ1–4GlcNAc-) core structures, poly-N-acetyllactosamines (poly-LacNAc), are known to represent the backbone or branched epitopes on many N-linked and O-linked glycans—where they participate in intercellular signaling via binding to various lectins, such as selectins or galectins (Leppaenen *et al.*, 2002; Niemela *et al.*, 1998), inflammation processes, and cancer (Hakomori, 2001; Ujita *et al.*, 1999). Type 1 chains are also present at the periphery of type 2 chains (Galβ1–4GlcNAc-), as well as in their sialylated and/or fucosylated forms. Chemical syntheses of poly-LacNAc have been developed using a variety of advanced synthetic strategies, yet the methods always involve tedious multiple protection/de-protection steps (Aly *et al.*, 2000; Buskas *et al.*, 2005; Misra *et al.*, 2001; Mong *et al.*, 2003). A limited number of structures were synthesized in small amounts using glycosyltransferases as an alternative approach to the chemical synthesis (Bårström *et al.*, 2000; Bintein *et al.*, 2003; Koeller and Wong, 2000; Zeng and Uzawa, 2005). In Scheme 1 (Fig. 1) and in support of Table I, we describe several enzymatic routes to synthesize blood group type 1 and type 2 oligosaccharides and derivatives thereof.

Notes

1. Elongation of type 1 and type 2 terminal sequences with β1–3GlcNAc-T. Previous specificity studies of the β3GlcNAcT (Table I, entry a) reveal that about 7% to 8% of enzyme activity is found with type 2 N-acetyllactosamine and 4% to 5% with type 1 lacto-N-biose relative to lactose, respectively. Still, with increased reaction times and additional enzymes (10 to 20 U/mole acceptor), we have demonstrated that β3GlcNAcT is very useful for preparative synthesis of various carbohydrate structures, including type 2 poly-LacNAc and type 1 poly-LacNAc (see Fig. 1).

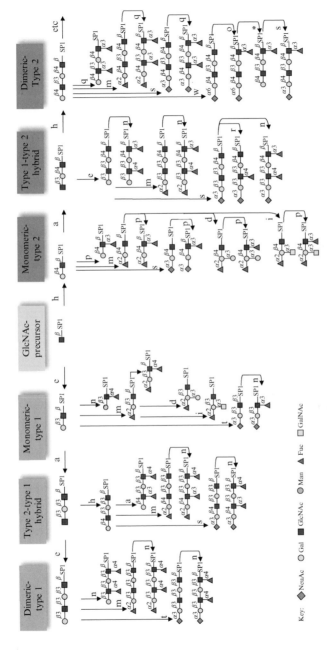

FIG. 1. SCHEME 1. Enzymatic synthesis of sialylated and fucosylated poly-*N*-acetyllactosamines. See Table I for enzyme conditions (entries a–z). (See color insert.)

Ganglio-Oligosaccharide Synthesis

Gangliosides are glycolipids that comprise a structurally diverse set of sialylated molecules that are found in most cells but are particularly abundant in neuronal tissues. They have been found to act as receptors for growth factors, toxins, and viruses and to facilitate the attachment of human melanoma and neuroblastoma cells. Specific gangliosides are also present in early stages of human neural development and affect major cellular processes, including proliferation, differentiation, survival, and apoptosis. They are also well-known tumor-associated antigens, and active immunization using gangliosides may suppress melanoma growth (Gagnon and Saragovi, 2002; Hakomori and Zhang, 1997; Kannagi *et al.*, 2004; Ravindranath *et al.*, 2000; Schnaar, 2000; Svennerholm, 2001). One of the crucial steps to synthesize the gangliosides is the introduction of sialic acid. Our enzymatic approach has significantly reduced the complications and labor for synthesis of these sialylated glycans from years to weeks of experimental work (Blixt *et al.*, 2005). We summarize our chemoenzymatic approach for synthesis of the ganglio-oligosaccharide family in Scheme 2 (Fig. 2) and Table I (Blixt *et al.*, 2005).

Notes

1. α2-8-sialylation using cst-II (Table I, entry u). With excess amounts of CMP-Neu5Ac, the cst-II demonstrates extensive α2-8 multi-sialylation activities. Therefore, the synthesis of GD3-, GT3-, and tetra-sialyllacto-oligosaccharides required carefully controlled conditions using restricted amounts of CMP-Neu5Ac. 2-azidoethyllactoside was elongated with Cst-II and 2.5 equivalents of CMP-Neu5Ac to produce 2.4 g (34 %) of GD3 and

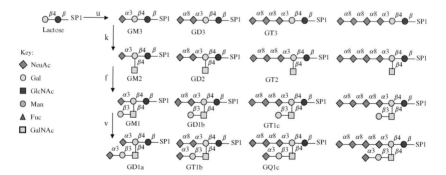

FIG. 2. SCHEME 2. Enzymatic synthesis of ganglio-oligosaccharides. See Table I for enzyme conditions (entries a–z). (See color insert.)

0.39 g (6 %) of GT3 (see Fig. 2). GM3 was completely consumed and converted to GD3 and GT3 using this molar ratio of CMP-Neu5Ac donor. By increasing the molar ratio of CMP-Neu5Ac to 4 equivalent, the reaction was driven to greater than an 80% conversion of the newly formed GD3 to generate larger amounts of GT3 (1.0 g, 20%). Smaller amounts of the tetra-sialic acid $(Sia\alpha2-8)_3Sia\alpha2-3lactose$ (0.25 g, 4%) were also isolated, and traces of higher sialylated fractions were also detected but not purified. Products were subject to ion-exchange chromatography and loaded onto a Dowex 1 × 8-400 formate ion exchange resin (30 × 2.5 cm). Compound GM3 oligosaccharide and free Neu5Ac were in the void fractions, whereas GD3 oligosaccharide, GT3 oligosaccharide, and nucleotides absorbed to the resin. The column was washed with water (500 ml) followed by elution with a gradient of aqueous sodium formate (20 to 80 mM [pH, 7.5], 2.5 l). Fractions containing separated compounds were lyophilized and further purified by size-exclusion chromatography to give pure material (>90%) per nuclear magnetic resonance (NMR) and mass spectrometry analysis.

2. β1-3-galactosylation using cgtB (Table I, entry f). Branched GalN-Acβ1-3 oligosaccharides could be further elongated to the Galβ1-3 GalNAcβ1-4 branch by using the β1-3GalT (cgtB), UDP-glucose (2 equiv), and GalNAcE in high yields. Initiating the reaction at high concentrations of acceptor substrates (>40 mM) is critical for enzyme activity. After 48 hrs' reaction at room temperature, the reaction was purified as described in the General Isolation and Purification section.

O-Glycans

The most common O-linked carbohydrates are based on core structures represented by the Tn-(GalNAcα1-1Thr/Ser) and T-(Galβ1-3GalNAcα1-1Thr/Ser) antigens (Scheme 3; Fig. 3). Sialylated versions of these O-antigens are expressed at low levels by many normal tissues, but they become highly expressed in many types of human malignancies, including colon, breast,

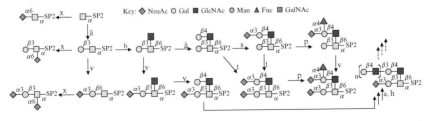

FIG. 3. SCHEME 3. Synthesis of O-glycans. See Table I for enzyme conditions (entries a–z). (See color insert.)

pancreas, ovary, stomach, and lung adenocarcinomas, as well as myelogenous leukemias (Dabelsteen, 1996; Itzkowitz *et al.*, 1989; Mori *et al.*, 1995). On the other hand, *O*-glycans based on the Core-2 epitope (Galβ1-3[GlcNAcβ1-6] GalNAcα1-1Thr/Ser) are expressed in normal tissues. We have been synthesizing various compounds using the key enzymes chST6GalNAc-I and Core-2-β1-6GlcNAcT on acceptors containing the threonine aglycon (see Fig. 3 and Table I).

Notes

1. α2-6-sialylation using chST6GalNAc-I (Table I, entry x). The enzymatic activity was previously evaluated on a set of small acceptor molecules (Blixt *et al.*, 2002), and it was found that an absolute requirement for enzymatic activity is that the anomeric position on GalNAc be α-linked to threonine. Thus, *O*-linked sialosides terminating with a protected threonine could successfully be synthesized on a gram-scale reactions (see Fig. 3). To be able to attach these compounds to other functional groups, the *N*-acetyl protecting group on threonine could be substituted with a biotin derivative before enzymatic extension with chST6GalNAc-I.

Other Glycans

We have demonstrated efficient strategies for generating *N*-acetyllactosamines, ganglio-oligosaccharides, and *O*-glycans. Our enzymatic approach has also been used to expand the library with other glycans as well, such as the globoside series, chemically and enzymatically sulfated glycans, and in the modification of isolated *N*-glycans. Due to space limitations, these categories are only briefly summarized later.

Globoside Glycans

The P blood group antigens are glycan structures displayed by membrane-associated glycosphingolipids present on red cells and on other tissues (Hellberg *et al.*, 2002; Ziegler *et al.*, 2004) . The P1 antigen is formed by the addition of a galactose in an α1,4-linkage to the paragloboside by the P1 α1,4galactosyltransferase, forming the pentasaccharide Galα1-4Galβ1-4GlcNAcβ1-3Galβ1-4Glc. The physiological functions of the P blood group antigens are not known, but these molecules have been implicated in the pathophysiology of urinary tract infections and parvovirus infections (Yang *et al.*, 1994). We have successfully generated various structures of the globoside series using the key enzymes α1,4galactosyltransferase-GalE (Table I, entry g) and β1,3-*N*-acetylgalactosaminyltransferase (Table I, entry d).

Sulfated Glycans

In addition to the glycan structural linkage diversity, sulfation also possesses a large amount of biological information. A number of proteoglycans, glycoproteins, and glycolipids contain sulfated carbohydrates. Their sulfate groups provide a negative charge and play a role in a specific molecular recognition process (Honke and Taniguchi, 2002). We have synthesized a number of sulfated disaccharides of Galβ1-3GlcANc, Galβ1-4GlcNAc, and Galβ1-4Glc at various positions and further performed enzymatic elongation with sialyltransferases and fucosyltransferases from Table I (to be published elsewhere). In addition, we have also begun expressing the sulfotransferases KSGal6ST (Fukuta *et al.*, 1997; Torii *et al.*, 2000) and GlcNAc6ST (Uchimura *et al.*, 1998) for sulfation of poly-LacNAc.

N-*Glycans*

Isolation and purification of natural glycans are complicated and tedious. Nevertheless, it is important to include them in existing and future glycan libraries. Several recent reports demonstrate that isolated N-glycans can be efficiently modified with various glycosyltransferases (Kajihara *et al.*, 2005). We have also taken on this task to expand our library with milligram quantities of isolated and enzymatically diversified N-glycans using the produced enzymes in Table I.

References

Aly, M. R. E., Ibrahim, E.-S. I., El-Ashry, El-S., H. E., and Schmidt, R. R. (2000). Synthesis of lacto-*N*-neohexaose and lacto-*N*-neooctaose using the dimethylmaleoyl moiety as an amino protective group. *Eur. J. Org. Chem.* **2000**, 319–326.

Auge, C., and Crout, D. H. (1997). Chemoenzymatic synthesis of carbohydrates. *Carbohydr. Res.* **305**, 307–312.

Bårström, M., Bengtsson, M., Blixt, O., and Norberg, T. (2000). New derivatives of reducing oligosaccharides and their use in enzymatic reactions: Efficient synthesis of sialyl Lewis a and sialyl dimeric Lewis x glycoconjugates. *Carbohydr. Res.* **328**, 525–531.

Bartolozzi, A., and Seeberger, P. H. (2001). New approaches to the chemical synthesis of bioactive oligosaccharides. *Curr. Opin. Struct. Biol.* **11**, 587–592.

Bintein, F., Auge, C., and Lubineau, A. (2003). Chemo-enzymatic synthesis of a divalent sialyl Lewis(x) ligand with restricted flexibility. *Carbohydr. Res.* **338**, 1163–1173.

Blixt, O., Allin, K., Pereira, L., Datta, A., and Paulson, J. C. (2002). Efficient chemo-enzymatic synthesis of *O*-linked sialyl oligosaccharides. *J. Am. Chem. Soc.* **124**, 5739–5746.

Blixt, O., Brown, J., Schur, M. J., Wakarchuk, W., and Paulson, J. C. (2001). Efficient preparation of natural and synthetic galactosides with a recombinant beta-1,4-galactosyltransferase-/UDP-4'-gal epimerase fusion protein. *J. Org. Chem.* **66**, 2442–2448.

Blixt, O., Collins, B. E., van den Nieuwenhof, I. M., Crocker, P. R., and Paulson, J. C. (2003). Sialoside specificity of the Siglec family assessed using novel multivalent probes: Identification of potent inhibitors of myelin-associated glycoprotein. *J. Biol. Chem.* **278**, 31007–31019.

Blixt, O., Head, S., Mondala, T., Scanlan, C., Huflejt, M. E., Alvarez, R., Bryan, M. C., Fazio, F., Calarese, D., Stevens, J., Razi, N., Stevens, D. J., Skehel, J. J., van Die, I., Burton, D. R., Wilson, I. A., Cummings, R., Bovin, N., Wong, C. H., and Paulson, J. C. (2004). Printed covalent glycan array for ligand profiling of diverse glycan binding proteins. *Proc. Natl. Acad. Sci. USA.* **101,** 17033–17038.

Blixt, O., and Razi, N. (2004). Strategies for synthesis of an oligosaccharide library using a chemoenzymatic approach. *In* "Synthesis of Carbohydrates through Biotechnology" (P. G. Wang and Y. Ichikawa, eds.), Vol. 873, pp. 93–112. American Chemical Society, Washington, DC.

Blixt, O., Vasiliu, D., Allin, K., Jacobsen, N., Warnock, D., Razi, N., Paulson, J. C., Bernatchez, S., Gilbert, M., and Wakarchuk, W. (2005). Chemoenzymatic synthesis of 2-azidoethyl-ganglio-oligosaccharides GD3, GT3, GM2, GD2, GT2, GM1, and GD1a. *Carbohydr. Res.* **340,** 1963–1972.

Boons, G.-J., and Demchenko, A. V. (2000). Recent advances in *O*-sialylation. *Chem. Rev. (Washington, D. C.)* **100,** 4539–4565.

Buskas, T., Li, Y., and Boons, G.-J. (2005). Synthesis of a dimeric Lewis antigen and the evaluation of the epitope specificity of antibodies elicited in mice. *Chemistry J.* **11,** 5457–5467.

Dabelsteen, E. (1996). Cell surface carbohydrates as prognostic markers in human carcinomas. *J. Pathol.* **179,** 358–369.

Eklind, K., Gustafsson, R., Tiden, A. K., Norberg, T., and Aberg, P. M. (1996). Large-scale synthesis of a Lewis B tetrasaccharide derivative, its acrylamide copolymer, and related di- and trisaccharides for use in adhesion inhibition studies with *Helicobacter pylori. J. Carbohydr. Chem.* **15,** 1161–1178.

Endo, T., Koizumi, S., Tabata, K., Kakita, S., and Ozaki, A. (1999). Large-scale production of *N*-acetyllactosamine through bacterial coupling. *Carbohydr. Res.* **316,** 179–183.

Fukuta, M., Inazawa, J., Torii, T., Tsuzuki, K., Shimada, E., and Habuchi, O. (1997). Molecular cloning and characterization of human keratan sulfate Gal-6-sulfotransferase. *J. Biol. Chem.* **272,** 32321–32328.

Gagnon, M., and Saragovi, H. U. (2002). Gangliosides: Therapeutic agents or therapeutic targets? *In* "Expert Opinion on Therapeutic Patents," Vol. 12, pp. 1215–1223. Ashley Publications, Ltd., London, England.

Garegg, P. J. (2004). Synthesis and reactions of glycosides. *Adv. Carbohydr. Chem. Biochem.* **59,** 69–134.

Hakomori, S. (2001). Tumor associated carbohydrate antigens defining tumor malignancy: Basis for development of anti-cancer vaccines. *Adv. Exp. Med. Biol.* **491,** 369–402.

Hakomori, S.-I., and Zhang, Y. (1997). Glycosphingolipid antigens and cancer therapy. *Chem. Biol.* **4,** 97–104.

Hanson, S., Best, M., Bryan, M. C., and Wong, C. H. (2004). Chemoenzymatic synthesis of oligosaccharides and glycoproteins. *Trends Biochem. Sci.* **29,** 656–663.

Hellberg, A., Poole, J., and Olsson, M. L. (2002). Molecular basis of the globoside-deficient P(k) blood group phenotype: Identification of four inactivating mutations in the UDP-*N*-acetylgalactosamine: Globotriaosylceramide 3-beta-*N*-acetylgalactosaminyltransferase gene. *J. Biol. Chem.* **277,** 29455–29459.

Honke, K., and Taniguchi, N. (2002). Sulfotransferases and sulfated oligosaccharides. *Med. Res. Rev.* **22,** 637–654.

Itzkowitz, S. H., Yuan, M., Montgomery, C. K., Kjeldsen, T., Takahashi, H. K., and Bigbee, W. L. (1989). Expression of Tn, sialosyl-Tn, and T antigens in human colon cancer. *Cancer Res.* **49,** 197–204.

Johnson, K. F. (1999). Synthesis of oligosaccharides by bacterial enzymes. *Glycoconjugate J.* **16**, 141–146.

Kajihara, Y., Yamamoto, N., Miyazaki, T., and Sato, H. (2005). Synthesis of diverse asparagine linked oligosaccharides and synthesis of sialylglycopeptide on solid phase. *Curr. Med. Chem.* **12**, 527–550.

Kannagi, R., Izawa, M., Koike, T., Miyazaki, K., and Kimura, N. (2004). Carbohydrate-mediated cell adhesion in cancer metastasis and angiogenesis. *Cancer Sci.* **95**, 377–384.

Koeller, K. M., and Wong, C.-H. (2001). Enzymes for chemical synthesis. *Nature (London)* **409**, 232–240.

Koeller, K. M., and Wong, C. H. (2000). Chemoenzymatic synthesis of sialyl-trimeric-Lewis x. *Chemistry* **6**, 1243–1251.

Lee, H. K., Scanlan, C. N., Huang, C. Y., Chang, A. Y., Calarese, D. A., Dwek, R. A., Rudd, P. M., Burton, D. R., Wilson, I. A., and Wong, C. H. (2004). Reactivity-based one-pot synthesis of oligomannoses: Defining antigens recognized by 2G12, a broadly neutralizing anti–HIV-1 antibody. *Angew Chem. Int. Ed. Engl.* **43**, 1000–1003.

Leppaenen, A., Penttilae, L., Renkonen, O., McEver, R. P., and Cummings, R. D. (2002). Glycosulfopeptides with *O*-glycans containing sialylated and polyfucosylated polylactosamine bind with low affinity to P-selectin. *J. Biol. Chem.* **277**, 39749–39759.

Misra, A. K., Fukuda, M., and Hindsaul, O. (2001). Efficient synthesis of lacosaminylated core-2 *O*-glycans. *Bioorg. Med. Chem. Lett.* **11**, 2667–2669.

Mong, T. K.-K., Huang, C.-Y., and Wong, C.-H. (2003). A new reactivity based one-pot synthesis of N-acetyllactosamine oligomers. *J. Org. Chem.* **68**, 2135–2142.

Mori, S., Suzushima, H., Nishikawa, K., Miyake, H., Yonemura, Y., Tsuji, N., Kawaguchi, T., Asou, N., Kawakita, M., and Takatsuki, K. (1995). Smoldering gamma delta T-cell granular lymphocytic leukemia associated with pure red cell aplasia. *Acta Haematologica* **94**, 32–35.

Nicolaou, K. C., and Mitchell, H. J. (2001). Adventures in carbohydrate chemistry: New synthetic technologies, chemical synthesis, molecular design, and chemical miology. *Angew Chem. Int. Ed. Engl.* **40**, 1576–1624.

Niemela, R., Natunen, J., Majuri, M. L., Maaheimo, H., Helin, J., Lowe, J. B., Renkonen, O., and Renkonen, R. (1998). Complementary acceptor and site specificities of Fuc-TIV and Fuc-TVII allow effective biosynthesis of sialyl-TriLex and related polylactosamines present on glycoprotein counterreceptors of selectins. *J. Biol. Chem.* **273**, 4021–4026.

Ramakrishnan, B., and Qasba, P. K. (2002). Structure-based design of beta 1,4-galactosyltransferase I (beta 4Gal-T1) with equally efficient *N*-acetylgalactosaminyltransferase activity: Point mutation broadens beta 4Gal-T1 donor specificity. *J. Biol. Chem.* **277**, 20833–20839.

Ravindranath, M. H., Gonzales, A. M., Nishimoto, K., Tam, W.-Y., Soh, D., and Morton, D. L. (2000). Immunology of gangliosides. *Indian J. Exper. Biol.* **38**, 301–312.

Schnaar, R. L. (2000). Glycobiology of the nervous system. In "Carbohydrates in Chemistry and Biology" (B. Ernst, G. W. Hart, and P. Sanay, eds.), Vol. 4, pp. 1013–1027. Wiley-VCH, Weinheim, New York.

Schwientek, T., Yeh, J. C., Levery, S. B., Keck, B., Merkx, G., van Kessel, A. G., Fukuda, M., and Clausen, H. (2000). Control of *O*-glycan branch formation: Molecular cloning and characterization of a novel thymus-associated core 2 beta1, 6-*N*-acetylglucosaminyltransferase. *J. Biol. Chem.* **275**, 11106–11113.

Sharon, N., and Lis, H. (1993). Carbohydrates in cell recognition. *Sci. Am.* **268**, 82–89.

Svennerholm, L. (2001). Identification of the accumulated ganglioside. *Adv. Genetics* **44**, 33–41.

Tanaka, H., Adachi, M., Tsukamoto, H., Ikeda, T., Yamada, H., and Takahashi, T. (2002). Synthesis of di-branched heptasaccharide by one-pot glycosylation using seven independent building blocks. *Org. Lett.* **4**, 4213–4216.

Torii, T., Fukuta, M., and Habuchi, O. (2000). Sulfation of sialyl *N*-acetyllactosamine oligosaccharides and fetuin oligosaccharides by keratan sulfate Gal-6-sulfotransferase. *Glycobiology* **10,** 203–211.

Uchimura, K., Muramatsu, H., Kadomatsu, K., Fan, Q. W., Kurosawa, N., Mitsuoka, C., Kannagi, R., Habuchi, O., and Muramatsu, T. (1998). Molecular cloning and characterization of an *N*-acetylglucosamine-6-*O*-sulfotransferase. *J. Biol. Chem.* **273,** 22577–22583.

Ujita, M., McAuliffe, J., Hindsgaul, O., Sasaki, K., Fukuda, M. N., and Fukuda, M. (1999). Poly-*N*-acetyllactosamine synthesis in branched *N*-glycans is controlled by complemental branch specificity of i-extension enzyme and .beta.1,4-galactosyltransferase I. *J. Biol. Chem.* **274,** 16717–16726.

Vasiliu, D., Razi, N., Zhang, Y., Allin, K., Jacobsen, N., Liu, X., Hoffman, J., Bouhorov, O., and Blixt, O. (2006). Large-scale chemoenzymatic synthesis of blood group and tumor associated poly-*N*-acetyllactosamine antigens. *Carbohydr. Res.* **341,** 1447–1457.

Yang, Z., Bergstrom, J., and Karlsson, K. A. (1994). Glycoproteins with Gal alpha 4Gal are absent from human erythrocyte membranes, indicating that glycolipids are the sole carriers of blood group P activities. *J. Biol. Chem.* **269,** 14620–14624.

Ye, X. S., and Wong, C. H. (2000). Anomeric reactivity-based one-pot oligosaccharide synthesis: A rapid route to oligosaccharide libraries. *J. Org. Chem.* **65,** 2410–2431.

Zeng, X., and Uzawa, H. (2005). *Carbohydr. Res.* **340,** 2469–2475.

Ziegler, T., Jacobsohn, N., and Funfstuck, R. (2004). Correlation between blood group phenotype and virulence properties of *Escherichia coli* in patients with chronic urinary tract infection. *Int. J. Antimicrob. Agents* **24**(Suppl. 1), S70–S75.

[10] Glycoconjugate Vaccines Against *Haemophilus influenzae* Type b

By Violeta Fernandez Santana, Luis Peña Icart, Michel Beurret, Lourdes Costa, and Vicente Verez Bencomo

Overview

Glycoconjugation technology leads to the development of prophylactic vaccines against infectious diseases. Several vaccines were introduced starting in the 1980s to fight *Haemophilus influenzae* type b. Their compositions are very diverse, from those using the capsular polysaccharide almost intact before the modification, through oligosaccharide fragments obtained by fragmentation of the capsular polysaccharide, to the most recent, which contain oligosaccharides prepared by chemical synthesis. We describe here some examples that illustrate the methods used in the development of such vaccines.

Glycoconjugates composed of a capsular polysaccharide or its oligosaccharide fragments covalently linked to a protein carrier have been shown to be excellent tools as preventive vaccines against infectious diseases (Jennings and Pon, 1996; Pozsgay, 2000). The first vaccine of this type was introduced for

METHODS IN ENZYMOLOGY, VOL. 415
0076-6879/06 $35.00
DOI: 10.1016/S0076-6879(06)15010-7

the immunization of children against *Haemophilus influenzae* type b (Hib) in the 1990s and proved highly efficient in preventing bacterial meningitis and pneumonia caused by this organism (Peltola, 2000). The success of conjugate vaccines against Hib prompted further development of other glycoconjugate vaccines against meningococcus and pneumococcus (Lindberg, 1999).

The capsular polysaccharide of Hib, a polymer of ribose-ribitol phosphate (PRP; Fig. 1), is suitable for the application of a wide variety of modification and glycoconjugation procedures. As a result, several glycoconjugate vaccines are presently available against Hib. All of them demonstrated high efficacy in clinical trials (Robbins *et al.*, 1996). Existing vaccines could be divided into three types according to the size and the source of PRP: (1) native polysaccharide (nPRP); (2) oligosaccharides obtained by fragmentation of oPRP; and (3) oligosaccharides obtained by chemical synthesis (sPRP).

For the manufacture of conjugates of type 1, the full-size or slightly depolymerized polysaccharide is randomly activated at several points for conjugation to a protein carrier. As a result, the conjugates resemble a high molecular mass lattice formed by several cross-linked protein and polysaccharide molecules. Several of the currently commercialized anti-Hib conjugate vaccines could be included in this group. For example, the polysaccharide activated with cyanogen bromide, followed by substitution with adipic acid dihydrazide (ADH; Fig. 2) and conjugation to tetanus (Chu *et al.*, 1983) or diphtheria toxoids mediated by 1-ethyl-3 (3-dimethylaminopropyl) carbodiimide (EDC), is the basis for vaccines manufactured by Sanofi-Pasteur (formerly Aventis), Connaught, and Glaxo-SmithKline. In the vaccine manufactured by Merck, the polysaccharide is activated by carbonyldiimidazole, followed by reaction with butane diamine and coupling to thiolated *Neisseria meningitis* outer membrane protein complex (Marburg *et al.*, 1986).

A second type of conjugate vaccines uses a rather smaller oligosaccharide fragment obtained by fragmentation of the capsular polysaccharide and

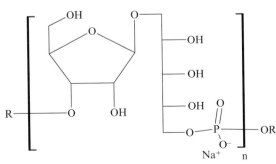

FIG. 1. Structure of the *Haemophilus influenzae* type b capsular polysaccharide.

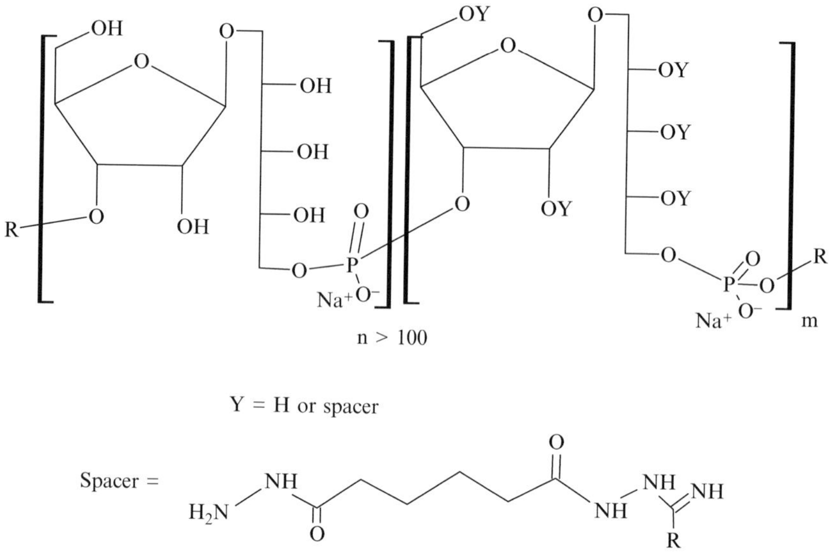

FIG. 2. Structure of PRP modified by cyanogen bromide followed by ADH.

further activation at one end. One of the commercially available vaccines is manufactured by periodic oxidation of the capsular polysaccharide (Anderson et al., 1986). The resulting oligosaccharide fragments have aldehyde groups at both ends (Fig. 3) and are coupled to CRM_{197} diphtheria mutant toxin by reductive amination with sodium cyanoborohydride. In another example, the polysaccharide is fragmented by acid hydrolysis (Costantino et al., 1999), and the resulting oligosaccharide is aminated through its ribose moiety with ADH, followed by coupling to CRM_{197} mediated by EDC.

An additional new type of conjugate was recently introduced (Vérez-Bencomo et al., 2004), using synthetic oligosaccharide fragments that reproduce the structure of the capsular polysaccharide. The oligosaccharide is obtained already with a spacer (Fig. 4) that is further activated, and it is coupled to tetanus toxoid (TTd) using maleimido-thiol chemistry.

At the molecular level, conjugates of types 1 and 2–3 are very different. This difference did not affect their clinical behavior that, otherwise, is quite similar (Granoff and Holmes, 1991) but strongly influences the quality control strategy. The following are the most important parameters to be defined in the glycoconjugate: molecular integrity, carbohydrate-protein ratio, unconjugated PRP or protein, and residual reagents.

The molecular integrity of the glycoconjugates is easily analyzed by size exclusion chromatography (SEC)–high-performance liquid chromatography

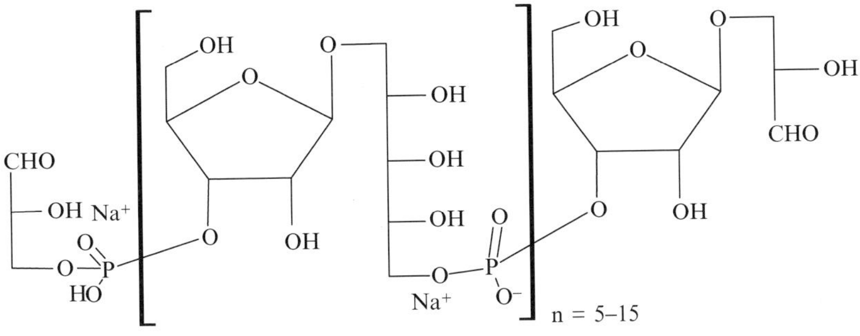

FIG. 3. Structure of oligosaccharides obtained by periodic oxidation of PRP.

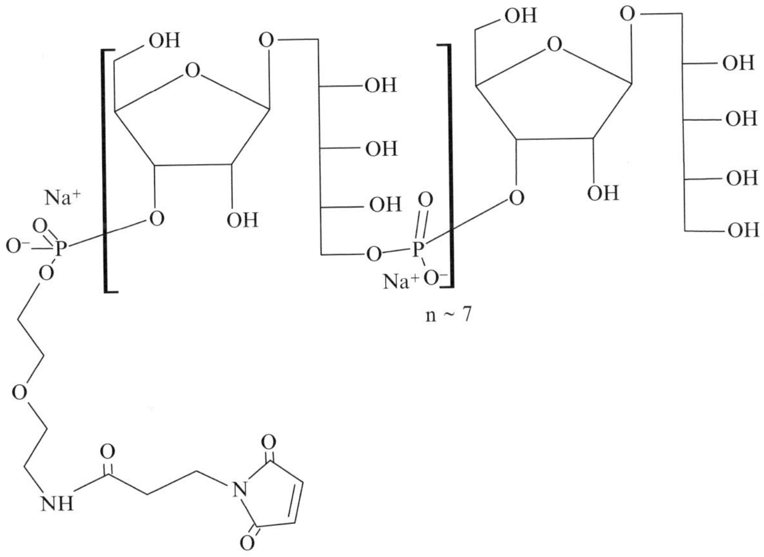

FIG. 4. Structure of the synthetic oligosaccharides employed in the manufacture of conjugated vaccine.

(HPLC) (Plumb and Yost, 1996). Columns like TSKgel 5000 PW (Tosoh) could be very efficiently used for oligosaccharide conjugates of either type 2 or 3. A typical chromatogram is shown in Fig. 5B. Glycoconjugates of type 1 are better analyzed with an OHpak 805 column (Showa Denko, Fig. 5A).

The integrity of the oligosaccharide either after modification or in glycoconjugates could be ascertained by ^{1}H–nuclear magnetic resonance (NMR) spectroscopy as shown in Fig. 6.

The carbohydrate-protein ratio is usually defined by colorimetric methods. Protein determination is generally performed by the method of

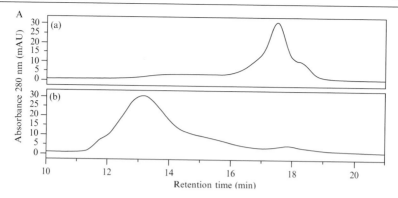

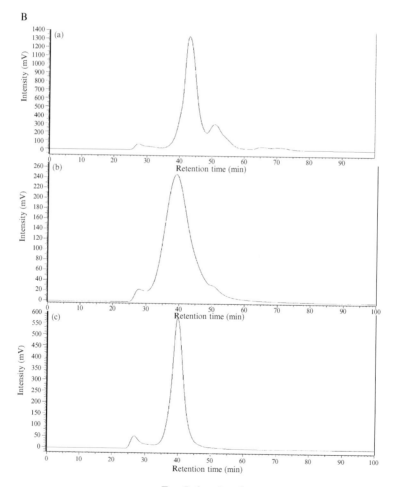

FIG. 5. (*continued*)

Lowry *et al.* (1951) unless reagents or buffer components prevent its use. Alternatives are the bicinchoninic acid (Smith *et al.*, 1985) and Peterson (1979) methods. PRP determination is generally performed by the orcinol method (Ashwell, 1957), with ribose as the standard. High-performance anion exchange chromatography with pulsed amperometric detection can be used very efficiently for the quantification of PRP through ribose-ribitol-phosphate (Tsai *et al.*, 1994), or ribitol (Bardotti *et al.*, 2000). The residual reagents can be monitored by a variety of methods including chromatography but also colorimetric assays. For example, dithiotreitol can be quantified by the Ellman (1958) method, and ADH by the trinitrobenzene sulfonic acid method (Habeeb, 1966).

Unconjugated PRP can be quantified after separation by ultrafiltration in conjugates of types 2 and 3. The task is more complex for separation of full-size polysaccharide in conjugates of type 1, but this can be accomplished, for example, after selective acid precipitation by deoxycholate-HCl (Guo *et al.*, 1998; Lei *et al.*, 2000).

Unconjugated protein can be monitored by SEC-HPLC on a TSKgel 5000 PW column for conjugates 2 and 3, and on an OHpak 805 column for conjugate 1 (see, for example, Fig. 5).

We selected three examples: (1) conjugation of the native polysaccharide through cyanogen-bromide-ADH modification followed by EDC-mediated amidation; (2) periodic oxidation of the native polysaccharide followed by reductive amination with sodium cyanoborohydride; and (3) activation of synthetic oligosaccharide with maleimidopropionic acid followed by conjugation to thiolated TTd. These examples (see Figs. 2, 3, and 4) illustrate methods currently used in glycoconjugate vaccine manufacture. The following description includes the methods for carbohydrate activation, the conjugation procedure for TTd, and some of the analytical results obtained for characterization of the conjugates.

Experimental Procedures

General Methods

In all the procedures described, the removal of low-molecular-mass reagents and purification of the conjugates usually exploits the differences in molecular mass between the product and reagents. The process could be

---

FIG. 5. (A) SEC-HPLC chromatograms of TTd and PRP conjugate of type 1. (a) TTd, OHpak 805, (b) nPRP-TTd, OHpak 805. (B) SEC-HPLC chromatograms of TTd and PRP conjugates of type 2 and 3. (a) TTd, TSK-5000, (b) oPRP-TTd, TSK-5000, (c) sPRP-TTd, TSK-5000.

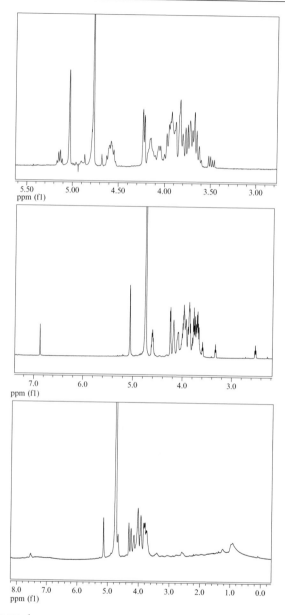

FIG. 6. 250 MHz ^{1}H-NMR spectrum of oPRP, sPRP and glycoconjugate sPRP-TTd.

performed either by SEC, hydrophobic interaction chromatography, or ultrafiltration. For simplicity, we uniformed most experiments and used only ultrafiltration on pressure cells with cellulose acetate membrane. In all cases, TTd was first buffer-exchanged to the buffer used in the conjugation process.

Methods for Modification-Activation of PRP

1. *Direct activation of nPRP by cyanogen bromide-adipic acid dihydrazide procedure.* nPRP (20 mg) was chilled to about 2° and then brought to a pH of 10 to 11 by the addition of carbonate-bicarbonate buffer (Kohn and Wilchek, 1986). A solution of cyanogen bromide (CNBr; 20 mg, 0.2 mmol; 5 M in acetonitrile) was then added while the mixture was stirred. The reaction proceeded for about 10 min, at which point the pH was lowered to a value of 8 to 9. A pre-chilled solution of ADH (360 mg, 2.1 mmol) was then added. Both the CNBr and the ADH reactions were monitored with a pH meter, optionally combined with a pH-stat apparatus (Metrohm) to control the addition of reagents. After about 16 h, the resulting reaction mixture was ultrafiltered on a 5- to 30-kDa cutoff membrane until free ADH was below the detectable level (Habeeb, 1966). The ADH-modified PRP (PRP-ADH) was then buffer-exchanged to a pH of 4 to 6. This solution could be stored safely at $-20°$ for several months. The yield measured 18 mg.

2. *Periodic oxidation for direct synthesis of activated PRP oligosaccharides (oPRP).* To a solution of nPRP (10 mg) in water (1 ml) was added sodium periodate ($NaIO_4$, 2.4 mg), and the mixture was stirred in the dark for 20 min. Glycerol (1 μl) was added to quench the reaction, and the stirring was continued for another 20 min. The solution was diafiltered against water (cellulose acetate membrane, 1-kDa cutoff membrane). A sample of oxidized PRP oligosaccharides (oPRP) was lyophilized and analyzed by NMR spectroscopy. A series of signals were detected in two regions: δ 3.5 to 3.8 ppm and 5.15 to 5.17 ppm. The yield was 8 mg.

3. *Activation of synthetic PRP oligosaccharide (sPRP).* sPRP (20 mg) was dissolved in anhydrous dimethylsulfoxide (4 ml), and to this solution N-hydroxysuccinimidyl 3-maleimidopropionate (5.4 mg, 20 μmol) was added. The stirring was continued for 12 h. The reaction mixture was then transferred to a centrifuge tube, and dioxane (16 ml) was added dropwise. The mixture was centrifuged and the liquid was discarded. The solid was redissolved in water and lyophilized. The yield was 16.8 mg.

Conjugation to Tetanus Toxoid

1. *Preparation of nPRP-TTd by EDC-mediated conjugation.* To a solution of TTd (15 mg; 0.1 μmol; 1.5 ml; pH, 4 to 6) was added a solution

of PRP-ADH (15 mg; 1.5 ml; pH, 4 to 6), while the solution was gently stirring, at about 4°. EDC (15 mg, 0.08 mmol) was then added, and the reaction was allowed to proceed for about 6 h. Aliquots were removed at regular intervals to monitor the reaction progress by SEC-HPLC (using OHpak 805). When the reaction was complete, which was indicated by the near-total disappearance of the unconjugated protein carrier, it was stopped by neutralization. The resulting solution was purified by gel permeation chromatography on Sepharose CL-4B (Chu et al., 1983). The void volume fractions containing the purified nPRP-TTd conjugate were pooled and then buffer-exchanged to a pH of 6.5 to 7.5. This solution could be stored safely at 0° to 5° for several months. A PRP to TTd ratio in the 0.3 to 0.6 (w/w) range is optimal as vaccine for human use. The level of unconjugated PRP-ADH was also measured (Guo et al., 1998) and usually found to be in the 5% to 20% range. A human vaccine should not contain more than 20% free PRP.

2. *Preparation of oPRP-TTd from oxidized oligosaccharides by reductive amination.* oPRP (8 mg) was added to a solution of TTd (12 mg, 0.08 μmol) in PBS (pH 7.0, 2,6 ml), and then sodium cyanoborohydride (5 mg) was added. The mixture was stirred at 37° for 3 days. The solution was then diafiltered against PBS (pH, 7.2; 30 kDa cutoff membrane).

The conjugate was stable in buffer solution at 4° for months. The carbohydrate–protein ratio could be controlled by the molar ratio of reactants. Usually, a ratio on the order of 0.3 is optimal as vaccine for human use. The level of unconjugated PRP was also measured by ultrafiltration and quantificationandusually found to be below 5%. A sample was lyophilized and analyzed by NMR spectroscopy and SEC-HPLC (using TSK-5000) with PBS as eluent.

3. *Preparation of sPRP-TTd by maleimido-thiol chemistry* (Fernandez-Santana et al., 2004). Under N_2 atmosphere, a solution of N-hydroxysuccinimide dithiopropionate (1.6 mg, 4 μmol) in dimethylsulfoxide (50 μl) was added to a solution of TTd (23 mg, 0.15 μmol) in PBS (pH, 8; with EDTA 1.86 g/l, 5.0 ml). After 2 hrs, dithiotreitol (19.3 mg, 25 mmol/l) was added (under N_2 gas), and the mixture was stirred at 4° for 1 hr. The resulting solution was diafiltered using N_2 as the pressure source (pH, 7.2; regenerated cellulose membrane; 30 kDa cutoff membrane). Protein and SH content was analyzed by the Lowry (Lowry et al., 1951) and Ellman (1958) methods, respectively. A 20- to 25-molar substitution was usually attained under this condition. A solution of activated sPRP (16.8 mg) previously dissolved in PBS (pH, 7.2; 0.4 ml) was added under N_2 atmosphere. The resulting solution was gently stirred for several hours at 4 to 8°. The reaction was then diafiltered against PBS (pH, 7.2; cellulose acetate membrane; 30 kDa cutoff membrane).

The conjugate was stable in buffer solution at 4° for months. The carbohydrate-protein ratio could be controlled by the molar ratio of reactants. The level of unconjugated PRP was also measured after ultrafiltration and was usually found to be below 5%. A ratio of 0.32 to 0.48 was confirmed in clinical evaluation to be optimal as vaccine for human use (Verez-Bencomo et al., 2004).

References

Anderson, P. W., Pichichero, M. E., Insel, R. A., Betts, R., Eby, R., and Smith, D. H. (1986). Vaccines consisting of periodate-cleaved oligosaccharides from the capsule of Haemophilus influenzae type b coupled to a protein carrier: Structural and temporal requirements for priming in the human infant. J. Immunol. 137, 1181–1186.

Ashwell, G. (1957). Colorimetric analysis of sugars. Methods Enzymol. III, 73–105.

Bardotti, A., Ravenscroft, N., Ricci, S., D'Ascenzi, S., Guarnieri, V., Averani, G., and Constantino, P. (2000). Quantitative determination of saccharide in Haemophilus influenzae type b glycoconjugate vaccines, alone and in combination with DPT, by use of high-performance anion-exchange chromatography with pulsed amperometric detection. Vaccine 18, 1982–1993.

Chu, C.-Y., Schneerson, R., Robbins, J. B., and Rastogi, S. C. (1983). Further studies on the immunogenicity of Haemophilus influenzae type b and pneumococcal type 6A polysaccharide-protein conjugates. Infect. Immunol. 40, 245–256.

Costantino, P., Norelli, F., Giannozzi, A., D'Ascenzi, S., Bartoloni, A., Kaur, S., Tang, D., Seid, R., Viti, S., Paffetti, R., Bigio, M., Pennatini, C., Averani, G., Guarnieri, V., Gallo, E., Ravenscroft, N., Lazzeroni, C., Rappuoli, R., and Ceccarini, C. (1999). Size fractionation of bacterial capsular polysaccharides for their use in conjugate vaccines. Vaccine 17, 1251–1263.

Ellman, G. L. (1958). A colorimetric method for determining low concentrations of mercaptans. Arch. Biochem. Biophys. 74, 443–450.

Fernandez-Santana, V., Cardoso, F., Rodriguez, A., Carmenate, T., Peña, L., Valdes, Y., Hardy, E., Mawas, F., Heynngnezz, L., Rodríguez, M. C., Figueroa, I., Chang, J., Toledo, M. E., Musacchio, A., Hernandez, I., Izquierdo, M., Cosme, K., Roy, R., and Verez-Bencomo, V. (2004). Antigenicity and Immunogenicity of a Synthetic Oligosaccharide-Protein Conjugate Vaccine against Haemophilus influenzae Type b. Infect. Immunol. 72, 7115–7123.

Granoff, D. M., and Holmes, S. J. (1991). Comparative immunogenicity of Haemophilus influenzae type b polysaccharide-protein conjugate vaccine. Vaccine 9, S30–S34.

Guo, Y.-Y., Anderson, R., McIver, J., Gupta, R. K., and Siber, G. R. (1998). A simple and rapid method for measuring unconjugated capsular polysaccharide (PRP) of Haemophilus influenzae type b in PRP-tetanus toxoid conjugate vaccine. Biologicals 26, 33–38.

Habeeb, A. F. S. A. (1966). Determination of free amino groups in proteins by trinitrobenzenesulfonic acid. Anal. Biochem. 14, 328–336.

Jennings, H. H., and Pon, R. A. (1996). Polysaccharides and glycoconjugates as human vaccines. In "Polysaccharides in Medicinal Applications" (S. Dimitriu, ed.), pp. 473–479. Marcel Dekker, New York.

Kohn, J., and Wilchek, M. (1986). The use of cyanogen bromide and other novel cyanylating reagents for the activation of polysaccharide resins. Appl. Biochem. Biotechnol. 9, 285–305.

Lei, Q. P., Lamb, D. H., Heller, R., and Pietrobon, P. (2000). Quantitation of low level unconjugated polysaccharide in tetanus toxoid-conjugate vaccine by HPAEC/PAD following rapid separation by deoxycholate-HCl. *J. Pharmaceut. Biomed. Anal.* **21,** 1087–1091.

Lindberg, A. A. (1999). Glycoprotein conjugate vaccine. *Vaccine* **17,** 28–36.

Lowry, O. H., Rosebrough, N. J., Farr, A. L., and Randall, R. J. (1951). Protein measurement with the Folin phenol reagent. *J. Biol. Chem.* **193,** 265–275.

Marburg, S., Jorn, D., Tolman, R. L., Arison, B., McCauley, J., Kniskern, P. J., Hagopian, A., and Vella, P. P. (1986). Bimolecular chemistry of macromolecules: Synthesis of bacterial polysaccharide conjugates with *Neisseria meningitidis* membrane protein. *J. Am. Chem. Soc.* **108,** 5282–5287.

Peltola, H. (2000). Worldwide *Haemophilus influenzae* type b disease at the beginning of the 21st century: Global analysis of the disease burden 25 years after the use of the polysaccharide vaccine and a decade after the advent of conjugates. *Clin. Microbiol. Rev.* **13,** 302–317.

Peterson, G. L. (1979). Review of the Folin phenol protein quantitation method of Lowry, Rosebrough, Farr and Randall. *Anal. Biochem.* **100,** 201–220.

Plumb, J. E., and Yost, S. E. (1996). Molecular size characterization of *Haemophilus influenzae* type b polysaccharide-protein conjugate vaccines. *Vaccine* **14,** 399–404.

Pozsgay, V. (2000). Oligosaccharide-protein conjugates as vaccine candidate against bacteria. *Adv. Carbohydr. Chem. Biochem.* **26,** 153–198.

Robbins, J. B., Schneerson, R., Anderson, P., and Smith, D. H. (1996). Prevention of systemic infections, especially meningitis, caused by *Haemophilus influenzae* type b: Impact on public health and implications for other polysaccharide-based vaccines. *J. Am. Chem. Soc.* **276,** 1181–1185.

Smith, P. K., Krohn, R. I., Hermanson, G. T., Mallia, A. K., Gartner, F. H., Provenzano, M. D., Fujimoto, E. K., Goeke, N. M., Olson, B. J., and Klenk, D. C. (1985). Measurement of protein using bicinchoninic acid. *Anal. Biochem.* **150,** 76–85.

Tsai, C.-M., Gu, X.-X., and Byrd, R. A. (1994). Quantification of polysaccharide in *Haemophilus influenzae* type b conjugate and polysaccharide vaccines by high-performance anion-exchange chromatography with pulsed amperometric detection. *Vaccine* **12,** 700–706.

Vérez-Bencomo, V., Fernández-Santana, V., Hardy, E., Toledo, M. E., Rodríguez, M. C., Heynngnezz, L., Rodríguez, A., Baly, A., Herrera, L., Izquierdo, M., Villar, A., Valdés, Y., Cosme, K., Deler, M. L., Montane, M., García, E., Ramos, A., Aguilar, A., Medina, E., Toraño, G., Sosa, I., Hernández, I., Martínez, R., Muzachio, A., Carmenates, A., Costa, L., Cardoso, F., Campa, C., Díaz, M., and Roy, R. (2004). A synthetic conjugate polysaccharide vaccine against *Haemophilus influenzae* type b. *Science* **305,** 522–525.

[11]   Assay of Human Gastric Mucin as a Natural Antibiotic Against *Helicobacter pylori*

By Minoru Fukuda, Masatomo Kawakubo, Yuki Ito, Motohiro Kobayashi, Heeseob Lee, and Jun Nakayama

Abstract

Helicobacter pylori infects more than half of the world's population and is considered a leading cause of peptic ulcer and gastric carcinoma. Although a large number of persons are infected with *H. pylori*, only a limited number of those infected (approximately 3%) develop peptic ulcers and gastric carcinoma. The progression of the disease is restricted by deeper portion of the gastric mucosa, and in many persons glandular atrophy appears to be prevented by mucins secreted in the deeper portion of the mucosa. Recent studies have shown that this inhibitory activity is at least partly due to the expression of $\alpha 1,4$-*N*-acetylglucosamine residues attached to the mucin (MUC6) in the deeper portion of the mucosa. $\alpha 1,4$-*N*-acetylglucosamine residues inhibit cholesterol α-glucosyltransferase, the product of which constitutes a major component of *H. pylori* cell wall. This inhibitory activity is thus regarded as a natural antibiotic function. This chapter describes the assay for antibiotic activity of MUC6 mucin against *H. pylori* infection and growth as well as inhibition by $\alpha 1,4$-*N*-acetylglucosamine–capped mucin-type oligosaccharides.

Overview

Helicobacter pylori is a gram-negative bacteria causing gastric diseases such as chronic gastritis, peptic ulcer, and gastric cancer. It was first isolated by Marshall and Warren (1984) in Australia. Indeed, Marshall discovered the pathogenicity of *H. pylori* by drinking the bacteria to give himself some of the symptoms of acute gastritis. *H. pylori* provides one of the most exemplified cases in which bacterial infection leads to inflammation and cancer. For their contribution, Marshall and Warren received the Nobel Prize in Physiology or Medicine in 2005.

H. pylori infects more than 50% of the world's population. In early stages of infection, *H. pylori* organisms colonize the superficial layer of gastric mucosa (Hidaka *et al.*, 2001). If untreated, this type of chronic gastritis is occasionally associated with glandular atrophy (Sipponen and Hyvarinen, 1993). In the early stages of *H. pylori* infection and colonization, the bacteria facilitate the process by adhesion of BabA on the bacterium to Lewis b

METHODS IN ENZYMOLOGY, VOL. 415
0076-6879/06 $35.00
DOI: 10.1016/S0076-6879(06)15011-9

blood antigen secreted from surface mucous cells (Ilver *et al.*, 1998). In the advanced stages of infection, SabA on *H. pylori* binds to sialyl dimeric Lewis X glycolipid, expressed in surface mucous cells under inflammatory stress (Mahdavi *et al.*, 2002; see also Aspholm *et al.*, 2006).

During chronic active gastritis, lymphoplasmacytic infiltration takes place in the lamina propria. This inflammatory response is followed by intestinal metaplasia, dysplasia, and eventually gastric adenocarcinoma (Nomura *et al.*, 1991; Parsonnet *et al.*, 1991). We showed that the progression of the disease is highly correlated with the formation of peripheral lymph node addressin (PNAd), which expresses L-selectin ligands, 6-sulfo sialyl Lewis X (Kobayashi *et al.*, 2004). By comparing different stages of *H. pylori*–induced inflammation, we found that the abundance of PNAd was highly correlated with disease progression. Importantly, PNAd formed in high endothelial venules (HEV)-like vessels that disappear once *H. pylori* is eradicated by antibiotics, indicating that *H. pylori* directly induces HEV-like vessels. HEV plays an essential role in lymphocyte recruitment in lymphocyte circulation. These results thus indicate that *H. pylori*–induced inflammation is facilitated by the formation of HEV-like vessels. Indeed, strong lymphocyte recruitment was observed in infected tissue (Kobayashi *et al.*, 2004).

One of the major observations in the gastric mucosa infected by *H. pylori* is that this microbe is largely associated with surface mucous cell–derived mucin and rarely found in the deeper portion of the gastric mucosa (Hidaka *et al.*, 2001). Taking into account that only a small fraction (an estimated 3%) of infected persons will develop advanced stages of the disease such as peptic ulcer and gastric carcinoma (Peek and Blaser, 2002), it is possible that mucin secreted from deeper layer of the gastric mucosa plays a protective role against *H. pylori*.

A dual histochemical staining termed *galactose oxidase-cold thionin Schiff-paradoxical Concanavalin A staining* distinguished two distinct types of gastric mucous layer (Ota *et al.*, 1991). The gland mucous cells such as mucous neck cells and pyloric gland cells located in deeper portion of the mucosa were found to react with monoclonal antibodies specific to α1,4-*N*-acetylglucosamine (α4GlcNAc) residues (Nakayama *et al.*, 1999). It was then hypothesized that α4GlcNAc residues in the gland mucous cell may be responsible for preventing *H. pylori* from attacking the deeper portions of the mucosa. Since α1,4-*N*-acetylglucosaminyltransferase (α4GnT) was cloned by an expression cloning strategy (Nakayama *et al.*, 1999), it was possible to test the hypotheses. The results, as described later, demonstrate that α4GlcNAc-capping structures of *O*-glycans inhibit *H. pylori* growth and function as an antibiotic against *H. pylori* (Kawakubo *et al.*, 2004).

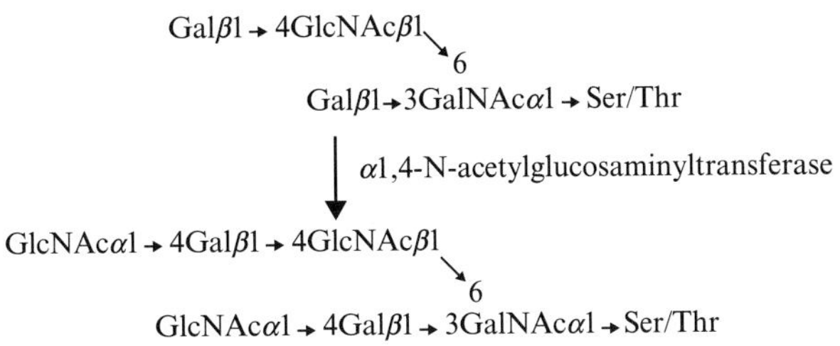

$$\text{Gal}\beta1 \rightarrow 4\text{GlcNAc}\beta1$$
$$\searrow 6$$
$$\text{Gal}\beta1 \rightarrow 3\text{GalNAc}\alpha1 \rightarrow \text{Ser/Thr}$$

$$\downarrow \alpha1,4\text{-N-acetylglucosaminyltransferase}$$

$$\text{GlcNAc}\alpha1 \rightarrow 4\text{Gal}\beta1 \rightarrow 4\text{GlcNAc}\beta1$$
$$\searrow 6$$
$$\text{GlcNAc}\alpha1 \rightarrow 4\text{Gal}\beta1 \rightarrow 3\text{GalNAc}\alpha1 \rightarrow \text{Ser/Thr}$$

FIG. 1. Biosynthesis of $\alpha1,4$-GlcNAc–capped O-glycans. Core 2–branched O-glycans serve as acceptor substrates for $\alpha1,4$-N-acetylglucosaminyltransferase ($\alpha4$GnT). $\alpha1,4$-GlcNAc–capping structure can be added to both core 2–branch and core 1 side-chains (Nakayama *et al.*, 1999).

In this chapter, we first describe the method for assaying *H. pylori* growth and adhesion to human gastric adenocarcinoma cells, which were transfected to express $\alpha4$GlcNAc (Fig. 1). Second, we will describe production of recombinant mucin-type glycoproteins expressing $\alpha4$GlcNAc. Third, we will describe the purification of two human mucin types, glycoprotein MUC5AC and MUC6, which represent mucins derived from the surface mucous cells and pyloric gland cells, respectively (Nordman *et al.*, 2002). Fourth, we will describe assay for *H. pylori* growth in the presence of those recombinant or natural mucin glycoproteins. Fifth, we will describe the quantitation of cholesterol α-glucoside in the cell wall and its biosynthesis in the presence of mucin glycoprotein expressing $\alpha4$GlcNAc. We then describe assays for *H. pylori* cholesterol α-glucosyltransferase activity in the presence of recombinant mucin glycoproteins.

Culture of *H. pylori* with AGS Cells Stably Expressing $\alpha4$GlcNAc-Capping Structure

Gastric adenocarcinoma AGS cells are commonly used for studies of interaction with *H. pylori* (Higashi *et al.*, 2002). Stable transfection of pcDNAI-$\alpha4$GnT resulted in AGS cells expressing $\alpha4$GlcNAc-capping structures, which were detected by immunofluorescent staining using $\alpha4$GlcNAc-specific HIK1083 antibody (Kanto Chemical, Tokyo, Japan) (Ishihara *et al.*, 1996), establishing the AGS-$\alpha4$GnT line.

The effects of $\alpha4$GlcNAc-capping structure on *H. pylori* were examined by overlaying *H. pylori* on mock-transfected control AGS cells and AGS- $\alpha4$GnT cells. Initially, 1.5×10^5 cells/ml of control AGS and AGS-α4GnT are plated in six-well plates containing cover glasses. After 24 hrs, 1×10^7 cells/ml of *H. pylori* were introduced to those six-well plates and

cultured for an additional 8 or 24 hrs. The cells were then fixed in 20% buffered formalin and incubated with antibodies against *H. pylori* (Dako, Glostrup, Denmark) and mouse monoclonal antibody HIK1083 specific to α4GlcNAc structure (Ishihara *et al.*, 1996). After washing with phosphate-buffered saline (PBS), rhodamine-labeled anti-rabbit immunoglobulins (for anti–*H. pylori* antibody) and fluorescein isothiocyanate-labeled anti-mouse IgM (for the HIK1083 antibody) were added. Cover glasses were then mounted on glass slides using Vectashield (Vector Laboratories, Burlingame, CA) and viewed under a Zeiss laser confocal microscope. Alternatively, a fluorescence microscope could also be used. Previous results showed that there is no significant difference of the numbers of *H. pylori* organisms attached to AGS-α4GnT cells and control AGS cells after 8 hrs of incubation. However, the number of *H. pylori* and AGS-α4GnT cells was dramatically decreased after 24 hrs of incubation. Moreover, it was apparent that control AGS cells were damaged as visualized under Nomarski optics, whereas AGS-α4GnT cells were not.

To determine how AGS cells are affected by *H. pylori*, the numbers of intact AGS or AGS-α4GnT cells were estimated by CellTiter 96 Aqueous one solution-cell proliferation assay kit (Promega, Madison, MI). AGS-α4GnT and mock-transfected AGS cells were plated at a density of 2×10^4 cells per well in 96-well plates. After culturing for 20 hrs, 1×10^6 cells of *H. pylori* were added to each well, and AGS cells and *H. pylori* cells were cocultured up to 4 days. The assay was performed every 24 hrs in triplicate; that is, the color reaction was developed after removing *H. pylori* with PBS washing and replacing the culture medium with 120 μl of reaction buffer containing 20 μl of MTS reaction mixture (Promega) and 100 μl of Dulbecco's Modified Eagle Medium. The reaction was terminated by adding 25 μl of 10% SDS to the reaction mixture, and absorbance was measured at 490 nm using a microplate spectrophotometer with 650 nm as reference. The viability of control AGS cells was significantly reduced after the third day, whereas AGS-α4GnT cells were fully viable up to 4 days. These results indicate that the presence of α4GlcNAc-capped *O*-glycans did not interfere the binding of *H. pylori* to AGS cells but protected the host cells against this microbe (Kawakubo *et al.*, 2004).

Expression of CD43 Expressing α4GlcNAc

Chinese hamster ovary (CHO) cells synthesize relatively simple mucin type *O*-glycans, and core 2–branched *O*-glycans or *O*-glycans containing α4GlcNAc are virtually absent in CHO cells (Sasaki *et al.*, 1987). To produce mucin type *O*-glycans, soluble CD43 (leukosialin) protein (sCD43) was used as a scaffold. Thus, DNA fragment encoding the entire extracellular domain

of CD43 (amino acid residues 20 to 254) was amplified by polymerase chain reaction (PCR) with KpnI- and XhoI-tagged primers using pRcCMV-leu as a template and then subcloned into KpnI and XhoI sites of pSecTag2 (Invitrogen, Carlsbad, CA). This vector harbors Igκ leader peptide followed by extracellular domain of CD43, a myc epitope, and $(His)_6$, thus forming pSecTag2-sCD43.

Alternatively, CD43 protein fused with the hinge and constant (Fc) regions of IgG can be used. For this preparation, CD43 cDNA was obtained using PCR and cDNA flanked by EcoRI (5'-end) and BglII (3'-end) sites. The BglII site was fused to an oligonucleotide sequence corresponding to amino acid residues 247 to 252. PCR product thus harbors the sequence that encompasses seven nucleotides upstream from the methionine initiation codon and amino acid residue right before the transmembrane region. In parallel, genomic DNA-encoding the human IgG hinge plus constant regions was excised by Bam HI/XbaI digestion of pcDM8-CD4 IgG (Aruffo et al., 1990). The EcoRI-BglII fragment of soluble CD43 cDNA and the BamHI-XbaI fragment of the human IgG hinge plus constant region were ligated into EcoRI and XbaI sites of pSRα vector, yielding pSRα-sCD43 · IgG (Sawada et al., 1994). pcDNAI vector harboring core2β1,6-N-acetylglucosaminyltransferase I (C2GnT-I) (Bierhuizen and Fukuda, 1992) and another pcDNAI vector harboring α4GnT (Nakayama et al., 1999) were isolated by expression cloning strategy as described previously.

To increase the efficiency of protein production derived from pcDNAI, which contains the polyoma large T–binding site, polyoma large T–expressing pSVE1-PyE can be introduced (Bierhuizen and Fukuda, 1992). pcDNAI plasmids are highly amplified in those cells expressing pSVE1-Py-E, and the amount of protein encoded in pcDNAI should be substantially increased. To increase α4GlcNAc-containing CD43 proteins, CHO mutant Lec2 cells can be used. Lec2 cells lack the CMP (cytidine monophosphate)-sialic acid transporter and thus lack Golgi sialylation (Deutscher et al., 1984). α4GnT and sialyltransferases compete for the same acceptor. In the absence of Golgi- sialylation, the yield of α4GlcNAc-capping O-glycans should be substantially increased.

Lec2 cell are thus transfected with pcDNAI-C2GnT, pSecTag2-sCD43, or psRα-sCD43 · IgG and pPSVE1-PE together with pcDNAI-α4GnT or pcDNAI vector using Lipofectamine 2000 (Invitrogen). Lec2 cells are cultured in α-MEM with 10% fetal calf serum. After 1 week of culture from transfection, the medium is concentrated using Centriprep YM-30 (Millipore, Bedford, MA). The amount of α4GlcNAc-capping structures can be estimated by an enzyme-linked immunosorbent assay (ELISA) using HIK1083 antibody. In this assay, one unit is defined as immunoreactivity equivalent to 1 mg (2.9 μmol) of p-nitrophenyl N-acetylglucosamine,

according to the manufacturer's instructions. To determine whether protein carrier influences inhibitory efficiency, CD34 expressing α4GlcNAc-capping structure is produced using pcDNA3-sCD34-IgG encoding a soluble CD34-IgG chimeric protein.

H. *pylori* (ATCC43504) was used in the previous work (Kawakubo *et al.*, 2004); H. *pylori* (ATCC700392) from American Type Culture Collection may be useful because its entire genome has been sequenced (Alm *et al.*, 1999).

Preparation of Human Gastric Mucins

It has been reported that MUC5AC and MUC6 are produced in surface mucus cells and gland mucous cells of the human stomach, respectively (Nordman *et al.*, 2002; Zhang *et al.*, 2001). Since mucins from gland mucous cells but not surface mucous cells contain α4GlcNAc-capping structures, it is critical to determine whether two different mucins have different inhibitory activity toward H. *pylori* growth. To purify these two mucins, the difference in carbohydrate-capping structures can be used. Normal-looking gastric mucosa distant from tumor is obtained from the patient after written informed consent has been obtained. The frozen material is thawed in the presence of a protease inhibitor cocktail, Complete (Roche Diagnostic), and minced into small pieces. The minced sample is homogenized in 3 ml (for 0.8 g wet mucosa) of 10 mM sodium phosphate buffer (pH, 6.5) containing 6 M guanidine hydrochloride, 5 mM Na$_2$EDTA, and 5 mM N-ethylmaleimide, followed by slow stirring overnight at 4°. The sample is then centrifuged at 10,000$\times g$ for 60 min, and the supernatant (3 ml) is incubated with 30 μl of 1 mM dithiothreitol (DTT) at 37° for 5 hrs and then alkylated with 25 mM (final concentration) iodoacetamide overnight in the dark at 4°. To remove reagents, the sample is dialyzed against saline, obtaining dissociated mucin proteins. Five hundred microliters of the reduced and alkylated sample is incubated with anti-MUC5A antibody 45M1 (Novocastra Laboratories, Newcastle upon Tyne, UK) using Seize X protein G immunoprecipitation kit (Pierce, Rockford, IL). The mucin sample is dissociated from the immune complex with ImmunoPure Gentle Ag/Ab Buffers (Pierce) and dialyzed against saline. It was shown previously that MUC5A, with α4GlcNAc-capping structure, is also produced by a few gastric pit cells adjacent to gland mucous cells (Zhang *et al.*, 2001) and that α4GlcNAc-capping structures can be recognized by *Griffonia simplicifolia* agglutinin-2 (GSA-2) (Lyer *et al.*, 1976). MUC5AC with terminal α4GlcNAc-capping structure is thus removed by applying a 45M1-bound fraction to GSA-2 agarose beads (EY Laboratories, San Mateo, CA) in 10 mM phosphate buffer (pH, 7.3)

containing 0.5 mM CaCl$_2$. MUC5AC without α4GlcNAc-capping structure is recovered as unbound fraction to GSA-2 agarose beads.

To prepare MUC6 derived from deeper regions of the mucosa—the gland mucous cells—the above-dissociated protein is incubated with GSA-2 agarose beads in 10 mM phosphate buffer (pH, 7.3) containing 0.5 mM CaCl$_2$ overnight at 4°. GSA-2 bound proteins are eluted with 0.4 M GlcNAc and immunoprecipitated with 45M1 antibody to remove MUC5AC with α4GlcNAc-capping structures. This protocol was devised because no appropriate antibody specific to MUC6 was available. The sample bound to GSA-2 but not anti-MUC5AC (45M1) antibody is dialyzed against saline and used as mucin glycoproteins (mostly MUC6) containing α4GlcNAc-capping structure. The amount of α4GlcNAc-capping structure is estimated by immunoreactivity using an ELISA kit with the HIK1083 antibody. The protein concentration of the previously mentioned MUC5AC and MUC6 samples can be estimated using a Micro BCA Protein Assay Kit (Pierce).

Bacterial Growth Assay

H. pylori is initially cultured in brucella broth (Becton Dickinson Microbiology Systems, Sparks, MD) supplemented with 10% horse serum. In certain instances, *H. pylori* is cultured in brain heart infusion (3.5 %) plus yeast extract (0.2%) (Becton Dickinson Microbiology Systems) supplemented with 10% fetal bovine serum (Rasko *et al.*, 2000). Diluted bacteria (1×10^7 cells/ml) are then cultured with brucella broth supplemented with 5% horse serum containing various amounts of soluble CD43 with or without α4GlcNAc-capping structures. In certain experiments, human gastric mucins with or without α4GlcNAc-capping structures prepared as described previously are added to test their inhibitory activity. *H. pylori* (1×10^7 cells/ml) is cultured in 96-well plates at 35° under 15% CO$_2$ up to 4 days. Bacterial growth is measured at OD 600 nm using a microplate spectrophotometer. Similarly, *p*-nitrophenyl α-*N*-acetylglucosamine (Toronto Chemicals, North York, Canada) is tested on the bacterial growth under the same conditions. Previous results indicate that α4GlcNAc-capped *O*-glycans inhibit the growth of *H. pylori* in a dose-dependent manner, whereas control *O*-glycans do not show such an inhibitory effect (Fig. 2) (Kawakubo *et al.*, 2004).

Cell Motility and Morphology

Motility and morphology of *H. pylori* cultured with brucella broth containing 31.2 mU/ml of sCD43 with capping α4GlcNAc or the same protein concentration of sCD43 without α4GlcNAc for 3 days are evaluated by taking time-lapse images at 1-sec intervals using a laser confocal

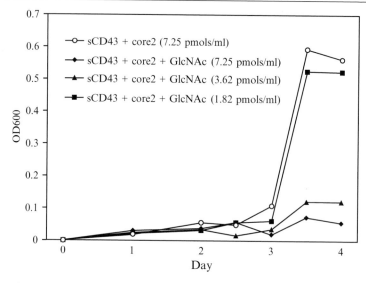

FIG. 2. *O*-Glycans with terminal α1,4-linked GlcNAc inhibit growth of *H. pylori*. The presence of a soluble CD43 containing α1,4-GlcNAc–capping structure inhibited *H. pylori* growth, whereas control-soluble CD43 lacking α1,4-GlcNAc did not (Kawakubo *et al.*, 2004).

microscope and by scanning electron micrographs. Previous results demonstrate that the motility of *H. pylori* in the presence of α4GlcNAc-capped *O*-glycans is dramatically suppressed, and abnormal morphology such as elongation and segmental narrowing is observed in these bacteria. By contrast, such abnormal findings are not found in *H. pylori* cultured with soluble CD43 without capping α4GlcNAc (Kawakubo *et al.*, 2004).

Detection of α-D-glucosyl Cholesterol by Mass Spectrometry

H. pylori contains cholesterol-α-D-glucopyranoside (CGL) as a major component of its cell wall (Haque *et al.*, 1996; Hirai *et al.*, 1995). Since the α4GlcNAc structure is similar to that of α-D-glucose, it is reasonable to test whether cholesterol-α-D-glucosyltransferase is inhibited by α4GlcNAc-capping structure. This hypothesis is also based on findings that the amino acid sequence of α4GnT (Nakayama *et al.*, 1999) is homologous to α1,4-galacto syltransferase that forms globosyl trisaccharide (Furukawa *et al.*, 2000; Keusch *et al.*, 2000; Steffensen *et al.*, 2000).

To determine whether mucin glycoproteins containing α4GlcNAc-capping structures actually inhibit cholesterol α-glucosyltransferase, soluble CD43 with or without α4GlcNAc structures is prepared as described

previously. *H. pylori* is first cultured in brucella broth supplemented with 10% horse serum and then cultured in brucella broth containing 5% horse serum and soluble CD43 with α4GlcNAc-capping structure or the same concentration of soluble CD43 without α4GlcNAc-capping structure.

After culturing at 35° for 2 days, *H. pylori* is harvested, washed in Ca^{2+}- and Mg^{2+}-free PBS three times, and resuspended in 1 ml of distilled water. Lipids are extracted from *H. pylori* cell pellets with 2 ml of chloroform-methanol mixture (2:1) at 4° overnight (Folch *et al.*, 1957). The extract, after filtering through filter paper, is dried under a nitrogen gas stream; the dried extract is dissolved in 4 ml of chloroform-methanol mixture (2:1), and 1 ml of water is further added. Lipids dissolved in the lower phase of the mixture are dried under a nitrogen stream and treated with 1 ml of 0.5 N NaOH in methanol at 50° for 1 hour. After neutralization with 6 N HCl, 1 ml of petroleum ether is added to the reaction tube. After removal of the upper phase, which contains polar lipids, 2 ml of petroleum ether is added to the lower phase. The dried lower phase is dissolved in chloroform-methanol-water (86:14:1), and 0.5 ml of another chloroform-methanol-water (3:48:47) is added. The recovered lower phase is then dried by a nitrogen gas stream and dissolved in 50 μl of chloroform.

For matrix-assisted laser desorption/ionization–time-of-flight (MALDI-TOF) mass spectrometry, 1 μl of the sample is added to 1 μl of 2.5-dihydroxybenzoic acid or 1 μl of *trans*-3-indoleacrylic acid, which is used as matrix (Taketomi and Sugiyama, 2000). Mass spectrum of α-glucosyl cholesterol is taken in positive and negative ion mode by MALDI-TOF mass spectrometry in a reflector mode with laser intensity of 2300. A two-point external calibration is performed, and an endogenously expressed phosphatidic acid in *H. pylori* (Hirai *et al.*, 1995) is used for internal standard. Our previous results indicate that CGL in the bacteria incubated with capping α4GlcNAc is reduced to 29.5% of *H. pylori* cultured without capping α4GlcNAc (Fig. 3) (Kawakubo *et al.*, 2004).

In Vitro Assay of Cholesterol α-Glucosyltransferase from H. pylori

Since cholesterol α-glucosyltransferase has not been characterized, the conditions for the enzymatic reaction were adapted from the studies of sterol glucosyltransferase from *Saccharomyces cerevisiae* and other lower organisms (Warnecke *et al.*, 1999). *H. pylori* is preincubated in brucella broth supplemented with 10% horse serum. The bacteria are then diluted at 5×10^7 cells/ml in brucella broth containing 5% of horse serum at 35° for 2 days. Two ml of *H. pylori* (5×10^8 cells/ml) is centrifuged, and the pellet is washed in PBS (Ca^{2+}- and Mg^{2+}-free) three times. *H. pylori* is then suspended in 1 ml of reaction buffer containing 100 mM Tris-HCl buffer

(pH, 7.5), 15% glycerol, 5 mM DTT, 200 μl Pefabloc (Merck), and 0.5 mg/ml of lysozyme, and this is incubated at 20° for 5 min. The samples are placed on ice and sonicated 10 times for 30 sec in an ultrasonic bath at 30-sec intervals. To the 80 μl enzyme solution made up of sonicated *H. pylori*, 5 μl of 8 mM cholesterol in ethanol, 7.2 μmol UDP-glucose in 5 μl, 1 μl of Triton CF-54, and 9 μl of 100 mM Tris-HCl buffer (pH 7.5) are added.

For the inhibition test, a soluble CD43 containing α4GlcNAc or control soluble CD43 is dissolved in the Tris-HCl buffer (pH, 7.5) and incubated in the same reaction mixture described previously at 30° for 3 hrs. The reaction is terminated by adding 900 μl of 0.45% NaCl solution and 4 ml of chloroform-methanol (2:1) mixture. The lower phase of the resultant mixture is filtered and dried under a nitrogen stream. The sample can be sequentially treated with various solutions as described previously to isolate α-glucosyl cholesterol. Then isolated lipids are subjected to MALDI-TOF mass spectrometry as described. Our previous results indicate that the synthesis of CGL is inhibited in the presence of soluble CD43 containing α4GlcNAc, whereas control-soluble CD43 does not affect the synthesis of CGL (see Fig. 3) (Kawakubo *et al.*, 2004).

The methods previously described provide solid evidence that the enzyme product is α-glucosyl cholesterol. On the other hand, it is cumbersome and is not suited for obtaining quantitative data. To achieve such a goal, we can use an alternative assay. This method is made on the basis of radioactive glucose incorporation from radioactive donor substrate (Lee *et al.*, unpublished results). The cell lysates prepared as described previously are incubated in the same reaction mixture, except that UDP-[^{3}H] glucose (0.05 μCi, 100,000 cpm, final concentration of 3.6 μM) is used as a donor substrate. The reaction is stopped by adding 100 mM HCl (final concentration) and the upper phase in ethylacetate-water partition is taken to determine incorporated radioactivity. A measurement of ^{3}H-glucose incorporated into cholesterol tells the extent of the enzymatic reaction. If necessary, the product is digested by α-glucosidase or β-glucosidase, confirming that it is α-glucosylated cholesterol.

Continuous Spectrophotometric Assay for Cholesterol α-Glucosyltransferases

To obtain kinetic data of cholesterol α-glucosyltransferases, it would be much better to continuously measure its enzymatic activity without interrupting the enzymatic reaction. This will be achieved by using coupling assay and spectrophotometer with temperature control. The reactions for coupling assay are summarized in Fig. 4. This assay contains three enzymes. The first reaction by cholesterol α-glycosyltransferase generates UDP from

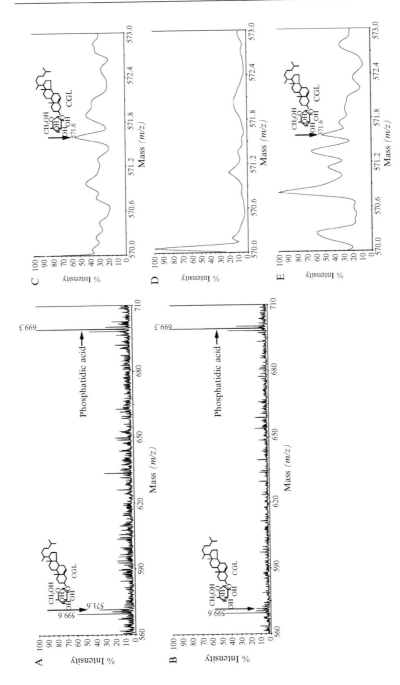

Fig. 3. MALDI-TOFF analysis of CGL biosynthesis in *H. pylori* after incubation with soluble CD43 with terminal α1,4-GlcNAc residue. (A) [M + Na]⁺, sodium-adducted CGL at m/z 571.6, was detected in the lipid fraction of *H. pylori* incubated with control-soluble

$$\text{UDP-glucose + cholesterol} \xrightarrow{\underset{\alpha\text{-glucosyltransferase}}{\text{Cholesterol}}} \text{cholesteryl-}\alpha\text{-o-glucose + UDP} \quad [1]$$

$$\text{UDP + phosphoenolpyruvate} \xrightarrow{\text{Pyruvate kinase}} \text{pyruvate + UTP} \quad [2]$$

$$\text{Pyruvate + NADH} \xrightarrow{\underset{\text{dehydrogenase}}{\text{Lactate}}} \text{lactate + NAD}^+ \quad [3]$$

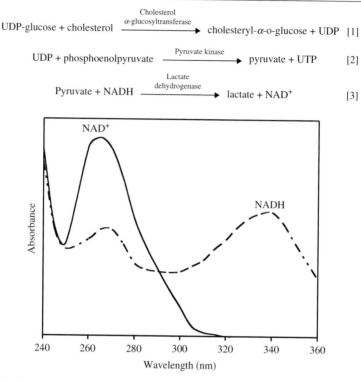

FIG. 4. Coupling assay for cholesterol α-glucosyltransferase. Cholesterol α-glucosyltransferase is incubated with UDP-Glc and cholesterol together with two enzymatic systems that convert formed UDP to UTP and then to NAD$^+$. The amount of NADH can be measured by a 96-well (corning costar 3696) ELISA system at 340 nm. The decrease of NADH is the measurement of the enzymatic activity.

UDP-glucose. UDP is then used by pyruvate kinase generating from phosphoenol pyruvate in the presence of NADH, that is used by lactate dehydrogenase, producing lactate + NAD$^+$. Because of this last reaction, the

CD43 (arrow). (B) CGL in *H. pylori* incubated with 4.0 mU/ml of αGlcNAc-capped soluble CD43 was reduced to 29.5% of the control experiment (arrow). In both (A) and (B), amounts of an endogenous standard, phosphatidic acid, are normalized at 100%. (C) MALDI-TOF mass spectrum of products synthesized from UDP-Glc and cholesterol by *H. pylori* cell lysates. CGL, [M + Na]$^+$ at m/z 571.6 is shown. (D and E) Mass spectrum of products synthesized from UDP-Glc and cholesterol by sonicated *H. pylori* in the presence of 50.0 mU/ml of αGlcNAc-capped soluble CD43 containing α1,4-GlcNAc structures (D) or control-soluble CD43 lacking α1,4-GlcNAc structures (E). Note that CGL was not synthesized in the presence of α1,4-GlcNAc-capped soluble CD43 in (D) (Kawakubo *et al.*, 2004).

consumption of NADH is proportional to production of cholesterol α-glucosyltransferases (Fitzgerald *et al.*, 1970; Gosselin *et al.*, 1994).

Materials

1. Pyruvate kinase (type II, rabbit muscle)
2. Lactate dehydrogenase (type III, rabbit muscle)
3. Phosphoenol pyruvate
4. NADH
5. UDP-glucose
6. 8 mM cholesterol dissolved in ethanol.

All of these reagents are purchased from Sigma.

Reaction Mixture

In 300 μl of reaction mixture:

1. 7.5 U pyruvate kinase
2. 15 U lactate dehydrogenase
3. 0.7 mM phosphoenol pyruvate
4. 0.3 mM NADH
5. 50 mM N-2-hydroxyethylpiperazine-N-2-ethanesulfonic acid (HEPES) (pH, 7)
6. 5 mM MnCl$_2$
7. 0.13% bovine serum albumin
8. 0.4 mM cholesterol (all in final concentration)
9. 100 mM UDP-Glc.

Assay

After preincubating at 37° for 10 min, 50 μl of the enzyme solution in HEPES (pH, 7) is added.

NADH has absorbance at 340 nm, and the absorbance at 340 nm is continuously monitored. Using a standard solution, the molarity of the NADH can be calculated. This method is useful in obtaining kinetic data. When low-molecular-weight compounds are screened for inhibiting cholesterol α-glucosyltransferase, one-time incubation may be used. This assay will be carried out continuously in 96-well. Since the enzyme has been cloned, this method will be become powerful.

Future Perspective

The studies described in this chapter have an important perspective since they provide a potentially novel treatment for *H. pylori* infection.

Although *H. pylori* infection can be treated with antibiotics such as amoxicillin and erythromycin together with acid-lowering drugs, *H. pylori* becomes resistant to these drugs in a minority of patients. It is thus critical to develop new treatments to overcome the increasing resistance of *H. pylori* against these drugs. This direction is supported by a recent report that porcine milk, which contains Lewis b and sialyl Lewis X as ligand of adhesions and most likely α4GlcNAc-capping structures, inhibits *H. pylori* growth (Gustafsson *et al.*, 2006). For this, synthetic oligosaccharides carrying α4GlcNAc-capping structures can be a therapeutic agent and may be used in eradicating *H. pylori* on the surface or close to the colonization beneath the surface of the gastric mucosa.

We have recently cloned cholesterol α-glucosyltransferase and found that this enzyme is unique to *Helicobacter* species (Lee *et al.*, submitted). By inhibiting cholesterol α-glucosyltransferase, it is thus feasible to eradicate *H. pylori* without causing diarrhea or other symptoms due to adverse effects to other organisms in the digestive tract. Moreover, cholesterol α-glucosyltransferase represents an entirely new target for drug development. It is expected that developing novel drugs, either specific oligosaccharides or small-molecule drugs, would be highly beneficial for the treatment of *H. pylori* infection and would thus prevent peptic ulcers and gastric carcinoma.

Acknowledgments

We thank all of our many colleagues and collaborators who have contributed to this work and Aleli Morse for organizing the manuscript. This chapter is largely based on our previously published article (Kawakubo *et al.*, 2004). The work in our laboratory was supported by NIH grant CA 33000 (to M.F.) and a Grant-in-Aid for Scientific Research on Priority Area 148082201 from the Ministry of Education, Culture, Sports, and Science and Technology of Japan (to J.N.).

References

Alm, R. A., Ling, L. S., Moir, D. T., King, B. L., Brown, E. D., Doig, P. C., Smith, D. R., Noonan, B., Guild, B. C., deJonge, B. L., Carmel, G., Tummino, P. J., Caruso, A., Uria-Nickelsen, M., Mills, D. M., Ives, C., Gibson, R., Merberg, D., Mills, S. D., Jiang, Q., Taylor, D. E., Vovis, G. F., and Trust, T. J. (1999). Genomic-sequence comparison of two unrelated isolates of the human gastric pathogen. *Helicobacter pylori. Nature* **397**, 176–180.

Aruffo, A., Stamenkovic, I., Melnick, M., Underhill, C. B., and Seed, B. (1990). CD44 is the principal cell surface receptor for hyaluronate. *Cell* **61**, 1303–1313.

Aspholm, M., Kalia, A., Ruhl, S., Schedin, S., Arnqvist, A., Lindén, S., Sjöström, R., Gerhard, M., Semino-Mora, C., Subois, A., Unemo, M., Danielsson, D., Teneberg, S., Lee, W.-K., Berg, D., and Borén, T. (2006). Helicobacter pylori adhesion to carbohydrates. *Methods Enzymol.* **417**, 289–335.

Bierhuizen, M. F., and Fukuda, M. (1992). Expression cloning of a cDNA encoding UDP-GlcNAc:Gal beta 1–3-GalNAc-R (GlcNAc to GalNAc) beta 1–6GlcNAc transferase by gene transfer into CHO cells expressing polyoma large tumor antigen. *Proc. Natl. Acad. Sci. USA* **89**, 9326–9330.

Deutscher, S. L., Nuwayhid, N., Stanley, P., Briles, E. I., and Hirschberg, C. B. (1984). Translocation across Golgi vesicle membranes: A CHO glycosylation mutant deficient in CMP-sialic acid transport. *Cell* **39**, 295–299.

Fitzgerald, D. K., Colvin, B., Mawal, R., and Ebner, K. E. (1970). Enzymic assay for galactosyl transferase activity of lactose synthetase and alpha-lactalbumin in purified and crude systems. *Anal. Biochem.* **36**, 43–61.

Folch, J., Lees, M., and Sloane Stanley, G. H. (1957). A simple method for the isolation and purification of total lipides from animal tissues. *J. Biol. Chem.* **226**, 497–509.

Furukawa, K., Iwamura, K., Uchikawa, M., Sojka, B. N., Wiels, J., Okajima, T., and Urano, T. (2000). Molecular basis for the p phenotype: Identification of distinct and multiple mutations in the alpha 1,4-galactosyltransferase gene in Swedish and Japanese individuals. *J. Biol. Chem.* **275**, 37752–37756.

Gosselin, S., Alhussaini, M., Streiff, M. B., Takabayashi, K., and Palcic, M. M. (1994). A continuous spectrophotometric assay for glycosyltransferases. *Anal. Biochem.* **220,** 92–97.

Gustafsson, A., Hultberg, A., Sjostrom, R., Kacskovics, I., Breimer, M. E., Boren, T., Hammarstrom, L., and Holgersson, J. (2006). Carbohydrate-dependent inhibition of *Helicobacter pylori* colonization using porcine milk. *Glycobiology* **16**, 1–10.

Haque, M., Hirai, Y., Yokota, K., Mori, N., Jahan, I., Ito, H., Hotta, H., Yano, I., Kanemasa, Y., and Oguma, K. (1996). Lipid profile of *Helicobacter* spp.: Presence of cholesteryl glucoside as a characteristic feature. *J. Bacteriol.* **178**, 2065–2070.

Hidaka, E., Ota, H., Hidaka, H., Hayama, M., Matsuzawa, K., Akamatsu, T., Nakayama, J., and Katsuyama, T. (2001). *Helicobacter pylori* and two ultrastructurally distinct layers of gastric mucous cell mucins in the surface mucous gel layer. *Gut* **49**, 474–480.

Higashi, H., Tsutsumi, R., Muto, S., Sugiyama, T., Azuma, T., Asaka, M., and Hatakeyama, M. (2002). SHP-2 tyrosine phosphatase as an intracellular target of *Helicobacter pylori* CagA protein. *Science* **295**, 683–686.

Hirai, Y., Haque, M., Yoshida, T., Yokota, K., Yasuda, T., and Oguma, K. (1995). Unique cholesteryl glucosides in *Helicobacter pylori:* Composition and structural analysis. *J. Bacteriol.* **177**, 5327–5333.

Ilver, D., Arnqvist, A., Ogren, J., Frick, I. M., Kersulyte, D., Incecik, E. T., Berg, D. E., Covacci, A., Engstrand, L., and Boren, T. (1998). *Helicobacter pylori* adhesin binding fucosylated histo-blood group antigens revealed by retagging. *Science* **279**, 373–377.

Ishihara, K., Kurihara, M., Goso, Y., Urata, T., Ota, H., Katsuyama, T., and Hotta, K. (1996). Peripheral alpha-linked *N*-acetylglucosamine on the carbohydrate moiety of mucin derived from mammalian gastric gland mucous cells: Epitope recognized by a newly characterized monoclonal antibody. *Biochem. J.* **318**(Pt. 2), 409–416.

Kawakubo, M., Ito, Y., Okimura, Y., Kobayashi, M., Sakura, K., Kasama, S., Fukuda, M. N., Fukuda, M., Katsuyama, T., and Nakayama, J. (2004). Natural antibiotic function of a human gastric mucin against *Helicobacter pylori* infection. *Science* **305**, 1003–1006.

Keusch, J. J., Manzella, S. M., Nyame, K. A., Cummings, R. D., and Baenziger, J. U. (2000). Cloning of Gb3 synthase, the key enzyme in globo-series glycosphingolipid synthesis, predicts a family of alpha 1,4-glycosyltransferases conserved in plants, insects, and mammals. *J. Biol. Chem.* **275**, 25315–25321.

Kobayashi, M., Mitoma, J., Nakamura, N., Katsuyama, T., Nakayama, J., and Fukuda, M. (2004). Induction of peripheral lymph node addressin in human gastric mucosa infected by *Helicobacter pylori. Proc. Natl. Acad. Sci. USA* **101**, 17807–17812.

Lee, H., Kobayashi, M., Wang, P., Nakayama, J., Seeberger, P. H., and Fukuda, M. (2006). Expression cloning of cholesterol α-glucosyltransferase, a unique enzyme that can be inhibited by natural antibiotic gastric mucin O-glycans, from *Helicobacter pylori. Biochem. Biophys. Res. Commun.* In press.

Lyer, P. N., Wilkinson, K. D., and Goldstein, L. J. (1976). An *N*-acetyl-D-glycosamine binding lectin from *Bandeiraea simplicifolia* seeds. *Arch. Biochem. Biophys.* **177**, 330–333.

Mahdavi, J., Sonden, B., Hurtig, M., Olfat, F. O., Forsberg, L., Roche, N., Angstrom, J., Larsson, T., Teneberg, S., Karlsson, K. A., Altraja, S., Wadstrom, T., Kersulyte, D., Berg, D. E., Dubois, A., Petersson, C., Magnusson, K. E., Norberg, T., Lindh, F., Lundskog, B. B., Arnqvists, A., Hammarstrom, L., and Boren, T. (2002). *Helicobacter pylori* SabA adhesin in persistent infection and chronic inflammation. *Science* **297,** 573–578.

Marshall, B. J., and Warren, J. R. (1984). Unidentified curved bacilli in the stomach of patients with gastritis and peptic ulceration. *Lancet* **1,** 1311–1315.

Nakayama, J., Yeh, J. C., Misra, A. K., Ito, S., Katsuyama, T., and Fukuda, M. (1999). Expression cloning of a human alpha1, 4-N-acetylglucosaminyltransferase that forms GlcNAcα1→4Galβ→R, a glycan specifically expressed in the gastric gland mucous cell-type mucin. *Proc. Natl. Acad. Sci. USA* **96,** 8991–8996.

Nomura, A., Stemmermann, G. N., Chyou, P. H., Kato, I., Perez-Perez, G. I., and Blaser, M. J. (1991). *Helicobacter pylori* infection and gastric carcinoma among Japanese Americans in Hawaii. *N. Engl. J. Med.* **325,** 1132–1136.

Nordman, H., Davies, J. R., Lindell, G., de Bolos, C., Real, F., and Carlstedt, I. (2002). Gastric MUC5AC and MUC6 are large oligomeric mucins that differ in size, glycosylation and tissue distribution. *Biochem. J.* **364,** 191–200.

Ota, H., Katsuyama, T., Ishii, K., Nakayama, J., Shiozawa, T., and Tsukahara, Y. (1991). A dual staining method for identifying mucins of different gastric epithelial mucous cells. *Histochem. J.* **23,** 22–28.

Parsonnet, J., Friedman, G. D., Vandersteen, D. P., Chang, Y., Vogelman, J. H., Orentreich, N., and Sibley, R. K. (1991). *Helicobacter pylori* infection and the risk of gastric carcinoma. *N. Engl. J. Med.* **325,** 1127–1131.

Peek, R. M., Jr., and Blaser, M. J. (2002). *Helicobacter pylori* and gastrointestinal tract adenocarcinomas. *Nat. Rev. Cancer* **2,** 28–37.

Rasko, D. A., Wang, G., Monteiro, M. A., Palcic, M. M., and Taylor, D. E. (2000). Synthesis of mono- and di-fucosylated type I Lewis blood group antigens by. *Helicobacter pylori. Eur. J. Biochem.* **267,** 6059–6066.

Sasaki, H., Bothner, B., Dell, A., and Fukuda, M. (1987). Carbohydrate structure of erythropoietin expressed in Chinese hamster ovary cells by a human erythropoietin cDNA. *J. Biol. Chem.* **262,** 12059–12076.

Sawada, R., Tsuboi, S., and Fukuda, M. (1994). Differential E-selectin–dependent adhesion efficiency in sublines of a human colon cancer exhibiting distinct metastatic potentials. *J. Biol. Chem.* **269,** 1425–1431.

Sipponen, P., and Hyvarinen, H. (1993). Role of *Helicobacter pylori* in the pathogenesis of gastritis, peptic ulcer and gastric cancer. *Scand. J. Gastroenterol. Suppl.* **196,** 3–6.

Steffensen, R., Carlier, K., Wiels, J., Levery, S. B., Stroud, M., Cedergren, B., Nilsson Sojka, B., Bennett, E. P., Jersild, C., and Clausen, H. (2000). Cloning and expression of the histo-blood group Pk UDP-galactose: Ga1beta-4G1cbeta1-cer alpha1, 4-galactosyltransferase: Molecular genetic basis of the p phenotype. *J. Biol. Chem.* **275,** 16723–16729.

Taketomi, T., and Sugiyama, E. (2000). Extraction and analysis of multiple sphingolipids from a single sample. *Methods Enzymol.* **312,** 80–101.

Warnecke, D., Erdmann, R., Fahl, A., Hube, B., Muller, F., Zank, T., Zahringer, U., and Heinz, E. (1999). Cloning and functional expression of UGT genes encoding sterol glucosyltransferases from *Saccharomyces cerevisiae, Candida albicans, Pichia pastoris,* and *Dictyostelium discoideum. J. Biol. Chem.* **274,** 13048–13059.

Zhang, M. X., Nakayama, J., Hidaka, E., Kubota, S., Yan, J., Ota, H., and Fukuda, M. (2001). Immunohistochemical demonstration of alpha1,4-*N*-acetylglucosaminyltransferase that forms GlcNAcalpha1,4Galbeta residues in human gastrointestinal mucosa. *J. Histochem. Cytochem.* **49,** 587–596.

[12] Molecular Contacts Between Antibiotics and the 30S Ribosomal Particle

By JULIA WIRMER and ERIC WESTHOF

Abstract

Crystal structures of complexes between ribosomal particles and antibiotics have pinned down very precisely the discrete binding sites of several classes of antibiotics inhibiting protein synthesis. The crystal structures of complexes between various antibiotics and ribosomal particles show definitively that ribosomal RNAs (rRNAs), rather than ribosomal proteins, are overwhelmingly targeted. The antibiotics are found at messenger RNA or transfer RNA binding sites and, most importantly, at pivot locations that are key for the structural rearrangements during the molecular mechanical steps in initiation, elongation, or termination of protein synthesis. We focus here on the 30S particle. Structurally, the antibiotics interact in many ways with RNA: (i) only with the phosphate groups (streptomycin); (ii) mainly with bases (hygromycin, spectinomycin); (iii) with a mixture of both (paromomycin, Geneticin); (iv) via magnesium ions (tetracycline) or a protein side chain (streptomycin). The antibiotics can mimic base stacking (pactamycin) or form pseudo–base pairing interactions with ribosomal bases (paromomycin and related aminoglycosides). Resistance strategies (mutations or methylations in rRNA or enzymatic modifications of the antibiotics) can generally be understood on the basis of the intermolecular contacts made between the antibiotics and rRNA residues in the crystal structures. In humans, toxicity of ribosomal antibiotics is most likely due, at least in part, to the sensitivity of mitochondrial ribosomes, since mitochondria evolved from a bacterial ancestor. Antibiotic families (e.g., aminoglycosides) form a set of invariant H-bonds to defined rRNA residues. When such residues are conserved in bacteria, but not in eukaryotes, resistance of eukaryotic ribosomes is observed. The structural knowledge, together with comparative genomic analysis, should allow for the development of new broad-spectrum antibiotics with higher selectivity toward bacterial ribosomes and less toxicity on eukaryotic cytoplasmic and mitochondrial ribosomes.

Overview

A vital step in protein biosynthesis is ribosomal polypeptide catalysis. It is therefore not surprising that the ribosome is the target of as much as half of the antibiotics known so far (Gale *et al.*, 1981). Different from the long-lived perception that small-molecule inhibitors interact with proteins

METHODS IN ENZYMOLOGY, VOL. 415
0076-6879/06 $35.00
DOI: 10.1016/S0076-6879(06)15012-0

only, these antibiotics target either RNA-protein complexes or directly target regions of ribosomal RNAs (rRNAs) (Brodersen *et al.*, 2000; Carter *et al.*, 2000; Hansen *et al.*, 2002, 2003; Ogle *et al.*, 2001, 2002, 2003; Steitz, 2005; Steitz and Moore, 2003; Wimberly *et al.*, 2000). Antibiotics binding to the ribosome belong to different types of small molecules such as macrolides, puromycin, thiostrepton, glycopeptides, tetracycline, and the aminoglycosides. In this review, we will concentrate on structural studies involving antibiotics, and especially aminoglycosides, binding to the small subunit of the bacterial ribosome, the 30S ribosomal particle. These are spectinomycin, edeine, tetracycline, pactamycin, and the aminoglycosides (Fig. 1).

Aminoglycosides are oligosaccharides containing a variable number of sugar rings and ammonium groups. The core of almost all the 30S-binding aminoglycosides is built around a streptamine derivative. The streptamine core is found in streptomycin, but the largest group of aminoglycoside antibiotics contains a central 2-deoxystreptamine (2-DOS) core (Fig. 1A and B, ring II; Fig. 1C and D, ring I). Hygromycin, apramycin, and neamine contain a terminal 2-DOS ring. Neamine is the simplest of the 2-DOS aminoglycosides binding to the ribosome because it contains only two rings, whereby ring I is attached to position 4 in ring II (2-DOS). The neomycin and paromomycin subclasses of aminoglycosides are 4,5-disubstituted on ring II of neamine, whereas the 4,6 disubstitution is found in the kanamycin and gentamicin families.

Since the main roles of the 30S ribosomal subunit consist in the decoding step (at the A-site) and in the translocation step (from A-site to the P-site), most antibiotics binding the 30S particle act by inhibiting either one or both of those recognition and mechanical processes. Those small but chemically complex molecules bind in critical regions where molecular dynamics and rearrangements control biological functions. The neamine-containing aminoglycosides stabilize the conformation of the A-site binding to the amino-acylated transfer RNA (tRNA) bound to a cognate messenger RNA (mRNA) codon and thereby introduce misreading of the mRNA (Pape *et al.*, 2000). In addition, it was recently shown that the best-studied aminoglycoside of this group, paromomycin, increases the affinity for the amino-acylated tRNA to the A-site by a factor of 210 and thereby inhibits translocation by a factor of 160 corresponding to ground-state stabilization (Peske *et al.*, 2004). Apramycin (Perzynski *et al.*, 1979) and hygromycin (Cabanas *et al.*, 1978; Eustice and Wilhelm, 1984a,b; Gonzalez *et al.*, 1978) primarily inhibit the translocation step during elongation and also to a minor extent introduce misreading of mRNA. Streptomycin interferes with decoding and thus induces misreading (Ruusala and Kurland, 1984). Tetracycline inhibits protein synthesis by preventing the binding of amino-acylated tRNA to the ribosome (Spahn

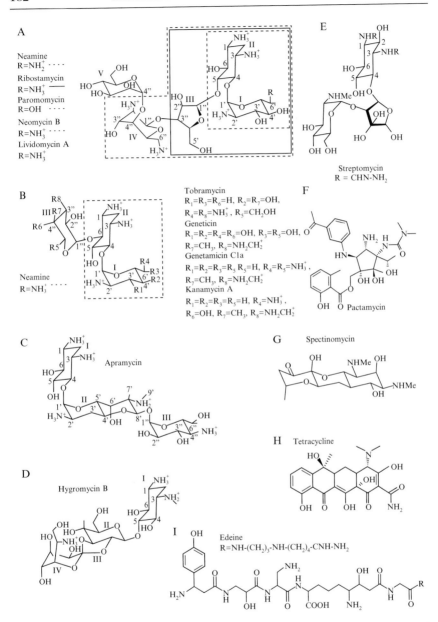

FIG. 1. Chemical structures of antibiotics binding to the 30S ribosomal particle. (A through E), Aminoglycosidic compounds; (F through I), non-aminoglycosidic compounds.

and Prescott, 1996). Pactamycin as well as edeine (Odon *et al.*, 1978) affect translation initiation in all organisms. At higher concentrations, pactamycin affects also the elongation in procaryotes (Cohen *et al.*, 1969). Spectinomycin inhibits translocation of peptidyl-tRNA (Bilgin *et al.*, 1990).

Binding of antibiotics occurs predominantly at the RNA component of the ribosome. First insights into the binding site of antibiotics stemmed from chemical footprinting experiments (Moazed and Noller, 1987; Woodcock *et al.*, 1991). The neamine-containing aminoglycosides bind to the decoding aminoacyl site (A-site) of the rRNA, on which neomycin, hygromycin, apramycin, and neamine reveal protections at A1408, G1491, and G1494 (Moazed and Noller, 1987; Purohit and Stern, 1994; Woodcock *et al.*, 1991). Hygromycin binds at the top of h44, close to the A-site, and protects residue G1494 from modification with dimethyl sulfate (DMS); modification of A1408, in contrast, is enhanced in the presence of the antibiotic (Moazed and Noller, 1987). Chemical footprinting (Moazed and Noller, 1987) and mutagenesis data (Melancon *et al.*, 1988; Montandon *et al.*, 1985; Pinard *et al.*, 1993) using streptomycin showed that residues 13, 526, 915, and 1490 are involved in binding. Again, binding occurs near the A-site (residue 1490 is protected) but also on three additional regions of the rRNA. Edeine strongly protects residues G693, G926, A794, and C795 as well as weakly A790, G791, and A1394; tetracycline protects A892, however it enhances modification for residues U1052 and C1054; the presence of spectinomycin protects C1063 and G1064 and enhances modification in residue G973 (Moazed and Noller, 1987). Pactamycin protects G693 and C795 (Woodcock *et al.*, 1991).

In 1994, Purohit and Stern showed that the whole ribosome is not necessary but that a reduced fragment of rRNA is sufficient to detect specific binding of neamine-containing aminoglycosides (Purohit and Stern, 1994). A first NMR structure of Puglisi and coworkers using a 27mer RNA revealed the basis for specificity as shape recognition, electrostatic interactions, and hydrogen bonds (Fourmy *et al.*, 1996). Furthermore, structural changes on aminoglycoside binding compared with the free RNA were observed based on chemical shift changes (Fourmy *et al.*, 1998).

Important insight into the mechanism of binding has come from high-resolution crystal structures (2.4 to 3.8 Å) of bacterial ribosomal particles complexed to various aminoglycosides (Brodersen *et al.*, 2000; Carter *et al.*, 2000; Pioletti *et al.*, 2001; Schlunzen *et al.*, 2001) and more detailed from crystal structures of oligonucleotides representing the minimal A-site in complex with several 2-DOS aminoglycosides (Francois *et al.*, 2004, 2005; Pfister *et al.*, 2003a; Vicens and Westhof, 2001, 2002, 2003a) (Table I).

TABLE I

X-RAY STRUCTURAL DATA OF ANTIBIOTICS COMPLEXED WITH EITHER THE 30S SUBUNIT OF THE BACTERIAL RIBOSOME OR WITH OLIGONUCLEOTIDES CONTAINING A MINIMAL A-SITE

Antibiotic	PDB-file	Method	Resolution	Publication
Hygromycin*	1HNZ	X-ray	3.3	(Brodersen et al., 2000)
Streptomycin*	1FJG	X-ray	3.0	(Carter et al., 2000)
Paromomycin*	1FJG	X-ray	3.0	(Carter et al., 2000)
Tetracycline*	1HNW	X-ray	3.4	(Brodersen et al., 2000)
Tetracycline*	1I97	X-ray	4.5	(Pioletti et al., 2001)
Spectinomycin*	1FJG	X-ray	3.0	(Carter et al., 2000)
Pactamycin*	1HNX	X-ray	3.4	(Brodersen et al., 2000)
Edeine*	1I95	X-ray	4.5	(Pioletti et al., 2001)
Paromomycin	1J7T	X-ray	2.5	(Vicens and Westhof, 2001)
Tobramycin	1LC4	X-ray	2.54	(Vicens and Westhof, 2002)
Geneticin	1MWL	X-ray	2.4	(Vicens and Westhof, 2003a)
Neomycin B	2ET4	X-ray	2.4	(Francois et al., 2005)
Neomycin B	2A04	X-ray	2.95	(Zhao et al., 2005)
Ribostamycin	2ET5	X-ray	2.2	(Francois et al., 2005)
Lividomycin A	2ESJ	X-ray	2.2	(Francois et al., 2005)
Neamine	2ET8	X-ray	2.5	(Francois et al., 2005)
Kanamycin A	2ESI	X-ray	3.0	(Francois et al., 2005)
Gentamicin C1A	2ET3	X-ray	2.8	(Francois et al., 2005)
Apramycin	1YRJ	X-ray	2.7	(Han et al., 2005)

*Indicates that the complex structure was solved in the context of the small ribosomal subunit; otherwise, the structure was solved using an oligonucleotide containing a minimal A-site.

Binding of Non-Aminoglycoside Antibiotics to the 30S Particle

Tetracycline

Two high-resolution X-ray structures of the 30S particle in complex with tetracycline are available (Brodersen et al., 2000; Pioletti et al., 2001). These structures and earlier biochemical data reveal that there is a primary binding site as well as several secondary binding sites for tetracycline in the 30S particle (Fig. 2A,B). Here, the earlier structure will be discussed (Brodersen et al., 2000). Tetracycline is a flat fused ring system; one site is hydrophilic due to the presence of several hydroxyl and keto groups, the other face is hydrophobic (Fig. 1H). The primary binding site of tetracycline is located near the A-site; most contacts are formed with the irregular shallow/minor groove of h34 (1196–1200, 1053–1056), and additional contacts are found with h31 (964–967). Binding occurs only with the backbone

of the RNA. Direct hydrogen bonds could be identified between the hydrophilic side of tetracycline and the phosphate oxygens of G1053, C1054, C1195, U1196, G1198, and G966, the 5′OH of C1054 and the 2′OH groups of C1195 and A965. In addition, stacking interactions between C1054 and U1196 and the hydrophobic face of the antibiotic are observed. Further stabilization is obtained by a magnesium ion bridging the hydrophilic side of tetracycline to O1P(G1197), O2P(G1198), and O2P(C1054). This magnesium ion is also present in the structure in the absence of tetracycline (Carter *et al.*, 2000). Overall, the conformations of the 30S particle near the primary binding site in the presence and absence of tetracycline are very similar. However, binding of tetracycline at this side clashes with A-site tRNA and thus explains the absence of tRNA binding to the A-site in the presence of the antibiotic by steric hindrance.

In contrast to the primary binding site, binding at the secondary site between h27 and h11 involves contacts with the bases (i.e., A892, C893, and U244). As in the first binding site, contacts are only observed between the RNA and the antibiotic. The comparison with the other available structure (Pioletti *et al.*, 2001) reveals the great variety of binding motifs of tetracycline. Six tetracycline molecules are bound to the 30S ribosome. The primary binding site agrees with the primary binding site observed in the earlier structure (Brodersen *et al.*, 2000). Also, the secondary binding site in the earlier paper is observed (called Tet5). In addition, one tetracycline molecule (Tet2) that does not interact with the rRNA but only with the protein S4 is observed. Tet3 interacts with the 16S RNA only (h40), whereas Tet4 interacts with the RNA (h29 and h43) as well as with a ribosomal protein (S9). The last identified tetracycline molecule (i.e., Tet6) interacts with RNA (i.e., h28, h29) and proteins (i.e., S7 and S9) and, as in the primary binding site, a magnesium ion is involved in binding. The absence of a magnesium ion in four of the six binding sites is surprising, since crystal structures of tetracycline bound to biomacromolecules always displayed a magnesium ion in the vicinity of the antibiotic (Hinrichs *et al.*, 1994; Kisker *et al.*, 1995; Orth *et al.*, 1999, 2000). The structure of the secondary binding sites (tet2-tet6) cannot be related to any particular mechanism of ribosomal function. It is possible that the binding of tetracycline blocks ribosomal assembly of new particles and/or inhibits required dynamics.

Spectinomycin

The spectinomycin binding site is located in the shallow/minor groove at one end of h34 (Fig. 2C) (Carter *et al.*, 2000). Spectinomycin is a very rigid molecule. Binding is a rather sequence specific, involving H-bridges

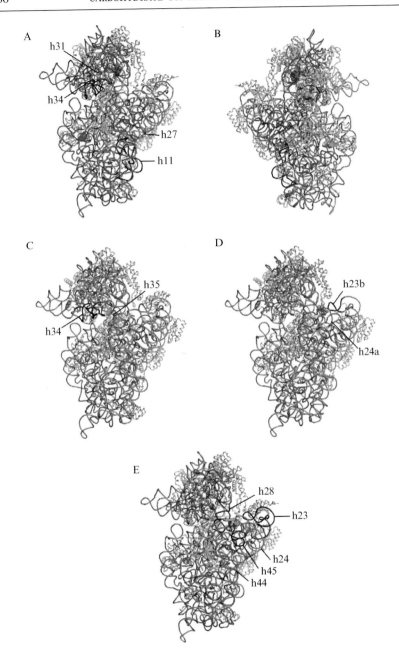

with O2(C1066), O6 and N7(G1064), and N3(C1192)—all of which are in h34. In addition, one H-bond with 2′OH(G1068) in h35 is observed. Mutagenesis studies confirmed the importance of G1064 and C1192 for spectinomycin binding. Furthermore, mutations in the ribosomal protein S5 have been identified that confer resistance to spectinomycin. Although S5 is not involved in binding, it is very close to the antibiotic (within 5 Å) and is thought to stabilize the conformation of the whole region. The mechanism of action of this antibiotic is not entirely clear yet. One likely explanation would be the inhibition of ribosome dynamics during translocation.

Pactamycin

Pactamycin binds in the central domain of the 16S rRNA to h23b, h24a, and the ribosomal protein S7 (Fig. 2D) (Brodersen *et al.*, 2000). In contrast to tetracycline and spectinomycin, pactamycin is a flexible molecule containing two outer aromatic rings and one inner five-member ring. In the structure with the 30S particle, the molecule is bent and mimics a dinucleotide with the outer two rings stacking on each other and on nucleotide G693 of the rRNA (h23b) and the inner ring, mimicking the sugar phosphate. Hydrogen bonds to the antibiotic are observed with N6 (A694) and O6(G693) (h23b), with 3′OH(C788), N4(C795), and O2 as well as O4′ (C796) (h24a) and with Gly81 of the ribosomal protein S7. Thus, contacts with the RNA are base specific but also occur with the backbone. The crystal structure confirms earlier protection studies proposing G693 and C695 to be involved in antibiotic binding (Woodcock *et al.*, 1991). Mutagenesis revealed A694G, C795U, and C796U as mutations, conferring resistance against the antibiotic (Mankin, 1997). The presence of pactamycin in the 30S particle displaces the mRNA in the E-site of the ribosome by approximately 12 Å. This explains the effect on translation initiation, since a productive formation of the initiation complex can be excluded in the presence of the antibiotic.

FIG. 2. Three-dimensional structures of antibiotics bound to the 30S ribosomal particle. RNA is colored grey, black, and blue; proteins are colored green; antibiotics are indicated as spheres. (A and B), Tetracycline (1I97.pdb (Pioletti *et al.*, 2001)); the six tetracycline molecules binding to the subunit are shown (tet1, red; tet2, cyan; tet3, blue; tet4, magenta; tet5, yellow; tet6, orange). Helices in contact with tet1 and tet5 are indicated (h31, blue; h34, black; h27, blue; h11, black). (C) Spectinomycin (1FJG.pdb (Carter *et al.*, 2000)); helices in contact with the antibiotic are indicated (h34, black; h35, blue). (D) Pactamycin (1HNX.pdb (Brodersen *et al.*, 2000)) h23b (black) and h24a (blue) are indicated. (E) Edeine (1I95.pdb (Pioletti *et al.*, 2001)); interacting helices are indicated (h28, black; h44, blue; h45, black; h24, blue; h23, black). (See color insert.)

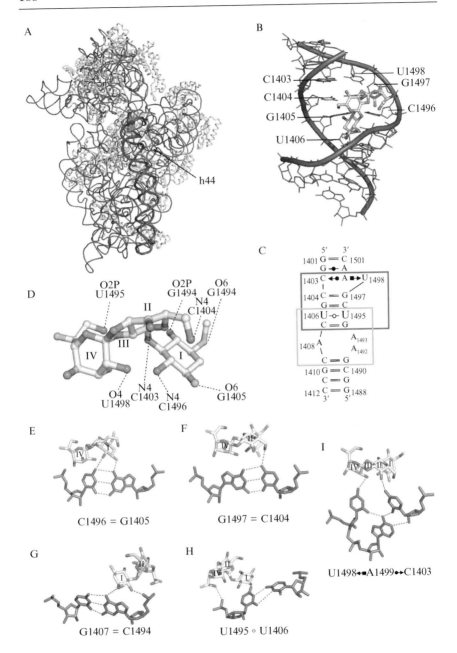

Edeine

Edeine is a peptide-like antibiotic with a tyrosine-like head and a long tail. In the X-ray structure (Pioletti *et al.*, 2001), edeine is located near the E-site and interacts with h28, h44, and h45 of the 16S rRNA only (Fig. 2E). N2(G926) (h28) forms a H-bond with the hydroxyl group of the tyrosine-like head of edeine, mimicking a base pair. Further interactions involve the phosphate groups of U1498 (h44) and G1505 (h45) and mainly the sugars of residues A790, C791, and A792 (h24). The latter interactions lead to the distortion of h24, inducing the formation of a cross-helix hydrogen bridge between C795 (h24) and G693 (h23). This conformational change is thought to be responsible for the inhibitory effect of edeine on protein synthesis.

Binding of Aminoglycoside Antibiotics to the 30S Ribosomal Particle

Hygromycin

Structural insights into hygromycin binding to the ribosome were revealed by the crystal structure of hygromycin bound to the 30S subunit of the *Thermus thermophilus* ribosome (Brodersen *et al.*, 2000). Hygromycin binds at the top of the rRNA helix h44 above the A-site (Fig. 3). The antibiotic is located in the deep/major groove of the RNA, close to the helical axis. Nucleotides in the region from 1490 to 1500 and 1400 to 1410 are contacted; weak contacts with the mRNA can also be observed. Half of the contacts to the rRNA are made with the nucleobases. Hygromycin binding is thus sequence specific. Comparison of the complex with the structure in absence of hygromycin reveals that the antibiotic does not induce significant changes in the structure of the rRNA (Brodersen *et al.*, 2000). Detailed interactions with the rRNA are as follows: Ring I forms hydrogen bonds with the backbone (O2P) and the base (O6) of G1494. In addition, ring I interacts with the phosphate of U1495 of the adaptable (see later) U1495oU1406 base pair and with N4 (C1496) and O6(G1405).

FIG. 3. (A) 3D structure of hygromycin bound to the 30S ribosome (1HNZ.pdb (Brodersen *et al.*, 2000)). Hygromycin is shown as red spheres. (B) Close-up view of the hygromycin binding site (helix h44 residues 1401–1410 and 1490–1501). Hygromycin is shown as ball and sticks. (C) Secondary structure diagram of parts of h44 that interact with hygromycin (red boxed area) and paromomycin (green boxed area). Tertiary interactions are indicated using the Leontis-Westhof nomenclature (Leontis and Westhof, 2001). (D through I) Interactions of hygromycin with the 30S particle. Hydrogen bonds with donor-acceptor distance of less than 3.5 Å are shown. (See color insert.)

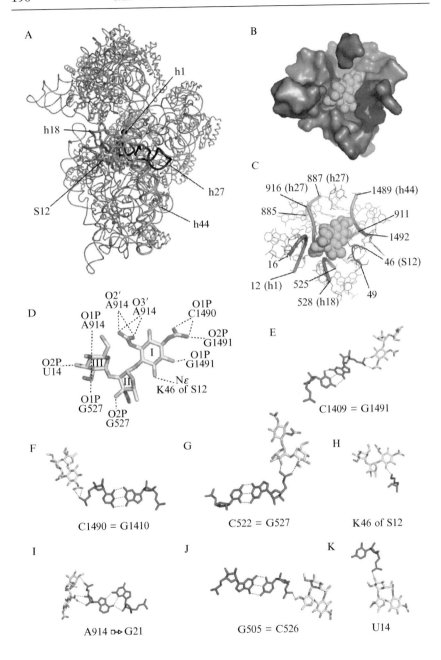

A

h1

h18

h27

S12

h44

B

C

887 (h27)

916 (h27)   1489 (h44)

885   911

                           1492

16   46 (S12)

12 (h1)   525   49

528 (h18)

D

O2′
A914   O3′
        A914   O1P
                     C1490

O1P   O2P
A914   I   G1491

O2P   O1P
U14   III   G1491

                II   Nε
O1P   K46 of S12
G527   O2P
        G527

E

C1409 = G1491

F

C1490 = G1410

G

C522 = G527

H

K46 of S12

I

A914 ⇨ G21

J

G505 = C526

K

U14

Ring II interacts with N4(C1403) and N4(C1404), and ring IV interacts with U1498.

The importance of these contacts is supported by mutational studies. A U1495C mutation is found in some eukaryotes. This mutation abolishes binding due to replacement of O4, which forms a strong H-bond with N1 of hygromycin (Spangler and Blackburn, 1985). Mutation of the G1491=C1409 base pair, which is not near the binding site, abolishes binding due to an allosteric effect (De Stasio et al., 1989). The disruption of the base pair geometry of the highly conserved U1406oU1495 base pair (in eukaryotes and in prokaryotes) by a U1406C replacement also confers hygromycin resistance with a relative resistance (RR) of 64 compared to the wild type (Pfister et al., 2003b). RR of 32 is observed on a C1496U mutation. This mutation disrupts the interaction of N4 of the nucleobase with O3 of ring I in the antibiotic. In addition, the C1496=G1405 base pair is disrupted by the mutation. This change in the base pair geometry may also affect the H-bond between G1405 and hygromycin and contribute to the RR (Pfister et al., 2003b). The effect of a U1498C mutation has a RR value of 8 to 16, smaller than the previous described mutations. This mutation abolishes the H-bond between O4 of the base and a hydroxyl group on ring IV of hygromycin. In contrast to the previous two mutants, a mutation of U1498 probably has no effect on the ribosome conformation because only very few contacts are formed between the base and the rest of the ribosomal subunit, explaining the moderate change in RR (Pfister et al., 2003b). The inhibition of translocation by the antibiotic is currently attributed to its binding position at the top of h44. This part is implicated in ribosome movement during translocation (Frank and Agrawal, 2000), which could be restricted by hygromycin binding.

Streptomycin

The structure of streptomycin binding to the ribosome is shown in Fig. 4 (Carter et al., 2000). Streptomycin binds to helices 1, 18, 27, and 44 and specifically to residues C1490, G1491, A914, G526, G527, and U14. In addition, K46 of the ribosomal protein S12 builds a hydrogen bond with streptomycin. In opposition to the other aminoglycosides (see previous and

---

FIG. 4. (A) Three-dimensional structure of streptomycin bound to the 30S ribosome (1FJG.pdb (Carter et al., 2000)). Streptomycin is shown as red spheres, proteins are green, and RNA is grey. (B) Surface of the binding pocket showing the region within 8 Å of streptomycin. (C) Residues and helices within 8 Å of streptomycin. (D through K) Interactions of streptomycin with the 30S particle. Hydrogen bonds with donor-acceptor distance of less than 3.5 Å are shown. (See color insert.)

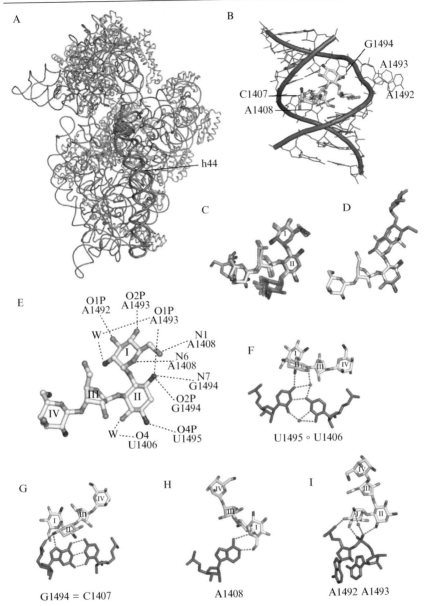

FIG. 5. (A) Three-dimensional structure of paromomycin bound to the 30S ribosome (1FJG.pdb (Carter *et al.*, 2000)). Paromomycin is shown as red spheres, proteins are green, and RNA is grey. (B) paromomycin binding site taken from a structure presenting the minimal A-site (1J7T.pdb (Vicens and Westhof, 2001)). (C) Superposition of aminoglycoside antibiotics containing the neamine moiety onto paromomycin (yellow). Color coding:

later sections), contacts with the rRNA are solely formed between the phosphate backbone and the antibiotic. This is also in sharp contrast to a recently selected 40mer streptomycin binding aptamer in which recognition is performed in a sequence-specific fashion mainly via contacts with the base moieties (Tereshko *et al.*, 2003). Specificity of the recognition in the complex of the 30S particle and streptomycin is achieved by the formation of a pocket allowing the accommodation of streptomycin (Fig. 4C).

Mutations conferring resistance against streptomycin are found in the 530 loop and around nucleotide 912 (Frattali *et al.*, 1990; Powers and Noller, 1991). Recently, a mutation in the decoding center, A1408G, has been identified as conferring resistance against the antibiotic (Gregory *et al.*, 2005). This is surprising since A1408 makes no direct contacts with the antibiotic and a general conformational change altering the binding pocket is not very likely. Interestingly, however, the A1408G mutation is a frequent mutation observed in mutants resistant to aminoglycosides (see later). It has been proposed that the antibiotic alters the conformational equilibrium occurring in the A-site during tRNA selection (Gregory *et al.*, 2005).

Neamine-Containing Aminoglycosides and Apramycin

Neamine (4), paromomycin (4,5), neomycin B (4,5), lividomycin A (4,5), ribostamycin (4,5), tobramycin (4,6), gentamicin C1a (4,6), geneticin (4,6), kanamycin A (4,6), and apramycin all have a common binding site and mechanism of action. Paromomycin is the only antibiotic of this group that has been crystallized in the context of the 30S ribosome (Carter *et al.*, 2000; Murphy and Ramakrishnan, 2004; Murphy *et al.*, 2004; Ogle *et al.*, 2001, 2002). The antibiotic binds to the upper part of h44 and only contacts with the rRNA are observed (see later). All other available structures have been crystallized using an oligonucleotide construct representing the minimal A-site. The structures of paromomycin binding to the 30S particle and to the oligonucleotide (Vicens and Westhof, 2001) superimpose very well (<1 Å). This confirms the validity of the structures obtained using minimal A-site oligonucleotides (see Table I).

Common to all the mentioned antibiotics is the 2-DOS moiety, which is linked on position 4 with position 1' in Ring I (ring II in apramycin). All

---

Geneticin (cyan), gentamicin A (magenta), kanamycin A (green), lividomycin (deep-salmon), neamine (grey), neomycin B (slate-blue), ribostamycin (orange), and tobramycin (marine). For pdb codes and references, see Table I. (D) Superpositon of apramycin (1YRJ.pdb (Han *et al.*, 2005), blue-green) onto paromomycin (yellow). (E through I) Common interactions of ring I and II of neamine-containing aminoglycosides; paromomycin is shown as example. Common hydrogen bonds with donor-acceptor distance of less than 3.5 Å are shown. (See color insert.)

structures of these compounds have some common features, which will be summarized here.

The antibiotics bind in the deep/major groove of RNA, and the binding is mediated by direct as well as by water-bridged hydrogen bonds (Fig. 5). All base pairs constituting the A-site are Watson-Crick type base pairs except for the UoU, which is bifurcated. Ring I of the neamine-containing aminoglycosides stacks, despite the non-planarity of the sugar ring, against G1491 and intercalates, leading to the formation of a pseudo–base pair between O5′ and R6′ (R6′ being OH in paromomycin, lividomycin A, and Geneticin and NH_3^+ in neomycin B, ribostamycin, neamine, tobramycin kanamycin A, and gentamicin C1A) of ring I and the Watson-Crick edge of A1408 in all structures. These crucial interactions are also formed in apramycin where there are the only hydrogen bonds made to this ring. Hydrogen bonds in ring I of the neamine-containing aminoglycosides are, in addition, formed with the phosphate backbone of A1492 and A1493 by O3′ and/or O4′ and/or water-bridged hydrogen bonds with N2′ and/or R6′ (but in gentamicin C1A, Geneticin). The interaction with the phosphate backbone, the intercalation of ring I, and, even more importantly, the base pairing-like interaction with A1408 lead to the bulging out of adenines A1492 and A1493. These bulged-out base pairs make contacts to Watson-Crick base pairs in neighboring oligonucleotides, mimicking interactions with the cognate tRNA-rRNA helix in the ribosome ternary complex.

Three conserved hydrogen bonds with bases are observed in ring II (ring I in apramycin; *ring numbering* in the following refers to the neamine-containing aminoglycosides). N3 (ring II) forms a hydrogen bond with N7 (G1494), O6 (ring II) contacts O4(1406) via a water molecule (only 4,5 aminoglycosides and apramycin), and N1(ring II) forms a hydrogen bond with O4(U1495). U1495 is part of the conserved U1406oU1495 base pair. Further common interactions of ring II are direct or water-bridged (only lividomycin) H bonds to the phosphate backbone (O1P or O2P) of A1493 and G1494.

Depending on the aminoglycoside, different additional contacts are observed in the complexes. The total number of contacts for ring I with the RNA varies between two (gentamicin) and seven (neamine and tobramycin) of which two to four are residue-specific contacts to bases. For ring II, between four (gentamicin) and 10 (neomycin) contacts are observed of which two to seven contacts are sequence-specific hydrogen bonds with bases. Taking into consideration all rings together between 9 (gentamicin, kanamycin) and 24 (lividomycin), contacts between the aminoglycoside and the RNA are observed. For a more specific list of all contacts, see supplementary material (Francois *et al.*, 2005).

Occupancy Varies Between the Complexes

A single type of oligonucleotide has been used in each of the previously mentioned X-ray structures (with the exception of 2A04.pdb, (Zhao *et al.*, 2005)). Because the A-site is asymmetric, this leads to the observation of two A-sites in the crystals that are separated by four G=C base pairs. Paromomycin (4,5), neomycin B (4,5), lividomycin A (4,5), tobramycin (4,6), gentamicin C1a (4,6), and Geneticin (4,6) form what could be called standard complexes with the ribosomal A-site. All of these antibiotics are composed of at least three rings. In these standard complexes, one antibiotic molecule is bound in each of the two A-sites of the oligonucleotide representing the minimal A-site. In each of the complexes, one of the A-sites is better defined (lower B-factor) than the other. In contrast, in the crystal structure of the complexes between kanamycin and the oligonucleotide, the A-site with the lower B-factor contains two kanamycin molecules, whereas the better defined A-site contains one antibiotic molecule as observed in the standard complexes (Francois *et al.*, 2005). One of the two antibiotics in the A-site occupied by two molecules is bound in the standard binding mode as described previously, whereas the other is bound in a different manner. In the ribostamycin complex, two antibiotic molecules are found in both binding sites. Again, one of the two is in the standard binding mode, whereas the second binds in a different mode. This second mode is different for the kanamycin and the ribostamycin complexes; hence, the name *unspecific* was chosen. Lastly, the complex with neamine has to be mentioned. Neamine is the smallest of the crystallized antibiotics and is composed of only two core rings. Only one of the two A-sites of the oligonucleotide is occupied by the aminoglycoside, and the other site is empty. In addition, the conformation of the U1406 in the occupied A-site differs from the conformation in all other complexes: U1406 is bulging out, instead of forming a base pair with U1495. U1495, in consequence, forms a base pair (*cis* Watson-Crick/Hoogsteen) with A1492 of a neighboring duplex.

Binding of Neamine-Containing Aminoglycosides to Mutant Ribosomes

The importance of the previously mentioned conserved contacts (see Fig. 5) for binding of the aminoglycosides to the ribosome was tested by determining minimal inhibitory concentrations using bacteria carrying homogeneous populations of mutant ribosomes (Hobbie *et al.*, 2005, 2006a,b; Pfister *et al.*, 2003a, 2005) and has been summarized in Table II.

Stacking between ring I and G1491 was tested by G1491A, G1491C, G1491U, C1409G, and C1409U mutations. Only small to moderate (up to RR of 128) changes in mean inhibitory concentration (MIC) values

TABLE II
Minimal Inhibitory Concentrations in μG/ML for Different Aminoglycosides

Mutation	4,6 Kanamycins		4,6 Gentamicins		4,5 ring I 6'-NH3		Neamine***	4,5 ring I 6'-OH	
	Tobramycin	Kanamycin A	Geneticin	Gentamicin	Paromomycin	Lividomycin***		Neomycin	Ribostamycin
WT	2	0.5–1	16–32	1	1*	2	64	1–2*	8
1408G	>1024	>1024	64–128	>1024	64*	64	>1024	>1024*	>1024
1406A	128	64–128	>1024	1024	1–2**	1	64	2–4**	8
1406C	128–256	32–64	>1024	128–256	64**	8–16	256–512	4**	32–64
1495A	128	512	>1024	32	256–512**	>1024	256–512	16–32**	>1024
1495C	8–16	4	64	4	256**	64–128	256–512	8**	128
1491A	4	2	16	2	32*	256–512	64–128	4*	128
1491C	16	16–32	256–512	16–32	512–1024*	>1024	512–1024	32*	1024
1491U	64	128	>1024	32	512*	>1024	1024	8–16*	1024
1409G	8	16–32	1024	2	32*	64–128	128	4*	512
1409U	8–16	16	128	8	8*	8–16	256	1*	32–64
1406C–1495A	>1024	>1024	>1024	>1024	>1024**	>1024	>1024	128**	>1024
1406C–1495G	256	>1024	>1024	>1024	2–4***	8–16	256	0.5***	32
1408G–1491U	>1024*		>1024*	>1024*	1024*	>1024	>1024	>1024*	>1024
1408G–1409G	>1024*		>1024*	>1024*	>1024*	>1024	>1024	>1024*	>1024
1406C–1408G	>1024**		>1024**	>1024**	>1024**	>1024	>1024	>1024**	>1024

Not marked, taken from (Hobbie et al., 2005).
*Taken from (Pfister et al., 2005).
**Taken from (Pfister et al., 2003a).
***Taken from (Hobbie et al., 2006a).

occurring with mutation are observed, with the exception of the G1491U mutant. Surprisingly, there are little differences between the 4,5 disubstituted and the 4,6 disubstituted aminoglycosides. However, the G1491U mutant differentiates between aminoglycosides carrying a 6'OH or a 6'NH_3 group on ring I. For the 6'OH 4,5 disubstituted aminoglycosides (i.e., paromomycin, lividomycin), a significant increase of the RR value is observed. These differences can be attributed to individual contacts whose discussion is beyond the focus of this review (Hobbie et al., 2006b; Pfister et al., 2005).

All aminoglycosides carrying a 6'NH_3 group rely on the pseudo–base-pair interaction made between ring I and A1408, which is reflected by MIC values higher than 1024 μg/ml for A1408G. In contrast, aminoglycosides carrying a 6'OH group such as Geneticin, paromomycin, and lividomycin can compensate for the loss of the interaction with A1408 by becoming H-bond acceptor from N1 or N2 of A1408G as exhibited by Pfister et al. (2003a).

The crystal structures previously discussed mentioned the adaptability of the conserved U1406oU1495 base pair; furthermore, an H-bond is formed between the N1 of ring I and O4 of U1495. A mutation of U1406 to A confers a moderate to high increase of RR to 4,6 disubstituted aminoglycosides, though this mutation does not effect the 4,5 disubstituted aminoglycosides. A model of the resulting U1495-A1406 base pair shows that appropriate positioning of ring III of the 4,6 aminoglycosides would not be possible in such a geometry (Pfister et al., 2003a). The U1406C mutant confers similar resistance with the exception of paromomycin, which is moderately effected by the mutation.

The 4,6 disubstituted aminoglycosides differentiate between a U1495 to A and C mutation. Although the U1495A mutant shows moderate to high increase of RR, only low levels are observed with the U1495C mutant. The U1495A mutations alters the base pair in a way that prevents proper positioning of ring II and ring III.

Drastic changes in drug resistance are observed in the A1495A/C mutants for the paromomycin subgroup antibiotics, whereas RR of the neamine subclass does not change significantly. These differences are attributed to compensation by stronger binding of the NH_3^+ carrying neamine subclass antibiotics to the RNA.

The double mutations 1406C–1495A and 1406C–1495G lead to drug resistance against all 4,6 disubstituted aminoglycosides due to change in base pair geometry preventing the accommodation of ring III. In contrast, only the 1406C–1495A mutant confers resistance to the 4,5 disubstituted antibiotics whereas the activity is restored in the 1406C–1495G mutant.

A number of double mutants changing the pseudo–base pair with A1408 and either the stacking interaction with 1491 or the geometry of the UoU base pair lead to loss of drug susceptibility with all antibiotics, confirming the possibility of compensation by one another in the single-point mutants.

"Off" state   "On" state

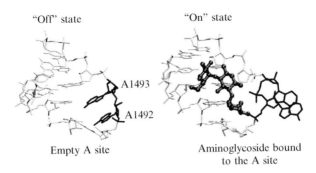

Empty A site   Aminoglycoside bound
                            to the A site

FIG. 6. Views of the two states of the A-site. (left) A-site in the "off" state as observed in the 30S ribosomal subunit (PDB ID 1J5E) (Wimberly *et al.*, 2000). (Right) A-site in the "on" state as observed in the A-site/tobramycin complex (PDB ID 1LC4) (Vicens and Westhof, 2002). The "on" state is the state normally induced and stabilized by binding of cognate tRNA to a codon.

Mechanism of Action of Neamine-Containing Aminoglycosides

Fidelity of protein synthesis is ensured by specific interactions of A1492 and A1493 of the ribosomal A-site with the first two of the three base pairs formed by the cognate codon–anticodon interaction (Ogle *et al.*, 2001). If a specific interaction occurs, the A-site changes its conformation from the "off" state (with A1492 and A1493 folded into the shallow groove of the A-site) to the "on" state (with A1492 and A1493 fully bulged out from the A-site (Fig. 6)) (Ogle *et al.*, 2001, 2002; Wimberly *et al.*, 2000). This structural change of the A-site to the "on" state provokes the transition of the whole ribosome from an open to a closed form that is stabilized by contacts involving the cognate tRNA and the ribosome (Ogle *et al.*, 2002, 2003).

2-DOS aminoglycosides such as paromomycin (4,6-DOS linked) and tobramycin (4,5-DOS linked) interfere with decoding (Moazed and Noller, 1987) by locking the A-site in the "on" conformation (see Fig. 6) (Carter *et al.*, 2000; Vicens and Westhof, 2001, 2003b) and, by doing so, they also pay for a part of the energetic cost associated with the tRNA-dependent ribosome closure (Ogle *et al.*, 2002, 2003). As a consequence, the ribosome loses its ability to discriminate cognate versus non-cognate tRNA–mRNA associations (Ogle *et al.*, 2001, 2002; Rodnina and Wintermeyer, 2001; Shandrick *et al.*, 2004).

Acknowledgment

J. W. was supported by a fellowship from the German academic exchange service (DAAD).

References

Bilgin, N., Richter, A. A., Ehrenberg, M., Dahlberg, A. E., and Kurland, C. G. (1990). Ribosomal RNA and protein mutants resistant to spectinomycin. *EMBO J.* **9**, 735–739.

Brodersen, D. E., Clemons, W. M., Jr., Carter, A. P., Morgan-Warren, R. J., Wimberly, B. T., and Ramakrishnan, V. (2000). The structural basis for the action of the antibiotics tetracycline, pactamycin, and hygromycin B on the 30S ribosomal subunit. *Cell* **103**, 1143–1154.

Cabanas, M. J., Vazquez, D., and Modolell, J. (1978). Inhibition of ribosomal translocation by aminoglycoside antibiotics. *Biochem. Biophys. Res. Commun.* **83**, 991–997.

Carter, A. P., Clemons, W. M., Brodersen, D. E., Morgan-Warren, R. J., Wimberly, B. T., and Ramakrishnan, V. (2000). Functional insights from the structure of the 30S ribosomal subunit and its interactions with antibiotics. *Nature* **407**, 340–348.

Cohen, L. B., Goldberg, I. H., and Herner, A. E. (1969). Inhibition by pactamycin of the initiation of protein synthesis: Effect on the 30S ribosomal subunit. *Biochemistry* **8**, 1327–1335.

De Stasio, E. A., Moazed, D., Noller, H. F., and Dahlberg, A. E. (1989). Mutations in 16S ribosomal RNA disrupt antibiotic—RNA interactions. *EMBO J.* **8**, 1213–1216.

Eustice, D. C., and Wilhelm, J. M. (1984a). Fidelity of the eukaryotic codon-anticodon interaction: Interference by aminoglycoside antibiotics. *Biochemistry* **23**, 1462–1467.

Eustice, D. C., and Wilhelm, J. M. (1984b). Mechanisms of action of aminoglycoside antibiotics in eucaryotic protein synthesis. *Antimicrob. Agents Chemother.* **26**, 53–60.

Fourmy, D., Recht, M. I., Blanchard, S. C., and Puglisi, J. D. (1996). Structure of the A site of *Escherichia coli* 16S ribosomal RNA complexed with an aminoglycoside antibiotic. *Science* **274**, 1367–1371.

Fourmy, D., Yoshizawa, S., and Puglisi, J. D. (1998). Paromomycin binding induces a local conformational change in the A-site of 16 S rRNA. *J. Mol. Biol.* **277**, 333–345.

Francois, B., Russell, R. J., Murray, J. B., Aboul-ela, F., Masquida, B., Vicens, Q., and Westhof, E. (2005). Crystal structures of complexes between aminoglycosides and decoding A site oligonucleotides: Role of the number of rings and positive charges in the specific binding leading to miscoding. *Nucleic Acids Res.* **33**, 5677–5690.

Francois, B., Szychowski, J., Adhikari, S. S., Pachamuthu, K., Swayze, E. E., Griffey, R. H., Migawa, M. T., Westhof, E., and Hanessian, S. (2004). Antibacterial aminoglycosides with a modified mode of binding to the ribosomal-RNA decoding site. *Angew Chem. Int. Ed. Engl.* **43**, 6735–6738.

Frank, J., and Agrawal, R. K. (2000). A ratchet-like inter-subunit reorganization of the ribosome during translocation. *Nature* **406**, 318–322.

Frattali, A. L., Flynn, M. K., De Stasio, E. A., and Dahlberg, A. E. (1990). Effects of mutagenesis of C912 in the streptomycin binding region of *Escherichia coli* 16S ribosomal RNA. *Biochim. Biophys. Acta* **1050**, 27–33.

Gale, E. F., Cundliffe, E., Reynolds, P. E., Richmond, M. H., and Warning, M. J. (1981). *In* "The Molecular Basis of Antibiotic Action," pp. 278–379. Wiley, London.

Gonzalez, A., Jimenez, A., Vazquez, D., Davies, J. E., and Schindler, D. (1978). Studies on the mode of action of hygromycin B, an inhibitor of translocation in eukaryotes. *Biochim. Biophys. Acta* **521**, 459–469.

Gregory, S. T., Carr, J. F., and Dahlberg, A. E. (2005). A mutation in the decoding center of Thermus thermophilus 16S rRNA suggests a novel mechanism of streptomycin resistance. *J. Bacteriol.* **187**, 2200–2202.

Han, Q., Zhao, Q., Fish, S., Simonsen, K. B., Vourloumis, D., Froelich, J. M., Wall, D., and Hermann, T. (2005). Molecular recognition by glycoside pseudo base pairs and triples in an apramycin-RNA complex. *Angewandte Chemie-International Edition* **44**, 2694–2700.

Hansen, J. L., Ippolito, J. A., Ban, N., Nissen, P., Moore, P. B., and Steitz, T. A. (2002). The structures of four macrolide antibiotics bound to the large ribosomal subunit. *Mol. Cell* **10**, 117–128.

Hansen, J. L., Moore, P. B., and Steitz, T. A. (2003). Structures of five antibiotics bound at the peptidyl transferase center of the large ribosomal subunit. *J. Mol. Biol.* **330**, 1061–1075.

Hinrichs, W., Kisker, C., Duvel, M., Muller, A., Tovar, K., Hillen, W., and Saenger, W. (1994). Structure of the Tet repressor-tetracycline complex and regulation of antibiotic resistance. *Science* **264**, 418–420.

Hobbie, S., Pfister, P., Francois, B., Westhof, E., and Bottger, E. C. (2006a). Binding of neomycin-class aminoglycoside antibiotics to mutant ribosomes with alterations in the A-site of 16S rRNA. *Antimicrob. Agents Chemother.* **50**, 1489–1496.

Hobbie, S. N., Bruell, C., Kalapala, S., Akshay, S., Schmidt, S., Pfister, P., and Bottger, E. C. (2006b). A genetic model to investigate drug-target interactions at the ribosomal decoding site. *Biochimie* In Press.

Hobbie, S. N., Pfister, P., Brull, C., Westhof, E., and Bottger, E. C. (2005). Analysis of the contribution of individual substituents in 4,6-aminoglycoside-ribosome interaction. *Antimicrob. Agents Chemother.* **49**, 5112–5118.

Kisker, C., Hinrichs, W., Tovar, K., Hillen, W., and Saenger, W. (1995). The complex formed between Tet repressor and tetracycline-Mg2+ reveals mechanism of antibiotic resistance. *J. Mol. Biol.* **247**, 260–280.

Leontis, N. B., and Westhof, E. (2001). Geometric nomenclature and classification of RNA base pairs. *RNA* **7**, 499–512.

Mankin, A. S. (1997). Pactamycin resistance mutations in functional sites of 16 S rRNA. *J. Mol. Biol.* **274**, 8–15.

Melancon, P., Lemieux, C., and Brakier-Gingras, L. (1988). A mutation in the 530 loop of *Escherichia coli* 16S ribosomal RNA causes resistance to streptomycin. *Nucleic Acids Res.* **16**, 9631–9639.

Moazed, D., and Noller, H. F. (1987). Interaction of antibiotics with functional sites in 16S ribosomal RNA. *Nature* **327**, 389–394.

Montandon, P. E., Nicolas, P., Schurmann, P., and Stutz, E. (1985). Streptomycin-resistance of *Euglena gracilis* chloroplasts: Identification of a point mutation in the 16S rRNA gene in an invariant position. *Nucleic Acids Res.* **13**, 4299–4310.

Murphy, F. V. T., and Ramakrishnan, V. (2004). Structure of a purine-purine wobble base pair in the decoding center of the ribosome. *Nat. Struct. Mol. Biol.* **11**, 1251–1252.

Murphy, F. V. T., Ramakrishnan, V., Malkiewicz, A., and Agris, P. F. (2004). The role of modifications in codon discrimination by tRNA(Lys)UUU. *Nat. Struct. Mol. Biol.* **11**, 1186–1191.

Odon, O. W., Kramer, G., Henderson, A. B., Pinphanichakarn, P., and Hardesty, B. (1978). GTP hydrolysis during methionyl-tRNAf binding to 40 S ribosomal subunits and the site of edeine inhibition. *J. Biol. Chem.* **253**, 1807–1816.

Ogle, J. M., Brodersen, D. E., Clemons, W. M., Jr., Tarry, M. J., Carter, A. P., and Ramakrishnan, V. (2001). Recognition of cognate transfer RNA by the 30S ribosomal subunit. *Science* **292**, 897–902.

Ogle, J. M., Carter, A. P., and Ramakrishnan, V. (2003). Insights into the decoding mechanism from recent ribosome structures. *Trends Biochem. Sci.* **28**, 259–266.

Ogle, J. M., Murphy, F. V., Tarry, M. J., and Ramakrishnan, V. (2002). Selection of tRNA by the ribosome requires a transition from an open to a closed form. *Cell* **111**, 721–732.

Orth, P., Saenger, W., and Hinrichs, W. (1999). Tetracycline-chelated Mg2+ ion initiates helix unwinding in Tet repressor induction. *Biochemistry* **38**, 191–198.

Orth, P., Schnappinger, D., Hillen, W., Saenger, W., and Hinrichs, W. (2000). Structural basis of gene regulation by the tetracycline inducible Tet repressor-operator system. *Nat. Struct. Biol.* **7**, 215–219.

Pape, T., Wintermeyer, W., and Rodnina, M. V. (2000). Conformational switch in the decoding region of 16S rRNA during aminoacyl-tRNA selection on the ribosome. *Nat. Struct. Biol.* **7**, 104–107.

Perzynski, S., Cannon, M., Cundliffe, E., Chahwala, S. B., and Davies, J. (1979). Effects of apramycin, a novel aminoglycoside antibiotic on bacterial protein synthesis. *Eur. J. Biochem.* **99**, 623–628.

Peske, F., Savelsbergh, A., Katunin, V. I., Rodnina, M. V., and Wintermeyer, W. (2004). Conformational changes of the small ribosomal subunit during elongation factor G–dependent tRNA-mRNA translocation. *J. Mol. Biol.* **343**, 1183–1194.

Pfister, P., Hobbie, S., Brull, C., Corti, N., Vasella, A., Westhof, E., and Bottger, E. C. (2005). Mutagenesis of 16S rRNA C1409-G1491 base-pair differentiates between 6'OH and 6'NH3+ aminoglycosides. *J. Mol. Biol.* **346**, 467–475.

Pfister, P., Hobbie, S., Vicens, Q., Bottger, E. C., and Westhof, E. (2003a). The molecular basis for A-site mutations conferring aminoglycoside resistance: Relationship between ribosomal susceptibility and X-ray crystal structures. *Chembiochem.* **4**, 1078–1088.

Pfister, P., Risch, M., Brodersen, D. E., and Bottger, E. C. (2003b). Role of 16S rRNA helix 44 in ribosomal resistance to hygromycin B. *Antimicrob. Agents Chemother.* **47**, 1496–1502.

Pinard, R., Payant, C., Melancon, P., and Brakier-Gingras, L. (1993). The 5' proximal helix of 16S rRNA is involved in the binding of streptomycin to the ribosome. *FASEB J.* **7**, 173–176.

Pioletti, M., Schlunzen, F., Harms, J., Zarivach, R., Gluhmann, M., Avila, H., Bashan, A., Bartels, H., Auerbach, T., Jacobi, C., Hartsch, T., Yonath, A., and Franceschi, F. (2001). Crystal structures of complexes of the small ribosomal subunit with tetracycline, edeine and IF3. *EMBO J.* **20**, 1829–1839.

Powers, T., and Noller, H. F. (1991). A functional pseudoknot in 16S ribosomal RNA. *EMBO J.* **10**, 2203–2214.

Purohit, P., and Stern, S. (1994). Interactions of a small RNA with antibiotic and RNA ligands of the 30S subunit. *Nature* **370**, 659–662.

Rodnina, M. V., and Wintermeyer, W. (2001). Fidelity of aminoacyl-tRNA selection on the ribosome: Kinetic and structural mechanisms. *Annu. Rev. Biochem.* **70**, 415–435.

Ruusala, T., and Kurland, C. G. (1984). Streptomycin preferentially perturbs ribosomal proofreading. *Mol. Gen. Genet.* **198**, 100–104.

Schlunzen, F., Zarivach, R., Harms, J., Bashan, A., Tocilj, A., Albrecht, R., Yonath, A., and Franceschi, F. (2001). Structural basis for the interaction of antibiotics with the peptidyl transferase centre in eubacteria. *Nature* **413**, 814–821.

Shandrick, S., Zhao, Q., Han, Q., Ayida, B. K., Takahashi, M., Winters, G. C., Simonsen, K. B., Vourloumis, D., and Hermann, T. (2004). Monitoring molecular recognition of the ribosomal decoding site. *Angew Chem. Int. Ed. Engl.* **43**, 3177–3182.

Spahn, C. M., and Prescott, C. D. (1996). Throwing a spanner in the works: antibiotics and the translation apparatus. *J. Mol. Med.* **74**, 423–439.

Spangler, E. A., and Blackburn, E. H. (1985). The nucleotide sequence of the 17S ribosomal RNA gene of *Tetrahymena thermophila* and the identification of point mutations resulting in resistance to the antibiotics paromomycin and hygromycin. *J. Biol. Chem.* **260**, 6334–6340.

Steitz, T. A. (2005). On the structural basis of peptide-bond formation and antibiotic resistance from atomic structures of the large ribosomal subunit. *FEBS Lett.* **579**, 955–958.

Steitz, T. A., and Moore, P. B. (2003). RNA, the first macromolecular catalyst: The ribosome is a ribozyme. *Trends Biochem. Sci.* **28**, 411–418.

Tereshko, V., Skripkin, E., and Patel, D. J. (2003). Encapsulating streptomycin within a small 40-mer RNA. *Chem. Biol.* **10**, 175–187.

Vicens, Q., and Westhof, E. (2001). Crystal structure of paromomycin docked into the eubacterial ribosomal decoding A site. *Structure (Camb)* **9,** 647–658.

Vicens, Q., and Westhof, E. (2002). Crystal structure of a complex between the aminoglycoside tobramycin and an oligonucleotide containing the ribosomal decoding a site. *Chem. Biol.* **9,** 747–755.

Vicens, Q., and Westhof, E. (2003a). Crystal structure of Geneticin bound to a bacterial 16S ribosomal RNA A site oligonucleotide. *J. Mol. Biol.* **326,** 1175–1188.

Vicens, Q., and Westhof, E. (2003b). RNA as a drug target: The case of aminoglycosides. *Chembiochem.* **4,** 1018–1023.

Wimberly, B. T., Brodersen, D. E., Clemons, W. M., Jr., Morgan-Warren, R. J., Carter, A. P., Vonrhein, C., Hartsch, T., and Ramakrishnan, V. (2000). Structure of the 30S ribosomal subunit. *Nature* **407,** 327–339.

Woodcock, J., Moazed, D., Cannon, M., Davies, J., and Noller, H. F. (1991). Interaction of antibiotics with A- and P-site–specific bases in 16S ribosomal RNA. *EMBO J.* **10,** 3099–3103.

Zhao, F., Zhao, Q., Blount, K. F., Han, Q., Tor, Y., and Hermann, T. (2005). Molecular recognition of RNA by neomycin and a restricted neomycin derivative. *Angew Chem. Int. Ed. Engl.* **44,** 5329–5334.

[13] Mechanism-Based Inhibitors to Probe Transitional States of Glycoside Hydrolases

By Hiroshi Hinou, Masaki Kurogochi, and Shin-Ichiro Nishimura

Abstract

Recent structural and kinetic studies indicate that glycosidases (glycoside hydrolases) change the peripheral structure of their catalytic sites dynamically to trim glycan structures. Inhibitors that label specific amino acid residues in the active site of these enzymes based on its mechanism of action are powerful tools to probe such a hidden transitional state. This chapter describes methods of mechanism-based irreversible inhibitors having fluorescence tags, including synthesis, inhibitory assay, rapid separation of the peptides containing labeled residues using antibody column, and proteomic analysis of key amino acid residues using matrix-assisted laser desorption/ionization–time-of-flight (TOF)/TOF mass spectrometry.

Overview

Glycosylation and deglycosylation reactions are two crucial processes for the modifications of protein and lipid functions that greatly influence various molecular recognition events including bacterial/viral infections,

METHODS IN ENZYMOLOGY, VOL. 415   0076-6879/06 $35.00
Copyright 2006, Elsevier Inc. All rights reserved.   DOI: 10.1016/S0076-6879(06)15013-2

cell adhesion in inflammation, immune response, cellular differentiation, development, regulation, and many other intercellular communication and signal transductions (Dwek, 1996). Glycoside hydrolases (GHs; e.g., carbohydrases, glycosidases) take part in such processes by trimming glycan chains of glycoconjugates. Therefore, precise analysis of the structures, functions, and catalytic mechanism of GHs has become one of the most important steps to investigate systematic biosynthesis and metabolism of glycoconjugates.

Moreover, it is evident that fine structural information on the active site of GHs permits rational design and efficient synthesis of potential modulators or inhibitors such as sugar-based therapeutic reagents. Mechanism-based inhibitors called *suicide substrates* are versatile tools to elucidate functional amino acid residue in active sites based on the catalytic mechanism of GHs. For example, Withers' group established practical methods for the identification of the catalytic nucleophile in retention-type glycosidases using 2-deoxy-2-fluoro glycosides (Withers *et al.*, 1987) in combination with Edman degradation (Withers *et al.*, 1990) or electrospray ionization mass spectrometry (MS) (Hekmat *et al.*, 2003; Withers and Aebersold, 1995). Halomethyl-substituted aryl groups (Loudon and Koshland, 1970) are one of the potential options to be employed as useful suicide substrates both for retention- and inversion-type GHs and other enzymes such as glycosyltransferases (Takaya *et al.*, 2005). Danzin and coworkers developed an irreversible inhibitor of glucosidase, using the difluoromethylated aryl group as a novel aglycon (Halazy *et al.*, 1990). Ichikawa and Ichikawa (2001) developed novel biotin-conjugated suicide substrates for the isolation and characterization of *N*-acetyl-β-D-glucosaminidase. Janda *et al.* (1997) have reported an excellent application of this tagging strategy for chemical selection of a catalytic antibody having β-galactosidase–like activity from a phage display library. Recently, we developed new methods to catch and identify the amino acid residues that involved in a dynamic nature of the enzymatic reactions by changing a peripheral structure of the active site (Chavas *et al.*, 2005) using 4-*N*-dansyl-2-difluoromethylphenyl aglycone, which was designed for the synthesis of mechanism-based fluorescent labeling reagents (Hinou *et al.*, 2005; Kurogochi *et al.*, 2004). 4-*N*-dansyl-2-difluoromethylphenyl β-galactoside revealed that this type of inhibitor inactivated both retaining- and inverting-type glycosidases, and the labeled dansyl groups greatly facilitated further isolation by antibody column, and subsequent matrix-assisted laser desorption/ionization–time-of-flight (MALDI-TOF)/TOF MS for identification of the labeled amino acids (Fig. 1; Kurogochi *et al.*, 2004). This chapter introduces the synthetic procedure and analytical protocols of a novel mechanism-based inhibitor designated for neuraminidases.

A

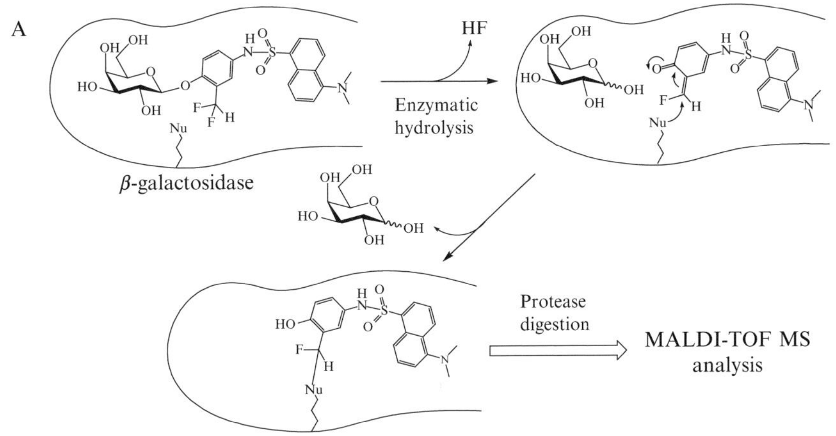

B

Xanthomonas manihotis	^{51}IHFQRIPRAYWKDRLQKARA^{70}
Bacillus circulans	^{251}AESVVETLEEILKQNGSVNF^{270}
E. coli	^{521}IKKWLSLPGETRPLILCEYA^{540}

FIG. 1. (A) Chemical structure of the mechanism-based fluorescence labeling reagent, and the plausible mechanism of the specific nucleophilic labeling in an irreversible manner. (B) Sequences surrounding the labeled nucleophile region of β-galactosidases.

Characterization of *Vibrio cholerae* Neuraminidase by Using a Mechanism-Based Fluorescent Labeling Reagent

Vibrio cholerae neuraminidase (VCNA) plays a significant role in the pathogenesis of cholera by removing sialic acid residues from higher-order gangliosides to an unmasked GM1, the essential receptor for cholera toxin. (Galen *et al.*, 1992) A novel mechanism-based fluorescent labeling reagent, 5-acetamido-2-(4-*N*-5-dimethylaminonaphthalene-1-sulfonyl-2-difluoromethylphenyl)-3,5-dideoxy-D-glycero-α-D-galacto-2-nonulopyranosonic acid (compound 1; Fig. 2), becomes a unique irreversible inhibitor of VCNA. As described in the experimental section, characterization of an inactivated VCNA was carried out basically by combining antibody affinity chromatography and subsequent MALDI-TOF/TOF MS analysis of the isolated peptide obtained from the tryptic hydrolysate of the inactivated enzyme. It was clearly revealed that the Asp-576 and Arg-577 residues, which are located within the sequence ^{576}*DR*FF^{579}, were specifically labeled by this suicide-type fluorescent substrate. Neither Asp-576 nor Arg-577 has ever been known to contribute to a specific residue in the rigid and highly conserved active site of VCNA investigated by crystallographic

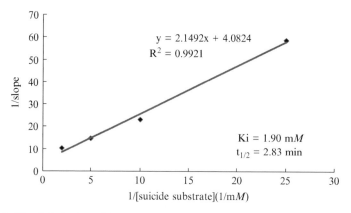

FIG. 2. Kinetic behavior of the compound **1**. To a solution of VCNA and 0–1.5 mM of compound **1** was added *p*-nitrophenyl *N*-acetylneuraminic acid (final concentration of 5 mM), and changes in the hydrolytic activity of the enzyme were determined by measuring the relative intensity at 410 nm.

analysis, suggesting that a flexible β-turn structure containing this sequence may have a crucial role in the dynamic nature of substrate recognition and catalytic action by VCNA. High-throughput proteomic analysis of the mechanism-based tagged neuraminidases by combined use of anti-dansyl antibody column chromatography and MALDI-TOF/TOF MS was proved to become a facile and general method both for identifying the candidates of the catalytic amino acids and for rapid characterization of new bacterial/viral neuraminidases (Taylor, 1996) from unknown pathogens and human neuraminidases as possible targets for diagnosis and therapy of colon cancer (Kakugata *et al.*, 2002).

Synthesis of Suicide Substrate of Neuraminidase 1

Compound 1 (Fig. 3) was synthesized from commercially available *N*-acetylneuraminic acid (Fig. 3, compound 2) according to the procedure shown in Fig. 3. A well-known glycosyl donor, methyl 5-acetamido-4,7,8, 9-tetra-*O*-acetyl-2-chloro-2,3,5-trideoxy-β-D-glycero-D-galacto-2-nonulopyranosonate (Kuhn *et al.*, 1966), was coupled with 2-hydroxy-5-nitrobenzaldehyde in the presence of diisopropylethyleneamine (Kuboki *et al.*, 1998) to afford α-glycoside 3 in 72.3%. Treatment of compound 3 with diethylamino sulfur trifluoride (Driguez *et al.*, 1992; Halazy *et al.*, 1990) furnishing intermediate compound 4 (67.2%) led to the desired compound 1 after de-*O*-acetylation, saponification of the methyl ester, reduction of the *p*-nitro

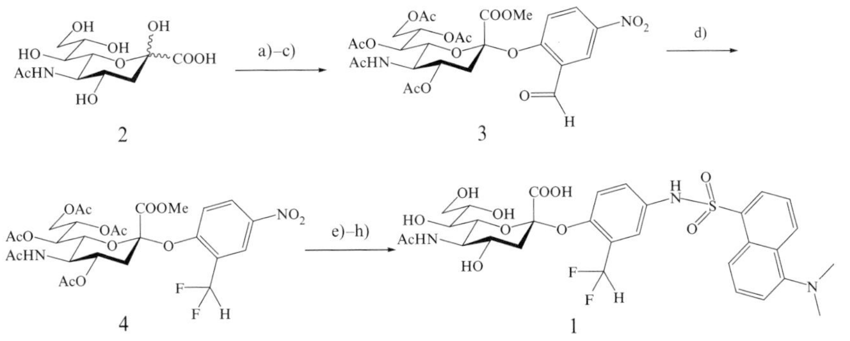

FIG. 3. Synthetic route of the mechanism-based fluorescence labeling reagent. (a) Dowex 50W(H⁺)/ MeOH, (b) CH₃COCl, (c) 2-hydroxy-5-nitrobenzaldehyde/CH₃CN, 3 steps, 72.7%); (d) dimethylamino sulfur trifluoride, 0°, (67.2%); (e) 1 M NaOMe/MeOH, (f) 1 M NaOMe/MeOH/H₂O, (g) 10% Pd(OH)₂/c, H₂, (h) dansyl chloride/Et₃N/MeOH, (4 steps, 45.2%).

group, and subsequent dansylation (4 steps, 45.2%) (Kurogochi *et al.*, 2004). The purity and chemical structures of the compounds synthesized herein were ascertained by nuclear magnetic resonance and high-resolution MS.

Spectral data for 5-acetamido-2-(4-*N*-5-dimethylaminonaphthalene-1-sulfonyl-2-difluoromethylphenyl)-3,5-dideoxy-D-glycero-α-D-galacto-2-nonulopyranosonic acid (1) were as following: ¹H nuclear magnetic resonance (D₂O) δ 8.44 (d, 1 H, J = 8.8 Hz, aromatic), 8.29 (d, 1 H, J = 8.7 Hz, aromatic), 8.15 (d, 1 H, J = 6.9 Hz, aromatic), 7.68 (t, 1 H, aromatic), 7.57 (t, 1 H, aromatic), 7.40 (d, 1 H, J = 7.6 Hz, aromatic), 7.13 (d, 1 H, J = 8.6 Hz, aromatic), 7.02 (brd, 1 H, aromatic), 6.97 (brs, 1 H, aromatic), 6.87 (t, 1 H, J = 55.2 Hz, CHF₂), 3.84 to 3.54 (m, 7 H, H-4, H-5, H-6, H-7, H-8, H-9a, H-9b), 2.85 (s, 6 H, NMe₂), 2.76 (dd, 1 H, J = 4.5, 12.6 Hz, H-3a), 2.03 (s, 3 H, NHAc), 1.78 (t, 1 H, J = 12.0 Hz, H-3b); FABHRMS calcd. For C₃₀H₃₆F₂N₃O₁₁S [M+H]⁺: 684.2033. Found 684.2032.

Inhibition and Specific Labeling of VCNA

An ultraviolet/visible spectrometer at 410 nm was used to measure irreversible inhibitions of α2,3,6,8-neuraminidase from *V. cholerae* by a suicide substrate (see Fig. 3, compound 1) in the presence of 4-nitrophenyl *N*-acetylneuraminic acid as a competitive substrate. The inhibition constants (K_i) and the half-life for inhibition ($t_{1/2}$) for the enzyme were determined by incubating the enzyme in 100 mM citrate phosphate buffer (pH, 5.0) containing various concentrations (0 to 1.0 mM) of the suicide substrate at 37°.

Aliquots (2 μl) of the inactivation mixture were removed at different time intervals (preincubation time, 0, 15, and 30 min) and diluted into a reaction cell containing 4-nitorophenyl N-acetylneuraminic acid solution (final concentration, 5 mM, 50 μl). Residual enzyme activity was determined by monitoring the rate of release of 4-nitrophenol with an ultraviolet/visible spectrometer (410 nm) after the addition of 0.1 M glycine-NaOH buffer (pH, 11) to stop the enzymatic activity. Slopes of plots of the natural algorithm of residual enzyme activity versus preincubation time yielded pseudo–first-order rate constants for inactivation at each suicide substrate concentration. Values for Ki (irreversible inhibitor constant) and half-life ($t_{\{1/2\}}$) of inhibition were calculated from plots of the reciprocal of the pseudo–first-order rate constants versus the reciprocal of the suicide substrate concentration. The Ki and $t_{\{1/2\}}$ (Kitz and Wilson, 1962) were calculated to be 1.90 mM and 2.83 min, respectively (see Fig. 2).

Mechanism-Based Fluorescent Labeling of Neuraminidase

α2,3,6,8-Neuraminidase from *V. cholerae* (50 μg) was incubated with or without the suicide substrate 1 (see Fig. 3) (5 μg; final concentration, 45 μM) at 37° in 200 μl of 100 mM citrate phosphate buffer (pH, 5.0). The reaction mixture was subjected directly to purification by small gel filtration column (Sephadex G-25, ϕ 8 $\times$ 20 mm, 1 ml) using 100 mM ammonium bicarbonate buffer (pH, 8.0) to remove the excess suicide substrate. Next, the labeled enzyme was concentrated by centrifugal evaporation system to approximately 50 μg/10 μl in 100 mM ammonium bicarbonate. To the previously described solution was added 10 mM HCl solution (90 μl) and pepsin (5 μg), and the mixture was incubated overnight at 37°. Then, the digests were subjected to affinity purification using anti-dansyl antibody column.

Purification of the Fluorescence-Labeled Peptide by Anti-Dansyl Antibody Column

Anti-dansyl rabbit immunoglobulin (Ig) G (0.5 ml, 1 mg/ml) was dialyzed at 4° for 10 min against a solution of 0.1 M sodium bicarbonate containing 0.1 M NaCl using Centricon YM-30 (Millipore) (5 $\times$ 3 ml, 10,000 rpm). Then, the anti-dansyl rabbit IgG solution was mixed with 0.5 ml of Affinity-Gel 10 (Bio-Rad) according to the manufacturer's instructions, and the mixture was packed into a 0.5 $\times$ 2-cm column (Bradford, 1976). The binding capacity (1.18 nmole/0.98 mg of IgG) of this antibody column was tested with (4-N-dansyl-phenyl)-β-D-galactopyranoside (Kurogochi *et al.*, 2004). A typical procedure for the isolation of a fluorescence-labeled peptide fragment generated by

peptidase treatment was as follows: A sample solution (50 μl) was allowed to pass through the anti-dansyl rabbit IgG column at room temperature with a flow rate of 0.1 ml/min. After washing with 3 ml of 100 mM Hepes-NaOH solution (pH, 7.5) containing 100 mM NaCl, peptides adsorbed onto the antibody column were eluted with 10% (v/v) acetic acid:water, and the fractions containing the desired materials were concentrated using centrifugal evaporation.

MALDI-TOF MS and MS/MS Finger Printing Analysis of the
    Labeled Peptide

VCNA inactivated (labeled) with an irreversible inhibitor was subjected to peptidase treatment to give rise to the complex peptide fragments observed in MALDI-TOF MS, as shown in Fig. 4A. It is evident that the signals resulting from the fluorescence-labeled peptide ([M + H]$^+$ = 1328.49, [M + Na]$^+$ = 1350.47, and [M + Ka]$^+$ = 1366.45) cannot be directly identified if the peptide mixture was analyzed, owing to the complexity of signals of other peptides found in the upper spectra. However, the MALDI-TOF MS of the peptide readily purified by the anti-dansyl antibody column exhibited clear and simple signals due to the ions of a labeled peptide fragment, as indicated in Fig. 4B.

The parent ion of 1328.49 *m/z* from the fluorescence-labeled peptide obtained by the peptidase treatment was subsequently applied for further LIFT-TOF/TOF fragmentation analysis to yield a spectrum having a series of

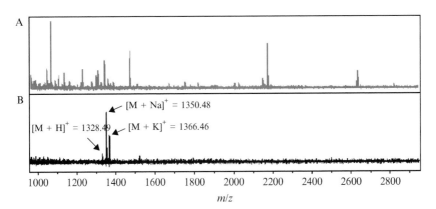

FIG. 4. MALDI-TOF mass spectra of fluorescence-labeled peptides. Peptide fragments produced by digestion with pepsin were analyzed. (A) Mixtures of peptides derived from labeled enzyme; (B) Peptide isolated by using anti-dansyl antibody column chromatography. (See color insert.)

b-ions (488.13, 1036.34, 1116.41, and 1310.48) and y-ions (1328.49, 841.38, 313.38, and 166.09), as shown in Fig. 5. These product ions and fragmentation from two dansylated amino acids allowed us to conclude that this peptide with calculated molecular mass ($[M + H]^+ = 1328.48$) was sequenced as ^{576}DRFF^{579}, in which Asp576 and Arg577 were dansylated. Surprisingly, it has been known that neither Asp576 nor Arg577 was assigned as crucial amino acid residues located in the catalytic site of VCNA (Moustafa *et al.*, 2004).

The three-dimensional (3-D) structure of VCNA (see Fig. 5) (Moustafa *et al.*, 2004) indicated that both Asp576 and Arg577 residues seemed to be located away (approximately 20 Å) from the catalytic pocket of VCNA. It has been postulated that a solvent-exposed Asp250 located close (4.7 Å) to the anomeric carbon, one of the well-conserved residues in the rigid active site, seems to be a key amino acid residue that acts as a proton donor or a stabilizer of a proton-donating water molecule (Crennell *et al.*, 1994). Since a pair of Asp320 (14.3 Å) and Arg321 (16.9 Å) can be seen between the active site and the flexible β-turn involving Asp576 and Arg577, the labeling of these two amino acid residues by compound 1 seems to be quite a specific and interesting result. Considering the unstable nature of the reactive intermediate of fluorinated quinone methide generated by hydrolysis, one may speculate that Asp576 and Arg577 could participate in the catalytic mechanism of the hydrolysis of compound 1 only when VCNA conducts a movement of the flexible β-turn structure by about 20 Å, bringing Asp576 and Arg577 to the active site.

At present, we cannot determine the mechanism of unusual dual labeling observed in the flexible β-turn, and further structural and mutagenesis studies may be required to provide a more sufficient and plausible answer to explain this phenomenon. However, Wakatsuki *et al.* also demonstrated in the crystallographic studies on human cytosolic sialidase (Neu2) (Chavas *et al.*, 2005) that the loops containing Glu111 and the catalytic Asp46 residues located far away from the active site are disordered in the apo form but upon binding of DANA (2-deoxy-2,3-dehydro- *N*-acetylneuraminic acid) become ordered to adopt two short α-helices and come in close contact with the inhibitor and cover the catalytic site, illustrating the dynamic nature of substrate recognition and catalytic action by sialidase. Taken together, these results may suggest that the use of the flexible loop or β-turn structure containing catalytic amino acids located away from the catalytic crevice seems to be feasible for adopting sialidase for substrates with different size or characteristics.

Three-dimensional structural information of enzymes at atomic resolution obtained by crystallographic analysis can often be used for conclusive identification of the active site amino acid residues. When the 3-D structure

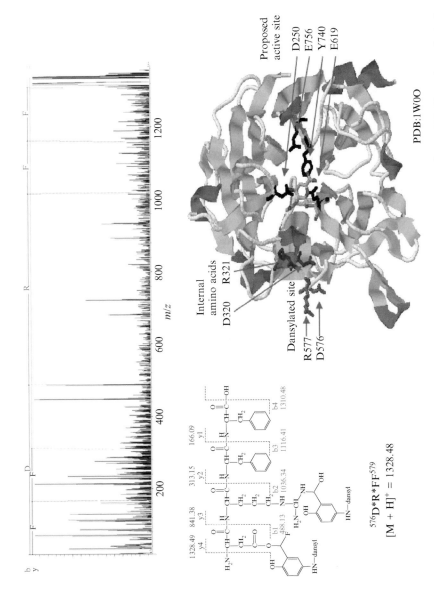

Fig. 5. MALDI-TOF/TOF mass spectrum of the dansylated peptide fragment (m/z 1328.49) and topological assignment of the identified sequence [576]DRFF[579], which contains dansylated Asp576 and Arg577. (See color insert.)

of an enzyme/inhibitor or enzyme/substrate complex is available, the residues important to catalysis, either directly as an acid–base, as nucleophilic catalysts, or less directly through binding of the substrate in its ground state or at the transition state, will be apparent. However, static 3-D information is still not useful for investigating a dynamic reaction mechanism of the enzymatic action. Even when the identities of the amino acid residues in close spatial proximity to the substrate are known, their specific functions in the catalysis frequently cannot be predicted. Therefore, verification of functional roles in the catalysis of the specific residues postulated by 3-D structural analysis requires further kinetic analyses of mutants, preferably in conjunction with studies using specific mechanism-based labeling reagents.

References

Bradford, M. M. (1976). A rapid and sensitive method for the quantification of microgram quantities of protein utilizing the principle of protein-dye binding. *Anal. Biochem.* **72,** 248–254.

Chavas, L. M. G., Tringali, C., Fusi, P., Venerando, B., Tettamanti, G., Kato, R., Monti, E., and Wakatsuki, S. (2005). Crystal structure of the human cytosolic sialidase Neu2: Evidence for the dynamic nature of substrate recognition. *J. Biol. Chem.* **280,** 469–475.

Crennell, S., Garman, E., Laver, G., Vimr, E., and Taylor, G. (1994). Crystal structure of *Vibrio cholerae* neuraminidase reveals dual lectin-like domains in addition to the catalytic domain. *Structure* **2,** 535–544.

Driguez, P.-A., Barrere, B., Chantegrel, B., Deshayes, C., Doutheau, A., and Quash, G. (1992). Synthesis of sodium salt of *ortho*-(difluoromethyl)phenyl-α-ketoside of *N*-acetylneuraminic acid: A mechanism-based inhibitor of *Clostridium perfringens* neuraminidase. *Bioorg. Med. Chem. Lett.* **2,** 1361–1366.

Dwek, R. A. (1996). Glycobiology: Toward understanding the function of sugars. *Chem. Rev.* **96,** 683–720.

Galen, J. E., Ketley, J. M., Fasano, A., Richardson, S. H., Wasserman, S. S., and Kaper, J. B. (1992). Role of *Vibrio cholerae* neuraminidase in the function of cholera toxin. *Infect. Immunol.* **60,** 406–415.

Halazy, S., Berges, V., Ehrhard, A., and Danzin, C. (1990). *Ortho*- and *para*-(difluoromethyl) aryl-β-D-glucosides: A new class of enzyme-activated irreversible inhibitors of β-glucosidases. *Bioorg. Chem.* **18,** 330–344.

Hekmat, O., Tokuyasu, K., and Withers, S. G. (2003). Subsite structure of the endo-type chitin deacetylase from a Deuteromycete, *Colletotrichum lindemuthianum*: An investigation using steady-state kinetic analysis and MS. *Biochemical J.* **374,** 369–380.

Hinou, H., Kurogochi, M., Shimizu, H., and Nishimura, S.-I. (2005). Characterization of *Vibrio cholerae* neuraminidase by a novel mechanism-based fluorescent labeling reagent. *Biochemistry* **44,** 11669–11675.

Ichikawa, M., and Ichikawa, Y. (2001). A mechanism-based affinity-labeling agent for possible use in isolating *N*-acetylglucosaminedase. *Bioorg. Med. Chem. Lett.* **11,** 1769–1773.

Janda, D. K., Lo, L.-C., Lo, C.-H. L., Sim, M.-M., Wang, R., Wong, C.-H., and Larner, R. A. (1997). Chemical selection for catalysis in combinatorial antibody libraries. *Science* **275,** 945–948.

Kakugata, Y., Wada, T., Yamaguchi, K., Yamanami, H., Ouchi, K., Sato, I., and Miyagi, T. (2002). Up-regulation of plasma membrane-associated ganglioside sialidase (Neu3) in human colon cancer and its involvement in apoptosis suppression. *Pro. Nat. Acad. Sci.* **99,** 10718–10723.

Kitz, R., and Wilson, I. B. (1962). Esters of methanesulfonic acid as irreversible inhibitors of acetylcholinesterase. *J. Biol. Chem.* **237,** 3245–3249.

Kuboki, A., Sekiguchi, T., Sugai, T., and Ohta, H. (1998). A facile access to aryl a-sialosides: The combination of a volatile amine base and acetonitrile in glycosidation of Sialosyl chlorides. *Synlett.* 479–482.

Kuhn, R., Lutz, P., and Mac Donald, P. L. (1966). Synthesis of the anomer sialic acid-methylketoside. *Chem. Ber.* **99,** 611–617.

Kurogochi, M., Nishimura, S.-I., and Lee, Y. C. (2004). Mechanism-based fluorescent labeling of β-galactosidases: An efficient method in proteomics for glycoside hydrolases. *J. Biol. Chem.* **279,** 44704–44712.

Loudon, G. M., and Koshland, D. E., Jr. (1970). The chemistry of a reporter group: 2-hydroxy-5-nitrobenzyl bromide. *J. Biol. Chem.* **245,** 2247–2254.

Moustafa, I., Connaris, H., Taylor, M., Zaitsev, V., Wilson, J. C., Kiefel, M. J., Itzstein, M., and Taylor, G. (2004). Sialic acid recognition by *Vibrio cholerae* neuraminidase. *J. Biol. Chem.* **279,** 40819–40826.

Takaya, K., Nagahori, N., Kurogochi, M., Furuike, T., Miura, N., Monde, K., Lee, Y. C., and Nishimura, S.-I. (2005). Rational design, synthesis, and characterization of novel inhibitors for human β1,4-galactosyltransferase. *J. Med. Chem.* **48,** 6054–6065.

Taylor, G. (1996). Sialidases: Structures, biological significance and therapeutic potential. *Curr. Opin. Struct. Biol.* **6,** 830–837.

Withers, S. G., and Aebersold, R. (1995). Approaches to labeling and identification of active site residues in glycosidase. *Protein Sci.* **4,** 361–372.

Withers, S. G., Street, I., Bird, P., and Dolphin, D. H. (1987). 2-Deoxy-2-fluoroglucosides: A novel class of mechanism-based glucosidase inhibitors. *J. Am. Chem. Soc.* **109,** 7530–7531.

Withers, S. G., Warren, R. A. J., Street, I. P., Rupitz, K., Kempton, J. B., and Aebersold, R. (1990). Unequivocal demonstration of the involvement of a glutamate residue as a nucleophile in the mechanism of a "retaining" glycosidase. *J. Am. Chem. Soc.* **112,** 5887–5889.

[14] Regulating Cell Surface Glycosylation with a Small-Molecule Switch

By Danielle H. Dube, Christopher L. de Graffenried, and Jennifer J. Kohler

Abstract

Correct localization of Golgi-resident enzymes is essential for the formation of specific glycan epitopes. In this chapter, we describe a method to control the localization, and thus the activity, of an individual glycosyltransferase by administration of a small molecule. Our method takes advantage of the modularity of most Golgi-resident enzymes, which are composed of localization and catalytic domains. These domains can be physically separated and fused to the small molecule binding proteins FRB and FKBP, which dimerize in the presence of rapamycin. In this way, rapamycin serves as a "switch" for enzyme activity.

Overview

In eukaryotes, complex carbohydrate assembly is accomplished by the coordinate action of Golgi-resident enzymes. The responsible enzymes, which include glycosyltransferases, glycosidases, sulfotransferases, deacetylases, and at least one epimerase (Krause *et al.*, 2005), are spatially arranged within the compartments of the Golgi to create an assembly line for glycan biosynthesis (Opat *et al.*, 2001; Varki, 1998). As proteins destined for secretion or presentation on the cell surface traffic through the Golgi, they become modified sequentially by these enzymes. Correct localization of a set of Golgi-resident enzymes within this enzymatic array is essential for the formation of specific glycan epitopes (de Graffenried and Bertozzi, 2004; Osman *et al.*, 1996; Skrincosky *et al.*, 1997).

An important goal for glycobiology is to establish the set of Golgi-resident enzymes responsible for production of particular glycans. This task is complicated both by the non–template-directed nature of glycan biosynthesis and the functional redundancy of many Golgi-resident enzymes. For example, within the fucosyltransferase family there are six enzymes that are able to catalyze the addition of fucose to GlcNAc via an $\alpha1,3$ linkage (de Vries *et al.*, 2001a). Despite their common ability to catalyze this reaction *in vitro*, these enzymes appear to modify distinct substrates in cells (Huang *et al.*, 2002). Establishing a clear link between glycosyltransferases and their glycan

METHODS IN ENZYMOLOGY, VOL. 415
0076-6879/06 $35.00
DOI: 10.1016/S0076-6879(06)15014-4

products is even more challenging in whole organisms, where the deletion of one enzyme often causes a compensatory up-regulation of related family members—or worse—results in embryonic lethality (Lowe and Marth, 2003). Compensation can cause predicted phenotypic changes to be subtle or absent. An embryonic lethal phenotype, which results from the fact that some glycosyltransferases are essential in early development, compromises researchers' attempts to study the role of these enzymes at later developmental stages.

Chemical approaches to regulation of enzyme activity offer an alternative to genetic methods (Mitchison, 1994; Specht and Shokat, 2002). Chemical tools enable temporal control, can be exquisitely selective, and provide the added feature of dose-dependent control (crucial in studies of developmental processes). As a result, the use of chemical tools to manipulate glycosyltransferases could circumvent embryonic lethality, compensation, and functional redundancy. In a quest to de-convolute cellular roles of Golgi-resident enzymes, we decided to develop a method in which the cellular activity of a single Golgi-resident enzyme is controlled by the administration of a small molecule. The design of active site inhibitors represented one potential route to small-molecule control, but in the case of carbohydrate biosynthetic enzymes, two essential inhibitor traits—cell permeability and specificity— are particularly difficult to achieve (Rath et al., 2004). Rather than pursue active site inhibitors, we turned to another salient functional characteristic of these enzymes, their localization to specific sites within the secretory pathway. The fact that localization is a critical determinant of Golgi-resident enzyme activity suggested that control of localization could provide a potential mechanism for regulating glycosyltransferase activity. This article describes a general method that exploits enzyme localization to achieve time-dependent and dose-dependent control of the cellular activity of an *individual* glycosyltransferase.

Experimental Design

Approximately 250 human genes encoding Golgi-resident enzymes have been identified (CAZY database); most exhibit a similar arrangement of functional elements. They are type II transmembrane proteins composed of a short amino terminal tail that extends into the cytoplasm, a hydrophobic transmembrane region that spans the Golgi membrane, and a large carboxy terminal catalytic domain located within the lumen of the Golgi (Fig. 1) (Munro, 1998). A loosely defined stem region joins the transmembrane and catalytic domains.

Our method relies on the modularity often observed in Golgi-resident enzymes (Colley, 1997; Munro, 1998). The amino terminal residues comprising the cytoplasmic, transmembrane, and stem regions of the enzyme

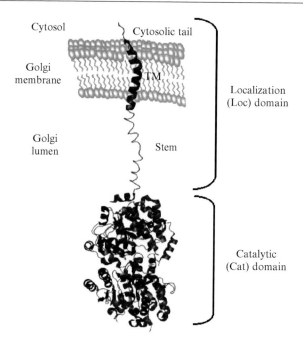

FIG. 1. Anatomy of a Golgi-resident enzyme. Most Golgi-resident enzymes, such as glycosyltransferases, are type II transmembrane proteins. These proteins are modular in nature; they are composed of catalytic (Cat) and localization (Loc) domains that, in general, can be separated at the level of primary sequence. As its name suggests, the Cat domain contains the catalytic activity of the enzyme. The Loc domain, which contains an N-terminal cytoplasmic tail, single-pass transmembrane (TM) region, and lumenal stem region, controls the localization of the enzyme within the secretory pathway.

constitute the localization domain (Loc domain) and are often both necessary and sufficient for localization (Milland et al., 2002; Yang et al., 1996). An isolated Loc domain is retained at the enzyme's normal location within the Golgi but lacks the ability to catalyze the reaction. Similarly, only the C-terminal catalytic domain (Cat domain) is necessary for catalysis (Xu et al., 1996). A detached Cat domain maintains the ability to catalyze the enzyme's reaction but will not be retained at a location where it has access to appropriate substrates. We have demonstrated that a small molecule can be used to regulate association of the Loc and Cat domains: the presence of this molecule results in the retention of the Cat domain at the site of the Loc domain and reconstitution of the enzyme's normal activity (de Graffenried et al., 2004; Kohler and Bertozzi, 2003; Kohler et al., 2004).

To control association of the two protein domains, we take advantage of the rapamycin-mediated heterodimerization (Choi et al., 1996) of FKBP

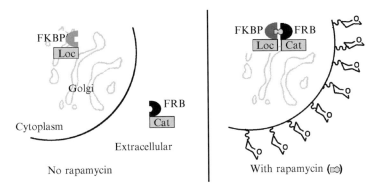

F<small>IG</small>. 2. Rapamycin-induced glycosyltransferase activity. Loc and Cat domains of glycosyltransferases can be physically separated and fused to the small molecule binding proteins FKBP and FRB, which dimerize in the presence of rapamycin. In the presence of rapamycin, the enzyme is reassembled and its glycan product appears on the cell surface. In the absence of rapamycin, the Cat domain is secreted and the glycan product is not observed.

(Standaert *et al.*, 1990) and FRB (Brown *et al.*, 1994). This ternary interaction has been used widely to mediate the dimerization of protein fragments (Belshaw *et al.*, 1996; Clemons, 1999) and to control a variety of biological events including signal transduction (Crabtree and Schreiber, 1996) and transcription (Liberles *et al.*, 1997). In our system, the localization and catalytic domains are separated into two polypeptides, each of which is fused to FRB or FKBP. The addition of rapamycin causes the dimerization of FRB and FKBP, bringing together the localization and catalytic domains and reconstituting the active enzyme (Fig. 2). In this way, the addition of a small molecule results in an increase in glycan production.

Previous experiments using rapamycin-induced dimerization of FRB and FKBP have demonstrated that multiple FRB or FKBP domains in fusion proteins lead to enhanced rapamycin response (Rivera *et al.*, 1996; Spencer *et al.*, 1993). In addition, Stankunas *et al.* (2003) reported that the T2098L mutant of FRB is susceptible to degradation and confers this susceptibility to the protein to which it is fused. Guided by these precedents, we have investigated the use of multiple repeats of FRB and FKBP, and of the T2098L FRB mutant, in our chimeric constructs.

Identification of Localization and Catalytic Domains

Our first target for small-molecule control was the Golgi-resident enzyme fucosyltransferase VII (FUT7) (Natsuka *et al.*, 1994; Sasaki *et al.*, 1994). Previous work had identified that amino acids 1 through 51 were

capable of dictating correct localization within the secretory pathway (Grabenhorst and Conradt, 1999). This region can be grafted onto the catalytic domain of another fucosyltransferase (i.e., FUT4), thereby producing a chimeric protein whose localization is identical to that of FUT7. Literature precedent also facilitated the identification of the amino acids necessary and sufficient for catalysis. de Vries *et al.* (2001b) showed that amino acids 39 through 342 can be expressed independently of the N-terminal region and comprise a folded and active catalytic domain. Guided by these results, we used amino acids 1 through 51 for our localization domain constructs and 39 through 342 for our catalytic domain constructs (Kohler and Bertozzi, 2003).

For many other Golgi-resident enzymes, the localization and catalytic domains have not yet been delineated. We have found that sequence alignments of related enzymes provide a useful guide in the identification of the functional domains (de Graffenried *et al.*, 2004; Kohler *et al.*, 2004). In general, enzymes that catalyze the same chemical transformation exhibit high sequence similarity in the catalytic domain but show significant differences within the N-terminal localization domains.

Construction of Expression Vectors

We use a modular strategy for constructing plasmids that encode localization and catalytic domain chimeras. To allow for the introduction of a varied number of FKBP and/or FRB repeats, we take advantage of restriction enzymes (Xba I and Spe I) that generate compatible cohesive ends (Fig. 3). All plasmids also code for epitope tags (myc for localization domains and HA for catalytic domains) to facilitate analysis of chimera expression levels and their cellular localization. To follow, the procedure used to create plasmids encoding FUT7-based chimeras is outlined (Kohler and Bertozzi, 2003).

Localization Domain Plasmids

We used polymerase chain reaction (PCR) to amplify the localization domain of FUT7 and to introduce an Nhe I site at the 5′ end of the coding strand and an Xba I site at the 3′ end of the coding strand. This insert was ligated between the Nhe I and Xba I sites of the plasmid pcDNA3.1-Zeo to produce pcDNA3.1-Loc.

We also amplified, with PCR, DNA encoding FRB and FKBP and, in the process, introduced an Xba I site at 5′ ends of the coding strands and Spe I sites at the 3′ ends of the non-coding strands. These PCR products were cloned into the plasmid pCR4-Blunt-TOPO (Invitrogen) so that the

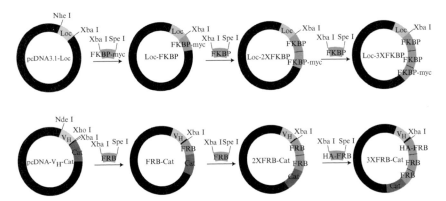

FIG. 3. Cloning strategy for producing chimeric proteins with multiple FRB and FKBP repeats. Each of our coding inserts was flanked by recognition sites for restriction enzymes (Xba I and Spe I) that generate compatible cohesive ends. We ligated the inserts sequentially into the expression plasmids that had been linearized with Xba I and regenerated an upstream Xba I site with each insertion.

FRB and FKBP inserts could be produced in large quantities for ligation into various plasmids. We produced two versions of the FRB and FKBP inserts: one version contained only the amino acid coding sequence lacking a stop codon (FRB and FKBP), and a second version contained the coding sequence followed by the sequence encoding the myc epitope tag followed by a stop codon (FRB-myc-stop and FKBP-myc-stop). For plasmids encoding FRB, we often use the T2098L mutant, which has been shown to be sensitive to proteolytic degradation in the absence of a small-molecule stabilizer (Stankunas *et al.*, 2003).

To construct the localization chimeras, pcDNA3.1-Loc was linearized by digestion with Xba I and ligated to the FRB-myc-stop or FKBP-myc-stop insert. Ligation regenerated the Xba I site between the Loc domain and FRB/FKBP. The resulting plasmids (Loc-FRB and Loc-FKBP) could be used directly or linearized again to allow for insertion of an additional FRB or FKBP insert, producing Loc-2XFRB and Loc-2XFKBP. Once again, the Xba I site between the Loc and FRB/FKBP domains was regenerated on ligation, allowing for the introduction of a third copy of FRB or FKBP and production of Loc-3XFRB and Loc-3XFKBP.

Catalytic Domain Plasmids

The plasmid pCMV/myc/ER was digested with Nde I and Xho I to excise the murine V_H chain signal peptide (Invitrogen), which was ligated into pcDNA3.1-Zeo, producing pcDNA3.1-V_H. The Cat domain of FUT7

(including the stop codon) was amplified by PCR with the concomitant introduction of the Xba I and Spe I restriction sites at the 5′ and 3′ ends of the coding strand, respectively. pcDNA3.1-V_H was linearized by digestion with Xba I and ligated to the Cat insert, producing pcDNA-V_H-Cat, in which the Xba I site is regenerated between the V_H domain and the Cat domain.

We used PCR to amplify DNA encoding FRB and FKBP and to introduce an Xba I site and the coding sequence for the HA epitope tag at 5′ ends of the coding strands and Spe I sites at the 3′ ends of the non-coding strands. These PCR products were cloned into the plasmid pCR4-Blunt-TOPO (Invitrogen) to produce a renewable source of the HA-FRB and HA-FKBP inserts. The HA-FRB and HA-FKBP inserts, along with the FRB and FKBP inserts described previously, were ligated into the pcDNA-V_H-Cat plasmid as described in the next paragraph.

To produce the plasmids FRB-CAT and FKBP-CAT, pcDNA-V_H-Cat was digested with Xba I and ligated to the HA-FRB or HA-FKBP insert. To produce the plasmids 2XFRB-CAT and 2XFRB-CAT, pcDNA-V_H-Cat was linearized with Xba I and ligated to the FRB or FKBP insert, regenerating the Xba I site between the V_H domain and FRB or FKBP. The resulting plasmids were linearized with Xba I and ligated to the HA-FRB or HA-FKBP insert, yielding 2XFRB-CAT and 2XFRB-CAT. To produce the plasmids 3XFRB-CAT and 3XFRB-CAT, pcDNA-V_H-Cat was linearized with Xba I and ligated to the FRB or FKBP insert, regenerating the Xba I site between the V_H domain and FRB or FKBP. Resultant plasmids were linearized with Xba I and ligated to the FRB or FKBP insert, regenerating the Xba I site between the V_H domain and the two copies of FRB or FKBP. The resulting plasmids were digested with Xba I and ligated to the HA-FRB or HA-FKBP insert, yielding 3XFRB-CAT and 3XFRB-CAT.

Activity Assays of Enzyme Reconstitution

We used flow cytometry to detect rapamycin-induced changes in cell surface glycosylation (Kohler and Bertozzi, 2003) and Western blot analysis to monitor rapamycin-induced changes in the glycosylation of secreted proteins (de Graffenried et al., 2004).

Flow Cytometry Analysis of Cell Surface Glycoproteins

Flow cytometry can be used to measure an enzyme's cellular activity and its response to a small-molecule regulator. Analysis requires the availability of an antibody or lectin with specificity for the cell surface glycan of interest. These protein probes can be used in conjunction with fluorescent

tags to quantitatively detect the appearance of a particular glycan on the surface of an individual cell. Following is a protocol for the detection of FUT7 activity in Chinese hamster ovary (CHO) cells using the HECA-452 antibody, which detects FUT7's product, sialyl Lewis X (sLe^X). This protocol can be modified for the detection of other glycans (de Graffenried et al., 2004; Kohler et al., 2004).

Day 1: Plate Cells

1. Trypsinize CHO cells growing in culture plates.
2. Centrifuge (2500g) to remove excess trypsin and dead cells.
3. Resuspend in F12 nutrient mixture (HAM) with L-glutamine (Gibco) containing 10% fetal bovine serum (HyClone), 100 units/ml penicillin, and 0.1 mg/ml streptomycin.
4. Place 5×10^5 cells in each well of a six-well plate (10 cm^2 surface area) and add additional media to a final volume of 2 ml per well.

Day 2: Transfection

1. For transfection, use LipofectAMINE PLUS reagent (Gibco-BRL) and follow the manufacturer's recommendations.
2. Unsupplemented OptiMEM I (Gibco) is used as media during the transformation. Equivalent amounts of localization and catalytic domain plasmids are used for each transfection.
3. For control experiments lacking either the localization or catalytic plasmid, an "empty" expression vector (pcDNA3.1) should be substituted for the expression plasmid.
4. Allow transfections to proceed for 3 to 5 hrs, then remove transfection media and replace with F12-HAM.

Day 3: Split Cells and Add Rapamycin

1. Wash cells two times with Dubecco's (1× phosphate-buffered saline [PBS], Gibco) to remove cell debris.
2. Remove cells from the plate by trypsinization and count live cells.
3. Pellet cells by centrifugation.
4. Resuspend in F12-HAM media and seed cells at a density of 1×10^5 cells per well of a 12-well plate (4 cm^2 surface area/well).
5. Dissolve rapamycin in ethanol to make a 1-mM stock solution (stable at $-20°$ indefinitely). The rapamycin stock solution can be further diluted in F12-HAM media to achieve the desired final concentration. We have observed that 200 nM rapamycin is usually sufficient to achieve maximal signal within 18 hrs (see later).

6. Incubate the cells with or without rapamycin for 18 hrs.
7. To measure the values in triplicate, prepare three wells for each set of conditions.

Day 4: Flow Cytometry

1. Wash the cells twice with $1\times$ PBS, lacking Mg^{2+} and Ca^{2+}.
2. To remove cells from the plate, add 0.2 ml of $1\times$ PBS containing 1 mM ethylenediamine tetraacetic acid (and lacking Mg^{2+} and Ca^{2+}) to each well and incubate at 37° for 20 min.
3. After the incubation period, it may be necessary to strike the plate against the side of a counter to dislodge the cells. Transfer the cells from each well to a single well of a 96-well V-bottom plate.
4. Keep the samples on ice for all subsequent steps.
5. Pellet the cells at 2500g, and then decant the supernatant.
6. Wash cells two times by resuspending in fluorescence-activated cell sorter (FACS) buffer ($1\times$ PBS, 1 mg/ml bovine serum albumin [BSA], 0.2% NaN_3), pelleting at 2500g, and decanting the supernatant.
7. Incubate the cells with the primary detection agent, in this case, biotinylated HECA-452 (BD PharMingen), which recognizes the tetrasaccharide sLe^X.
8. Add the antibody to FACS buffer to achieve a final dilution of 1:25.
9. Add 50 μl of the diluted antibody to each well, resuspend the cells, and incubate for 1 hr.
10. Pellet the cells and decant the supernatant.
11. Wash the cells two times with FACS buffer (as previously).
12. Incubate cells with the secondary detection reagent, in this case streptavidin-tricolor (Caltag Laboratories).
13. Add the streptavidin to FACS buffer to achieve a final dilution of 1:50.
14. Add 50 μl of diluted streptavidin to each well, resuspend the cells, and incubate for 0.5 hr.
15. Pellet the cells and decant the supernatant, then wash two more times with FACS buffer.
16. Resuspend each well in 150 μl of FACS buffer.
17. Measure the fluorescence of at least 10,000 live cells on the appropriate channel of a flow cytometer.

Western Blot Analysis of Secreted Glycoproteins

We use Western blot analysis to measure the rapamycin-induced cellular activity of an enzyme that acts on secreted substrates. For example, we assessed the activity of the sulfotransferase GlcNAc6ST-2 by analyzing the

formation of the MECA-79 epitope on the secreted glycoprotein GlyCAM-immunoglobulin (Ig). Cells transfected with localization and catalytic domain chimeras of GlcNAc6ST-2 and the substrate GlyCAM-Ig are split into several plates. One day after transfection, the cells are washed with PBS and then incubated with OptiMEM media containing rapamycin (0 to 200 nM) for 4 days.

Purification of Secreted GlyCAM-Ig

The conditioned media are collected, clarified by centrifugation, and concentrated to 1.5 ml. The concentrate is then incubated with 0.5 ml of a slurry of protein A-Sepharose (Zymed Laboratories, 50% by volume) overnight at 4° with shaking. Beads are washed with 10 ml of PBS containing 0.05% Tween-20. The GlyCAM-Ig chimera is eluted from the beads with 1.5 ml of 100 mM glycine (pH, 2.9), then neutralized with 150 μl of 1 M Tris (pH, 8). Each sample is then lyophilized and resuspended in water.

Western Blot of Purified Protein

The total protein content of each sample is determined using the Bradford assay (Bio-Rad) with BSA as the standard. Equivalent quantities of protein are loaded on a 12% SDS (sodium dodecyl sulfate) polyacrylamide gel. Separated proteins are transferred to nitrocellulose, which is blocked overnight at 4° with blocking buffer (10% nonfat dried milk, 0.05% Tween-20 in PBS). Nitrocellulose is incubated with MECA-79 overnight at 4°, followed by goat anti-mouse IgM conjugated to horseradish peroxidase. To verify equivalent loading, an identical nitrocellulose membrane is prepared, but it is probed with donkey anti-human IgG conjugated to horseradish peroxidase. The blots are developed by using SuperSignal West Pico chemiluminescent substrate (Pierce) and Kodak film. We observe an increase in sulfation of GlyCAM-Ig in the presence of increasing concentrations of rapamycin, indicating that rapamycin can modulate GlcNAc6ST-2 activity on a secreted glycoprotein in a tunable fashion.

Variables Affecting Enzyme Activity

To determine the concentration dependence of small molecule–mediated changes in glycosylation, we transfected cells with the FUT7-based plasmids Loc-FKBP and FRB-Cat and incubated them with various concentrations of rapamycin. We observed increasing sLe^X levels with higher doses of rapamycin (Fig. 4). Maximal sLe^X expression was achieved with rapamycin concentrations of 200 nM or greater; we used 200 nM rapamycin as a standard for most subsequent experiments.

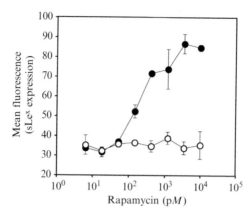

Fig. 4. FUT7 activity can be turned on in the presence of the small molecule rapamycin. Cells were transfected with FUT7-derived Loc-FKBP and FRB-Cat (filled circles) or with FUT7-derived Loc-FRB and FRB-Cat (open circles) in the presence of increasing amounts of rapamycin. The cells were then probed for sLe^X expression by flow cytometry.

In our studies of FUT7, we prepared a series of plasmids in which the localization domain was linked to one, two, or three copies of FRB or FKBP and another series in which the catalytic domain was linked to one, two, or three copies or FRB or FKBP. These plasmids were transfected pairwise into CHO cells, and sLe^X levels were measured in the absence and presence of 200 nM rapamycin. Figure 5 shows the fold-change in sLe^X expression with addition of rapamycin. Increasing the number of FRB or FKBP repeats attached to the localization domain decreased the rapamycin response: in every case, Loc-FRB and Loc-FKBP were the best localization domain plasmids. For the catalytic domain plasmids, FKBP-Cat provided a stronger rapamycin response than 2XFKBP-Cat or 3XFKBP-Cat. This trend was reversed for the FRB-Cat series; in this case, 3XFRB-Cat yielded the largest increase in glycan expression in response to rapamycin treatment. Thus, our empirical observation is that Loc-FKBP and 3XFRB-Cat form the best pair of plasmids for FUT7 reconstitution.

We also tested whether rapamycin-induced stabilization of the T2098L mutant of FRB against proteolysis is responsible for the signal enhancement that we observe in the presence of rapamycin. We transfected cells with the FUT7-derived plasmids Loc-FKBP and FRB-Cat and compared sLe^X expression for cells treated with 0 or 200 nM rapamycin. We observe the same increase in sLe^X expression, regardless of whether the FRB sequence is wild type or the T2098L mutant. These data suggest that localization, and

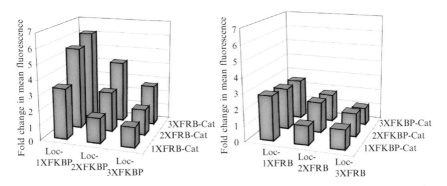

FIG. 5. The Loc-FKBP and 3XFRB-Cat pair provides the largest change in glycan expression in response to rapamycin. Cells were transfected with the indicated pairs of FUT7-derived plasmids. Half of the transfected cells were treated with 200 nM rapamycin; the other half were untreated. The fold change in mean fluorescence reported on the y-axis is the mean fluorescence of rapamycin-treated cells divided by the mean fluorescence of untreated cells.

not protein stabilization, is the primary mechanism of signal enhancement. We have not performed the same comparison for chimeras containing multiple repeats of FRB.

We hypothesized that controlling enzyme activity at the level of enzyme *localization*, rather than at the level of transcription or translation, might offer a rapid response. To quantify the time dependence of enzyme activation, cells should be transfected with Loc-FKBP and 3XFRB-Cat. One day after transfection, cells should be split into 60 samples. At 0.5- or 1-hr intervals, media should be removed from a pair of samples and replaced—one sample with plain media and one with media containing 200 nM rapamycin. After 25 hrs, glycan expression levels can be quantified by flow cytometry. We observe an increase in sLe^X expression within 1 hr of the addition of rapamycin (Fig. 6). Expression levels increase over time, reaching a plateau at about 18 hrs after rapamycin addition.

Although transient transfection with the localization and catalytic domain-encoding plasmids yields rapamycin-inducible changes in glycan expression, not all cells express the chimeric genes. Although we have not produced cell lines that stably express the chimeric genes, we hypothesize that they would yield a more robust rapamycin response. As a test of this hypothesis, we transiently transfected cells with the GlcNAc6ST-2–based plasmids Loc-FKBP and 3XFRB-Cat as well as a plasmid encoding green fluorescence protein (GFP). We gated our flow cytometry experiment on the cells that expressed high levels of GFP, with the assumption that they also express high levels of Loc-FKBP and 3XFRB-Cat (Fig. 7). Considering only

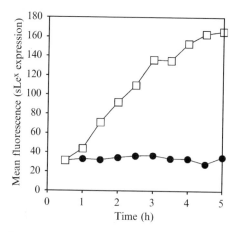

FIG. 6. Addition of rapamycin causes a rapid increase in sLe^X expression. Cells were transfected with FUT7-derived Loc-FKBP and 3XFRB-Cat, then split into pools and incubated in the absence of rapamycin for 2 days. The media was replaced with media containing 0 nM (filled circles) or 200 nM (open squares) rapamycin and the cells incubated for the time indicated (x-axis). sLe^x expression was measured by flow cytometry.

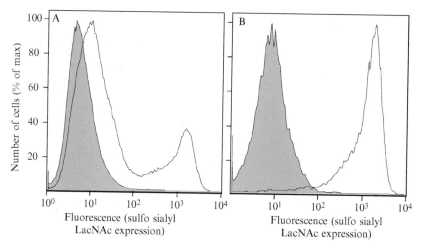

FIG. 7. Cells expressing high levels of chimeric proteins exhibit a switch-like response to rapamycin. Cells were transfected with the GlcNAc6ST-2 derived plasmids Loc-FKBP and 3XFRB-Cat and with a plasmid encoding GFP. Transfected cells were split and treated with 0 nM (grey fill) or 200 nM (white fill) rapamycin, then probed for sulfo sialyl LacNAc expression by flow cytometry. Histogram A shows data for all cells. Histogram B shows data for the subset of cells that express high levels of GFP, indicative of high transfection efficiency.

these high-expressing cells, we observe a switch-like response to rapamycin. Cells not treated with rapamycin produce low levels of sulfo sialyl LacNAc, GlcNAc6ST-2's product, whereas cells treated with 200 nM rapamycin all display high levels of the sulfated glycan.

Location of Catalytic Domains

In our design of the conditional glycosylation system, we predict that in the absence of rapamycin the localization domain will be retained in the Golgi, whereas the catalytic domain will be secreted into the extracellular space. With the addition of rapamycin, we expect that the localization domain will remain at the same site in the secretory pathway and that the catalytic domain will be retained there as well. Next, we outline two experimental techniques that can be used to track the location and amounts of catalytic domain chimeric proteins observed in the absence and presence of rapamycin.

Flow Cytometry

Two-dimensional flow cytometry can be used to detect a correlation between intracellular levels of the catalytic domain chimera and cell surface expression of its glycan product. We have applied this technique to cells transfected with the FUT7-based plasmids Loc-FKBP and 3XFRB-Cat, as follows. Cells are stained for sLe^X using biotinylated HECA-452 and streptavidin-tricolor, as described previously. Cells are fixed and washed using the Cytofix/Cytoperm Kit (BD Biosciences), according to kit instructions. Cells are probed for HA levels by incubation with 2 μl of FITC-labeled anti-HA (Covance Research Products, Denver, PA) in 50 μl BD Perm/Wash buffer for 30 min at 4°. Cells are washed twice with BD Perm/Wash buffer and then resuspended in FACS buffer before fluorescence is measured on the appropriate channels of a flow cytometer. We observe that the addition of rapamycin causes increases in both HA and sLe^X levels. Furthermore, these increases are correlated: cells with higher HA levels tend to also have higher levels of sLe^X.

Immunofluorescence Microscopy

Immunofluorescence microscopy can be used to determine the subcellular localization of the HA-tagged catalytic domain chimeras by comparison to various marker proteins. To label the HA-tagged chimeras, we use mouse anti-HA (3:1000 dilution, Covance Research Products) as the primary antibody and Alexa546-conjugated goat anti-mouse (1:1000 dilution, Molecular Probes) as the secondary antibody. To label the Golgi, we use rabbit anti-giantin (1:500 dilution, Covance Research Products) as the

primary antibody and Alexa647-conjugated goat anti-rabbit (1:1000 dilution, Molecular Probes) as the secondary antibody. In addition, it is possible to use anti-glycan antibodies to simultaneously detect the enzyme's cell surface product (de Graffenried *et al.*, 2004).

CHO cells are transfected with Loc-FKBP and 3XFRB-Cat, as described previously. One day after transfection, cells are seeded on slides mounted with tissue culture wells (Lab-tek, Nalge Nunc Incorporated) and incubated with or without rapamycin (200 n*M*). After 2 days, the cells are washed three times with PBS and then fixed in 3% paraformaldehyde in PBS (20 min at room temperature). After three more PBS washes, cells are permeabilized with PBS containing 1% BSA and 0.1% Triton X-100 (5 min at room temperature). Cells are blocked in PBS containing 1% BSA (20 min at room temperature), then incubated with the primary antibodies in PBS containing 1% BSA (2 hrs at room temperature). Cells are washed with PBS three times, then they are blocked for an additional 20 min before incubation with the secondary antibodies in PBS containing 1% BSA (1 hr at room temperature). Cells are washed three times with PBS, then three times with water. They are mounted using Vectashield with DAPI (Vector Laboratories, Inc.) and visualized by de-convolution microscopy.

In the absence of rapamycin, we observe diffuse HA staining; with rapamycin treatment, HA staining is observed in punctate juxtanuclear structures that colocalize with Golgi markers. Microscopy results are consistent with rapamycin-induced retention of the catalytic domain in the Golgi.

Applications and Future Directions

We have summarized herein a method to regulate cell surface glycosylation with a small molecule. To date, we have successfully used the small molecule rapamycin to regulate the activity of two fucosyltransferases (FUT1 and FUT7) and two sulfotransferases (GlcNAc6ST-1 and GlcNAc6ST-2) in cells. Although we have found that we have the most success with enzymes that generate terminal modifications, we expect that other Golgi-resident enzymes will be amenable to this strategy.

The dearth of tools available to establish the set of Golgi-resident enzymes responsible for the production of particular glycans inspired us to create this technology. Our "conditional knockout" system is uniquely poised to study the roles of glycosyltransferases in an animal model. This application may be facilitated by new enhancements to the FRB-rapamycin-FKBP system, which have improved its utility, especially in animal experiments (Stankunas *et al.*, 2003). In addition, other small molecule–controlled protein–protein interactions have been reported recently (Bayle *et al.*, 2006).

Furthermore, we expect that our system will be easily adaptable to alternative dimerization reagents, possibly allowing for simultaneous control of more than one Golgi-resident enzyme.

References

Bayle, J. H., Grimley, J. S., Stankunas, K., Gestwicki, J. E., Wandless, T. J., and Crabtree, G. R. (2006). Rapamycin analogs with differential binding specificity permit orthogonal control of protein activity. *Chem. Biol.* **13,** 99–107.

Belshaw, P. J., Ho, S. N., Crabtree, G. R., and Schreiber, S. L. (1996). Controlling protein association and subcellular localization with a synthetic ligand that induces heterodimerization of proteins. *Proc. Natl. Acad. Sci. USA* **93,** 4604–4607.

Brown, E. J., Albers, M. W., Shin, T. B., Ichikawa, K., Keith, C. T., Lane, W. S., and Schreiber, S. L. (1994). A mammalian protein targeted by G1-arresting rapamycin-receptor complex. *Nature* **369,** 756–758.

CAZY database, Carboydrate Active enZYmes database. Available at http://afmb.cnrs-mrs.fr/CAZY/.

Choi, J. W., Chen, J., Schreiber, S. L., and Clardy, J. (1996). Structure of the FKBP12-rapamycin complex interacting with the binding domain of human FRAP. *Science* **273,** 239–242.

Clemons, P. A. (1999). Design and discovery of protein dimerizers. *Curr. Opin. Chem. Biol.* **3,** 112–115.

Colley, K. J. (1997). Golgi localization of glycosyltransferases: More questions than answers. *Glycobiology* **7,** 1–13.

Crabtree, G. R., and Schreiber, S. L. (1996). Three-part inventions: Intracellular signaling and induced proximity. *Trends Biochem. Sci.* **21,** 418–422.

de Graffenried, C. L., and Bertozzi, C. R. (2004). The roles of enzyme localization and complex formation in glycan assembly within the Golgi apparatus. *Curr. Op. Cell. Biol.* **16,** 356–363.

de Graffenried, C. L., Laughlin, S. T., Kohler, J. J., and Bertozzi, C. R. (2004). A small-molecule switch for Golgi sulfotransferases. *Proc. Natl. Acad. Sci. USA* **101,** 16715–16720.

de Vries, T., Knegtel, R. M. A., Holmes, E. H., and Macher, B. A. (2001a). Fucosyltransferases: Structure/function studies. *Glycobiology* **11,** 119R–128R.

de Vries, T., Storm, J., Rotteveel, F., Verdonk, G., van Duin, M., van den Eijnden, D. H., Joziasse, D. H., and Bunschoten, H. (2001b). Production of soluble human α1,3-fucosyltransferase (FucT VII) by membrane targeting and *in vivo* proteolysis. *Glycobiology* **11,** 711–717.

Grabenhorst, E., and Conradt, H. S. (1999). The cytoplasmic, transmembrane, and stem regions of glycosyltransferases specify their *in vivo* functional sublocalization and stability in the Golgi. *J. Biol. Chem.* **274,** 36107–36116.

Huang, M. C., Laskowska, A., Vestweber, D., and Wild, M. K. (2002). The α(1,3)-fucosyltransferase Fuc-TIV, but not Fuc-TVII, generates sialyl Lewis X-like epitopes preferentially on glycolipids. *J. Biol. Chem.* **277,** 47786–47795.

Kohler, J. J., and Bertozzi, C. R. (2003). Regulating cell surface glycosylation by small molecule control of enzyme localization. *Chem. Biol.* **10,** 1303–1311.

Kohler, J. J., Czlapinski, J. L., Laughlin, S. T., Schelle, M. L., de Graffenried, C. L., and Bertozzi, C. R. (2004). Directing flux in glycan biosynthetic pathways with a small molecule switch. *Chembiochem* **5,** 1455–1458.

Krause, S., Hinderlich, S., Amsili, S., Horstkorte, R., Wiendl, H., Argov, Z., Mitrani-Rosenbaum, S., and Lochmuller, H. (2005). Localization of UDP-GlcNAc 2-epimerase/ManAc kinase (GNE) in the Golgi complex and the nucleus of mammalian cells. *Exp. Cell Res.* **304**, 365–379.

Liberles, S. D., Diver, S. T., Austin, D. J., and Schreiber, S. L. (1997). Inducible gene expression and protein translocation using nontoxic ligands identified by a mammalian three-hybrid screen. *Proc. Natl. Acad. Sci. USA* **94**, 7825–7830.

Lowe, J. B., and Marth, J. D. (2003). A genetic approach to mammalian glycan function. *Annu. Rev. Biochem.* **72**, 643–691.

Milland, J., Russell, S. M., Dodson, H. C., McKenzie, I. F. C., and Sandrin, M. S. (2002). The cytoplasmic tail of α1,3-galactosyltransferase inhibits Golgi localization of the full-length enzyme. *J. Biol. Chem.* **277**, 10374–10378.

Mitchison, T. J. (1994). Toward a pharmacological genetics. *Chem. Biol.* **1**, 3–6.

Munro, S. (1998). Localization of proteins to the Golgi apparatus. *Trends Cell Biol.* **8**, 11–15.

Natsuka, S., Gersten, K. M., Zenita, K., Kannagi, R., and Lowe, J. B. (1994). Molecular cloning of a cDNA encoding a novel human leukocyte α1,3fucosyltransferase capable of synthesizing the sialyl Lewis X determinant. *J. Biol. Chem.* **269**, 16789–16794.

Opat, A. S., van Vliet, C., and Gleeson, P. A. (2001). Trafficking and localisation of resident Golgi glycosylation enzymes. *Biochimie* **83**, 763–773.

Osman, N., McKenzie, I. F. C., Mouhtouris, E., and Sandrin, M. S. (1996). Switching amino-terminal cytoplasmic domains of α(1,2)fucosyltransferase and a(1,3)galactosyltransferase alters the expression of H substance and Gal α(1,3)Gal. *J. Biol. Chem.* **271**, 33105–33109.

Rath, V. L., Verdugo, D., and Hemmerich, S. (2004). Sulfotransferase structural biology and inhibitor discovery. *Drug Disc. Today* **9**, 1003–1011.

Rivera, V. M., Clackson, T., Natesan, S., Pollock, R., Amara, J. F., Keenan, T., Magari, S. R., Phillips, T., Courage, N. L., Cerasoli, F., Holt, D. A., and Gilman, M. (1996). A humanized system for pharmacologic control of gene expression. *Nat. Med.* **2**, 1028–1032.

Sasaki, K., Kurata, K., Funayama, K., Nagata, M., Watanabe, E., Ohta, S., Hanai, N., and Nishi, T. (1994). Expression cloning of a novel α1,3-fucosyltransferase that is involved in biosynthesis of the sialyl Lewis X carbohydrate determinants in leukocytes. *J. Biol. Chem.* **269**, 14730–14737.

Skrincosky, D., Kain, R., ElBattari, A., Exner, M., Kerjaschki, D., and Fukuda, M. (1997). Altered golgi localization of core 2 β-1,6-*N*-acetylglucosaminyltransferase leads to decreased synthesis of branched *O*-glycans. *J. Biol. Chem.* **272**, 22695–22702.

Specht, K. M., and Shokat, K. M. (2002). The emerging power of chemical genetics. *Curr. Opin. Cell Biol.* **14**, 155–159.

Spencer, D. M., Wandless, T. J., Schreiber, S. L., and Crabtree, G. R. (1993). Controlling signal transduction with synthetic ligands. *Science* **262**, 1019–1024.

Standaert, R. F., Galat, A., Verdine, G. L., and Schreiber, S. L. (1990). Molecular cloning and overexpression of the human FK506-binding protein FKBP. *Nature* **346**, 671–674.

Stankunas, K., Bayle, J. H., Gestwicki, J. E., Lin, Y. M., Wandless, T. J., and Crabtree, G. R. (2003). Conditional protein alleles using knockin mice and a chemical inducer of dimerization. *Mol. Cell* **12**, 1615–1624.

Varki, A. (1998). Factors controlling the glycosylation potential of the Golgi apparatus. *Trends Cell Biol.* **8**, 34–40.

Xu, Z., Vo, L., and Macher, B. A. (1996). Structure function analysis of human α1, 3-fucosyltransferase. *J. Biol. Chem.* **271**, 8818–8823.

Yang, W., Pepperkok, R., Bender, P., Kreis, T. E., and Storrie, B. (1996). Modification of the cytoplasmic domain affects the subcellular localization of Golgi glycosyltransferases. *Eur. J. Cell Biol.* **71**, 53–61.

[15] Metabolic Labeling of Glycans with Azido Sugars for Visualization and Glycoproteomics

By SCOTT T. LAUGHLIN, NICHOLAS J. AGARD, JEREMY M. BASKIN,
ISAAC S. CARRICO, PAMELA V. CHANG, ANJALI S. GANGULI,
MATTHEW J. HANGAUER, ANDERSON LO,
JENNIFER A. PRESCHER, and CAROLYN R. BERTOZZI

Abstract

The staggering complexity of glycans renders their analysis extraordinarily difficult, particularly in living systems. A recently developed technology, termed *metabolic oligosaccharide engineering,* enables glycan labeling with probes for visualization in cells and living animals, and enrichment of specific glycoconjugate types for proteomic analysis. This technology involves metabolic labeling of glycans with a specifically reactive, abiotic functional group, the azide. Azido sugars are fed to cells and integrated by the glycan biosynthetic machinery into various glycoconjugates. The azido sugars are then covalently tagged, either *ex vivo* or *in vivo*, using one of two azide-specific chemistries: the Staudinger ligation, or the strain-promoted [3+2] cycloaddition. These reactions can be used to tag glycans with imaging probes or epitope tags, thus enabling the visualization or enrichment of glycoconjugates. Applications to noninvasive imaging and glycoproteomic analyses are discussed.

Introduction

Glycans are involved in virtually all aspects of life, including development (Okajima and Irvine, 2002), cancer (Borsig *et al.*, 2001), diabetes (Ohtsubo *et al.*, 2005), inflammation (Phillips *et al.*, 1990), and host–pathogen interactions (Wei *et al.*, 2003). Yet, despite their ubiquitous nature, little is known about the functions of glycans relative to life's other crucial biomolecules (i.e., DNA, RNA, and protein). This disparity has likely arisen from a difference in biosynthetic programs. Whereas DNA, RNA, and protein are synthesized in a template-defined manner (which has been elegantly exploited to aid their study), carbohydrates are constructed stochastically by an array of enzymes in the secretory pathway. These enzymes include the glycosyltransferases for the addition of monosaccharides, glycosidases for their removal, as well as various kinases, phosphorylases, sulfotransferases, and sulfatases for further glycan modification. The

METHODS IN ENZYMOLOGY, VOL. 415   0076-6879/06 $35.00
       DOI: 10.1016/S0076-6879(06)15015-6

resulting mature glycan can possess considerable heterogeneity, which, together with the "non–template driven" synthesis, precludes the use of standard genetic techniques for *in vivo* analysis. Thus, our knowledge of glycans has been advanced primarily though *in vitro* biochemistry, where physiological relevance is a frequent concern. Glycan and glycoconjugate analysis would benefit from technologies that function both within its stochastic biosynthetic regime and in the demanding environment of a living organism.

We have developed a technique for labeling glycans within cells or organisms with a chemical reporter group. Termed *metabolic oligosaccharide engineering,* an unnatural monosaccharide with a subtle structural modification is processed by the cell's biosynthetic pathways and incorporated into glycoconjugates analogously to the natural sugar (Fig. 1). Reutter and coworkers pioneered the technology by replacing the acetyl group in *N*-acetylmannosamine with longer acyl chain homologues (Kayser *et al.*, 1992), and it has since been expanded to include specifically reactive functional groups such as the ketone, the thiol, and this chapter's focus—the azide (Mahal *et al.*, 1997; Sampathkumar *et al.*, 2006; Saxon and Bertozzi, 2000). Metabolic oligosaccharide engineering has been used in cell culture experiments to label glycans containing sialic acid (Saxon and Bertozzi, 2000), *N*-acetylgalactosamine (GalNAc) (Hang *et al.*, 2003), or *N*-acetylglucosamine (GlcNAc) (Vocadlo *et al.*, 2003) with their corresponding "azido sugars." However, it was not until recently that metabolic oligosaccharide engineering was used to introduce azides into glycans within a living animal (Scheme 1). In 2004, the *N*-acetylmannosamine (ManNAc) analogue *N*-azidoacetylmannosamine (ManNAz) was incorporated into murine glycans as *N*-azidoacetyl sialic acid (SiaNAz) (Prescher *et al.*, 2004). Even more recently, the GalNAc analogue *N*-azidoacetylgalactosamine (GalNAz) was used to label *O*-linked glycans in mice (Dube *et al.*, 2006).

After its metabolic installation, the azido sugar can be chemically reacted by exploiting the unique reactivity of the azide via the Staudinger ligation with phosphines or the strain-promoted [3+2] cycloaddition with cyclooctynes (Fig. 2) (Agard *et al.*, 2004; Saxon and Bertozzi, 2000). These reactions proceed selectively in cell lysates, on the surfaces of living cells, and in organisms including mice and *Caenorhabditis elegans,* and they afford the opportunity to attach a plethora of chemical probes to azide-containing glycans (Laughlin *et al.*, unpublished data; Prescher *et al.*, 2004; Prescher *et al.*, unpublished data). Epitope tags such as the small molecule biotin and various peptides (e.g., FLAG, myc, and His_6) facilitate detection *ex vivo* and enrichment by immunoprecipitation. A fusion of the FLAG and His_6 tags enables two orthogonal purification steps to be performed and has been pivotal in reducing sample complexity for glycoproteomic applications (Carrico *et al.*, unpublished data). Finally, a cornucopia of fluorescent

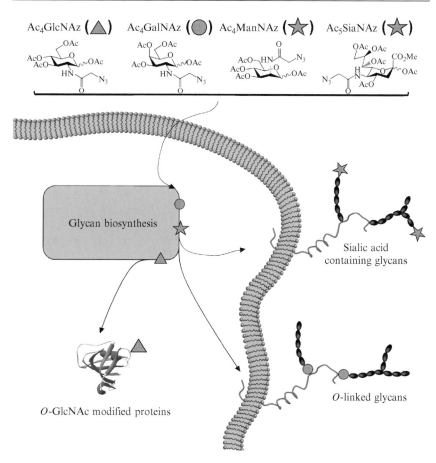

FIG. 1. Metabolic oligosaccharide engineering. Peracetylated *N*-azidoacetylglucosamine (Ac₄GlcNAz) and *N*-azidoacetylgalactosamine (Ac₄GalNAz) are taken up by the cell, deacetylated, and converted to the activated sugar. Peracetylated *N*-azidoacetylmannosamine (Ac₄ManNAz) is taken up by the cell, deacetylated, transformed into sialic acid, and subsequently converted into the activated sugar. The peracetylated methyl ester of *N*-azidoacetyl sialic acid (Ac₅SiaNAz) is taken up by the cell, deacetylated and saponified, and converted into the activated sugar. GlcNAz is attached to *O*-GlcNAc–modified proteins; GalNAz is appended to serine or threonine residues to initiate mucin-type *O*-glycan biosynthesis, and both ManNAz and SiaNAz are incorporated as sialic acid on the periphery of *N*- and *O*-linked glycans. Protection of the sugar hydroxyl groups as acetate esters is essential for cell permeability. Ac₄GlcNAz: 1,3,4,6-tetra-*O*-acetyl-*N*-azidoacetyl-α,β-D-glucosamine; Ac₄GalNAz: 1,3,4,6-tetra-*O*-acetyl-*N*-Azidoacetyl-α,β-D-galactosamine; Ac₄ManNAz: 1,3,4,6-tetra-*O*-acetyl-*N*-Azidoacetyl-α,β-D-mannosamine; Ac₅SiaNAz: 2,4,7,8,9-penta-*O*-acetyl-*N*-5-azidoacetamido-3,5-dideoxy-D-*glycero*-α,β-D-*galacto*-non-2-ulosonic-1-methyl ester.

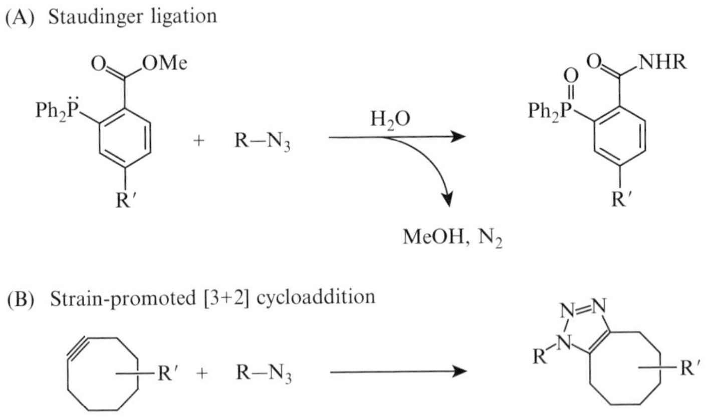

(A) Staudinger ligation

(B) Strain-promoted [3+2] cycloaddition

FIG. 2. The *in vivo* azido-ligation reactions. (A) In the Staudinger ligation, a phosphine reacts with the azide to form an intermediate that is captured by the adjacent ester to yield a ligation product and the phosphine oxide. (B) The strain-promoted [3+2] cycloaddition uses ring strain to promote a Huisgen-type [3+2] cycloaddition with the azide to afford the ligation product. R, Metabolic substrate; R', assay probe (e.g., epitope tag).

molecules enables the direct detection of azido species in cell lysates, on cell surfaces, and in living animals (Chang *et al.*, unpublished data; Hangauer *et al.*, unpublished data; Lemieux *et al.*, 2003; Prescher *et al.*, unpublished data). This chapter will describe procedures for metabolic oligosaccharide engineering experiments in cell culture and in living animals.

Materials and Methods

All chemical reagents were obtained from commercial suppliers and used without further purification unless otherwise noted. Matrix-assisted laser desorption/ionization (MALDI) mass spectra were obtained at the University of California Berkeley mass spectrometry laboratory. High-performance liquid chromatography (HPLC) was performed with a Rainin Dynamax SD-200 system (Varian, Palo Alto, CA) with detection at 220 nm. Protease inhibitor tablets were purchased from Roche. Detergent-compatible protein concentration assay kit and Biogel P-10 were obtained from Bio-Rad. The α-FLAG M2 conjugates, α-FLAG M2 agarose, penicillin, and streptomycin were purchased from Sigma. Fluorescein isothiocyanate (FITC)-conjugated mouse IgG$_1$ isotype control was purchased from BD PharMingen. Cell culture media were purchased from Invitrogen Life Technologies. Fetal calf serum was purchased from HyClone.

All flasks and plates for cell culture were purchased from Fisher Scientific. Flow cytometry data were acquired with a FACSCalibur flow cytometer (BD Biosciences Immunocytometry Systems) equipped with a 488-nm argon laser. Es1e/Es1e, B6D2F1, or C57BL/6 mice were obtained from the Jackson Laboratory or Charles River Laboratories, fed standard chow, and kept on a 12-hr light cycle. Sterile 26.5- or 30-gauge needles were used for intraperitoneal cavity or tail vein injection, respectively.

Synthetic Methods

The syntheses of the peracetylated azido sugars (i.e., Ac$_4$GlcNAz, Ac$_4$GalNAz, Ac$_4$ManNAz) were performed essentially as described by Luchansky et al. (2003). The following procedure describes an improved version of the Ac$_4$ManNAz synthesis. The same procedure can be used to synthesize Ac$_4$GlcNAz, and Ac$_4$GalNAz.

ManNAz (N-Azidoacetyl-α,β-D-mannosamine)

Azidoacetic acid was synthesized as previously described (Luchansky et al., 2003). D-mannosamine hydrochloride (1.5 g, 7.0 mmol) was added to a solution of azidoacetic acid (0.98 g, 9.7 mmol) in methanol (70 ml). Triethylamine (2.5 ml, 17 mmol) was added, and the reaction mixture was stirred for 5 min at room temperature. The solution was cooled to 0°, and N-hydroxybenzotriazole (0.86 g, 7.0 mmol) was added, followed by 1-[3-(dimethylamino)propyl]-3-ethylcarbodiimide hydrochloride (2.68 g, 14.0 mmol). The reaction was allowed to warm to room temperature overnight, at which point TLC (thin layer chromatography) analysis with ceric ammonium molybdate stain indicated that the reaction was complete. The solution was concentrated and the crude ManNAz was purified by silica gel chromatography, eluting with methanol-dichloromethane (1:9, v/v) before acetylation.

Ac$_4$ManNAz (1,3,4,6-tetra-O-acetyl-N-azidoacetyl-α,β-D-mannosamine)

Acetic anhydride (1.0 ml, 11 mmol) was added to a solution of ManNAz (0.025 g, 0.095 mmol) in pyridine (2 ml), and the reaction mixture was stirred overnight at room temperature. The solution was concentrated, resuspended in CH$_2$Cl$_2$, and washed with 1 M HCl, saturated NaHCO$_3$, and saturated NaCl. The organic phase was dried over Na$_2$SO$_4$, filtered, and concentrated. The crude material was purified by silica gel chromatography, eluting with hexanes-ethyl acetate (2:1, v/v). The fractions containing product were concentrated to yield Ac$_4$ManNAz (0.039 g, 95% yield). This material was further purified by recrystallization from

hexanes-ethyl acetate or by reversed-phase HPLC eluting with a gradient of CH_3CN (5% to 40%) and H_2O. ^{1}H NMR (500 MHz, $CDCl_3$): δ 1.98 (3H, s), 1.99 (3H, s), 2.05 (6H, s), 2.10 (6H, s), 2.11 (3H, s), 2.17 (3H, s), 3.81 (1H, ddd, J=2.5, 4.6, 9.7), 4.01–4.15 (7H, m), 4.20–4.26 (2H, m), 4.61 (1H, ddd, J=1.9, 4.2, 9.3), 4.72 (1H, ddd, J=1.6, 3.9, 9.0), 5.05 (1H, dd, J=3.9, 9.9), 5.15 (1H, app t, J=9.8), 5.21 (1H, app t, J=10.1), 5.33 (1H, dd, J=4.3, 10.2), 5.88 (1H, d, J=1.6), 6.03 (1H, d, J=1.8), 6.59 (1H, d, J=9.3), 6.65 (1H, d, J=9.0) ppm. ^{13}C NMR (500 MHz, $CDCl_3$): δ 20.5, 20.6, 20.6, 20.6, 20.6, 20.7, 20.7, 20.8, 49.2, 49.7, 52.3, 52.5, 61.6, 61.7, 64.9, 65.1, 68.8, 70.2, 71.4, 73.3, 90.2, 91.3, 166.8, 167.3, 168.1, 168.3, 169.5, 169.5, 169.6, 170.1, 170.1, 170.5 ppm. FAB-HRMS calculated for $C_{16}H_{22}LiN_4O_{10}$ [M + Li] [+] 437.1496, found 437.1496.

Ac₄GlcNAz (1,3,4,6-tetra-O-acetyl-N-azidoacetyl-α,β-D-glucosamine)

^{1}H NMR (500 MHz, $CDCl_3$, mixture of anomers 1:1 α:β): δ 2.03 (3H, s), 2.04 (3H, s), 2.05 (3H, s), 2.06 (3H, s), 2.09 (6H, s), 2.12 (3H, s), 2.21 (3H, s), 3.83 (1H, ddd, J=2.3, 5.3, 9.8), 3.92 (2H, s), 3.94 (2H, s), 4.04 (1H, ddd, J= 2.3, 6.1, 12.3), 4.08 (1H, dd, J=2.3, 12.5), 4.14 (1H, dd, J=2.2, 12.5), 4.18–4.24 (2H, m), 4.25–4.31 (1H, m), 4.47 (1H, ddd, J=3.3, 8.6, 10.9), 5.14 (1H, app t, J=9.7), 5.25 (3H, m), 5.78 (1H, d, J=8.7), 6.20 (1H, d, J=3.7), 6.43 (2H, d, J=8.8) ppm. ^{13}C NMR (125 MHz, $CDCl_3$, mixture of anomers 1:1 α:β): δ 20.5, 20.5, 20.5, 20.6, 20.6, 20.6, 20.8, 20.8, 51.1, 52.3, 52.5, 52.9, 61.4, 61.6, 67.3, 67.9, 69.7, 70.2, 72.1, 72.7, 90.2, 92.0, 167.0, 167.2, 168.7, 169.1, 169.3, 169.3, 170.6, 170.6, 170.8, 171.4 ppm. FAB-HRMS calculated for $C_{16}H_{22}LiN_4O_{10}$ [M + Li][+] 437.1496, found 437.1504.

Ac₄GalNAz (1,3,4,6-tetra-O-acetyl-N-azidoacetyl-α,β-D-galactosamine)

^{1}H NMR (500 MHz, $CDCl_3$): δ 2.01 (3H, s), 2.02 (3H, s), 2.02 (3H, s), 2.04 (3H, s), 2.12 (3H, s), 2.16 (3H, s), 2.16 (3H, s), 2.17 (3H, s), 3.92 (2H, s), 3.94 (2H, s), 4.07 (2H, m), 4.08 (1H, m), 4.14 (2H, m), 4.25 (1H, app t, J=6.7), 4.36 (1H, app dt, J=9.1, 11.2), 4.68 (1H, m), 5.21 (1H, dd, J=3.3, 11.3), 5.24 (1H, dd, J=3.2, 11.6), 5.39 (1H, app d, J=2.6), 5.44 (1H, app d, J=3.2), 5.79 (1H, d, J=8.8), 6.21 (1H, d, J=2.0), 6.28 (1H, d, J=9.0), 6.38 (1H, d, J=9.5) ppm. ^{13}C NMR (125 MHz, $CDCl_3$): δ 20.6, 20.6, 20.6, 20.7, 20.7, 20.7, 20.9, 20.9, 47.0, 50.1, 52.5, 52.6, 61.2, 61.3, 66.3, 66.6, 67.6, 68.7, 70.0, 71.9, 90.9, 92.6, 167.0, 167.2, 168.9, 169.4, 170.1, 170.2, 170.4, 170.4, 170.5, 171.1 ppm. FAB-HRMS calculated for $C_{16}H_{22}LiN_4O_{10}$ [M + Li][+] 437.1496, found 437.1508.

Phosphine-FLAG (Table I, 1a) was synthesized as previously described (Luchansky *et al.*, 2003). The following procedure describes the synthesis of 2-diphenylphosphanyl-terephthalic acid 1-methyl ester 4-pentafluoro-phenyl ester (phosphine-PFP), an intermediate in the syntheses of 1b, 1c, 1d, and 1e (see Table I).

TABLE I
PHOSPHINE AND CYCLOOCTYNE DETECTION AND ENRICHMENT PROBES

Reactive scaffold	Detection/enrichment probe

for the Staudinger ligation

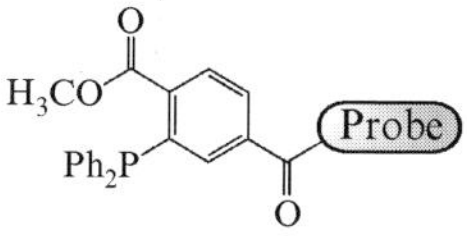

1a (Probe) =   -DYKDDDDK   (FLAG)

1b   -DYKDDDDKHHHHHH (FLAG-His$_6$)

1c   (Fluorescein)

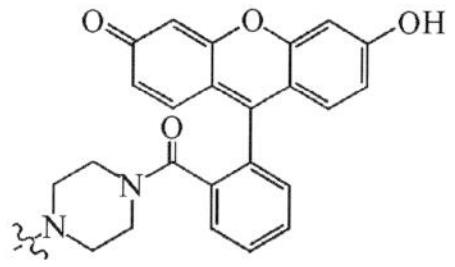

1d   (Rhodamine)

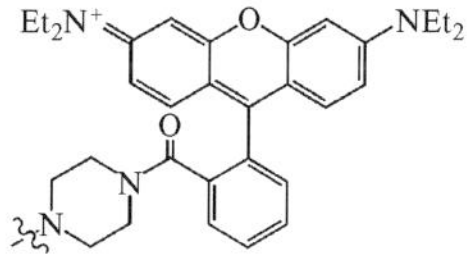

1e   (Cy5.5)

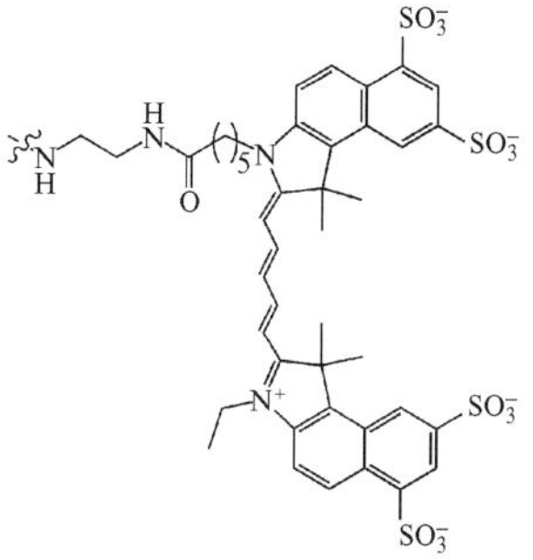

for the strain-promoted cycloaddition

(Probe) =

(FLAG)
DYKDDDDK

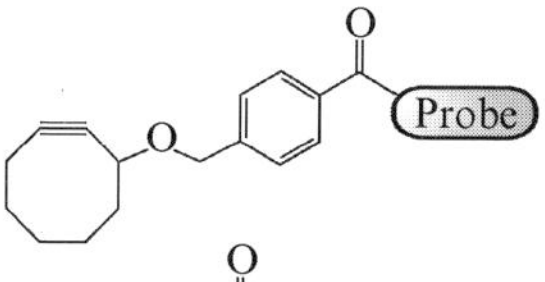

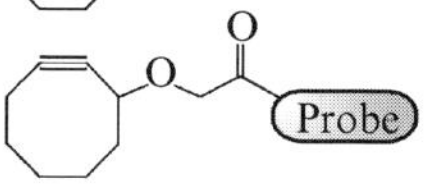

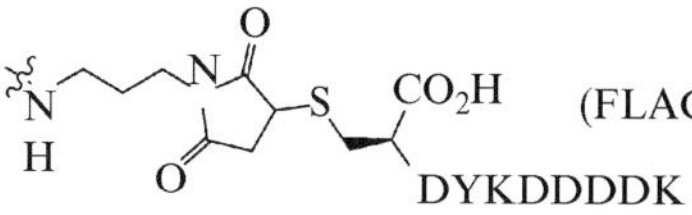

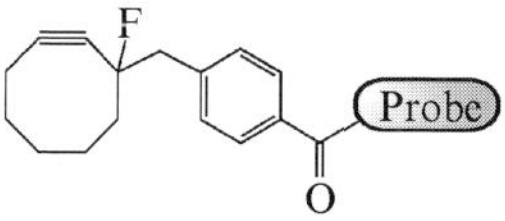

1-Methyl-2-iodoterephthalate

A solution of NaNO$_2$ (9.00 g, 0.132 mol) in 50 ml of H$_2$O was added dropwise to 1-methyl-2-aminoterephthalate (25.0 g, 0.128 mol) in 250 ml of cold, concentrated HCl, resulting in the evolution of a small amount of orange gas. The reaction mixture was stirred for 30 min at room temperature and then filtered through Celite into a solution of KI (215 g, 1.25 mol) in 350 ml of H$_2$O. The resulting dark red solution was stirred for 1 hr and then diluted with ethyl acetate (250 ml) and washed with saturated Na$_2$SO$_3$ (2 × 25 ml), followed by water (2 × 50 ml) and saturated NaCl (1 × 50 ml). The combined aqueous layers were back-extracted with ethyl acetate (50 ml). The organic layers were dried over Na$_2$SO$_4$ and concentrated. The crude product was recrystallized from MeOH and H$_2$O to afford 26.8 g of a yellow solid (69%). ^{1}H NMR (300 MHz, CDCl$_3$): δ 3.97 (s, 3H), 7.83 (d, J = 8.1, 1H), 8.12 (dd, J = 1.5, 8.1, 1H), 8.69 (d, J = 1.5, 1H); ^{13}C NMR (75 MHz, CDCl$_3$): δ 52.9, 129.4, 130.5, 132.5, 140.1, 142.6, 166.6, 169.44, 203.6.

1-Methyl-2-diphenylphosphinoterephthalate

A solution of 1-methyl-2-iodoterephthalate (26.2 g, 85.6 mmol) in dry MeOH (260 ml) was treated with triethylamine (23.9 ml, 171 mmol) and a catalytic amount of palladium acetate (192 mg, 0.856 mmol). The mixture was degassed *in vacuo,* and diphenylphosphine (14.9 ml, 85.6 mmol) was added under an atmosphere of argon. The resulting solution was heated at reflux overnight and then cooled to room temperature and concentrated. The crude residue was dissolved in 500 ml of a 1:1 CH$_2$Cl$_2$:H$_2$O mixture and the layers were separated. The organic layer was washed with 1 M HCl (1 × 100 ml) and concentrated. The crude product was purified by flash chromatography on silica gel (8:1 CHCl$_3$:MeOH) to yield a yellow solid (26.2 g, 69%). ^{1}H NMR (400 MHz, CDCl$_3$): δ 3.75 (s, 3H), 7.27–7.31 (m, 10H), 7.65 (m, 1H), 8.01–8.03 (m, 2H); ^{13}C NMR (100 MHz, CDCl$_3$): δ 52.3, 128.6, 128.7, 129.0, 129.7, 130.5, 133.7, 133.9, 135.5, 137.0, 166.7; ^{31}P NMR (160 MHz, CDCl$_3$): δ −4.36; HRMS (FAB): Calculated for C$_{21}$H$_{18}$O$_4$P [M + H]$^+$ 365.0943, found 365.0942.

Phosphine-PFP (2-diphenylphosphanyl-terephthalic acid 1-methyl ester
  4-pentafluorophenyl ester)

To a solution of 1-methyl-2-diphenylphosphinoterephthalate (2.53 g, 6.86 mmol) in (tetrahydrofolate) THF (50 ml) was added triethylamine, and the mixture was degassed *in vacuo.* Pentafluorophenyl trifluoroacetate was added dropwise over 2 min. After 3 hrs, the solvents were removed by rotary evaporation and the crude product was purified by flash chromatography on silica gel (95:5 hexanes:ethyl acetate) to yield a yellow oil (3.40 g,

92%). IR (thin film): 3350, 3072, 3003, 2953, 2920, 2850, 1767, 1726 cm^{-1}.
^{1}H NMR (400 MHz, CDCl$_3$): δ 3.77 (s, 3H), 7.29–7.37 (m, 10H), 7.72–7.73
(m, 1H), 8.17–8.18 (m, 2H); ^{13}C NMR (100 MHz, CDCl$_3$): δ 52.5, 128.6,
128.7, 129.2, 129.5, 130.9, 133.7, 133.9, 136.1, 136.5, 136.6, 139.4, 139.6,
142.1, 142.5, 153.8, 161.7, 166.5; ^{31}P NMR (160 MHz, CDCl$_3$): δ −3.96;
^{19}F NMR (376 MHz, CDCl$_3$): δ −161.2 (app t, 2, J = 18.8), −156.7 (app t, 1,
J = 18.8), − 151.4 (app d, 2, J = 18.8); HRMS (FAB): Calculated for
C$_{27}$H$_{17}$F$_5$O$_4$P [M + H]$^+$ 531.0785, found 531.0772.

Phosphine-FLAG-His$_6$ (1b)

The FLAG-His$_6$ peptide was synthesized on Wang resin using standard manual Fmoc-based peptide synthesis. After the final deblocking of
the N-terminal aspartic acid, five equivalents of phosphine-PFP and N, N-
diisopropylethylamine were added as a solution in dimethylformamide
(DMF); the resin was shaken in a sealed vessel for 3 hrs, and it was
then washed extensively with DMF, CH$_2$Cl$_2$, and acetic acid, sequentially.
Removal of side-chain protecting groups and cleavage from the resin were
accomplished by treatment with 95% trifluoroacetic acid, 2.5% triisopro-
pylsilane, and 2.5% water for 2 hrs. Evaporation of most of the cleavage
solution afforded a viscous orange liquid, from which the product was
precipitated using diethylether. This product was filtered, washed (3×) with
diethylether, and dried *in vacuo*. The product was dissolved in water and
purified by C$_{18}$ reversed-phase HPLC with a gradient of CH$_3$CN (0% to
10% over 1 min followed by 10% to 55% over 38 min; elution at ~14 min)
and H$_2$O. The purified product was lyophilized and stored under argon to
avoid phosphine oxidation. MALDI MS: Calculated for C$_{98}$H$_{117}$N$_{28}$O$_{29}$P
[M + H]$^+$ 2181.8351, found 2181.8834.

The syntheses of 1c and 1d (see Table I) were accomplished analogously
to 1a, essentially by coupling phosphine-PFP with a fluorophore piperazine
conjugate (Nguyen and Francis, 2003). The synthesis of 1e was accomplished
by condensation of the *N*-hydroxysuccinimide ester of Cy5.5 (Amersham
Biosciences) to phosphine-PFP via an ethylenediamine linker (Chang *et al.*,
unpublished results). The carboxylic acids of cyclooctynes 2, 3, and 4 (see
Table I) were synthesized as described (Agard *et al.*, 2004; Agard *et al.*,
unpublished results). Conversion to the pentafluorophenyl ester followed
by coupling to an amino maleimide linker and subsequent condensation
with FLAG-cysteine peptide yielded compounds 2, 3, and 4.

Metabolic Labeling of Glycans in Cell Culture

The azido sugar was added to an empty tissue culture dish from an
ethanol stock solution to afford a final concentration of 0 to 50 μM, and the
ethanol was evaporated before the addition of cells (Table II). Cells were

TABLE II
TYPICAL REAGENT CONCENTRATIONS FOR COMMON BIOLOGICAL APPLICATIONS

Reagent	Metabolic labeling		Ex vivo reaction		In vivo reaction
	Cell culture	Mice	Lysate analysis	Flow cytometry	
Ac$_4$GalNAz Ac$_4$ManNAz Ac$_4$GlcNAz	10–50 μM	300 mg/kg			
1a			250 μM	250 μM	24 mg in 150 μL H$_2$O
1b				N/A	N/A
1c			12.5 μM	N/A	
1d					Not optimized
1e				12.5 μM	
2					
3			100–200 μM		
4					

diluted to a concentration of 200,000 cells/ml and added to yield a desired confluence (approximately 2 million for a 10-cm dish). The cells were incubated with azido sugar for 1 to 3 days before analysis by flow cytometry or lysis followed by Western blot analysis of glycans.

Metabolic Labeling of Glycans in Mice

Adult mice were administered azido sugar (0 to 300 mg/kg, in ~200 μl of a 50 mg/ml stock solution in 70% aqueous dimethylsulfoxide) by injection into the intraperitoneal cavity once daily for 7 days. During the experiment, mice were monitored for signs of distress (e.g., sluggish movement, poor posture). Mice were euthanized for organ collection 12 to 24 hrs after the final azido sugar injection using a lethal dose of isoflurane anesthesia followed by cervical dislocation, or CO$_2$ inhalation. Standard dissection methods were used to isolate the liver, kidney, heart, spleen, gut, thymus, brain, lymph node, and serum (Suckow et al., 2000).

Reaction of Azido Sugars with Labeling Reagents in Cell Culture and in Tissue Lysates

For analysis of azido sugars in cellular glycoproteins by Western blot, the cells were transferred to a 15-ml conical tube, pelleted (~1500g, 3 min, 4°), resuspended in 10 ml of phosphate-buffered saline (PBS), and pelleted again. The cells were transferred to a 1.5-ml Eppendorf tube in 1 ml of

PBS, pelleted ($\sim$1500g, 3 min, 4°), and resuspended in $\sim$200 μl of lysis buffer (20 mM Tris [pH, 7.4], 150 mM NaCl, 1 mM ethylenediamine tetraacetic acid [EDTA], 1% NP40, protease inhibitor tablet) per 10-cm dish. The cells were lysed by sonication; insoluble debris was removed by centrifugation ($\sim$14,000 rpm, 10 min, 4°), and the supernatant was assayed for protein concentration (Bio-Rad, DC protein assay kit).

For analysis of tissue lysates from animals treated with azido sugars, the organs were added to ice-cold PBS ($\sim$5 ml, depending on organ size) immediately after dissection, rinsed with ice-cold PBS to remove blood, and added to 5 ml of ice-cold lysis buffer (10 mM Tris [pH, 7.4], 150 mM NaCl, 1% NP40, with Roche protease inhibitor cocktail). The organ was typically homogenized mechanically with approximately 15 strokes of Potter-Elvehjem homogenizer or until evenly homogenized. The homogenate was transferred to a 15-ml falcon tube and insoluble debris was pelleted (3700 g, 10 min, 4°). The supernatant was assayed for protein concentration using (Bio-Rad, DC protein assay kit).

For phosphine-based labeling reactions, the reagent was added to 10 to 100 μg of total protein-containing lysate at the concentrations indicated in Table II. The reaction was run at room temperature for 12 hrs, at which point the sample was analyzed using standard immunoblot techniques (Sambrook and Russell, 2001). For cyclooctyne-based labeling reactions, the reagents were also added directly to 10 to 100 μg of total protein-containing lysate at the concentrations dictated by Table II. However, heating the sample led to nonspecific protein labeling and, thus, it was necessary to destroy unreacted reagent before analysis by denaturing gel electrophoresis. This was accomplished by quenching the unreacted cyclooctyne with 2-azidoethanol or by dialysis (Agard et al., 2004).

Flow cytometry analysis of cells metabolically labeled with azido sugars in cell culture or in animals was performed as follows. Adherent cells in culture were lifted with a solution of 0.5 mM EDTA in PBS lacking calcium and magnesium ($\sim$2 ml, 10 min, 37°). The cells were then transferred to a 15-ml conical tube in 10 ml of PBS, pelleted by centrifugation ($\sim$1500g, 3 min, 4°), resuspended in a minimal amount of PBS ($\sim$1 ml), and added in triplicate to a 96-well V-bottom plate. For ex vivo splenocyte analysis by flow cytometry, murine splenocytes were isolated using a standard protocol (Suckow et al., 2000) and aliquoted into a 96-well V-bottom plate in triplicate. All subsequent steps were performed in the 96-well plate. The cells were pelleted ($\sim$1500g, 3 min, 4°) and the medium was decanted. The cells were resuspended in 200 μl of flow cytometry buffer (PBS with 1% fetal calf serum). Centrifugation was repeated and the cells were washed a second time. After pelleting, the supernatant was decanted; the cells were resuspended in an appropriate concentration of azide-reactive

probe (see Table II) in flow cytometry buffer and incubated for 1 to 1.5 hrs at room temperature or 37°. During each step, it is critical that the cells be completely resuspended to allow the reagents full access to azides.

After the labeling reaction, the cells were kept at 4° to slow internalization of modified glycans from the cell surface. The cells were pelleted ($\sim$1500g, 3 min, 4°), the supernatant was decanted, and the cells were resuspended in 200 μl of flow cytometry buffer. This wash cycle was repeated twice. If the azide-reactive probe contained a fluorophore for detection, the cells were submitted to six additional wash cycles and analyzed by flow cytometry. If a FLAG-tagged azide-reactive probe was used, the cells were incubated with FITC-conjugated α-FLAG (1:900 dilution) in flow cytometry buffer or the same amount of isotype-matched FITC-conjugated mouse IgG$_1$. The cells were incubated with the antibodies for 30 min on ice and then washed twice with 200 μl of flow cytometry buffer and transferred to 400 μl of flow cytometry buffer for analysis by flow cytometry.

Administration of Labeling Reagents *In Vivo*

Murine glycans were metabolically labeled with azido sugars *in vivo* as described previously. The mice were administered the desired azide-reactive probe in water or vehicle alone by intraperitoneal or intravenous injection 24 hrs after the final azido sugar injection (see Table II). After 1 to 3 hrs, the organs were harvested using standard dissection methods (Suckow *et al.*, 2000), and analysis by flow cytometry or Western blot was performed analogously as described previously.

Enrichment and Identification of Metabolically Labeled Glycans

Cell or tissues metabolically labeled with azido sugars were generated and homogenized as described previously. The lysate was reacted with compound 1b for 12 hrs at room temperature and unreacted compound 1b is removed by two rounds of size exclusion chromatography with Biogel P-10. The FLAG-containing proteins were enriched by immunoprecipitation with α-FLAG M2 agarose. For cell culture experiments, approximately 100 μl of the α-FLAG M2 agarose slurry was added for each 10 ml of cell culture media used. The ratio for metabolically labeled organs has not yet been optimized. The lysate was incubated with α-FLAG M2 agarose overnight at 4°. The beads were washed with 10 column volumes of each of the following solutions: 50 mM Tris (pH, 7.4), 300 mM NaCl, 1% triton X-100; 50 mM Tris (pH, 7.4), 1.3 M NaCl, 1% triton X-100; 50 mM Tris (pH, 7.4), 300 mM NaCl, 1% triton X-100, 1 M urea. The bound species were eluted with one column volume 8 M urea (pH, 8), 0.1 M Na$_2$PO$_4$, 10 mM Tris. For secondary capture with Ni^{2+}-NTA agarose, 0.1 times the eluant volume of a

1:1 solution of Ni^{2+}-NTA agarose was added and the mixture was incubated overnight at room temperature. The beads were washed with 10 column volumes of each of the following solutions: 8 M urea, 0.1 M Na_2PO_4, 10 mM Tris (pH, 8); 8 M urea, 0.1 M Na_2PO_4, 10 mM Tris (pH, 8), and 15 mM imidazole. The bound species are eluted with one column volume of 8 M urea, 0.1 M Na_2PO_4, 10 mM Tris (pH, 8), 250 mM imidazole.

Discussion

We have developed a method for detection and visualization of glycans that involves metabolic labeling with azido sugars followed by chemical labeling of the azides with probes or affinity reagents. The chemical labeling step can be executed with either phosphine probes via the Staudinger ligation or with cyclooctyne probes via the strain-promoted [3+2] cycloaddition. These chemoselective ligation reactions allow chemical tagging of azido sugars within complex biological samples ranging from cell lysates to living organisms.

A key issue in the implementation of this method is its sensitivity, which is governed by a combination of factors. The efficiency of metabolic labeling, overall abundance of the labeled species, and efficiency of the chemical tagging step all contribute. In our experience, labeling efficiency is highly dependent on the cell type (for cultured cells) or organ (for living animals) and on the pathway targeted. In some cultured cells, metabolic labeling with $Ac_4ManNAz$ results in up to 40% replacement of natural sialic acids with SiaNAz, whereas in others, only 4% labeling efficiency is observed (Luchansky $et\ al.$, 2004). These differences may reflect variations in the level of endogenous substrates that compete in the same metabolic pathway.

Labeling efficiencies and glycan quantities are governed by intrinsic properties of the biological system. But the efficiency of the chemical tagging step can be controlled to some extent by the choice of reactants. Ultimately, reactions with the fastest kinetics will provide the highest-sensitivity detection, since more azido sugars can be reacted in a given amount of time. By contrast, if the reaction is sluggish, sensitivity is compromised. Reaction kinetics are also important to consider for applications of the technique attempting the study of dynamic processes such as membrane glycan turnover.

The optimal labeling conditions are thus dependent on the application and the reactant's properties. For labeling of glycans in cell lysates with compound 1a, the reaction is typically run for 12 to 24 hrs at a concentration of 250 μM 1a and at room temperature. The duration of the reaction in this context is of little consequence, as most biological processes have

ceased. On the other hand, live cells in culture are typically limited to 1 to 1.5 hrs' incubation with 250 μM 1a to maintain cell viability in the isolated environment. Although this short incubation time does not permit completion of the reaction, it is typically adequate for the detection of azido sugar–labeled glycans. In mice, the reaction is typically limited to 3 hrs or less to minimize proteolysis of the FLAG peptide (Dube et al., 2006; Luchansky et al., 2003; Prescher et al., 2004).

Ideally, the reaction would proceed fast enough to study dynamic biological processes and to detect small changes in glycosylation on cells. Unfortunately, the phosphine reagents are not amenable to rate enhancement by chemical modification of the scaffold. Highly reactive phosphines are both synthetically challenging targets and prone to rapid oxidation in biological systems (Lin et al., 2005).

The cyclooctyne scaffold can be more easily modified to generate reactants with enhanced kinetics. The first-generation reagent (see Table I, compound 2) reacts selectively with azido sugars in cell lysates and on isolated cells and has been used for labeling of azide-labeled proteins on the surface of Escherichia coli (Agard et al., 2004; Prescher et al., unpublished results, Link et al., in press). Although the reaction kinetics of compound 2 with azides is similar to that of phosphines, the reagent suffers from poor water solubility and spontaneous decomposition. These liabilities have rendered the compound ineffective for tagging azido sugar–labeled glycans in vivo. Nevertheless, this first-generation cyclooctyne has functioned as a springboard for the creation of more effective reagents.

A more water-soluble cyclooctyne was developed by omitting the phenyl ring in compound 2 to yield compound 3. Like the first-generation reagent, compound 3 undergoes the strain-promoted [3+2] cycloaddition at a rate similar to the Staudinger ligation and reacts selectively in cell lysates and on isolated cells. However, unlike its predecessor, the improved solubility enables higher concentrations to be tested in vivo, where the compound has achieved modest success relative to the phosphine-based compounds (Prescher et al., unpublished results).

Cyclooctynes with improved kinetics would label azide-containing biomolecules with higher efficiency and, therefore, sensitivity. Thus, additional second-generation reagents boasting fluorine substitutions on the cyclooctyne scaffold have been synthesized and show progressive increases in reaction rate (Baskin et al., unpublished results). In vitro kinetics experiments revealed that the mono fluorinated compound 4 reacts 2-fold faster with model azides than its nonfluorinated relative compound 1, whereas a cyclooctyne with multiple fluorine substituents has a remarkable 40-fold increase in reaction rate over the original cyclooctyne (Baskin et al., unpublished results). These reactions remain selective for azides at room temperature and successfully

label azides in cell lysates and on cultured cells (Agard *et al.*, unpublished results).

Detection of Azide-Labeled Glycans

Whereas the phosphine or cyclooctyne scaffolds control the rate and the selectivity of the reaction, it is the probe that confers the powers of detection, enrichment, and identification. An epitope tag (e.g., FLAG, myc, His$_6$, biotin) offers the most sensitive means of detection (Fritze and Anderson, 2000). A more direct approach to visualization is to attach the imaging agent (e.g., a fluorophore) covalently to the reactive scaffold. These conjugates enable detection of azide-labeled proteins in a gel and on cultured cells without the need for secondary reagents, and they have potential use for noninvasive imaging in live animals. Phosphine-fluorescein (compound 1c) and -rhodamine (1d) derivatives have been synthesized and used to detect azide-labeled proteins in polyacrylamide gels (Chang *et al.*, unpublished results). Furthermore, compound 1d has shown azide-specific labeling in a living mouse (Prescher *et al.*, unpublished results). Although these first-generation fluorescent probes are paving the way for future *in vivo* imaging and pulse chase experiments, they have some fundamental limitations. The fluorescein and rhodamine fluorophores have excitation wavelengths with limited tissue penetrance that cause autofluorescence in animal tissues. The red-shifted excitation wavelengths of near infrared fluorophores, however, enjoy superior tissue penetrance and cause little autofluorescence (Massoud *et al.*, 2003). Cy5.5 is a near infrared fluorophore that has been used extensively in mice and will be implemented in the next generation *in vivo* imaging reagent (Bremer *et al.*, 2001; Weissleder *et al.*, 1999). Toward this end, a phosphine-Cy5.5 probe 1e was synthesized and used to visualize azide-labeled glycoproteins in gels and on cultured cells by flow cytometry; noninvasive *in vivo* imaging experiments are currently underway (Chang *et al.*, unpublished results).

A fundamental issue with the fluorescent probes is their high background labeling of live cell surfaces. Although repeated washing diminishes this background in cell culture, it is not possible in animals, so clearance of unreacted probe is determined by its inherent pharmacokinetic properties. Ideally, the fluorescence of the probe would be activated by the reaction with azides. As a proof of concept for this type of "smart" probe, Lemieux *et al.* (2003) synthesized a fluorogenic phosphine-dye conjugate based on a coumarin scaffold (Fig. 3A). This molecule is nonfluorescent in the reduced state because the phosphine lone pair exerts a quenching effect on the coumarin scaffold. However, on reaction with an azide, the phosphine is oxidized to the phosphine oxide, which activates fluorescence of the

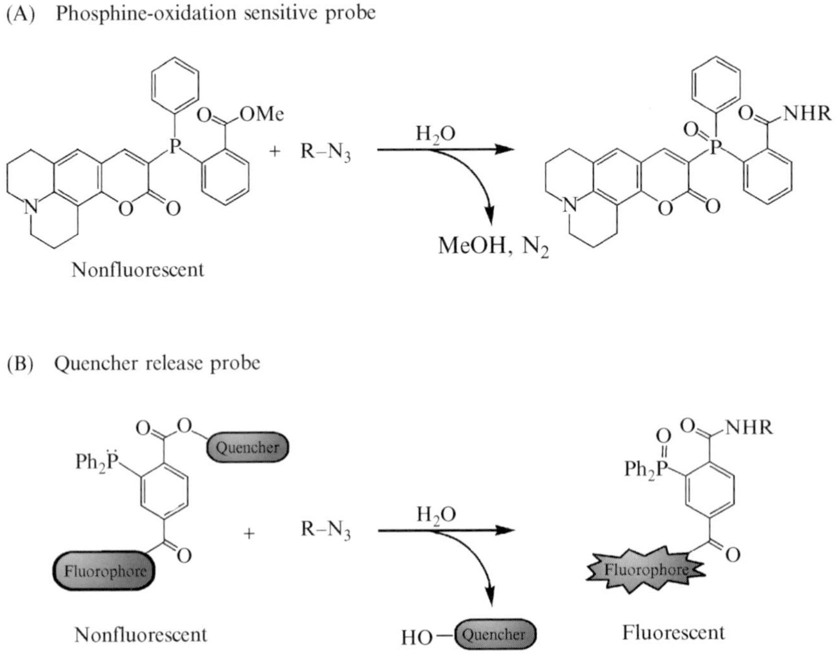

(A) Phosphine-oxidation sensitive probe

(B) Quencher release probe

FIG. 3. "Smart probes" based on the phosphine scaffold. (A) In the non-oxidized state, the phosphine lone pair quenches the fluorescence of the coumarin fluorophore. On oxidation of the phosphine by Staudinger ligation or nonspecific oxidation, the lone pair electrons are sequestered and the coumarin is brightly fluorescent. (B) A fluorophore attached to the phosphine scaffold is quenched by another organic dye with the appropriate spectral properties. On Staudinger ligation, but not nonspecific oxidation, ester cleavage removes the quencher and the molecule is brightly fluorescent.

coumarin dye. The molecule was shown to tag azide-labeled proteins in cell lysates, but the hydrophobic nature of the compound and its tendency to oxidize nonspecifically led to background fluorescence in the context of cultured cells. Efforts are currently underway to create new fluorogenic phosphines that have enhanced properties by exchanging the cleaved methyl ester with a FRET-quenching dye (Fig. 3B) (Hangauer et al., unpublished results).

Enrichment and Identification of Azide-Labeled Glycans

In addition to detection, an epitope tag facilitates the enrichment of azide-labeled glycoproteins by affinity purification. The FLAG tag is recognized by a commercially available antibody that can be used in

concert with solid support to achieve sample purification (Masters, 2004). Alternatively, the His_6 tag enables enrichment via Ni^{2+}-NTA agarose affinity purification (Hefti *et al.*, 2001).

The ability to enrich the azide-labeled population enables further identification by mass spectrometry (Fig. 4). The azide-labeled glycans are reacted with an epitope tag, purified from the general lysate, and submitted to proteomic analysis. To reduce sample complexity enough for a cogent

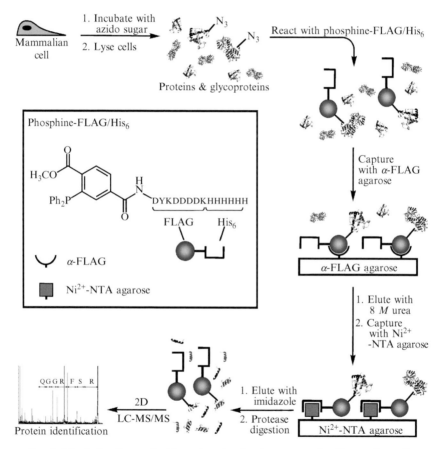

FIG. 4. Metabolic oligosaccharide engineering for glycoproteomics. Mammalian cells are labeled with azido sugar and harvested. Reaction of the azido glycans with a phosphine-FLAG/His_6 compound enables two orthogonal purification steps. The molecules are first immunoprecipitated with α-FLAG conjugated to agarose beads. The immunoprecipitate is then further purified using the His_6 epitope, yielding a sample suitable for two-dimensional liquid chromatography–mass spectrometry/mass spectrometry analysis.

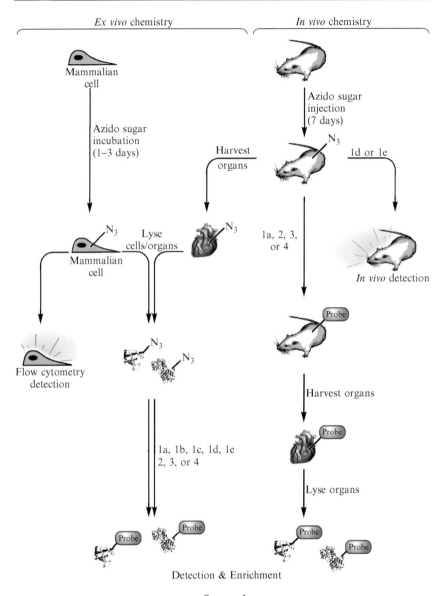

SCHEME 1.

analysis, multiple purification steps are usually necessary. Toward this end, Carrico *et al.* synthesized a combination FLAG-His$_6$ epitope tag that enables two orthogonal purification steps (Carrico *et al.*, unpublished

results). In the first purification by anti-FLAG immunoaffinity chromatography, the labeled species are enriched substantially, but the sample is often contaminated with the α-FLAG antibody. The second purification removes the antibody and any species with nonspecific affinity for the α-FLAG column. This technology has been used to identify azide-labeled glycans in Jurkat cells, and efforts are underway to identify glycans that are differentially expressed in cancer and other disease states.

Conclusion

It has become clear in recent years that understanding biological systems demands knowledge that extends beyond DNA, RNA, and protein into the arena of posttranslational modifications and thus glycans. Because such modifications are always at least one step removed from genetics and often exist as subtle protein modifications, glycans and their associated proteins are extraordinarily difficult to study. Yet, their detection and identification are crucial to our understanding. Metabolic labeling with azido sugars in concert with the Staudinger ligation or strain-promoted [3+2] cycloaddition provides a powerful means to study these glycans within their native environment. Notably, this technology is not restricted to the study of glycans. Other biomolecules (e.g., DNA, proteins, lipids) have been metabolically labeled with azides *in vitro*, and the move into a living system is only a matter of time (Comstock and Rajski, 2005; Kho *et al.*, 2004; Kiick *et al.*, 2002; Ryu and Schultz, 2006).

References

Agard, N. J., Prescher, J. A., and Bertozzi, C. R. (2004). A strain-promoted [3+2] azide-alkyne cycloaddition for covalent modification of biomolecules in living systems. *J. Am. Chem. Soc.* **127**, 15046–15047.
Borsig, L., Wong, R., Feramisco, J., Nadeau, D. R., Varki, N. M., and Varki, A. (2001). Heparin and cancer revisited: Mechanistic connections involving platelets, P-selectin, carcinoma mucins, and tumor metastasis. *Proc. Natl. Acad. Sci. USA* **98**, 3352–3357.
Bremer, C., Tung, C., and Weissleder, R. (2001). *In vivo* molecular target assessment of matrix metalloproteinase inhibition. *Nat. Med.* **7**, 743–748.
Chandra, R. A., Douglas, E. S., Mathies, R. A., Bertozzi, C. R., and Francis, M. B. (2005). Programmable cell adhesion encoded by DNA hybridization. *Angew. Chem. Int. Edit.* **45**, 896–901.
Comstock, L. R., and Rajski, S. R. (2005). Conversion of DNA methyltransferases into azidonucleosidyl transferases via synthetic cofactors. *Nucleic Acids Res.* **33**, 1644–1652.
Dube, D. H., Prescher, J. A., Quang, C. N., and Bertozzi, C. R. (2006). Probing mucin-type O-linked glycosylation in living animals. *Proc. Natl. Acad. Sci. USA* **103**, 4819–4824.
Eason, P. D., and Imperiali, B. (1999). A potent oligosaccharyl transferase inhibitor that crosses the intercellular endoplasmic reticulum membrane. *Biochemistry* **38**, 5430–5437.

Elbein, A. D. (1987). Inhibitors of the biosynthesis and processing of N-linked oligosaccharide chains. *Annu. Rev. Biochem.* **56,** 497–534.

Fritze, C. E., and Anderson, T. R. (2000). Epitope tagging: General method for tracking recombinant proteins. *Methods Enzymol.* **327,** 3–16.

Hang, H. C., Yu, C., Kato, D. L., and Bertozzi, C. R. (2003). A metabolic labeling approach toward proteomic analysis of mucin-type O-linked glycosylation. *Proc. Natl. Acad. Sci. USA* **100,** 14846–14851.

Hefti, M. H., Van Gug-Van der Toorn, C. J. G., Dixon, R., and Vervoort, J. (2001). A novel purification method for histidine-tagged proteins containing a thrombin cleavage site. *Anal. Biochem.* **295,** 180–185.

Hennet, T., Chui, D., Paulson, J. C., and Marth, J. D. (1998). Immune regulation by the ST6Gal sialyltransferase. *Proc. Natl. Acad. Sci. USA* **95,** 4504–4509.

Hennet, T., and Ellies, L. G. (1999). The remodeling of glycoconjugates in mice. *Biochim. Biophys. Acta* **1473,** 123–136.

Hwang, H., Olson, S. K., Esko, J. D., and Horvitz, H. R. (2003). *Caenorhabditis elegans* early embryogenesis and vulval morphogenesis require chondroitin biosynthesis. *Nature* **423,** 439–443.

Kayser, H., Zeitler, R., Kanicht, C., Grunow, D., Nuck, R., and Reutter, W. (1992). Biosynthesis of a nonphysiological sialic acid in different rat organs, using N-propanoyl-D-hexosamines as precursors. *J. Biol. Chem.* **267,** 16934–16938.

Kho, Y., Kim, S. C., Jiang, C., Barma, D., Kwon, S. W., Cheng, J., Jaunbergs, J., Weinbaum, C., Tamanoi, F., Falck, J., and Zhao, Y. (2004). A tagging-via-substrate technology for detection and proteomics of farnesylated proteins. *Proc. Natl. Acad. Sci. USA* **101,** 12479–12484.

Kiick, K. L., Saxon, E., Tirrell, D. A., and Bertozzi, C. R. (2002). Incorporation of azides into recombinant proteins for chemoselective modification by the standinger ligation. *Proc. Natl. Acad. Sci. USA* **99,** 19–24.

Lee, L. V., Mitchell, M. L., Huang, S., Forkin, V. V., Sharpless, K. B., and Wong, C. (2003). A potent and highly selective inhibitor of human a-1,3-fucosyltransferase via click chemistry. *J. Am. Chem. Soc.* **125,** 9588–9589.

Lemieux, G. A., De Graffenried, C. L., and Bertozzi, C. R. (2003). A fluorogenic dye activated by the Staudinger ligation. *J. Am. Chem. Soc.* **125,** 4708–4709.

Licha, K., Debus, N., Emig-Vollmer, S., Hofmann, B., Hasbach, M., Stibenz, D., Sydow, S., Schirner, M., Ebert, B., Petzelt, D., Buhrer, C., Semmler, W., and Tauber, R. (2005). Optical molecular imaging of lymph nodes using a targeted vascular contrast agent. *J. Biomed. Opt.* **10,** 41205.

Lin, F. L., Hoyt, H. M., van Halbeek, H., Bergman, R. G., and Bertozzi, C. R. (2005). Mechanistic investigation of the Staudinger Ligation. *J. Am. Chem. Soc.* **127,** 2686–2695.

Link, A. J., Vink, M. K., Agard, N. J., Prescher, J. A., Bertozzi, C. R., and Tirrell, D. A. (2006). Discovery of aminoacyl-tRNA synthetase activity through cell-surface display of non-canonical amino acids. *Proc. Natl. Acad. Sci.* **103,** 10180–10185.

Luchansky, S. J., Argade, S., Hayes, B. K., and Bertozzi, C. R. (2004). Metabolic functionalization of recombinant glycoproteins. *Biochemisty* **43,** 12358–12366.

Luchansky, S. J., Hang, H. C., Saxon, E., Grunwell, J. R., Yu, C., Dube, D. H., and Bertozzi, C. R. (2003). Constructing azide-labeled cell surfaces using polysaccharide biosynthetic pathways. *Methods Enzymol.* **362,** 249–274.

Mahal, L. K., Yarema, K. J., and Bertozzi, C. R. (1997). Engineering chemical reactivity on cell surfaces through oligosaccharide biosynthesis. *Science* **276,** 1125–1128.

Massoud, T. F., and Gambhir, S. S. (2003). Molecular imaging in living subject: Seeing fundamental biological processes in a new light. *Genes Dev.* **17,** 545–580.

Masters, S. C. (2004). Co-Immunoprecipitation from transfected cells. *Methods Mol. Biol.* **261,** 337–350.

McDonald, J. A., and Camenisch, T. D. (2002). Hyaluronan: Genetic insights into the complex biology of a simple polysaccharide. *Glycoconj. J.* **19,** 331–339.

Michelacci, Y. M. (2003). Collagens and proteoglycans of the corneal extracellular matrix. *Braz. J. Med. Biol. Res.* **36,** 1037–1046.

Nguyen, T., and Francis, M. B. (2003). Practical synthetic route to functionalized rhodamine dyes. *Org. Lett.* **18,** 3245–3248.

Ohtsubo, K., Takamatsu, S., Minowa, M. T., Yoshida, A., Takeuchi, M., and Marth, J. D. (2005). Dietary and genetic control of glucose transporter 2 glycosylation promotes insulin secretion in suppressing diabetes. *Cell* **123,** 1307–1321.

Okajima, T., and Irvine, K. D. (2002). Regulation of notch signaling by *O*-linked fucose. *Cell* **111,** 893–904.

Phillips, M. L., Nudelman, E., Gaeta, F. C. A., Perez, M., Singhal, A. K., Hakomori, S., and Paulson, J. C. (1990). ELAM-1 mediates cell adhesion by recognition of a carbohydrate ligand, sialyl-Le^x. *Science* **250,** 1130–1132.

Prescher, J. A., Dube, D. H., and Bertozzi, C. R. (2004). Chemical remodeling of cell surfaces in living animals. *Nature* **430,** 873–877.

Rose, M. C., and Voynow, J. A. (2006). Respiratory tract mucin genes and mucin glycoproteins in health and disease. *Physiol. Rev.* **86,** 245–278.

Ryu, Y., and Schultz, P. G. (2006). Efficient incorporation of unnatural amino acids into proteins in *Escherichia coli*. *Nat. Methods* **3,** 263–265.

Sambrook, J., and Russell, D. W. (2001). *In* "Molecular Cloning: A Laboratory Manual." Cold Spring Harbor Laboratory Press, Cold Spring Harbor, New York.

Sampathkumar, S. G., Li, A. V., Jones, M. B., Sun, Z., and Yarema, K. J. (2006). Metabolic installation of thiols into sialic acid modulates adhesion and stem cell biology. *Nat. Chem. Biol.* **2,** 149–152.

Saxon, E., and Bertozzi, C. R. (2000). Cell surface engineering by a modified Staudinger reaction. *Science* **287,** 2007–2010.

Schwarzkopf, M., Knobeloch, K., Rohde, E., Hinderlich, S., Wiechens, N., Lothar, L., Horak, I., Reutter, W., and Horstkorte, R. (2002). Sialylation is essential for early development in mice. *Proc. Natl. Acad. Sci. USA.* **99,** 5267–5270.

Suckow, M. A., Dannerman, P., and Brayton, C. (2000). *In* "The Laboratory Mouse." CRC Press, Boca Raton, Florida.

Vocadlo, D. J., Hang, H. C., Kim, E. J., Hanover, J. A., and Bertozzi, C. R. (2003). A chemical approach for identifying *O*-GlcNAc–modified proteins in cells. *Proc. Natl. Acad. Sci. USA* **100,** 9116–9121.

Wei, X., Decker, J. M., Wang, S., Hui, H., Kappes, J. C., Wu, X., Salazar-Gonzalez, J. F., Salazar, M. G., Kilby, J. M., Saag, M. S., Komarova, N. L., Nowak, M. A., Hahn, B. H., Kwong, P. D., and Shaw, G. M. (2003). Antibody neutralization and escape by HIV-1. *Nature* **422,** 307–312.

Weissleder, R., Tung, C. H., Mahmood, U., and Bogdanov, A., Jr. (1999). *In vivo* imaging of tumors with protease-activated near-infrared fluorescent probes. *Nat. Biotechnol.* **17,** 375–378.

Yeh, J., Hiraoka, N., Petryniak, B., Nakayama, J., Ellies, L. G., Rabuka, D., Hindsgaul, O., Marth, J. D., Lowe, J. B., and Fukuda, M. (2001). Novel sulfated lymphocyte homing receptors and their control by a core1 extension β1,3-*N*-acetylglucosaminyltransferase. *Cell* **105,** 957–969.

Section IV

Carbohydrate Ligand Specificity

[16] Functional Proteomic Profiling of Glycan-Processing Enzymes

By Keith A. Stubbs and David J. Vocadlo

Abstract

Glycoconjugates play critical roles in regulating cellular and organismal functions. Consequently, defining the relative levels of these glycoconjugates under varied physiological conditions is important. Thus identifying and understanding the regulation of the enzymes that process these glycoconjugates are essential steps in understanding the role of this "glycocode" in development and disease. Activity-based affinity reagents are useful tools for probing these enzymes and should facilitate the unraveling of proteomes. One advantage of activity-based affinity probes is that they can simultaneously reveal multiple enzymes having similar activities. These probes can also be used to enrich proteomes of interest, thereby facilitating identification and cloning of new carbohydrate-processing enzymes. Here we review the current state of activity-based affinity probes for profiling carbohydrate-processing enzymes, focusing on successes and limitations, general design features, and a specific example describing profiling of exoglycosidases from cell lysates.

Introduction

As genomes of organisms ranging from pathogens to humans are decoded at an accelerating pace, the field of proteomics is faced with the challenge of keeping step. One key difficulty is assigning biological functions to the growing list of thousands of putative proteins. Another challenge is the accurate quantitative analysis of changes in protein concentrations and activities that occur as a result of developmental changes or the onset of disease. Complicating both these issues, however, is the array of posttranslational modifications ornamenting proteins that can profoundly modulate their function. As well, the interaction of many proteins with various naturally occurring ligands mediates both their localization and, accordingly, their function.

The extensive posttranslational glycosylation of proteins (Varki, 1999) with a large variety of glycoforms is one such problem facing the field of proteomics. Defining the structures of unique and functionally important glycans in both prokaryotes and eukaryotes that mediate protein localization

METHODS IN ENZYMOLOGY, VOL. 415
Copyright 2006, Elsevier Inc. All rights reserved.
0076-6879/06 $35.00
DOI: 10.1016/S0076-6879(06)15016-8

or fulfill critical structural roles is a related challenge (Morelle and Michalski, 2005; Rabenstein, 2002). Together, these structures act as a complex code, decorating many proteins and mediating their interactions to dynamically modulate various cellular and organismal processes. The understanding of how, and where, carbohydrate motifs are attached to proteins and subsequently removed, as well as knowing which enzymes are responsible for these phenomena, are prerequisites for breaking this "glycocode." Such knowledge should provide improved understanding of the roles of glycan structures in health and disease. Compared with methods for identifying sites of glycosylation (Kaji *et al.*, 2003; Peter-Katalinic, 2005; Zhang *et al.*, 2003), however, the methods used to profile the activities of enzymes that install or cleave carbohydrates from glycan structures within proteomes and the discovery of new classes of these enzymes is greatly lagging.

The use of activity-based affinity probes, which have met with considerable success for profiling protease activities (Phillips and Bogyo, 2005; Speers and Cravatt, 2004), has only lately received attention for glycosyl-processing enzymes. Attempts to develop useful probes for these enzymes have met with mixed success. Therefore, a consideration of design features critical in developing versatile activity-based probes is warranted.

General Considerations in Developing Activity-Based Probes

The following is a list of general features that should be incorporated within the structure of an activity-based affinity probe to ensure its utility:

1. *Reporter Group:* A reporter group such as a fluorophore or biotin that permits rapid and sensitive detection and/or enrichment of labeled proteins should be incorporated into the molecule. Alternatively, an innocuous functional group that can be elaborated with a reporter group can be used.

2. *Recognition Element:* A critically important design feature is the inclusion of a moiety that can be specifically recognized by the enzyme(s) of interest. This element is usually a substrate analogue or a potent inhibitor of the enzyme.

3. *Linker Moiety:* A linker molecule between the reporter group and the recognition element is usually incorporated to increase solubility of the molecule in water and to ensure that the reporter group does not interfere with binding to the target enzyme. One commonly used series of linkers is ethylene glycol chains of various lengths.

4. *Reactive Group:* A reactive group that forms a covalent linkage between the probe and the enzyme(s) of interest under appropriate conditions is required. Such a group may be an electrophilic moiety, a photoreactive

cross-linker, or a mechanism-based component incorporated as part of the recognition element.

The general approach to using activity-based affinity probes is to simply add them to the cell lysate(s) of interest and incubate the mixture for a period of time (Fig. 1). Once incubation and cross-linking are complete, the probe should have formed a covalent bond to enzymes binding the recognition element. Choosing an appropriate combination of the correct recognition element and appropriately tuned reactive group is crucial to the success of the experiment. Binding between the recognition element and the target enzyme(s) should be tight; otherwise, a high concentration of the affinity probe will be required for efficient labeling. Such high probe concentrations may cause problems if the reactive group can cross-react nonspecifically. This problem of cross-reactivity is why affinity tags bearing inherently reactive groups appended to low-affinity ligands are often not useful for labeling complex proteomes.

The successfully labeled cell lysate can then be processed in different ways, depending on the aims of the experiment. The first and probably most

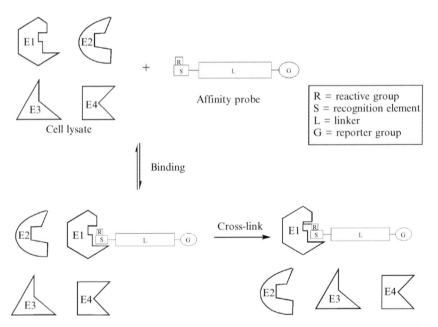

FIG. 1. A general mechanism by which activity-based probes can be used to label the active sites of targeted enzymes. Reversible binding of the probe is followed by the formation of an irreversible covalent linkage.

common treatment is to analyze the cell lysate by Western-blot using an antibody or protein recognizing the reporter group. Biotin is commonly used and offers some advantages, but highly fluorescent reporters, such as rhodamine or fluorescein, obviate the need to carry out blotting and probing steps and can be two orders of magnitude more sensitive (Liu *et al.*, 1999). Fluorescent gel-based methods are useful for rapid analysis of changes in enzyme expression between samples or for preliminary identification of tissues containing enzyme activities of interest. However, because there is no enrichment of labeled proteins, the detection of low-abundance proteins can be challenging. The second treatment is to carry out affinity purification of the enzymes of interest. Enriching the sample in this way facilitates identifying the enzyme with the activity that is sought. Using biotin as a reporter group offers some advantage since immobilized monomeric avidin can be used to enrich the sample in biotin-tagged enzymes. After enrichment, labeled proteins can be analyzed using tandem mass spectrometry or *N*-terminal sequencing to facilitate their cloning and detailed characterization.

Specific Case: Carbohydrate Processing Enzymes

Developing probes directed against specific carbohydrate-processing enzymes is complicated in that, unlike proteases, which commonly have a cleft-like active site, many carbohydrate-processing enzymes act on the termini of glycoconjugates and therefore have pocket-shaped active site architectures. Such sterically demanding active site structures hinder the development of useful probes since there is little space to accommodate reporter groups pendent to the recognition element. An exception are the endoglycosidases, which cleave glycan chains at internal glycosidic linkages. These enzymes tend to have more canyon-like active sites that can accommodate affinity probes having a reported group appended to the nonreducing terminus of the recognition element as shown in probe **1** (Fig. 2). Affinity probe **1** has been used successfully to probe a cellular extract of *Cellulomonas fimi* for glycanase activity (Hekmat *et al.*, 2005; Williams *et al.*, 2006). An elegant feature of this probe is that, since it is mechanism based, it is entirely specific for the enzymes of interest. Proteins identified will accordingly have the expected enzyme activity (Hekmat *et al.*, 2005). Such probes will likely prove widely useful for identifying proteins having desirable properties with this enzyme activity. Of course, since many glycoconjugates involved in biological recognition are not repeating oligomeric structures, there are few endoglycosidases within eukaryotes, and such an approach will be of limited use in profiling glycoside hydrolases from these proteomes.

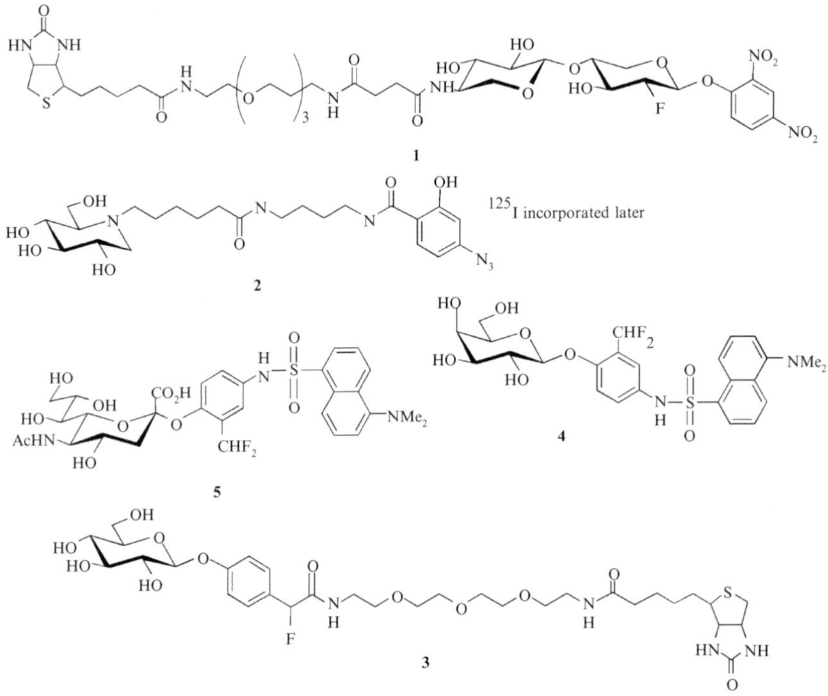

FIG. 2. Types of activity-based affinity probes already used in the literature.

Azasugars, which have long been used as inhibitors of glycosidases (Ganem, 1996), are another example of a specific recognition element that has been exploited for activity-based profiling. The 1-deoxynojirimycin (DNJ) derivative **2** (see Fig. 2) bearing a photoaffinity label as the reactive moiety has been used successfully for the active site-directed labeling of glucosidase I (Romaniouk *et al.*, 2004), an enzyme involved in the quality control of properly folded glycoproteins (Herscovics, 1999). By first labeling derivative **2** with radioactive iodine and then treating cell lysates with this radioactive probe, it appears that glucosidase I is labeled reasonably selectively, albeit perhaps at low efficiency. One intelligent feature of this approach was to use a high-affinity competitive inhibitor to direct the photoaffinity labeling, thereby reducing the need for high concentrations of material and so reducing nonspecific labeling. Accordingly, such a general strategy is a potentially viable route to profiling other glycosidases but will require further optimization. Disadvantages of using derivative **2**,

however, is the lengthy synthetic sequence required to prepare DNJ derivatives and the reliance on radioactivity.

A different approach to affinity-based labeling of glycosidases using
suicide substrates has been developed. This general strategy relies on the
substrate specificity of the glycosidases that are targeted. Enzymatic hydrolysis of the suicide glycoside (Fig. 3) liberates a reactive quinone methide
that can label any nucleophilic group encountered. Suicide substrates **3**
and **4** have been used as probes for β-glucosidases (Tsai *et al.*, 2002) and
β-galactosidases (Kurogochi *et al.*, 2004), respectively. A closely related
approach using suicide sialoside **5**, which has a fluorescent tag as the
reporter group, was tested as an activity-based probe for *Vibrio cholerae*
neuraminidase (Hinou *et al.*, 2005). Two advantages of these probes are
that they are comparatively easy to prepare and they can be used with both

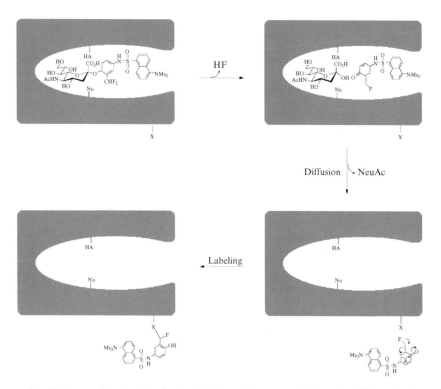

FIG. 3. Proposed mechanism for the labeling of the neuraminidase from *Vibrio cholerae*
using affinity probe 3. After the enzymatic activation of the probe, the fluorescent quinone
methide is free to diffuse away and may label any nucleophilic functional group in the target
or, it labeling is slow, another protein within the cell lysate.

mechanistic classes of glycosidase—retaining and inverting. One problem with these suicide probes, however, is that the highly reactive quinone methide can diffuse out of the enzyme active site and label residues indiscriminately (Hinou et al., 2005). For example, when using probe 5 to label pure V. cholerae neuraminidase, an enzymic residue approximately 20 Å away from the enzyme active site was tagged (Hinou et al., 2005). As a consequence, these compounds have not yet proved useful for labeling enzymes from cell lysates (Tsai et al., 2002) since they can label other proteins present in the mixture. Although promising, further development and optimization is required before these probes will be useful for activity-based profiling of complex proteomes.

Given the varied structures of glycosidases and the advantages and disadvantages inherent with these varied approaches, it appears that no one strategy is likely to be universally effective, and, consequently, there is considerable room and desire for the introduction of new and improved methods.

Case Study: Probing Exoglycosidases in Cell Lysates

As described previously, methods for profiling exoglycosidases are particularly challenging as a consequence of their constrained active site architectures. One advantage that can be exploited for developing probes for retaining exoglycosidases, however, is that their catalytic mechanism involves the transient formation of a covalent glycosyl-enzyme intermediate. These intermediates can be stabilized so they persist using either deoxy sugars (Roeser and Legler, 1981) or, more commonly, deoxy-fluorosugars (Wicki et al., 2002; Withers et al., 1988) as mechanism-based inactivators. The inclusion of a fluorine atom at the 2 or 5 position of the pyranose ring has the effect of destabilizing the two transition states that flank the covalent glycosyl-enzyme intermediate. Using these fluorosugars, the catalytic residues of many enzymic nucleophiles have been identified (Vocadlo and Withers, 2000; Vocadlo et al., 2001; Wicki et al., 2002; Withers et al., 1988). Because both enzyme-catalyzed steps are slowed by the presence of the fluorine group, a key structural requirement of these inactivators is that a very good leaving group be incorporated at the anomeric center. In this way, the intermediate is kinetically accessible since the very good leaving group compensates for the presence of the fluorine, yet the breakdown of the intermediate is very slow since the excellent leaving group is no longer present. The net effect is that these fluorosugars act both as a recognition element and reactive group, since a stable covalent bond is formed between the enzymic nucleophile and fluorosugar. All that remains to be resolved is how to attach the reporter group. As described earlier,

exoglycosidases have a pocket-shaped active site in which extensive contacts are formed between the protein and all substrate hydroxyl groups, making incorporation of a bulky reporter group such as biotin impractical.

Recently, several selective chemoselective ligations have been developed that make use of bioorthogonal functional groups. The Staudinger ligation (Saxon and Bertozzi, 2000), the copper catalyzed Huisgen cycloaddition (Rostovtsev et al., 2002; Tornoe et al., 2002), and a newer strain-promoted Huisgen ligation (Agard et al., 2004) all exploit the azide functionality and use either a triphenyl phosphine methyl ester, an alkyne, or a strained alkyne as the respective complementary reactant. These chemoselective ligations can be carried out under physiological conditions and are so exquisitely selective that other biologically relevant functional groups do not interfere. Any of these functional group pairs can be appended to various affinity probes or ligands as desired. By appending the azide group onto the fluorosugar, binding of the saccharide moiety by exoglycosidases is still possible since the azide tag is sterically unobtrusive. This azide group can be ligated to a phosphine ester or alkyne bearing a reporter group at any later stage. Therefore, the reporter group does not have to be initially present within the probe molecule. The enzyme can first be labeled by the affinity probe bearing the azide group and then the tagged enzyme can be elaborated with the reporter group through the ligation reaction. Using this approach, the first successful profiling of exoglycosidase activity from a complex proteome has been demonstrated (Vocadlo and Bertozzi, 2004).

Purified LacZ from *Escherichia coli* was selected as a preliminary model since it is known this enzyme is inactivated by a 2-deoxy-2-fluoro-β-galactoside (Gebler et al., 1992). Several other retaining exoglycosidases were also studied to demonstrate the generality of this approach. All six retaining β-galactosidases, which come from three different families of glycoside hydrolases, could be readily tagged and detected using a gel-based assay. To demonstrate the utility of this strategy in profiling a complex proteome, the study was extended successfully to probe *E. coli* cell lysates (Fig. 4).

The chemical synthesis of such azido-deoxy-fluorosugars as the one used here, 6-azido-2,6-dideoxy-2-fluoro-β-D-galactopyranosyl fluoride 6, is not trivial. Careful choice of which position the azide moiety should be appended is critical to ensure the targeted glycosidases are able to recognize and bind to the affinity probe (Vocadlo and Bertozzi, 2004). In this regard, the hydroxymethyl group of the sugar ring is a logical place to install the azide moiety; substitution of the hydroxyl group should not be too drastic a steric change, and free rotation of the methylene unit will allow it to adopt the most comfortable orientation. The high-resolution crystal structure case

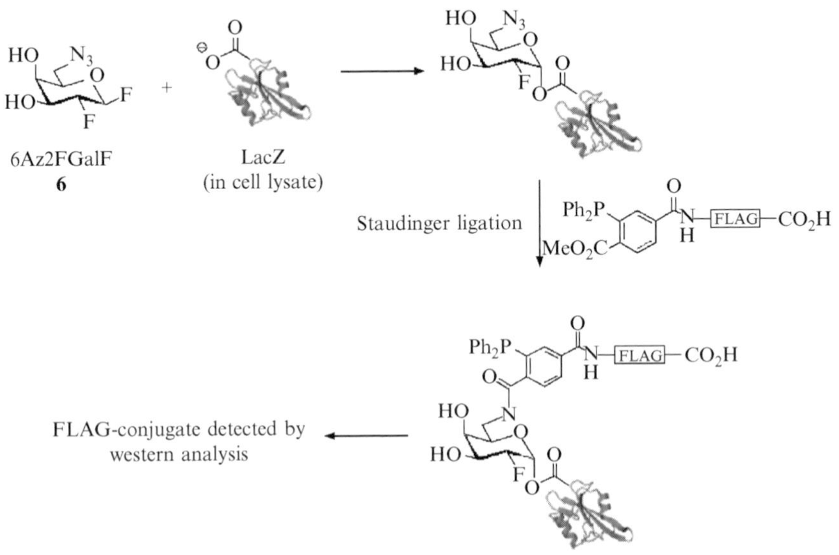

FIG. 4. The activity-based approach using the Staudinger ligation. The glycosidase is labeled due to the 2-fluoro moiety on the substrate 6Az2FGalF. Once the compound is covalently bound, attachment of the affinity label through the azide group is achieved.

of the 2-fluoroglycosyl enzyme complex of *E. coli* LacZ (Juers *et al.*, 2001) reveals a sterically congested environment at the 6-position. This enzyme therefore serves as a demanding test case, and success here suggests the method will be portable to other enzyme systems. Before embarking on a lengthy synthesis of such an azido-affinity probe, however, it is advantageous to examine the effect of a pendent azide group on enzymatic processing of a substrate analogue, which can be more readily prepared. If the enzyme can process the substrate, then it is likely that the analogous substitution made in the inactivator will also be tolerated. Accordingly, in the example described here, 4-nitrophenyl 6-azido-6-deoxy-β-D-galactopyranoside (Fig. 5) was prepared and tested.

Using this compound as a substrate of LacZ, saturation kinetics are not observed (Fig. 5), indicating that the azide group hinders binding to the β-galactosidase. The value of the second-order rate constant $V_{max}/K_m[E_o]$ for the azido-substrate is 12,000-fold lower than for the unsubstituted substrate (see Fig. 5). Although this difference may appear large, it is not a great concern for successful implementation of this activity-based proteomics profiling approach since the target enzymes only need to react once with the probe to form a covalent linkage and multiple turnovers are not

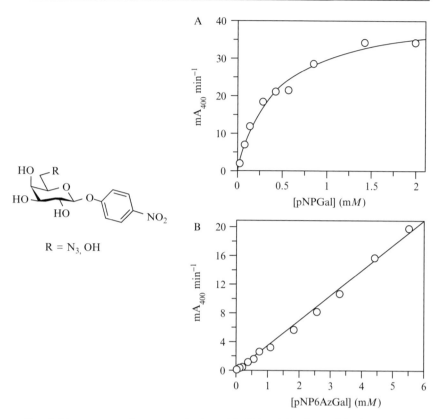

FIG. 5. Enzyme assays for a normal galactoside (A) and for the model substrate bearing the azide moiety at C-6 (B). (A) The Michaelis-Menten plot for the substrate R = OH. (B) The Michaelis-Menten plot for the substrate R = N₃.

required. Armed with the knowledge that the azide substitution is tolerated by the enzyme, the synthesis of the probe can be undertaken with confidence. To briefly summarize the synthetic route used to prepare probe 6, the 4-chlorothiophenyl glycoside is prepared from the glycosyl bromide after the fluorine atom has already been incorporated at C-2. This approach facilitates early installation of the azide at C-6 and then allows straightforward generation of the glycosyl fluoride at a later stage, either directly from the thioglycoside or from the readily accessed hemiacetal (Fig. 6).

Regardless of what route is selected to prepare the desired probe, it should be concise and robust, with key steps of functional group installation being facile. After synthesis of the probe, the inactivation and ligation can be carried out.

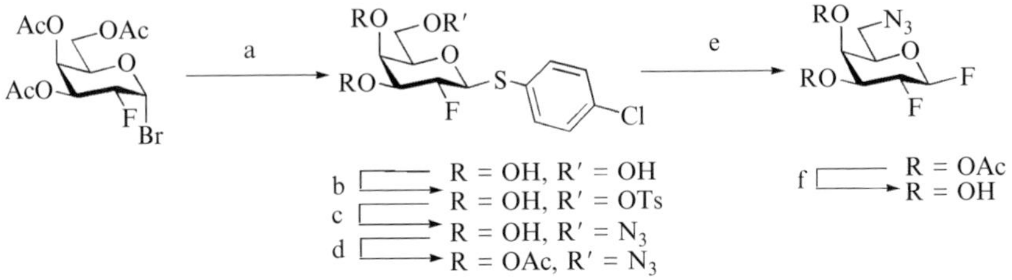

FIG. 6. (a) i. TBAHS, pClSPh, 1 M NaOH, CH$_2$Cl$_2$, ii. NaOMe, MeOH, iii. Amberlyst IR-120(H$^+$). (b) TsCl, pyridine. (c) NaN$_3$, DMF, 80°. (d) Ac$_2$O, pyridine. (e) i. NBS, Acetone, H$_2$O, ii. DAST, THF, CH$_2$Cl$_2$, –40° – RT. (f) i. NaOMe, MeOH, Amberlyst IR-120(H$^+$).

Step 1: Inactivation of Enzymes in the Complex Proteome

Because both the inactivation and ligation reactions are extraordinarily selective, high concentrations of probe can be used to speed the process. To prevent proteolytic degradation of the sample during handling, appropriate protease inhibitors should be added to the cell lysates immediately after cells are broken open. The fluorosugar probe is then added and the inactivation is allowed to proceed for as long as desired. The inactivation mixture can be comfortably left overnight at room temperature or incubating at 37°. To verify the inactivation is complete, an enzymatic assay using an appropriate substrate can be carried out.

Step 2: Chemoselective Ligation Reaction

After the inactivation reaction is complete, excess probe should be removed either by dialysis at 4°, by desalting column or by repeated buffer exchange. Before the ligation reaction, it may be useful to denature the proteins since the azide moiety within the folded enzyme active site might be occluded and therefore unable to efficiently ligate. Proteins should be denatured at room temperature using an acidic solution of denaturing concentrations of guanidine or urea and a low concentration of a monothiol such as β-mercaptoethanol. Dithiothreitol and phosphines should be avoided because these can reduce the azide functionality. Also, maintaining acidic conditions reduces susceptibility of the covalent glycosyl-enzyme adduct to cleavage. Any of the various chemoselective ligation reactions described previously can be used, but the Staudinger ligation, using 250 μM phosphine-FLAG reagent (Saxon and Bertozzi, 2000), works very well to elaborate the inactivated enzyme with a FLAG-peptide (e.g., DYKDD DDK) as the reporter group. The ligation is essentially complete after 6 hours.

Step 3: Gel-Base Profiling of the Proteome

After the ligation, a loading dye lacking thiols is added, and the mixture is then loaded directly onto a polyacrylamide gel. It is important to not heat the sample before electrophoresis since the glycosyl-enzyme adduct is a labile acylal ester that may be cleaved by nucleophiles at elevated temperatures. Also, all subsequent electrophoresis and blotting should be carried out at 4°. After electrophoresis, blotting to nitrocellulose using standard conditions is carried out. Analysis of the Western blot using appropriate protein or antibody conjugates should reveal that the probe has selectively labeled the target enzymes within the cell lysate. The use of appropriate controls in the experiment greatly facilitates trouble-shooting and reliable interpretation of the results. The controls also inform on the selectivity of the antibodies used and serve as a guide to the sensitivity of the protocol.

The Western blot analyses (Fig. 7) demonstrate that this profiling strategy for β-galactosidases in cell lysates is both selective and sensitive. These results show this methodology should prove generally applicable for evaluating the presence and concentration of various retaining exoglycosidases from

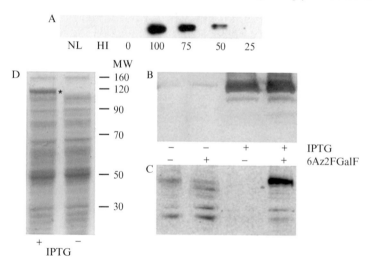

Fig. 7. (A) Western blot analysis showing the limit of detection using 6Az2FGalF in conjunction with the Staudinger ligation. NL, No 6Az2FGalF; HI, heat-inactivated LacZ (100 ng); 0, 0 ng LacZ; 100, 100 ng LacZ; 75, 75 ng LacZ; 50, 50 ng LacZ; 25, 25 ng LacZ. After inactivation, the samples were labeled with phosphine-FLAG and then analyzed by Western blot using anti–FLAG-HRP. (B) Western blot analysis of LacZ in the lysates from cells grown in the presence or absence of IPTG. The blot was probed using a mouse anti-LacZ monoclonal antibody followed by an anti-mouse immunoglobulin G–HRP conjugate. (C) The Western blot analysis shown in B, stripped and probed using anti-FLAG-HRP. (D) SDS-PAGE analysis of cell lysates from cultures of *E. coli* K-12, either induced with 0.1 m*M* IPTG or uninduced.

different families of glycoside hydrolases in a wide range of proteomes from different organisms.

Conclusion

Glycoconjugates play critical roles in regulating cellular and organismal functions. Consequently, defining the relative levels of these glycoconjugates under varied physiological conditions is important. Identifying and understanding the regulation of the enzymes that process these glycoconjugates is an essential step in understanding the role of the "glycocode" in development and disease. Activity-based affinity reagents are useful tools for probing these enzymes and should facilitate the unraveling of proteomes. One advantage of activity-based affinity probes is that they can be designed to target specific classes of enzymes. Accordingly, they can simultaneously reveal multiple enzymes having similar activities. These probes can also be used to enrich proteomes of interest, thereby facilitating identification and cloning of new carbohydrate-processing enzymes.

One potential application of probes for glycan-processing enzymes could be to address these enzymes in living systems. For example, highly selective probes could be used to evaluate the efficacy and selectivity of competitive inhibitors in living tissues, a more informative setting than *in vitro* assays using purified enzymes. If these inhibitors potently block the function of some enzymes in the cellular environment, the binding of the active site–directed affinity probes will be prevented and the specificity of the inhibitor will be revealed by changes to the profile. Such a strategy would be particularly useful for evaluating the selectivity of competitive inhibitors toward functionally related enzymes within tissues of interest.

Given both the structural and mechanistic diversity of glycan processing enzymes, it is most likely that no one type of affinity probe will be universally effective for investigating these enzymes within complex proteomes. Accordingly, there is considerable room and desire for the introduction of new and improved activity-based proteomics methods for this large and rapidly growing superfamily of enzymes (Davies *et al.*, 2005).

Experimental Procedures for Labeling *E. coli* Cell Lysates Using Probe 6

Cell cultures (6×5 ml) of *E. coli* K-12 were grown to an OD_{600} value of 0.6 and were either induced with 0.1 mM IPTG (isopropyl-β-D-thiogalactopyranoside) or untreated. After growing the cells at room temperature overnight to stationary phase, the cells were harvested by centrifugation ($5000 \times g$) and frozen at $-80°$ in a solution (500 μl) of Tris buffer (pH, 7.4)

containing 10% glycerol, 1 mM PMSF (phenylmethylsulfonyl fluoride), and hen egg white lysozyme (0.1 mg/ml). The suspension of frozen cells was then sonicated at 4° (4 × 30 sec; power level, 2 to 3; microtip, Misonix, Ultrasonic Processor XL). The crude lysate was clarified by centrifugation at 14,000 rpm in an Eppendorf 5415C microcentrifuge for 20 min at 4°. The supernatant was collected, and these samples were used as the complex proteome sample. The relative activities of induced and uninduced cultures were measured using an aliquot (100 μl) of a 50-fold dilution of culture lysate in a solution of 100 mM Tris (pH, 7.4) containing 6.6 mM pNPGal and 10% glycerol in a final volume of 850 μl at room temperature, and it was determined that uninduced cultures contained less than 1% β-galactosidase activity (0.0003 A$_{400}$/min/μl of cell lysate) compared with induced cultures (0.055 A$_{400}$/min/μl of cell lysate). An aliquot (5 μl) of each sample was subject to SDS-PAGE analysis (precast 10% or 12% Tris-HCl polyacrylamide gels, Bio-Rad) using coomassie stain to verify differences in LacZ protein expression levels between induced and uninduced cultures. An aliquot of each of the cell lysate samples (90 μl) was treated with a solution (10 μl) of 6Az2FGalF (526 mM, final [6Az2FGalF] = 52.6 mM) in phosphate-buffered saline (PBS) containing 50% DMF and incubated overnight at 37°. The inactivated samples were then dialyzed overnight (Pierce Slide-A-Lyzer Mini Dialysis Unit, 10,000 MWCO) at 4°. After dialysis, solutions of 1 M (pH, 2.0) sodium phosphate (20 μl), β-mercaptoethanol in water (100 mM), PMSF in ethanol (100 mM), and saturated urea (one third of the total volume) were added to yield a solution with a pH of approximately 3.5 containing 4 mM β-mercaptoethanol and 2 mM PMSF. One volume of a solution of FLAG-phosphine in water (500 μM) was added (250 μM final concentration), and the mixture was allowed to incubate overnight at room temperature. The sample was concentrated (AmiCon, microconcentrator 10,000 MWCO) to a final volume of approximately 20 μl, and an aliquot of the sample was mixed with SDS-PAGE loading buffer. Without heating, the sample was loaded onto precast 10% or 12% Tris-HCl polyacrylamide gels (Bio-Rad). After electrophoresis, the samples were electroblotted to nitrocellulose membrane (0.45 μm, Bio-Rad). Transfer was verified by visual inspection of the transfer of prestained markers (Benchmark, Invitrogen). The membrane was blocked using PBS containing 5% low-fat dry powdered milk, containing 0.1% Tween-20 (blocking buffer) for 1 hr at room temperature or overnight at 4°. The blocking solution was decanted, and a solution of blocking buffer A containing either anti-β-galactosidase mouse immunoglobulin G mAb (1:5000, Promega) or anti-FLAG-HRP mAb conjugate (1:3500, Sigma) was added. Membranes were incubated at room temperature for 1 hr or overnight at 4°, after which the blocking solution was decanted and the membrane was

rinsed with PBS containing 0.1% Tween-20 (wash buffer). Membranes were then rinsed for 2 × 5 min and 2 × 20 min with wash buffer. For detection of LacZ, the membrane was incubated in blocking solution for 1 hr at room temperature, and, after washing, the membrane was incubated with a secondary goat anti-mouse–HRP conjugate (1:10,000, Zymed Laboratories) for 1 hr at room temperature or 4° overnight in blocking solution. The membrane was washed, and detection of membrane-bound goat anti-mouse–HRP conjugate was accomplished as described later. Where necessary, membranes were stripped using 2% SDS, 100 mM β-mercaptoethanol, and 50 mM Tris (pH, 6.8) at 50° for 30 min and the reprobed using anti-FLAG–HRP mAb conjugate (1:3500, Sigma) in the blocking buffer. Control samples were treated in the same manner, except that specific reagents were replaced by buffer where appropriate. Detection of membrane-bound goat anti-mouse–HRP or anti-FLAG–HRP conjugates was accomplished with chemiluminescent detection using the SuperSignal West Pico Chemiluminescent Detection Kit (Pierce) and film (Kodak Biomax MR or X-OMAT).

References

Agard, N. J., Prescher, J. A., and Bertozzi, C. R. (2004). A strain-promoted [3 + 2] azide-alkyne cycloaddition for covalent modification of biomolecules in living systems. *J. Am. Chem. Soc.* **126,** 15046–15047.

Davies, G. J., Gloster, T. M., and Henrissat, B. (2005). Recent structural insights into the expanding world of carbohydrate-active enzymes. *Curr. Opin. Struct. Biol.* **15,** 637–645.

Ganem, B. (1996). Inhibitors of carbohydrate-processing enzymes: Design and synthesis of sugar-shaped heterocycles. *Acc. Chem. Res.* **29,** 340–347.

Gebler, J. C., Aebersold, R., and Withers, S. G. (1992). Glu-537, not Glu-461, is the nucleophile in the active site of (lac Z) beta-galactosidase from. *Escherichia coli. J. Biol. Chem.* **267,** 11126–11130.

Hekmat, O., Kim, Y. W., Williams, S. J., He, S., and Withers, S. G. (2005). Active-site peptide "fingerprinting" of glycosidases in complex mixtures by mass spectrometry. Discovery of a novel retaining β-1,4-glycanase in. *Cellulomonas fimi. J. Biol. Chem.* **280,** 35126–35135.

Herscovics, A. (1999). Importance of glycosidases in mammalian glycoprotein biosynthesis. *Biochim. Biophys. Acta* **1473,** 96–107.

Hinou, H., Kurogochi, M., Shimizu, H., and Nishimura, S. (2005). Characterization of *Vibrio cholerae* neuraminidase by a novel mechanism-based fluorescent labeling reagent. *Biochemistry* **44,** 11669–11675.

Juers, D. H., Heightman, T. D., Vasella, A., McCarter, J. D., Mackenzie, L., Withers, S. G., and Matthews, B. W. (2001). A structural view of the action of *Escherichia coli* (lacZ) β-galactosidase. *Biochemistry* **40,** 14781–14794.

Kaji, H., Saito, H., Yamauchi, Y., Shinkawa, T., Taoka, M., Hirabayashi, J., Kasai, K., Takahashi, N., and Isobe, T. (2003). Lectin affinity capture, isotope-coded tagging and mass spectrometry to identify *N*-linked glycoproteins. *Nat. Biotechnol.* **21,** 667–672.

Kurogochi, M., Nishimura, S., and Lee, Y. C. (2004). Mechanism-based fluorescent labeling of beta-galactosidases: An efficient method in proteomics for glycoside hydrolases. *J. Biol. Chem.* **279,** 44704–44712.

Liu, Y., Patricelli, M. P., and Cravatt, B. F. (1999). Activity-based protein profiling: The serine hydrolases. *Proc. Natl. Acad. Sci. USA.* **96,** 14694–14699.

Morelle, W., and Michalski, J. C. (2005). Glycomics and mass spectrometry. *Curr. Pharm. Des.* **11,** 2615–2645.

Peter-Katalinic, J. (2005). Methods in enzymology: *O*-glycosylation of proteins. *Methods Enzymol.* **405,** 139–171.

Phillips, C. I., and Bogyo, M. (2005). Proteomics meets microbiology: Technical advances in the global mapping of protein expression and function. *Cell Microbiol.* **7,** 1061–1076.

Rabenstein, D. L. (2002). Heparin and heparan sulfate: Structure and function. *Nat. Prod. Rep.* **19,** 312–331.

Roeser, K. R., and Legler, G. (1981). Role of sugar hydroxyl groups in glycoside hydrolysis: Cleavage mechanism of deoxyglucosides and related substrates by beta-glucosidase A3 from *Aspergillus wentii. Biochim. Biophys. Acta* **657,** 321–333.

Romaniouk, A. V., Silva, A., Feng, J., and Vijay, I. K. (2004). Synthesis of a novel photo-affinity derivative of 1-deoxynojirimycin for active site-directed labeling of glucosidase I. *Glycobiology* **14,** 301–310.

Rostovtsev, V. V., Green, L. G., Fokin, V. V., and Sharpless, K. B. (2002). A stepwise huisgen cycloaddition process: Copper(I)-catalyzed regioselective "ligation" of azides and terminal alkynes. *Angew. Chem. Int. Ed. Engl.* **41,** 2596–2599.

Saxon, E., and Bertozzi, C. R. (2000). Cell surface engineering by a modified Staudinger reaction. *Science* **287,** 2007–2010.

Speers, A. E., and Cravatt, B. F. (2004). Chemical strategies for activity-based proteomics. *Chem. Biol. Chem.* **5,** 41–47.

Tornoe, C. W., Christensen, C., and Meldal, M. (2002). Peptidotriazoles on solid phase: [1,2,3]-triazoles by regiospecific copper(I)-catalyzed 1,3-dipolar cycloadditions of terminal alkynes to azides. *J. Org. Chem.* **67,** 3057–3064.

Tsai, C. S., Li, Y. K., and Lo, L. C. (2002). Design and synthesis of activity probes for glycosidases. *Org. Lett.* **4,** 3607–3610.

Varki, A. (1999). *In* "Essentials of Glycobiology." Cold Spring Harbor Laboratory Press, Cold Spring Harbor, NY.

Vocadlo, D. J., and Bertozzi, C. R. (2004). A strategy for functional proteomic analysis of glycosidase activity from cell lysates. *Angew. Chem. Int. Ed. Engl.* **43,** 5338–5342.

Vocadlo, D. J., Davies, G. J., Laine, R., and Withers, S. G. (2001). Catalysis by hen egg-white lysozyme proceeds *via* a covalent intermediate. *Nature* **412,** 835–838.

Vocadlo, D. J., and Withers, S. G. (2000). Identification of active site residues in glycosidases by use of tandem mass spectrometry. *Methods Mol. Biol.* **146,** 203–222.

Wicki, J., Rose, D. R., and Withers, S. G. (2002). Trapping covalent intermediates on β-glycosidases. *Methods Enzymol.* **354,** 84–105.

Williams, S. J., Hekmat, O., and Withers, S. G. (2006). Synthesis and testing of mechanism-based protein-profiling probes for retaining *endo*-glycosidases. *Chem. Biol. Chem.* **7,** 116–124.

Withers, S. G., Rupitz, K., and Street, I. P. (1988). 2-Deoxy-2-fluoro-D-glycosyl fluorides: A new class of specific mechanism-based glycosidase inhibitors. *J. Biol. Chem.* **263,** 7929–7932.

Zhang, H., Li, X. J., Martin, D. B., and Aebersold, R. (2003). Identification and quantification of *N*-linked glycoproteins using hydrazide chemistry, stable isotope labeling and mass spectrometry. *Nat. Biotechnol.* **21,** 660–666.

[17] Oligosaccharide Microarrays to Map Interactions of Carbohydrates in Biological Systems

By Jose L. de Paz, Tim Horlacher, and Peter H. Seeberger

Abstract

Carbohydrate microarrays are becoming a standard tool for glycobiologists to screen large numbers of sugars and elucidate the role of carbohydrates in biological systems. This article describes detailed methods to prepare and use microarrays containing synthetic oligosaccharides as well as a summary of the biological information that can be obtained by using this technology. These methods use different linking chemistries to immobilize a wide range of synthetic oligosaccharides onto glass slides through the formation of a covalent bond. Therefore, this technology enables the elaborate study of a great variety of carbohydrate interactions.

Introduction

Carbohydrates, in the form of glycopeptides, glycolipids, glycosaminoglycans, or other glycoconjugates, have long been known to participate in a wide variety of biological processes (Dwek, 1996; Hakomori and Zhang, 1997). They are involved in viral entry, signal transduction, inflammation, cell–cell interactions, bacteria–host interactions, fertilization, and development. In the postgenomic era, glycomics (Bidlingmaier and Snyder, 2002; Doores *et al.*, 2006; Dube and Bertozzi, 2005; Feizi and Mulloy, 2003a; Kiessling and Cairo, 2002; Ratner *et al.*, 2004a), the functional study of carbohydrates in living organisms, has received increasing attention for biological research and biomedical applications (Seeberger and Werz, 2005; Werz and Seeberger, 2005). More rapid advances in the field of glycomics have been hindered by the complexity of the biomolecules involved. Oligosaccharides are structurally more complex than nucleic acids and proteins due to frequent branching and linkage diversity. Furthermore, the difficulty in isolating, characterizing, and synthesizing complex oligosaccharides has been a significant challenge to progress in the field.

In this context, carbohydrate microarrays (Feizi *et al.*, 2003b; Love and Seeberger, 2002; Shin *et al.*, 2005), carrying tens or hundreds of different sugars that are bound covalently or noncovalently in small spots on solid surfaces, are becoming a standard tool for glycobiologists to screen large numbers of sugars and elucidate the role of carbohydrates in biological

METHODS IN ENZYMOLOGY, VOL. 415

0076-6879/06 $35.00
DOI: 10.1016/S0076-6879(06)15017-X

systems. The miniaturized array methodology is particularly well suited for investigations in the field of glycomics because only tiny amounts of both analyte and ligand are required for one experiment, and several thousand binding events can be screened in parallel on a single glass slide.

In addition, carbohydrate microarrays are ideal to detect interactions that involve carbohydrates since the multivalent display of ligands on a surface (cluster effect, Mammen *et al.*, 1998) overcomes the relative weakness of these interactions by mimicking cell–cell interfaces. Moreover, carbohydrate microarrays possess a plethora of potential applications in glycomics, including the rapid determination of the binding profile of carbohydrate-binding proteins (Blixt *et al.*, 2004; Huang *et al.*, 2006), the detection of specific antibodies for the diagnosis of diseases (Manimala *et al.*, 2005; Thirumalapura *et al.*, 2005), the characterization of carbohydrate-cell recognition events (Nimrichter *et al.*, 2004), and the high-throughput screening of inhibitors to prevent carbohydrate-protein interactions (Bryan *et al.*, 2004).

In the past few years, our laboratory has prepared carbohydrate-containing slides to address biological problems including the identification of human immunodeficiency virus (HIV) vaccine candidate antigens (Adams *et al.*, 2004), the evaluation of aminoglycoside antibiotics (Disney and Seeberger, 2004a), the detection of pathogenic bacteria (Disney and Seeberger, 2004b), and the determination of the binding profile of heparin-binding proteins (De Paz *et al.*, 2006). This effort involved the development of linking chemistries that are compatible with the protecting group manipulations usually required for the synthesis of complex oligosaccharides. Here, we present several methods of attachment for synthetic oligosaccharides to glass slides through the formation of a covalent bond as well as a summary of the biological information that can be obtained by using this microarray technology.

Oligosaccharide Microarrays as Tools in HIV Glycobiology

Defining HIV envelope glycoprotein interactions with binding partners advances our understanding of the infectious process and provides a basis for the design of vaccines and agents that interfere with HIV entry (Adams *et al.*, 2004; Calarese *et al.*, 2005). Our group used carbohydrate and glycoprotein microarrays to analyze glycan-dependent interactions of two HIV-1 envelope glycoproteins, gp-120 and gp-41.

For this purpose, a series of high-mannose oligosaccharides (Fig. 1) was synthesized (Ratner *et al.*, 2002). These structures include the triantennary mannoside 1, as well as closely related substructures of this epitope, that is decorating the viral-surface envelope glycoproteins of HIV. The synthetic oligosaccharides contain a thiol-terminated ethylene glycol linker that permits their attachment onto maleimide-functionalized glass slides by

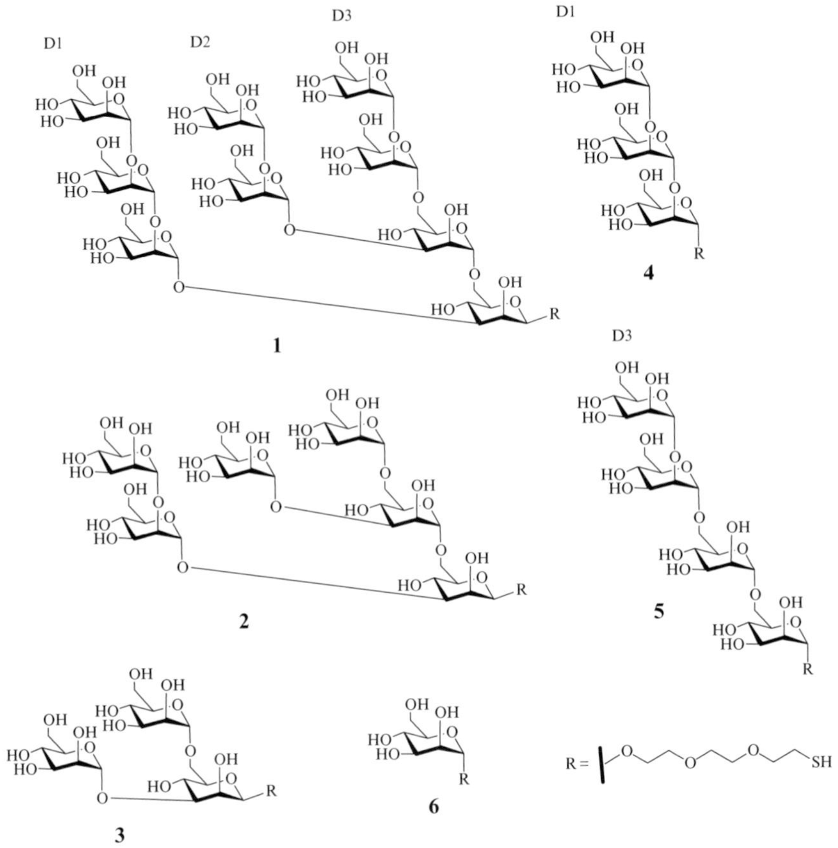

FIG. 1. A series of high-mannose oligosaccharides (1-6) containing a thiol linker for immobilization chemistry.

formation of a stable covalent bond (Ratner *et al.*, 2004b). We adopted two different strategies for the preparation of the high-mannose microarrays that differ in the density of the arrayed sugars. In one case, bovine serum albumin (BSA)-derivatized aldehyde slides were treated with succinimidyl 4-(*N*-maleimidomethyl)cyclohexane-1-carboxylate (SMCC) to present a low-density maleimide surface. Alternatively, amine-derivatized slides were directly modified with SMCC to prepare high-density maleimide surfaces (Fig. 2.)

Materials and Equipment

1. Milli-Q water
2. SuperAldehyde slides (TeleChem International)

3. Phosphate-buffered saline (PBS) buffer (pH, 7.4; 10 mM; 137 mM NaCl; 2.7 mM KCl)
4. BSA (Fluka)
5. Anhydrous dimethylformamide (DMF, Aldrich)
6. SMCC from Pierce Endogen
7. N,N-diisopropylethylamine (Aldrich)
8. Gamma Amino Propyl Silane (GAPS) slides (Corning)
9. Tris(carboxyethyl)phosphine hydrochloride (TCEP) from Pierce Endogen
10. 3-Mercaptopropionic acid
11. N-2-hydroxyethylpiperazine-N-2-ethanesulfonic (HEPES) buffer (50 mM, pH 7.5)
12. DNA array printer (MicroGrid TAS)
13. Hybridization coverslips (HybriSlip, Molecular Probes)
14. Fluorescence slide scanner (arrayWoRx Applied Precision)
15. Digital genome software (MolecularWare)
16. Ethylene-glycol-bis (succinimidylsuccinate) (Pierce Endogen).

Preparation of High-Mannose Microarrays

SuperAldehyde slides can be converted into maleimide slides by using the following protocol. The slides are immersed in PBS (pH, 7.4; 10 mM) containing BSA (1% w/v) and incubated overnight at room temperature. Then, the slides are rinsed twice with Nanopure water and twice with ethanol (95%) and dried under a stream of dry argon. Subsequently, the

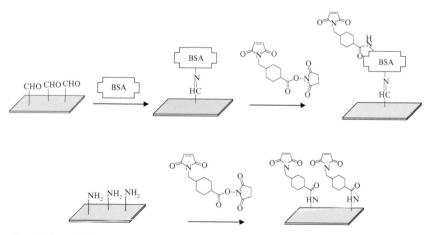

FIG. 2. Strategies for the preparation of high-mannose microarrays with different density.

BSA-coated slides are immersed in anhydrous DMF containing SMCC (4.3 mM) and N,N-diisopropylethylamine (100 mM). The slides are incubated in this solution for 24 hrs at room temperature, washed four times with ethanol (95%), and stored in a dry box until use. High-density maleimide slides can be prepared from amine-coated GAPS slides. The slides are incubated overnight at room temperature in a DMF solution containing SMCC (0.7 mM) and N,N-diisopropylethylamine (100 mM). Then, the slides are washed four times with methyl alcohol, dried under a stream of argon, and stored in a desiccator before use. These two functionalization methods offer different advantages. BSA-derivatized slides present a relatively low density of immobilized oligosaccharides and excellent resistance to nonspecific binding of proteins. GAPS slides permit high-density immobilization of oligosaccharides, allowing examination of carbohydrate clusters at the surface and presenting the sugar in a peptide-free context.

After preparing the maleimide functionalized surfaces, the thiol-containing sugars can be immobilized on the slides using the following protocol. First, the high-mannose oligosaccharides are treated with TCEP (1 equivalent) in PBS buffer (10 mM; pH, 7.4) for 1 hr at room temperature. Next, a standard DNA array printer is used to deliver as little as 1 nl of solutions containing carbohydrate that range in concentration from 5 mM μM (average spot diameter, 200 μm). Thereafter, the slides are stored in a humid chamber for 12 hrs at room temperature, washed twice with distilled water, and incubated for 1 hr in a solution of 3-mercaptopropionic acid (1 mM) or 2-(2-(2-mercaptoethoxy)ethoxy)ethanol (1 mM) in PBS buffer (10 mM) to quench all remaining maleimide groups. The slides are washed three times with distilled water, then twice with ethanol (95%), and are stored in a dry box until they are used for the binding experiments.

Hybridization Experiments

These high-mannose arrays have been used to determine the binding profile of four relevant gp-120 binding proteins: the dendritic cell lectin DC-SIGN, the HIV-inactivating proteins scytovirin and cyanovirin-N, and the human antibody 2G12. For this purpose, the synthetic oligosaccharides were printed across a wide range of concentrations to obtain the corresponding binding curves by plotting the fluorescence intensity against the concentration of the solution used for printing. For example, incubation of antibody 2G12 with the microarray revealed binding at spots corresponding to compounds 1, 2, 4, and 5 (see Fig. 1), but not to the branched trimannoside 3 or mannose 6. This result suggests that the Manα1–2Man linkage is necessary for recognition by 2G12. Analysis of scytovirin's carbohydrate binding profile revealed that only compounds 1 and 5 bound to this anti-HIV

protein, suggesting that the terminal α1-2 mannose linkage is necessary for the recognition of the α1-6 trimannoside substructure that constitutes the D3 arm (see Fig. 1). These binding experiments illustrate the potential of the microarray technology to pinpoint the structural requirements for carbohydrate–protein interactions, providing precious information for the development of an HIV vaccine.

For array incubation, the proteins are used at a concentration of approximately 20 μg/ml in HEPES buffer (50 mM; pH, 7.5) containing NaCl (0.1 M) and BSA (1% w/v). The protein solution is placed between array slides and hybridization coverslips and incubated for 30 min at room temperature. When fluorophore-labeled proteins are used, the slides are directly washed with HEPES buffer (50 mM; pH, 7.5) containing Tween-20 (1% v/v) and BSA (0.1% w/v), washed twice with distilled water, and then centrifuged for 5 min to ensure dryness. In cases in which secondary labeled antibodies are used, the slides are previously incubated with a solution of the antibody and then washed and dried as noted previously. Microarrays can be scanned by using a standard fluorescence slide scanner, and the spots can be quantified by using the appropriate software.

We have also included natural and modified glycoproteins, as well as neoglycoproteins on the arrays, to obtain additional information regarding the role of the polypeptide backbone in which glycans are presented to their binding partners.

Glycoproteins can be arrayed at high density on amine-reactive glass slides as follows: GAPS slides are treated with ethylene-glycol-bis(succinimidylsuccinate) (10 mM) in anhydrous DMF containing N,N-diisopropylethylamine (100 mM) for 24 hrs at room temperature. Subsequently, glycoproteins (100 μg/ml) in PBS buffer (10 mM, pH 7.4) are printed and incubated for 12 hrs in a humidity chamber, then washed twice with distilled water and incubated for 2 hrs with BSA (1% w/v) in PBS buffer to quench reactive succinimidyl groups. After rinsing three times with PBS, the arrays are dried under a stream of nitrogen and stored in a desiccator before use. These glycoprotein slides can be used in binding experiments as described previously.

Microarrays of Synthetic Heparin Oligosaccharides for
   High-Throughput Screening of Heparin–Protein Interactions

Heparin, an anticoagulant drug, is widely recognized to be a biologically important and chemically unique polysaccharide (Capila and Linhardt, 2002; Noti and Seeberger, 2005). Heparin is a highly sulfated, linear polymer that belongs to the family of glycosaminoglycans and participates in a plethora of biological processes by interaction with many proteins

(Casu and Lindahl, 2001; Conrad, 1998). For example, the heparin–antithrombin III (AT-III) interaction is responsible for heparin's anticoagulant activity. The interaction of heparin with fibroblast growth factors (FGF) is crucial for regulating the activity of these signaling polypeptides that are involved in angiogenesis, cell growth, and differentiation.

Heparin is predominantly formed by disaccharide repeating units of D-glucosamine (GlcN) and L-iduronic acid (IdoA), is linked by α1-4 glycosidic linkages, and typically contains sulfate groups at positions 2 and 6 of the GlcN unit and position 2 of the IdoA unit (Fig. 3). However, a number of structural variations of this trisulfated disaccharide exist and contribute to the microheterogeneity of heparin. The amino group of the glucosamine residue may be acetylated or unsubstituted. The uronic acid unit can also be D-glucuronic acid, O-sulfated or unsubstituted (see Fig. 3). This structural variability renders heparin an extremely challenging molecule to characterize and can be responsible for the interaction of heparin with a wide variety of proteins. The chemical complexity and heterogeneity of this polysaccharide can also explain the fact that, despite its widespread medical use, the structure–activity relationships of heparin are still poorly understood.

In this context, the use of microarrays of synthetic heparin oligosaccharides can substantially improve the understanding of heparin–protein interactions, opening an opportunity for the discovery of novel therapeutic interventions for a variety of disease states. We have recently reported (De Paz et al., 2006) the creation and use of microarrays containing synthetic heparin oligosaccharides (Orgueira et al., 2003). For this purpose, we have developed a novel linker strategy that is compatible with the protecting-group manipulations required for the synthesis of the highly sulfated oligosaccharides (De Paz et al., 2003; De Paz and Martin-Lomas, 2005) after solution or solid-phase assembly (Fig. 4). A series of fully

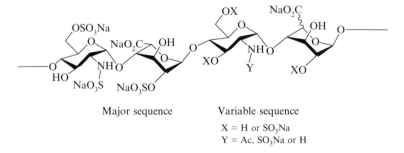

Major sequence   Variable sequence

X = H or SO$_3$Na
Y = Ac, SO$_3$Na or H

FIG. 3. Major and minor disaccharide repeating units in heparin.

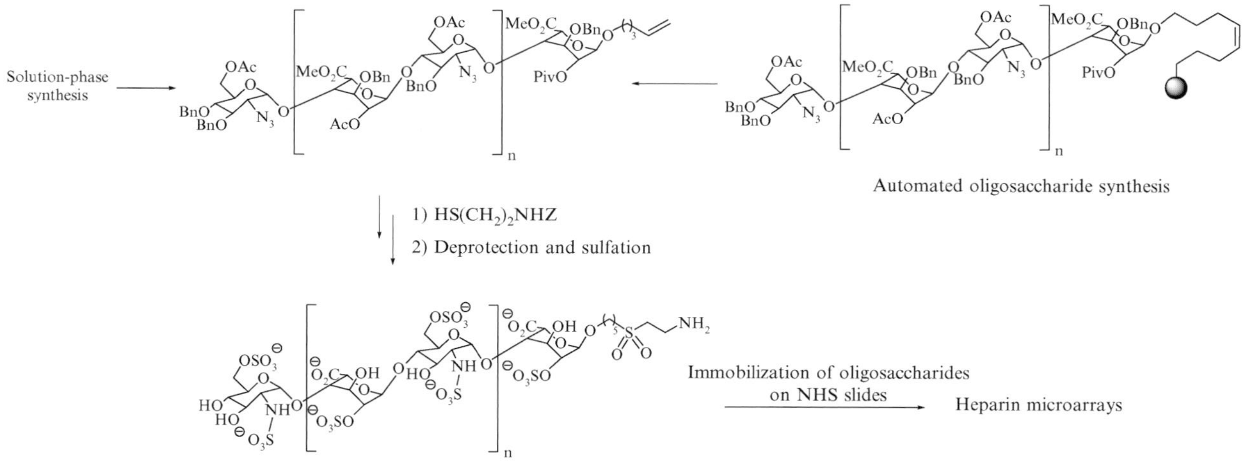

FIG. 4. General strategy for the preparation of microarrays containing synthetic heparin oligosaccharides. NHS, N-hydroxysuccinimide; Z, benzyloxycarbonyl.

protected oligosaccharides, precursors for the $GlcNSO_3(6\text{-}OSO_3)\text{-}IdoA$ $(2\text{-}OSO_3)$ repeating unit of the major sequence of heparin, were synthesized as pentenyl glycosides. Synthetic oligosaccharides obtained by automated solid-phase synthesis also can be released as terminal pentenyl glycosides (Plante *et al.*, 2001).

Strategic placement of an orthogonally protected amine linker was key to the success of the array construction. 2-(Benzyloxycarbonylamino)-1-ethanethiol was selected for the radical elongation of the pentenyl glycosides by using a catalytic amount of 2,2'-azobis(2-methylpropionitrile) at 75°. Treatment with lithium hydroperoxide and KOH hydrolyzed the acyl and methoxycarbonyl groups with simultaneous oxidation of sulfide into sulfone. Then, the introduction of the *O*-sulfate groups was achieved by treatment with SO_3. Py complex. Subsequently, the azide groups were transformed into the corresponding amines via Staudinger reduction. Finally, *N*-sulfation followed by hydrogenolysis afforded the synthetic heparin oligosaccharides ready for covalent immobilization onto commercially available *N*-hydroxysuccinimide (NHS) activated glass slides (see Fig. 4). The combination of amine-functionalized glycans with NHS-activated glass surfaces results in robust and reproducible covalent attachment of carbohydrates without modification of standard DNA printing protocols.

Materials and Equipment

1. Sodium phosphate buffer (pH, 9.0; 50 m*M*)
2. Automated arraying robot (Perkin Elmer)
3. NHS-activated slides (CodeLink, Amersham Biosciences)
4. FGF-1 and FGF-2 (PeproTech EC)
5. Anti–FGF-1 and anti–FGF-2 (Santa Cruz Biotechnology Inc)
6. AlexaFluor-546–labeled anti-rabbit immunoglobulin G (Molecular Probes)
7. PBS buffer (pH, 7.4; 10 m*M*; 137 m*M* NaCl; 2.7 m*M* KCl)
8. Hybridization coverslips (HybriSlip, Molecular Probes)
9. Milli-Q water
10. Fluorescence reader (LS400, Tecan)
11. Scan Array Express Software (Perkin Elmer).

Preparation of Heparin Arrays

The heparin microarrays are prepared by using the following protocol. The oligosaccharides are dissolved in sodium phosphate buffer (pH, 9.0; 50 m*M*) at concentrations ranging from 5 m*M* to 50 μM. An automated arraying robot is used to deliver 1 nl of sugar solutions onto NHS-activated

slides. All samples are printed in replicates of 15 for quantification purposes (average spot size, ~200 μm). Sodium phosphate buffer (pH, 9.0; 50 mM) as well as 2'-aminoethyl-2-acetamido-α-D-glucopyranoside (GlcNAc) and 2'-aminoethyl-β-D-galactopyranoside (Gal) can be used as negative controls. After printing, slides are immediately placed in a humidity chamber and incubated for 12 hrs. Arrays are washed three times with Nanopure water to remove the unbound carbohydrates from the surface. Remaining succinimidyl groups are quenched by placing slides in a solution preheated to 50° that contained ethanolamine (100 mM) in sodium phosphate buffer (pH, 9.0; 50 mM) for 1 hr. Slides are rinsed several times with distilled water, dried by centrifugation, and stored in a desiccator before binding experiments.

Incubation with Heparin-Binding Proteins

The utility of these heparin chips has been demonstrated by probing the carbohydrate affinity of two relevant heparin-binding growth factors, acidic FGF (FGF-1) and basic FGF (FGF-2) that are implicated in development and differentiation of several tumors (Pellegrini, 2001).

To perform the array incubation and analysis, FGF stock solutions are diluted to a concentration of 5 μg/ml with PBS buffer (pH, 7.5; 10 mM) containing BSA (1% w/v). FGF hybridization solution (60 to 80 μl) is placed between array slides and plain coverslips and incubated for 1 hr at room temperature. Slides are washed twice with PBS buffer (pH, 7.5; 10 mM) containing Tween-20 (1% v/v) and BSA (0.1% w/v), twice with Nanopure water, and then centrifuged for 5 min to ensure dryness. To detect bound FGF, arrays are incubated with anti-human FGF polyclonal antibody (5 μg/ml) and then washed as previously described. Finally, AlexaFluor-546–labeled anti-rabbit immunoglobulin G is used as secondary antibody and arrays are again washed as described previously. Heparin arrays are scanned by using a standard fluorescence reader, and spot intensities are integrated using appropriate software. The use of carbohydrate solutions at different concentrations allows for the construction of binding curves, by plotting the fluorescence intensity against the concentration. At subsaturating concentration levels, a carbohydrate binding profile can be obtained for a given protein by comparing the integrated fluorescence of different immobilized oligosaccharides.

Using this approach, we estimated the relative binding affinities of a series of heparin oligosaccharides (Fig. 5) to FGF-1 and FGF-2. Strong fluorescence signals at the tetra- and hexasaccharide positions were in agreement with previously reported data (De Paz et al., 2001; Raman et al., 2005). Interestingly, in the case of FGF-1, monosaccharide 8

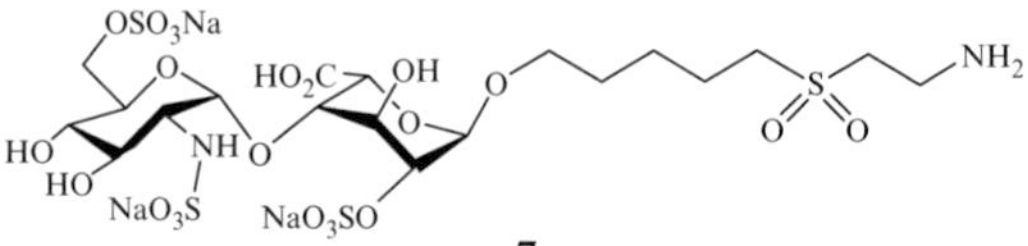

FIG. 5. A series of heparin oligosaccharides (7–11) ready for immobilization on a chip surface.

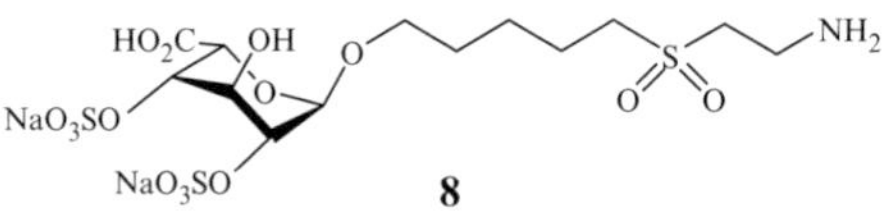

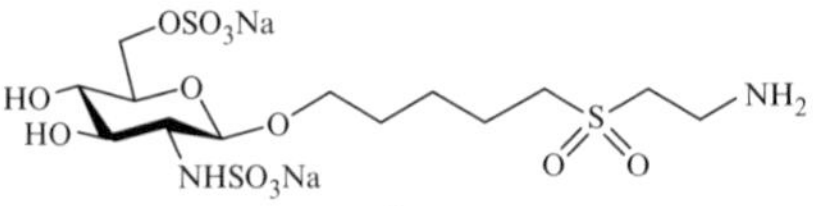

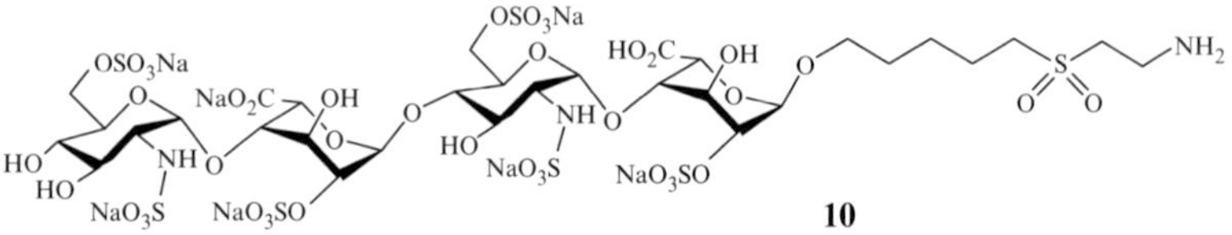

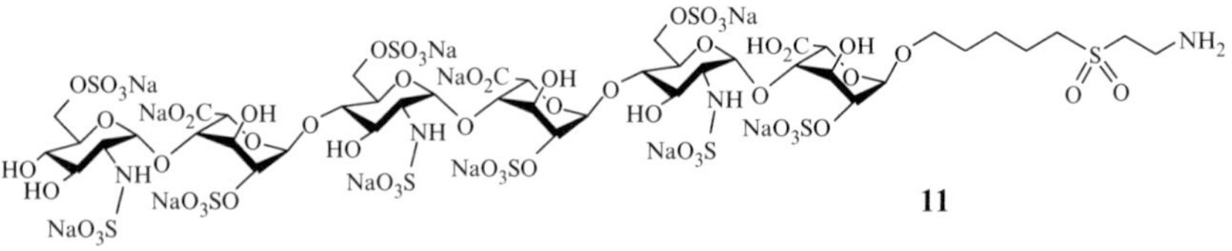

(see Fig. 5) exhibited a spot intensity comparable to that of longer oligosaccharides. The presence of a 2,4-O-sulfation pattern, not found in nature, may be responsible for this result. Indeed, compounds such as *myo*-inositol hexasulfate and sucrose octasulfate have been shown to bind FGF-1 with high affinity (Pineda-Lucena *et al.*, 1994; Zhu *et al.*, 1993). This result illustrates the possibility of employing the heparin microarrays to discover inhibitors for heparin-protein interactions.

Microarrays of Aminoglycosides

Aminoglycosides are a class of potent broad-range antibiotics. Their common structural motif contains an aminocyclitol ring with two or more amino sugars linked to it. Aminoglycosides bind to the A-site of 16S rRNA

of bacteria, thereby inhibiting bacterial protein translation (Fourmy *et al.*, 1996; Moazed and Noller, 1987). This type of antibiotics was first discovered in 1944; however, as for many other antibiotics, resistance to aminoglycosides is becoming an increasing concern (Jana and Deb, 2006; Walsh, 2000). In addition to 16S rRNA, aminoglycosides bind to many other cellular targets, including several RNA species and proteins (Morris *et al.*, 1996; Ren *et al.*, 2002; Walter *et al.*, 1999). The interactions with host proteins probably cause many of the often severe side effects known to be associated with aminoglycosides (Grovaerts *et al.*, 1990). However, the potential of aminoglycosides to interact with other specific RNA structures might lead to new application as potent drugs (Walter *et al.*, 1999).

Microarrays were used to test the binding of several interaction targets of aminoglycosides. Thereby, the advantages and drawbacks of this class of drugs can be explored in a convenient experimental setup. Aminoglycoside microarrays were used to study the interaction of aminoglycosides with the following: (1) 16S rRNA and other interesting RNA species to evaluate the microarray technique as a tool for high-throughput drug screening (Disney and Seeberger, 2004a); (2) potential cellular target proteins associated with side effects caused by aminoglycosides, thereby screening for more compatible aminoglycoside candidates (Disney and Seeberger, 2004a); and (3) bacterial enzymes modifying aminoglycosides, the most prominent mode of action leading to resistance to aminoglycosides (Disney *et al.*, 2004).

Materials and Equipment

1. Aminoglycosides (Fluka and Sigma)
2. Boc-β-Ala-Osu
3. N,N'-di(Boc)-N''-triflylguanidine
4. Trifluoroacetic acid (TFA)
5. GAPS II amine-coated slides (Corning)
6. Anhydrous DMF (Aldrich)
7. Disuccinimide (carbonate or polyethyleneglycol [PEG]) linker
8. N,N-diisopropyl amine (Aldrich)
9. BSA (Roche)
10. Ethanolamine
11. RNA/DNA synthesizer (Applied Biosystems)
12. Triisopropylsilyloxymethyl-protected RNA monomers (Glen Research)
13. Urea
14. Polyacrylamide gel (20%)
15. Milli-Q water
16. Sephadex NAP 25 prepacked column

17. Dithiothreitol
18. Spermidine
19. Nucleoside triphosphates (NTPs)
20. T7 RNA polymerase
21. Ribonuclease (RNase)-free DNase
22. Phenol
23. Chloroform
24. Sephadex G-10
25. Tween-20
26. SYBR green II nucleic acid stain (Molecular Probes)
27. Fluorescent succinimide ester (Molecular Probes)
28. Sephadex G-25
29. Glycerol
30. 2-Mercaptoethanol
31. Fluorescent slide scanner (ArrayWoRx, Applied Precision)
32. DNA array printer (MicroGrid TAS).

Stock Solutions

1. Sodium bicarbonate buffer (100 mM; pH, 8.8)
2. HEPES buffer (100 mM; pH, 7.5)
3. Tris buffer (1 M; pH, 7.5)
4. Tris/boric acid/ethylene diamine tetraacetic acid (TBE) buffer (89 mM Tris, 89 mM boric acid, 20 mM ethylenediamine tetraacetic acid [EDTA])
5. NaCl (1 M)
6. KCl (1 M)
7. MgCl$_2$ (1 M).

Preparation of Aminoglycoside-Bearing Microarrays

Aminoglycosides (Fig. 6) can be obtained from commercial sources. Neamine can be synthesized from neomycin by methanolysis. Guanidino-glycosides can be obtained from the corresponding aminoglycoside using the following protocol: Aminoglycosides are coupled with Boc-β-Ala-OSu, which serves as a linker for coupling to an amine reactive slide. Guanidiny-lation can be achieved by reacting with N,N'-di(Boc)-N''-triflylguanidine for 3 days. The reaction can be monitored by mass spectrometry. De-protection is conducted by treatment with TFA (50% in dichloromethane) (Fig. 7). All molecules can be linked to amine-reactive slides without further modification.

Several types of amine-reactive slides were tested before the experiments: aldehyde-coated slides, amine-coated slides functionalized with

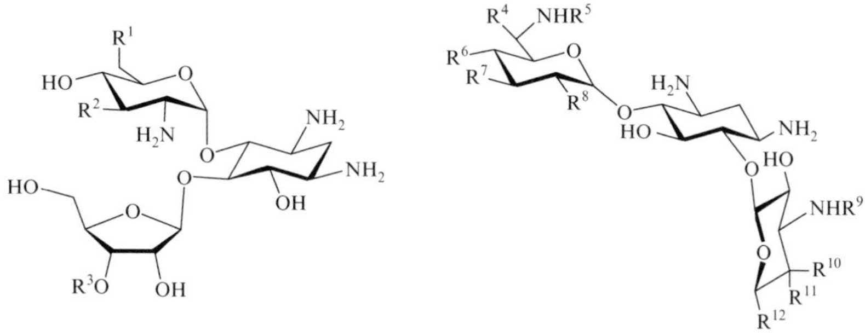

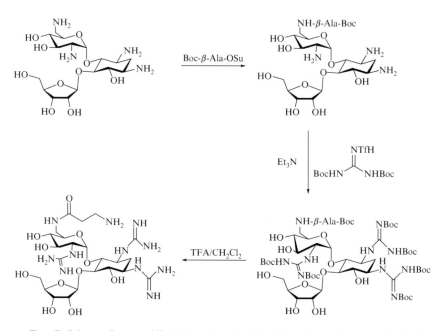

FIG. 6. Two major aminoglycoside core structures.

4,5-linked 2-deoxystreptamine derivatives

R^1 = OH, NH$_2$
R^2 = H, OH
R^3 = H, carbohydrates

4,6-linked 2-deoxystreptamine derivatives

R^4 = H, CH$_3$
R^5 = NH$_2$, NHCH$_3$
R^6 = H, OH
R^7 = H, OH
R^8 = H, OH, NH$_2$

R^9 = H, CH$_3$
R^{10} = H, OH
R^{11} = OH, CH$_3$
R^{12} = H, CH$_2$OH

FIG. 7. Scheme for guanidinylation of aminoglycosides. Boc, *t*-butoxycarbonyl; Su, succinimide; Tf, trifluoromethanesulfonyl; TFA, trifluoroacetic acid.

a disuccinimide PEG linker, and BSA-coated slides functionalized with a disuccinimide PEG linker. All types of slides work, but amine-coated slides functionalized with a disuccinimide PEG linker give the strongest signal to background.

GAPS II amine-coated slides are functionalized by submersion in DMF with disuccinimide (carbonate or PEG) linker (10 mM) and N,N-diisopropylamine (100 mM) for several hours. For coating slides with BSA, the slides treated with a disuccinimide carbonate linker are incubated for 12 hrs in sodium bicarbonate (100 mM; pH, 8.8) with BSA (1%). Afterward, slides are once again submerged in DMF with disuccinimide (carbonate or PEG) linker (10 mM) and N,N-diisopropylamine (100 mM). After each treatment, slides are washed several times with ethanol or methanol, centrifuged, and stored in a desiccator until further use.

Spots of approximately 2 nl of an aminoglycoside solution (5 mM) or derivative in DMF/H$_2$O (25% v/v) are arrayed onto glass slides using an automated arraying robot and slides are incubated in a humidity chamber overnight at room temperature. Slides are washed briefly with water to remove unreacted aminoglycosides and are quenched with ethanolamine (100 mM) and N,N-diisopropylamine (100 mM) in DMF for 3 hrs at room temperature. Then, slides are washed three times each with water and ethanol, dried by centrifugation, and stored in a desiccator.

Synthesis of RNA for Microarray Applications

RNA binding candidates were synthesized and tested for binding to various immobilized aminoglycosides (Fig. 8). Fluorescent-labeled oligo-nucleotides are synthesized on an automated RNA/DNA synthesizer using commercially obtained RNA monomers. The synthetic RNA contains triisopropylsilyloxymethyl esters as protection groups for the 2′ hydroxyl group that is de-protected according to the manufacturer's standard proto-col. After de-protection, RNA is purified by gel electrophoresis on a polyacrylamide gel (20%) with urea (8 M). The RNA band of interest is excised and the RNA extracted by stirring the gel slice in Nanopure water under RNase-free conditions (i.e., sterile). The RNA sample is desalted by chromatography on a Sephadex NAP 25 prepacked column. The RNA-containing samples are lyophylized, resuspended in sterile RNase-free water, and stored at −20°. The RNA concentration can be obtained by measuring the extinction of the RNA solution with an appropriate fluores-cent dye or other standard RNA-measuring methods. The RNA oligonu-cleotides mimicking the 16S rRNA A-site are refolded by incubation in a HEPES buffer (20 mM; pH, 7.5) with NaCl (0.2 M) for 60 min at 60° followed by slow cooling at room temperature.

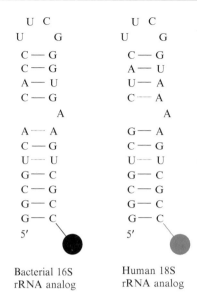

```
        U  C                    U  C
      U      G               U      G
        C — G                  C — G
        C — G                  A — U
        A — U                  U — A
        C — G                  C ···· A
              A                      A
      A ···· A               G — A
        C — G                  C — G
        U ···· U               U ···· U
        G — C                  G — C
        C — G                  C — G
        G — C                  G — C
        G — C                  G — C
      5'                       5'
```

Bacterial 16S   Human 18S
rRNA analog   rRNA analog

Fɪɢ. 8. Fluorescent-labeled 16S rRNA mimetics.

Alternatively, RNA can be synthesized by an *in vitro* transcription system, as in the case for the *Candida albicans* group I intron ribozyme (Disney *et al.*, 2001). The DNA for the corresponding RNA is cloned into a plasmid downstream of a T7 promoter. Plasmid (10 μg) bearing the transcription unit is linearized and buffered in 1 ml Tris/HCl (40 mM; pH, 7.5) with MgCl$_2$ (12 mM), dithiothreitol (10 mM), spermidine (4 mM), and NTPs (1 mM each). For transcription, T7 RNA polymerase (2000 U) is added and the mixture is incubated for 1 hr at 37°. After the reaction, RNase-free DNase (100 U) is added and the mixture is incubated for 20 min at 37°. The solution is phenol/chloroform extracted and precipitated with ethanol to purify the mixture. The precipitate is resuspended and desalted by size exclusion chromatography on a Sephadex G-10 column or by dialysis against sterile water. RNA-containing samples are collected, precipitated with ethanol, resuspended in RNase-free water, and stored at −20°. For renaturation of the *C. albicans* ribozyme, the ribozyme is heated for 10 min at 55° in HEPES buffer (50 mM; pH, 7.5) with KCl (135 mM) and MgCl$_2$ (10 mM) and cooled slowly to room temperature.

Hybridization of RNA to Aminoglycoside Microarrays

For hybridization, 10 μl of refolded rRNA mimetic (10 μM) is pipetted onto the microarray glass slide, covered with a glass cover slip,

and incubated for 1 hr at room temperature. Slides are washed with HEPES (20 mM; pH, 7.5) with NaCl (0.2 M) and optional Tween-20 (0.1%) for 5 min at room temperature, dipped into water, and dried by centrifugation. The back of the slide is cleaned by ethanol.

Hybridizing ribozymal RNA, 10 μl of refolded ribozyme (10 μM) in HEPES buffer (50 mM, pH 7.5) with KCl (135 mM) and MgCl$_2$ (10 mM) are incubated on the slide in a similar fashion. Afterward, slides are incubated in TBE buffer (89 mM Tris, 89 mM boric acid, 20 mM EDTA) with 1/10000 SYBR green II nucleic acid stain for 2 min. The stained microarrays are washed for 5 min with TBE with Tween-20 (0.1%), rinsed with water, and dried by centrifugation. Slides are scanned with a fluorescent microarray scanner and analyzed with MolecularWare software to quantify the binding intensities.

The experiments of bacterial rRNA mimetic and the *C. albicans* RNA showed specific binding of these RNAs to aminoglycosides immobilized on glass slides with little or no binding to control specimens. Comparison of the binding affinities of different aminoglycosides estimated by measuring the binding intensity to the respective target might aid the development of novel or improved antibiotics.

Hybridization of Proteins to Aminoglycoside Microarrays

Apart from RNA species, several proteins have been tested for binding to aminoglycosides. Therefore, proteins are labeled with succinimide ester bearing fluorescent dyes. In brief, 1 mg of protein is dissolved in 1 ml of sodium bicarbonate buffer (0.1 M; pH, 8.8) and 50 μl succinimide ester fluorescent dye (1 mg/50 μl) in DMF is added. The reaction is incubated for 2 hrs at room temperature and quenched by adding ethanolamine (0.1 M). The fluorescent-labeled protein is purified by size exclusion chromatography on a Sephadex G25 column. Fluorescent protein–containing fractions are pooled and stored in Tris/HCl (50 mM; pH, 7.5) and glycerol (20%) at $-20°$.

Proteins for hybridization are added to Tris/HCl (50 mM; pH, 7.5) with 2-mercaptoethanol (0.1%), NaCl (50 mM), and Tween-20 (0.01%) in an appropriate concentration. The slides are incubated with the protein hybridization solution for 1 hr at room temperature in the same fashion as for RNA hybridizations. Slides are washed with Tris/HCl (50 mM; pH, 7.5) with 2-mercaptoethanol (0.1%), NaCl (50 mM), and Tween-20 (0.1%) for 5 min at room temperature and dried. For detection and quantification, slides are scanned with a fluorescent microarray scanner and analyzed with MolecularWare software.

These binding studies showed that binding affinities of aminoglycosides for phospholipase C as well as for the Klenow fragment of the DNA

polymerase vary greatly between the derivatives. Since binding to host proteins is one of the main reasons for the side effects of aminoglycosides, this microarray system can help to select compatible aminoglycosides and hence avoid these adverse events. Interaction studies with modifying enzymes causing antibiotic resistance also yielded different affinities of aminoglycosides and their derivatives. This approach will help to develop and screen for aminoglycosides with strong binding to therapeutic targets and low affinity to resistance-causing enzymes.

Monosaccharide Microarrays to Screen Interactions of Bacteria and Carbohydrates

Animal cell surfaces are covered mainly by a layer of carbohydrates and proteins. These components mediate the interactions of the cell with the environment (Cook, 1995; Kubrycht and Sigler, 1997). Most viruses and many bacterial pathogens bind to the cell surface, many of them using carbohydrates of the cell surface as attachment sites (Corfield *et al.*, 2003; Karlsson, 1995). For several pathogens (e.g., *Helicobacter pylori, Escherichia coli*), the ability to bind carbohydrates displayed on human cells correlates at least in part with the pathogenicity of the bacteria, converting a harmless microbe into a pathogen (Hooper and Gordon, 2001).

In a novel approach, carbohydrate microarrays were used to study the interactions of *E. coli* with monosaccharides (Disney and Seeberger, 2004b). These investigations showed that *E. coli* cells bearing a mannose receptor bind specifically to mannose immobilized on microarrays. This finding can be exploited in several ways, using the binding assay to screen for attachment inhibitors that could serve as drugs, or to detect and recover the bacteria selectively in a mixture of cells. This technique constitutes a system to isolate and characterize pathogens by overcoming several of the drawbacks associated with traditional test systems for pathogens.

Materials and Equipment

1. NHS-activated slides (CodeLink, Amersham Biosciences)
2. SYTO 83 orange-fluorescent dye (Molecular Probes)
3. LB media
4. LB plates
5. Serum (Sigma)
6. Erythrocytes (Sigma)
7. Hybridization chamber (Grace Lab)
8. Fluorescent slide scanner (LS400, Tecan)
9. Automated arraying robot (Perkin Elmer).

Stock Solutions

1. Sodium phosphate buffer (300 mM; pH, 9.0)
2. Blocking buffer (50 mM sodium phosphate; 100 mM ethanolamine; pH, 9.0)
3. PBS buffer (10 mM; 137 mM NaCl; 2.7 mM KCl; pH, 7.4).

Bacterial Strains

1. *E. coli* strain ORN 178 (gift from Prof. Dr. P. Orndorf, University of North Carolina)
2. *E. coli* strain ORN 209 (gift from Prof. Dr. P. Orndorf, University of North Carolina).

Preparation of Carbohydrate Microarrays

Five different monosaccharides (12–16, Fig. 9) are synthesized bearing an ethanolamine linker at the reducing end of the carbohydrate (Disney and Seeberger, 2004b). For coupling to the slides, the derivatized mono-saccharides are solubilized in sodium phosphate buffer (50 mM; pH, 9.0).

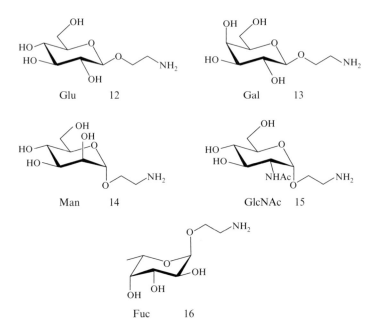

Fig. 9. Monosaccharides (12–16) for immobilization on amine-reactive slides.

By using an automated robot, 1 nl of the monosaccharides in the desired concentration is printed onto CodeLink slides. After printing, slides are sealed in a chamber with a slurry of sodium chloride leading to a constant humidity of approximately 70% and are incubated overnight for the coupling reaction. The slides are washed with water to remove uncoupled carbohydrates and unreacted groups are blocked with preheated (50°) blocking buffer for at least 30 min at 50°. After blocking, slides are washed several times with water and briefly submerged in ethanol. Finally, the slides are dried by centrifugation and stored in a desiccator until usage.

Hybridization of Bacteria with Carbohydrate Microarrays

For hybridization, E. coli cells are grown to an $OD_{660} = 1.0$ (corresponding to 10^8 cells/ml) at 37° in LB media with shaking at approximately 190 rpm. An appropriate amount of cells (i.e., 10^8 cells for binding and inhibition studies, various concentrations for detection limit determinations) are harvested by centrifugation and washed twice with PBS (pH, 7.2). The cells are resuspended in 1 ml of PBS with 50 μM SYTO 83 orange-fluorescent dye (Molecular Probes) and are shaken for 1 hr at room temperature in this solution. Dyed cells are centrifuged and washed twice with 1 ml of PBS.

The washed cells are resuspended in 800 μl PBS with $CaCl_2$ (1 mM) and $MnCl_2$ (1 mM) and are dispensed into a hybridization chamber attached to the microarray slide. For inhibition studies, inhibitors in various concentrations are added to this hybridization solution.

For the detection studies, 10^8 E. coli cells are harvested, and serum or erythrocytes are added and resuspended in PBS with $CaCl_2$ (1 mM) and $MnCl_2$ (1 mM). To this solution SYTO 83 dye is added, to a final concentration of 50 μM, and the solution is incubated for 30 min. This solution (800 μl) is applied into the hybridization chamber onto the microarrays without further purification.

For hybridization, each array with the attached and filled hybridization chamber is shaken for 1 hr at room temperature to facilitate binding of bacteria. Afterward, the chamber is removed and the arrays are dipped three times into each 50 ml of PBS with $CaCl_2$ (1 mM) and $MnCl_2$ (1 mM) and once into Nanopure H_2O. Slides are centrifuged at approximately 100×g to dry the array, and they are then analyzed with a fluorescent microarray scanner.

These studies demonstrated that as few as 10^5 E. coli cells can be detected in pure samples, and these bacteria can be still identified in mixtures with a 10-fold excess of contaminant.

To study the recovery of cells from microarray slides, E. coli bacteria (10^8 cells) are harvested, washed, stained with SYTO 62 dye, and hybridized with the microarray slides, and the slides are washed as previously described.

By scraping the positions on the array spotted with mannose with an inoculation loop and streaking the loop on a LB plate, *E. coli* bound to the slide can be transferred. After incubation of the plate at 37° overnight, colonies on the plate can be detected and used for further investigation.

These findings introduce the possibility of using carbohydrate microarrays as a detection system for pathogens by identification and isolation of suspicious bacteria that can be further examined (Mahal, 2004).

Conclusions

In summary, carbohydrate microarray experiments have clearly demonstrated the utility and advantages of microarrays. The interactions of various carbohydrates ranging from simple monosaccharides over glycosaminoglycans to complex oligosaccharide structures with a wide range of binding partners, including proteins, RNA, and even whole cells, were conveniently analyzed. The microarray system is flexible enough to investigate many different targets. Microarrays can be applied to study many aspects of binding events, such as the binding affinity of interactions partners, the inhibition of binding, and the screening of potential interaction partners. For cell interactions, the bound bacteria can be removed and used for further investigations.

In the emerging field of glycomics, microarrays have been proved to be the method of choice to study carbohydrate interactions because they can be used to investigate a broad range of possible binding partners. Furthermore, only very small amounts of carbohydrate are needed, thus overcoming the limitation of having to synthesize large amounts of complex sugars. Carbohydrates are displayed on the microarray slide, in a way mimicking the natural distribution on the cell surface. We expect the carbohydrate microarray technology to be essential to extending our current knowledge of carbohydrate functions in nature.

Acknowledgments

We thank ETH Zürich and the European Commission (Marie Curie Fellowship for J. L. P.) for financial support. We thank all present and past members of the Seeberger group and our collaborators who contributed to the results reported in this chapter.

References

Adams, E. W., Ratner, D. M., Bokesch, H. R., McMahon, J. B., O'Keefe, B. R., and Seeberger, P. H. (2004). Oligosaccharide and glycoprotein microarrays as tools in HIV glycobiology: Glycan-dependent gp120/protein interactions. *Chem. Biol.* **11,** 875–881.

Bidlingmaier, S., and Snyder, M. (2002). Carbohydrate analysis prepares to enter the "omics" era. *Chem. Biol.* **9,** 400–401.

Blixt, O., Head, S., Mondala, T., Scanlan, C., Huflejt, M. E., Alvarez, R., Bryan, M. C., Fazio, F., Calarese, D., Stevens, J., Razi, N., Stevens, D. J., Skehel, J. J., van Die, I., Burton, D. R., Wilson, I. A., Cummings, R., Wong, C.-H., and Paulson, J. C. (2004). Printed covalent glycan array for ligand profiling of diverse glycan binding proteins. *Proc. Nat. Acad. Sci. USA* **101**, 17033–17038.

Bryan, M. C., Lee, L. V., and Wong, C. H. (2004). High-throughput identification of fucosyl-transferase inhibitors using carbohydrate microarrays. *Bioorg. Med. Chem. Lett.* **14**, 3185–3188.

Calarese, D. A., Lee, H. K., Huang, C. Y., Best, M. D., Astronomo, R. D., Stanfield, R. L., Katinger, H., Burton, D. R., Wong, C. H., and Wilson, I. A. (2005). Dissection of the carbohydrate specificity of the broadly neutralizing anti–HIV-1 antibody 2G12. *Proc. Nat. Acad. Sci. USA* **102**, 13372–13377.

Capila, I., and Linhardt, R. J. (2002). Heparin-protein interactions. *Angew. Chem. Int. Ed.* **41**, 390–412.

Casu, B., and Lindahl, U. (2001). Structure and biological interactions of heparin and heparan sulfate. *Adv. Carbohydr. Chem. Biochem.* **57**, 159–256.

Conrad, H. E. (1998). "Heparin-Binding Proteins." Academic Press, San Diego, CA.

Cook, G. M. (1995). Glycobiology of the cell surface: The emergence of sugars as an important feature of the cell periphery. *Glycobiology* **5**, 449–458.

Corfield, A.P, Wiggins, R., Edwards, C., Myerscough, N., Warren, B. F., Soothill, P., Millar, M. R., and Horner, P. (2003). A sweet coating—how bacteria deal with sugars. *Adv. Exp. Med. Biol.* **535**, 3–15.

De Paz, J. L., Angulo, J., Lassaletta, J. M., Nieto, P. M., Redondo-Horcajo, M., Lozano, R. M., Gimenez-Gallego, G., and Martin-Lomas, M. (2001). The activation of fibroblast growth factors by heparin: Synthesis, structure, and biological activity of heparin-like oligosaccharides. *ChemBiochem.* **2**, 673–685.

De Paz, J. L., and Martin-Lomas, M. (2005). Synthesis and biological evaluation of a heparin-like hexasaccharide with the structural motifs for binding to FGF and FGFR. *Eur. J. Org. Chem.* **9**, 1849–1858.

De Paz, J. L., Noti, C., and Seeberger, P. H. (2006). Microarrays of synthetic heparin oligosaccharides. *J. Am. Chem. Soc.* **128**, 2766–2767.

De Paz, J. L., Ojeda, R., Reichardt, N., and Martin-Lomas, M. (2003). Some key experimental features of a modular synthesis of heparin-like oligosaccharides. *Eur. J. Org. Chem.* **17**, 3308–3324.

Disney, M. D., Haidaris, C. G., and Turner, D. H. (2001). Recognition elements for 5′exon substrate binding to the *Candida albicans* group I intron. *Biochemistry* **40**, 6507–6519.

Disney, M. D., Magnet, S., Blanchard, J. S., and Seeberger, P. H. (2004). Aminoglycoside microarrays to study antibiotic resistance. *Angew. Chem. Int. Ed.* **43**, 1591–1594.

Disney, M. D., and Seeberger, P. H. (2004a). Aminoglycoside microarrays to explore interactions of antibiotics with RNAs and proteins. *Chem. Eur. J.* **10**, 3308–3314.

Disney, M. D., and Seeberger, P. H. (2004b). The use of carbohydrate microarrays to study carbohydrate-cell interactions and to detect pathogens. *Chem. Biol.* **11**, 1701–1707.

Doores, K. J., Gamblin, D. P., and Davis, B. G. (2006). Exploring and exploiting the therapeutic potential of glycoconjugates. *Chem. Eur. J.* **12**, 656–665.

Dube, D. H., and Bertozzi, C. R. (2005). Glycans in cancer and inflammation: Potential for therapeutics and diagnostics. *Nat. Rev. Drug Discovery* **4**, 477–488.

Dwek, R. A. (1996). Glycobiology: Toward understanding the function of sugars. *Chem. Rev.* **96**, 683–720.

Feizi, T., Fazio, F., Chai, W., and Wong, C.-H. (2003b). Carbohydrate microarrays: A new set of technologies at the frontiers of glycomics. *Curr. Opin. Struct. Biol.* **13**, 637–645.

Feizi, T., and Mulloy, B. (2003a). Carbohydrates and glycoconjugates glycomics: The new era of carbohydrate biology. *Curr. Opin. Struc. Biol.* **13**, 602–604.

Fourmy, D., Recht, M. I., Blanchard, S. C., and Puglisi, J. D. (1996). Structure of the A site of *Escherichia coli* 16S ribosomal RNA complexed with an aminoglycoside antibiotic. *Science* **274**, 1367–1371.

Grovaerts, P. J., Claes, J., van de Heyning, P. H., Jorens, P. G., Marquet, J., and De Broe, M. E. (1990). Aminoglycoside-induced ototoxicity. *Toxicol. Lett.* **52**, 227–251.

Hakomori, S., and Zhang, Y. (1997). Glycosphingolipid antigens and cancer therapy. *Chem. Biol.* **4**, 97–104.

Hooper, L. V., and Gordon, J. I. (2001). Glycans as legislators of host-microbial interactions: Spanning the spectrum from symbiosis to pathogenicity. *Glycobiology* **11**, 1R–10R.

Huang, C. Y., Thayer, D. A., Chang, A. Y., Best, M. D., Hoffmann, J., Head, S., and Wong, C-H. (2006). Carbohydrate microarray for profiling the antibodies interacting with Globo H tumor antigen. *Proc. Nat. Acad. Sci. USA* **103**, 15–20.

Jana, S., and Deb, J. K. (2006). Molecular understanding of aminoglycoside action and resistance. *Appl. Microbiol. Biotechnol.* **70**, 140–150.

Karlsson, K. A. (1995). Microbial recognition of target-cell conjugates. *Curr. Opin. Struct. Biol.* **5**, 622–635.

Kiessling, L. L., and Cairo, C. W. (2002). Hitting the sweet spot. *Nat. Biotechnol.* **20**, 234–235.

Kubrycht, J., and Sigler, K. (1997). Animal membrane receptors and adhesive molecules. *Crit. Rev. Biotechnol.* **17**, 123–147.

Love, K. R., and Seeberger, P. H. (2002). Carbohydrate arrays as tools for glycomics. *Angew. Chem. Int. Ed.* **41**, 3583–3586.

Mahal, L. K. (2004). Catching bacteria with sugar. *Chem Biol.* **11**, 1602.

Manimala, J. C., Li, Z., Jain, A., VedBrat, S., and Gildersleeve, J. C. (2005). Carbohydrate array analysis of anti-Tn antibodies and lectins reveals unexpected specificities: Implications for diagnostic and vaccine development. *ChemBioChem.* **6**, 2229–2241.

Mammen, M., Choi, S. K., and Whitesides, G. M. (1998). Polyvalent interactions in biological systems: Implications for design and use of multivalent ligands and inhibitors. *Angew. Chem. Int. Ed.* **37**, 2754–2794.

Moazed, D., and Noller, H. F. (1987). Interaction of antibiotics with functional sites in 16S ribosomal RNA. *Nature* **327**, 389–394.

Morris, J. C., Ping-Sheng, L., Zhai, H. X., Shen, T. Y., and Mensa-Wilmot, K. (1996). Phosphatidylinositol phospholipase C is activated allosterically by the aminoglycoside G418: 2-deoxy-2-fluoro-scyllo-inositol-1-O-dodecylphosphonate and its analogs inhibit glycosylphosphatidylinositol phospholipase C. *J. Biol. Chem.* **271**, 15468–15477.

Nimrichter, L., Gargir, A., Gortler, M., Altstock, R. T., Shtevi, A., Weisshaus, O., Fire, E., Dotan, N., and Schnaar, R. L. (2004). Intact cell adhesion to glycan microarrays. *Glycobiology* **14**, 197–203.

Noti, C., and Seeberger, P. H. (2005). Chemical approaches to define the structure-activity relationship of heparin-like glycosaminoglycans. *Chem. Biol.* **12**, 731–756.

Orgueira, H. A., Bartolozzi, A., Schell, P., Litjens, R. E. J. N., Palmacci, E. R., and Seeberger, P. H. (2003). Modular synthesis of heparin oligosaccharides. *Chem. Eur. J.* **9**, 140–169.

Pellegrini, L. (2001). Role of heparan sulfate in fibroblast growth factor signalling: A structural view. *Curr. Opin. Struct. Biol.* **11**, 629–634.

Pineda-Lucena, A., Jimenez, M. A., Nieto, J. L., Santoro, J., Rico, M., and Giménez-Gallego, G. (1994). H-1-NMR Assignment and solution structure of human acidic fibroblast growth-factor activated by inositol hexasulfate. *J. Mol. Biol.* **242**, 81–98.

Plante, O. J., Palmacci, E. R., and Seeberger, P. H. (2001). Automated solid-phase synthesis of oligosaccharides. *Science* **291**, 1523–1527.

Raman, R., Sasisekharan, V., and Sasisekharan, R. (2005). Structural insights into biological roles of protein-glycosaminoglycan interactions. *Chem. Biol.* **12,** 267–277.

Ratner, D. M., Adams, E. W., Disney, M. D., and Seeberger, P. H. (2004a). Tools for glycomics: Mapping interactions of carbohydrates in biological systems. *Chem. Biol. Chem.* **5,** 1375–1383.

Ratner, D. M., Adams, E. W., Su, J., O'Keefe, B. R., Mrksich, M., and Seeberger, P. H. (2004b). Probing protein-carbohydrate interactions with microarrays of synthetic oligosaccharides. *Chem. Biol. Chem.* **5,** 379–383.

Ratner, D. M., Plante, O. J., and Seeberger, P. H. (2002). A linear synthesis of branched high-mannose oligosaccharides from the HIV-1 viral surface envelope glycoprotein gp120. *Eur. J. Org. Chem.* **5,** 826–833.

Ren, Y. G., Martinez, J., Kirsebom, L.A, and Virtanen, A. (2002). Inhibition of Klenow DNA polymerase and poly(A)-specific ribonuclease by aminoglycosides. *RNA* **8,** 1393–1400.

Seeberger, P. H., and Werz, D. B. (2005). Automated synthesis of oligosaccharides as a basis for drug discovery. *Nat. Rev. Drug Discovery* **4,** 751–763.

Shin, I., Park, S., and Lee, M. (2005). Carbohydrate microarrays: An advanced technology for functional studies of glycans. *Chem. Eur. J.* **11,** 2894–2901.

Thirumalapura, N. R., Morton, R. J., Ramachandran, A., and Malayer, J. R. (2005). Lipopolysaccharide microarrays for the detection of antibodies. *J. Immunol. Methods* **298,** 73–81.

Walsh, C. (2000). Molecular mechanisms that confer antibacterial drug resistance. *Nature* **406,** 775–781.

Walter, F., Vicens, Q., and Westhof, E. (1999). Aminoglycoside-RNA interactions. *Curr. Opin. Chem. Biol.* **3,** 694–704.

Werz, D. B., and Seeberger, P. H. (2005). Carbohydrates as the next frontier in pharmaceutical research. *Chem. Eur. J.* **11,** 3194–3206.

Zhu, X. T., Hsu, B. T., and Rees, D. C. (1993). Structural studies of the binding of the antiulcer drug sucrose octasulfate to acidic fibroblast growth-factor. *Structure* **1,** 27–34.

[18] Identification of Ligand Specificities for Glycan-Binding Proteins Using Glycan Arrays

By Richard A. Alvarez and Ola Blixt

Abstract

Protein–glycan interactions mediate diverse biological processes in cell communication and innate immunity. They involve the binding of a protein on one cell surface to a glycosylated protein or lipid on an opposing cell surface. Understanding the functional significance of these interactions is of major interest to the scientific community. Numerous studies have demonstrated the utility of solid-phase glycan arrays as a means of identifying

METHODS IN ENZYMOLOGY, VOL. 415   0076-6879/06 $35.00
DOI: 10.1016/S0076-6879(06)15018-1

glycan binding specificity. These approaches share a common format in that glycans are attached in some manner to a solid–phase surface and the glycan-binding protein (GBP) is presented in solution for binding analysis. Multiple options are available to investigators for detecting these interactions, but the most robust reporters are fluorescence based, and binding can be detected as relative fluorescence units. The solid-phase assays have been proved to be highly scalable and suitable for manufacturing processes. The latest innovations have led to the production of miniaturized arrays using microarray printing technology to produce glycan microarrays with the potential to spot and interrogate several thousands of unique glycans simultaneously. This chapter describes two glycan array platforms developed by the Consortium for Functional Glycomics, a microtiter plate-based array and a covalent printed array, and the analytic methods used to conduct binding specificity assays for a broad range of GBPs and organisms.

Overview

The measurement of protein–glycan binding interactions has taken on new meaning as the field of glycobiology (Taylor and Drickamer, 2003; Varki et al., 1999) and efforts to define the glycome have escalated (Bidlingmaier and Snyder, 2002; Perkel, 2002; Seeberger, 2005). Glycan-binding proteins (GBPs) and their glycan counter-receptors play important roles in mediating cell-to-cell recognition, cell adhesion, cell motility, and downstream signal processes in innate immunity and pathogen recognition (Sharon and Lis, 2004; Taylor and Drickamer, 2003). The development of high-throughput technologies to determine binding specificities is accelerating our understanding of these interactions (Disney and Seeberger, 2004; Feizi and Chai, 2004).

Of the many techniques used to measure specificity and affinity, solid-phase systems are the best adapted to high-throughput screening. A body of literature has emerged over the past few years, demonstrating the use of glycans immobilized on solid-phase substrates such as nitrocellulose, polystyrene, streptavidin, thin-layer silica (Feizi et al., 2003; Shin et al., 2005), and covalent (Angeloni et al., 2005; Bergh, et al., 2001; Blixt et al., 2004; Bryan et al., 2004; Dyukova et al., 2005; Mrksich, 2004; Park and Shin, 2002) and non-covalent platforms (Fukui et al., 2002; Ko et al., 2005; Wang et al., 2002; Willats et al., 2002). Binding specificity results of these studies have agreed with specificities previously determined in solution-based assays with a limited repertoire of glycans (Brewer et al., 1985; Gupta et al., 1996; Iglesias et al., 1982). The major advantage of the solid-phase

approach is that a broader range of glycans can be applied to and simultaneously interrogated on these platforms. As glycan libraries have expanded, new binding specificity information has begun to emerge (Blixt et al., 2004; Bochner et al., 2005; Guo et al., 2004; Palma et al., 2005; Schartz-Albiez and Kniep, 2005; Stevens et al., 2006; van Vliet et al., 2005). Thus, a bottleneck to the intrinsic value of these platforms is the production of new glycan structures to populate the array format. To fully exploit the glycan array technology, a mechanism for producing a larger and more diverse range of glycans is needed to yield biologically relevant results. The Consortium for Functional Glycomics (CFG) has addressed this need by constructing the most diverse library of compounds found on a solid-phase system (see Chapter 9, Chemoenzymatic Synthesis of Glycan Libraries). The CFG has used this extensive glycan array to generate binding specificity data for a broad range of GBPs and organisms.

Building on that understanding, the CFG has developed two generations of high-throughput arrays that are currently available to investigators. The consortium has also established two component cores to collaborate on the enablement of this technology. The Glycan Array Synthesis Core D has primary responsibility for the synthesis and printing of the glycan library that populates the CFG arrays. The Protein-Glycan Interaction Core H develops screening methods and conducts assays on the CFG arrays for projects approved by the Consortium Steering Committee. This chapter discusses the methods used in producing both the streptavidin-biotinylated plate glycan arrays and the printed covalent microarrays of the CFG and describes the methods of assay used in analyzing a broad range of lectins and organisms to determine glycan binding specificity.

The first-generation glycan array is built on 384-well microtiter plates coated with streptavidin to which a library of biotinylated glycans is attached. The second-generation glycan array is built using microarray printing technology to spot amine-reactive glycans on N-hydroxysuccinimide (NHS)-activated slides. Each array uses off-the-shelf technology and each can be analyzed using standard detections systems and software programs for microtiter and microarray applications.

Construction of the CFG Streptavidin-Biotin Glycan Array (Plate Glycan Array)

The materials needed for constructing a plate glycan array are as follows:

1. 384-well NeutrAvidin-coated microtiter plates (Pierce, Cat. No. 15513, NeutrAvidin HBC Black 384-Well Plates With SuperBlock Blocking Buffer)

2. Phosphate-buffered saline (PBS; pH, 7.4)
3. EP3 robotic liquid handling system (Perkin Elmer, optional)
4. Wallac Victor^2 1420 Multi-label Counter (Perkin Elmer, or similar fluorescence plate reader)
5. Embla 96/384 Well Washer (Molecular Devices, optional)
6. Biotinylated glycans.

Commercially produced 384-well streptavidin- and NeutrAvidin-coated microtiter plates are available from many sources (e.g., Pierce, Sigma, Greiner). Initial studies showed that streptavidin-coated plates produced excellent results but had variably high background signal. The background signal was significantly reduced using plates coated with NeutrAvidin, a deglycosylated form of streptavidin. To minimize background effect, the CFG plate array is constructed on black NeutrAvidin 384-well plates using a library of biotinylated glycans. The biotinylated glycans are produced by Core D. Each glycan is functionalized by attaching an LC-LC-Biotin (Pierce) moiety to the reducing sugar, yielding one biotin linked via spacer arm to each glycan (Blixt et al., 2003). The 384-well microtiter plates have a well volume of 100 μl and a biotin binding capacity of approximately 15 pmol biotin per well, as determined by a biotin inhibition assay. The NeutrAvidin is coated by the manufacturer to a volume of 50 μl in each well.

To construct the plate array, Core H applies an excess of biotinylated glycan (22.5 pmol) in 25 μl of PBS in replicates of n = 3 (plate array versions 1 and 2) or n = 4 (plate array version 3) to the wells. The glycans are incubated overnight at 4°, washed in PBS and stored long term at 4° in PBS with 0.05% sodium azide. The plates can also be manufactured using a robotic liquid handling system. Depending on the system used, multiple replicate plates can be manufactured by transferring each glycan from a source plate to multiple 384-well plates. The CFG system can manufacture nine sets of the plate array in one batch run of about 2 hrs. By adding excess biotinylated glycan to the capacity of the well, it is assumed to be saturated, and each well contains approximately equimolar concentrations of each glycan.

The plate-based glycan array has a number of attributes that make it especially useful for the development of assays. It can be produced manually, so sophisticated equipment is not required. A single 384-well plate can be subdivided to hold multiple assays run over time. For instance, in a quality control process used by Core H, each new glycan produced by Core D is assayed using fluorescein isothiocyanate (FITC)-labeled plant lectins to validate the expected specificity before addition to the full glycan array. The plate platform can also be used to verify binding seen in the primary screen by quickly rescreening primary hits in a mini-assay format. For this

application, biotinylated glycans can be absorbed to the streptavidin plate for as little as 1 hr before a binding assay is run.

The list of glycans present on the plate array can be found on the CFG Web site (http://www.functionalglycomics.org/static/consortium/). (*Note:* Each new version of the plate array is based on the number of 384-well plates used; that is, version one used 1 plate; version 2 used two plates, etc. As the glycan library grows, each succeeding version of the plate array retains the previous version's numbering for each glycan, with new glycans being added to the end of the list. In this way, historical data can be easily compared across time.)

Glycan Binding Assays on the CFG Plate Array

A diverse range of GBPs have been analyzed on the plate array using multiple detection strategies. These strategies are described in the section titled Fluorescence Reporters Used in Glycan Array Assays. (*Note:* Throughout this chapter, reference is made to FITC. This is meant to refer generally to any fluorescent compound that is used as a reporter for generating a signal to be read by a fluorescence reader. Investigators can use any of the many fluorescent compounds available that match the excitation and emission capabilities of their equipment.)

Buffers

1. Tris saline magnesium (TSM) Buffer (TSM): 20 mM Tris-HCl; pH, 7.4; 150 mM NaCl; 2 mM CaCl$_2$; 2 mM MgCl$_2$
2. TSM Wash Buffer (TSMW): TSM buffer + 0.05% Tween 20
3. TSM Binding Buffer (TSMBB): TSM buffer + 0.05% Tween 20 + 1% bovine serum albumin (BSA).

(*Note:* PBS can be substituted for TSM.)

Direct Binding Assays

In a direct binding assay, the lectin or organism is labeled directly with FITC and applied in a one-step incubation to the array.

1. Add 25 μl labeled lectin in TSMBB to each well and incubate 1 hr at room temperature protected from light.
2. Wash the plate three times with 100 μl TSMW using an automatic plate washer.
3. Add 25 μl TSM buffer to all wells before reading.
4. For negative controls, incubate 25 μl of FITC-labeled lectin to empty wells followed by standard washing.

5. For background controls, add 25 μl TSM buffer to empty wells after washing.
6. For positive controls, add FITC-labeled plant lectins to wells coated with known ligands and run in parallel with experimental lectin.
7. Read plate at 485/535 nm using a multimode fluorescence plate reader.

Indirect Binding Assays

Unlabeled lectins can be detected with FITC-labeled secondary antibodies that require additional incubation steps and washes.

1. After step 2 in direct binding assay procedure just described, add FITC-labeled secondary antibody in TSMBB and incubate for 20 min at room temperature.
2. Repeat with additional antibodies, as needed, by returning to step 2 in direct binding assay procedure.
3. Wash three times with 100 μl TSMW between each step.
4. Add 25 μl TSM buffer to each well and read the plate.

The optimal concentration of a lectin is based on its affinity for any given ligand in the array. Generally, a concentration of 10 to 30 μg/ml is a good starting point for binding assays in the plate array. Given the current number of wells in the CFG plate array, this requires 25 ml of a 30-μg/ml solution of a given lectin, or a total of 750 μg of protein, to run one assay. Optimal amounts of cells or organisms must be determined empirically. A good starting point is a solution containing 10^6 to 10^9 cells/ml.

Analysis of Binding Specificity on the CFG Plate Array

The average, standard deviation, and standard error of the mean for each glycan are determined for the four replicates. The signal-to-noise ratio (S/N) for each glycan is calculated by dividing the average relative fluorescence units (RFU) for each glycan by the average RFU of a control well that did not contain a glycan, but did contain the lectin and reporter. All glycans are then sorted and ranked by their S/N. Each glycan's S/N is then compared with the average S/N for the entire array (excluding positive controls) and ranked as follows: >3 times the average S/N is scored as a high affinity glycan; >2 times is scored as medium; and >1 times is scored as low (Bochner et al., 2005). The estimated working affinity range for the plate array is 100 μM or better, based on comparative studies using surface plasmon resonance (Bochner et al., 2005).

Construction of the CFG Covalent Printed Glycan Array (Printed Glycan Array)

Materials needed for constructing a printed glycan array are as follows:

1. Microarray pen-based spotting system and accessories
2. 96-well and 384-well polystyrene plates for master plate and source plates
3. Robotic liquid handling system for building source plates
4. NHS-activated slides (Schott, Nexterion Slide H)
5. Glycans with a functional amine-spacer arm.

The printed glycan array is constructed using a library of glycans and glycoproteins with an amine-reactive functionalized spacer arm (Blixt et al., 2004; see Chapter 9 of this volume). The glycans are attached to a slide coated with a proprietary matrix of polyethylene glycol containing high-density amine-reactive NHS (Schott, Nexterion Slide-H). The glycans are printed using a robotic 32-pin based spotting arrayer (Robotic LabWare Designs), which applies approximately 0.6 nl of amine-functionalized glycans in replicates of six at two concentrations (100 μM, saturating; 10 μM, subsaturating). The resulting array contains more than 4000 spotted ligands on each slide, with an upper capacity of about 10,000 spots. Because of the increased capacity, the printed array has the advantage of increasing the number of replicates and the number of glycan concentrations that can be used to analyze GBP binding specificity. By comparing GBP binding to saturated glycans with that seen to subsaturated glycans, the relative binding specificities to determine the preferred ligand can be evaluated.

For example, PAv2 glycan spots are printed in a metagrid format that contains four columns and eight rows of subgrids. Each subgrid contains eight columns and 16 rows of glycan spots. The first two columns of the metagrid are printed at 100-μM concentration; the last two columns of the metagrid are printed at 10-μM concentration. This metagrid layout is encoded into a spot-finding template within the image analysis software program and is used to analyze the results.

The CFG microarray facility can produce a batch run of 50 slides with more than 250 glycans printed in replicates of n = 6 at two concentrations in a 5-hr period. The slides retain their NHS activity during printing, which is conducted in a lint-free environmental chamber with regulated humidity. Once inactivated with buffer containing ethanolamine, the slides can be stored in a desiccated state for an indefinite period of time. All glycans incorporated onto the array are rigorously test printed at different concentrations to determine the concentration required for saturation. The test-printed glycans are also detected with a panel of lectins for verification.

Printing reproducibility is high, with a coefficient of variance of less than 1%. The CFG printed array (PA) has been produced in two public versions to date. PA version 1 contains 200 unique glycan structures; PA version 2 contains 264 structures. The list of glycans present on each can be found on the CFG Web site (http://www.functionalglycomics.org/static/consortium/).

Glycan Binding Assays on the Covalent Printed Array

The printed array uses the same buffers as the plate array. Each slide should be marked around the perimeter of the printed area because the glycans are not visible on the slide. They can be visualized by breathing across the slide to produce a fog that can reveal the printed grids (a halitosis array). Each new printed glycan array version is accompanied with an industry standard GAL file, which is a .txt file that lists the address of each spot and glycan on the array. The gridfile is used by image analysis software programs to assign the glycan structure to each spot during the analysis. Slides should be handled with powder-free gloves, and precautions should be taken to avoid touching the printed area of the slide.

The image of a GBP binding result may reveal that only a few ligands are bound specifically in the array. This can make placing the analysis template a challenge. To facilitate placement in the post-run analysis, the CFG has developed two approaches. Biotin spots can be incorporated during printing of the array into three of the four corners of each subarray and detected by incubating with streptavidin-FITC (0.4 ug/ml) after each experimental run. This will light up the corners of each grid and facilitate placement of the analysis template. An alternative approach is to incubate the array with a cocktail of FITC-labeled lectins, specific for terminal monosaccharides on the array, after the experimental sample has been captured (EY Laboratories, Inc.; Vector Laboratories). This will light up approximately 80% of the glycans on the array and reveal the printed layout of the subgrids for placement of the analysis template.

Direct Binding Assays

1. FITC-labeled lectin is diluted to assay concentration in 60 μl TSMBB and spun in microcentrifuge at maximum speed for 10 min at room temperature. In addition, the lectin can be applied in larger volumes, as described in the section titled Alternative Assay Conditions.
2. Fifty microliters of the labeled lectin is applied to the printed surface in a drop at the edge of the printed area.
3. A cover slip is applied, and the solution is spread over the entire printed area. (If bubbles form, they can usually be removed with gentle pressure applied to the coverslip.)

4. The slide is incubated, protected from light, in a humidified chamber for 1 hr at room temperature (a simple chamber can be made using a Petri dish filled with enough deionized water to cover the bottom and two pipets for elevating the slide).
5. The coverslip is removed and the slide washed by dipping four times each in TSMW, TSM, and deionized water.
6. The slide is spun dry using a tabletop slide centrifuge (Spectrafuge Mini, Labnet). The slide can also be dried under a gentle stream of air or nitrogen.
7. The image is read in a Perkin Elmer Microscanarray XL4000 scanner or other comparable microarray slide reader at em485/ex530.
8. Raw data results are generated in Excel format.

In the CFG process, the ImaGene data is uploaded to the database as .txt files, and database-driven presentation tools manage the output presented on the CFG Web site. All data are posted according to Consortium policy.

Indirect Binding Assays

Using a primary antibody—an antibody to a fusion tag or immunoglobulin chimera—will require additional incubations and washes for each.

1. After step 3 in direct binding assay procedure just described, the slide is washed and spun dry.
2. The secondary antibody is diluted into 60 μl TSMBB and spun in microcentrifuge at maximum speed for 10 min at room temperature
3. Fifty microliters of the supernatant is applied and a coverslip applied. The sample is incubated for 20 min at room temperature.
4. Proceed to step 4 in direct binding assay procedure.

Analyzing the Covalent Printed Array Results

To generate images with optimal resolution, the microarray scanning parameters are set to 10 μm resolution (about one tenth the average spot size of 100 μm); the appropriate fluorophore is chosen, and the scanning area is set to include the entire printed area on the slide. The photomultiplier gain is set initially to 90%, the laser power is set initially to 90%; the laser power is fixed, and the photomultiplier gain is allowed to vary. This will result in an uncompressed tiff image file of approximately 17 mb for the current PAv2 of the covalent printed array. Once the uncompressed tiff image has been collected, image analysis is performed using ImaGene

(V.6) image analysis software (GenePix, Molecular Devices, is another option). In ImaGene, a spot-finding template is applied to the image that defines the position and the spot area for each glycan is printed on the array. A gridfile (GAL file) is also used to assign a glycan to each address in the array. The ImaGene software uses these two files and a spot-finding algorithm to measure the signal for each spot on the array. The results are generated in a .txt tile. This file can be converted to an Excel workbook, and a macro can be generated to sort the results and determine the mean, standard deviation, and standard error of the mean for each glycan in the array. The processed results are presented as a table; a graph is generated, and the data can be sorted by average RFU to rank glycan hits.

Alternative Assay Conditions

The printed array platform conditions can be adjusted to suit the needs for screening lectins and organisms that do not fit the standard assay protocol.

Valency

Low-affinity interactions may require multivalent strategies in presentation of the lectin to the array (Blixt *et al.*, 2004). This effect can be enhanced by sequentially preincubating with multiple cross-species antibodies to build a multivalent presentation. Through the process of incubating with a series of secondary antibodies, a complex with multiple copies of the carbohydrate-binding domain can be assembled. By using FITC-labeled versions of each antibody, the overall signal can be enhanced.

Assays of Cells and Organisms

The same buffers and general incubation times and washes can be used to screen organisms on the array. They can be directly labeled using passive fluorescent dyes or detected using antibodies specific to proteins on the organism. Instead of applying a coverslip, the printed area can be encircled using a resistance pen (ImmunoEdge Pen, Vector). After rehydrating the slide, the edges of the printed area are marked with the pen and allowed to dry for 2 to 3 min. This will allow up to 1 ml of buffer (with Tween) to be pipetted onto the printed area. The slide can be gently rocked on a platform rocker during incubation. After the primary incubation, the slide can be washed as usual, and if a secondary antibody is used, it can be coverslipped as for the standard lectin-binding protocol.

An alternative approach is to place the slide face down in a Petri dish and then pipet the sample onto the floor of the dish. The sample will wick under the slide and come into contact with the printed area. The slide can be washed with successive changes of buffer in the Petri dish.

Other Detection Methods and Considerations

Detection can be achieved using GFP-expressing cells or organisms and passive fluorescent labeling dyes. The result can also be viewed under bright field and fluorescence microscopy directly on the slide to visualize the organism binding to spots and to capture images.

Fluorescence Reporters Used in Glycan Array Assays

Both array platforms lend themselves to multiple detection methods for analyzing binding specificity. Direct labeling of proteins or organisms via primary amine groups can be achieved with NHS ester labeling kits available from a number of commercial sources using the manufacturer's protocol

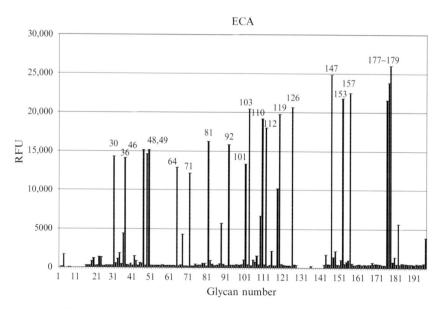

FIG. 1. Binding profile of *Erythrina cristagalli* agglutinin (ECA) assayed on the Consortium for Functional Glycomics plate glycan array. ECA was applied at a concentration of 30 μg/ml using the methods described in this chapter. RFU, relative fluorescence units.

TABLE I
THE BINDING SPECIFICITY OF *Erythrina cristagalli* AGGLUTININ (ECA) DETERMINED ON THE CFG PLATE GLYCAN
ARRAY AND RANKED BY S/N AS DESCRIBED IN THE METHODS

Glycan no.	Glycan	S/N	Rank
165	mixed biantennary with 2 galactose, with core α6Fucose (IgG pronase digest)	81.80	high
133	biantennary with 2 galactose (Asialo-Fibrinogen)	78.10	high
164	mixed biantennary with 1 galactose, with core α6Fuc (IgG pronase digest)	74.94	high
143	Galβ4GlcNAcβ6(Galβ 3)GalNAcα	70.75	high
139	Galβ4GlcNAcβ3GalNAcα	68.53	high
163	mixed biantennary with 0,1,2 galactose, with core α6Fuc (IgG pronase digest)	67.74	high
112	GalNAcβ4GlcNAcβ4Manα3(GalNAcβ4GlcNAcβ4Manα6)Manα4GlcNAcβ4GlcNAcβ-N	64.96	high
89	(Galb1-4GlcNAcβ)₂-3,6-GalNAcα	64.15	high
105	Fucα2Galβ4GlcNAcβ3Galβ4GlcNAc	61.98	high
96	Galβ4GlcNAcβ#PAA	60.34	high
98	GalNAcβ4GlcNAc#PAA	56.47	high
67	Galβ4(6-O-Su)GlcNAcβ	51.07	high
78	Galβ4GlcNAcβ6GalNAcα	49.84	high
32	Galβ4GlcNAcβ3Galβ4Glcβ	47.73	high
35	Galβ4GlcNAcβ3Galβ4GlcNAcβ 3Galβ4GlcNAcβ	47.58	high
34	Galβ4GlcNAcβ3Galβ4GlcNAcβ	45.82	high
16	Galβ4Glcβ	44.91	high
22	GalNAcβ4GlcNAcβ	44.35	high
87	Fucα2Galβ4GlcNAcβ	41.94	high
50	Galβ3(Galβ4GlcNAcβ6)GalNAcα1:T.P1	40.46	high
57	Galβ4GlcNAcβ6(Neu5Acα3Galβ3)GalNAcα1:T.P1	38.31	high
104	Fucα2Galβ4Glcβ	31.93	high
95	Galβ4Glcβ1#PAA	20.95	medium
74	Fucα2Galβ	18.04	low
169	Galβ3GlcNAcβ3Galβ4Glcβ	17.74	low
21	Galβ4GlcNAcβ	13.81	low
53	Galβ3(Neu5Acα3Galβ4GlcNAcβ6)GalNAcα1:T.P1	13.55	low
184	(Galβ1-4GlcNAcβ1-2Man)2-Man-GlcNAcβ1-4GlcNAcβ1	12.18	low

(Invitrogen Alexa Fluor 488 Protein Labeling Kit A-10235; Pierce, 46100). A variation of this approach is to label the protein with NHS-biotin and then incubate the biotinylated protein with streptavidin-FITC (Pierce, 212223; Invitrogen Molecular Probes, SA100-02). Similarly, FITC-labeling to sulfhydryl groups can be achieved using a commercially available thiol-reactive maleimide-labeling kits.

Indirect detection methods using primary antibodies and FITC-labeled secondary antibodies can be used. Primary antibodies generated by investigators from a variety of species, or from hybridomas, can be used with FITC-labeled secondary antibodies or with precomplexing methods. As a precaution, any primary antibody should be run against the array in a separate assay to determine whether there is any antibody-specific binding to the array. FITC-labeled antibodies are available from many manufacturers (Molecular Probes, Abcam, Bethyl Labs).

Proteins can be expressed with fusion tags (His, GST, Flag, MyC), which can be detected by the corresponding tag-specific antibody. Serum can be screened on the glycan array using FITC-labeled goat anti-human immunoglobulin (Ig) G, IgM, IgA, or combinations thereof, including subtype-specific immunoglobulins. In general, proteins can be screened

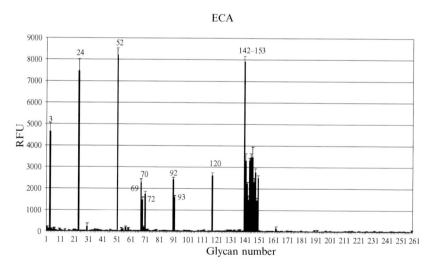

Fig. 2. Binding profile of *Erythrina cristagalli* agglutinin (ECA) assayed on the Consortium for Functional Glycomics printed glycan array with glycans printed at 100 μM concentration. ECA was applied at a concentration of 200 $\mu g/ml$ using the methods described in this chapter. RFU, relative fluorescence units.

on the printed array at concentrations of 200 μg/ml or less. Antibodies can be screened at concentrations ranging from 1 to 10 μg/ml for both platforms. If the concentration is not known, dilutions ranging from 1:25 to 1:100 are a good starting point for optimizing assay conditions.

Cells and organisms can be detected using passive fluorescent-labeling techniques. They can be stained with cell-permeant dyes that stain nucleic acids (Invitrogen, SYTO dyes, S32707). Green fluorescent protein-expressing cells and proteins can be detected using the direct binding methods described in this chapter.

TABLE II

THE BINDING SPECIFICITY OF *ERYTHRINA CRISTAGALLI* AGGLUTININ (ECA) DETERMINED ON THE CFG PRINTED GLYCAN ARRAY AND RANKED BY AVERAGE RFU AS DESCRIBED IN THE METHODS

Glycan no.	Glycan name	Avg. RFU	SEM
52	Galβ1-4GlcNAcβ1-2Manα1-3 (Galβ1-4GlcNAcβ1-2Manα1-6) Manβ1-4GlcNAcβ1-4GlcNAcβ-G	8217	323
143	Galβ1-4GlcNAcβ1-3 (Galβ1-4GlcNAcβ1-6)GalNAcα	7910	267
24	(Galβ1-4GlcNAcβ)₂-3,6-GalNAcα	7465	551
3	AGP-B (AGP ConA bound)	4639	398
149	Galβ1-4GlcNAcβ1-3Galβ1-4Glcβ	3463	477
148	Galβ1-4GlcNAcβ1-3Galβ1-4Glcβ	3454	187
144	Galβ1-4GlcNAcβ1-3GalNAcα	3298	355
147	Galβ1-4GlcNAcβ1-3Galβ1-4GlcNAcβ	3290	141
151	Galβ1-4GlcNAcβ1-6GalNAcα	2756	193
120	Galβ1-3(Galβ1-4GlcNAcβ1-6)GalNAcα	2594	151
153	Galβ1-4GlcNAcβ	2492	130
92	GalNAcβ1-4GlcNAcβ	2403	125
150	Galβ1-4GlcNAcβ1-6(Galβ1-3)GalNAcα	2305	172
69	Fucα1-2Galβ1-4GlcNAcβ1-3 Galβ1-4GlcNAc	2272	184
145	Galβ1-4GlcNAcβ1-3Galβ1-4(Fucα1-3) GlcNAcβ1-3Galβ1-4(Fucα1- 3)GlcNAcβ	2224	78
72	Fucα1-2Galβ1-4GlcNAcβ	1730	128
93	GalNAcβ1-4GlcNAcβ	1583	99
146	Galβ1-4GlcNAcβ1-3Galβ1-4 GlcNAcβ1-3Galβ1-4GlcNAcβ	1469	204
70	Fucα1-2Galβ1-4GlcNAcβ1-3 Galβ1-4GlcNAcβ1-3Galβ1-4GlcNAcβ	1465	131
152	Galβ1-4GlcNAcβ	1452	129

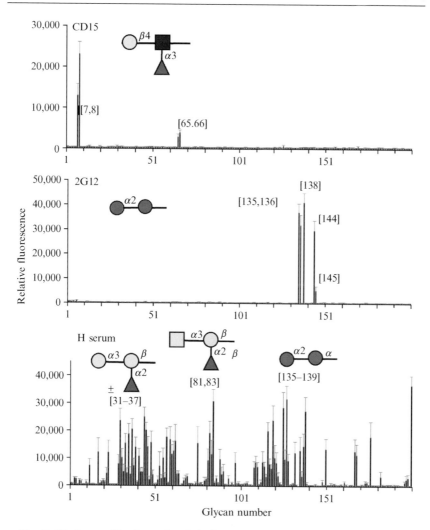

FIG. 3. Binding profiles for anti-carbohydrate antibodies assayed on the CFG printed glycan array. CD-15 (BD Biosciences clone H198, 30 μg/ml), 2G12 mAb precomplexed with goat anti-human IgG-FITC (15 μg/ml), and human serum (heat inactivated, 1:25 dilution) were applied to the CFG printed glycan array using the methods described in this chapter.

Example Data Output From the Plate and Printed Glycan Arrays

Erythrina cristagalli agglutinin (ECA) bound to glycans with terminal galactose, *N*-acetyllactosamine, and 2'fucosyllactosamine as expected

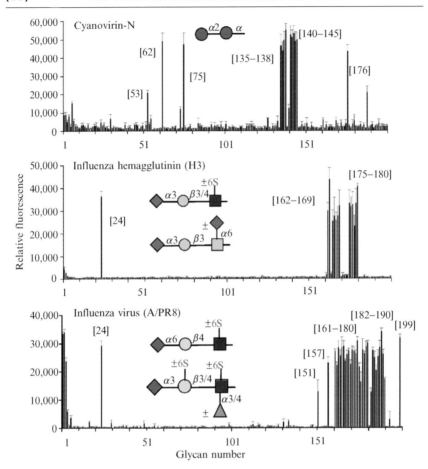

FIG. 4. Binding profiles for bacterial and viral glycan-binding proteins assayed on the CFG printed glycan array. Cyanovirin-N (30 μg/ml) detected with rabbit anti–cyanovirin-N (10 μg/ml) and goat anti-rabbit IgG-FITC (10 μg/ml), influenza recombinant hemagglutinin from duck/Ukraine/1/63(H3/N7) (150 μg/ml) precomplexed with mouse anti-HisTag IgG-Alexa488 (75 μg/ml) and anti-mouse IgG-Alexa488(35 μg/ml), and intact influenza virus A/Puerto Rico/8/34 (H1N1) (100 μg/ml, in the presence of 10 μM oseltamivir carboxylate) were applied to the CFG printed glycan array using the methods described in this chapter.

(Iglesias *et al.*, 1982) (Fig. 1; the corresponding processed results and glycan binding specificity for ECA are shown in Table I).

The binding of ECA on the CFG covalent printed array version 2 showed that binding was specific for glycans with terminal galactose,

N-acetyllactosamine, and 2'fucosyllactosamine (Fig. 2). Table II shows the corresponding results with glycans sorted by average RFU.

Binding of anti-carbohydrate antibodies (Fig. 3) (Blixt *et al.*, 2004, reprinted here with permission from Dr. Blixt and the *Proceedings of the National Academy of Sciences*) including anti-CD15, known to recognize the Lewis x antigen (Galβ1-4(Fucα1-3)GlcNac); anti-2G12, an anti-HIV monoclonal antibody known to bind high-mannose–type *N*-linked glycans on the major envelope protein of HIV gp120; and human serum, which exhibits a broad specificity to αGal/GalNAc, blood group antigens, and mannose fragments.

Cyanovirin-N was shown to bind glycans with terminal Manα1-2 residues as well as high-mannose structures with Manα1-2 termini in agreement with its known ability to block HIV-1 binding to gp120 (Blixt *et al.*, 2004, reprinted here with permission from Dr. Blixt and the PNAS). The bottom panels of Fig. 4 show binding of influenza recombinant hemagglutinin (H3) and whole influenza virus (A/PR8), respectively, to sialosides with differing specificities.

The CFG glycan arrays have been used to determine the binding specificity profiles for numerous samples, including GBPs, human sera, antibodies, viruses, fungal cells, sperm, bacteria, peptides, and others. All data generated for samples run on the CFG glycan arrays can be viewed on the CFG Web site. To learn more about screening samples on the CFG glycan array, please visit the Web site (http://www.functionalglycomics.org/static/consortium/).

References

Angeloni, S., Ridet, J. L., Kusy, N., Gao, H., Crevoisier, F., Guinchard, S., Kochhar, S., Sigrist, H., and Sprenger, N. (2005). Glycoprofiling with micro-arrays of glycoconjugates and lectins. *Glycobiology* **15**, 31–41.

Bergh, A., Magnusson, B. G., Ohlsson, J., Wellmar, U., and Nilsson, U. J. (2001). Didecyl squarate: A practical amino-reactive cross-linking reagent for neoglycoconjugate synthesis. *Glycoconj. J.* **18**, 615–621.

Bidlingmaier, S., and Snyder, M. (2002). Carbohydrate analysis prepares to enter the "Omics" era. *Chem. Biol.* **9**, 443–454.

Blixt, O., Collins, B. E., van den Nieuwenhof, I. M., Crocker, P. R., and Paulson, J. C. (2003). Sialoside specificity of the siglec family assessed using novel multivalent probes: Identification of potent inhibitors of myelin-associated glycoprotein. *J. Biol. Chem.* **278**, 31007–31019.

Blixt, O., Head, S., Mondala, T., Scanlan, C., Huflejt, M. E., Alvarez, R., Bryan, M. C., Fazio, F., Calarese, D., Stevens, J., Razi, N., Stevens, D. J., Skehel, J. J., van Die, I., Burton, D. R., Wilson, I. A., Cummings, R., Bovin, N., Wong, C. H., and Paulson, J. C. (2004).

Printed covalent glycan array for ligand profiling of diverse glycan binding proteins. *Proc. Natl. Acad. Sci.* **101,** 17033–17038.

Bochner, B. S., Alvarez, R. A., Mehta, P., Bovin, N. V., Blixt, O., White, J. R., and Schnaar, R. L. (2005). Glycan array screening reveals a candidate ligand for Siglec-8. *J. Biol. Chem.* **280,** 4307–4312.

Brewer, F., Bhattacharyya, L., Brown, R. D., and Koenig, S. H. (1985). Interactions of concanavalin A with a trimannosyl oligosaccharide fragment of complex and high mannose type glycopeptides. *Biochem. Biophys. Res. Commun.* **127,** 1066–10671.

Bryan, M. C., Fazio, F., Lee, H. K., Huang, C. Y., Chang, A. Y., Best, M. D., Calarese, D. A., Blixt, O., Paulson, J. C., and Burton, D. R. (2004). Covalent display of oligosaccharide arrays in microtiter plates. *J. Am. Chem. Soc.* **126,** 8640–8641.

Disney, M. D., and Seeberger, P. H. (2004). The use of carbohydrate microarrays to study carbohydrate-cell interactions and to detect pathogens. *Chem. Biol.* **11,** 1701–1707.

Dyukova, V. I., Dementieva, E. I., Zubtsov, D. A., Galanina, O. E., Bovin, N. V., and Rubina, A. Y. (2005). Hydrogel glycan microarrays. *Anal Biochem.* **347,** 94–105.

Feizi, T., Fazio, F., Wengang, C., and Wong, C.-H. (2003). Carbohydrate microarrays: A new set of technologies at the frontiers of glycomics. *Curr. Opin. Struct. Biol.* **13,** 637–645.

Fukui, S., Feizi, T., Galustian, C., Lawson, A. M., and Chai, W. (2002). Oligosaccharide microarrays for high-throughput detection and specificity assignments of carbohydrate-protein interactions. *Nat. Biotechnol.* **20,** 1011–1017.

Feizi, T., and Chai, W. (2004). Oligosaccharide microarrays to decipher the glyco code [review]. *Nat. Rev. Mol. Cell Biol.* **5,** 582–588.

Guo, Y., Feinberg, H., Conroy, E., Mitchell, D. A., Alvarez, R., Blixt, O., Taylor, M. E., Weis, W. I., and Drickamer, K. (2004). Structural basis for distinct ligand-binding and targeting properties of the receptors DC-SIGN and DC-SIGNR. *Nat. Struct. Mol. Biol.* **11,** 591–598.

Gupta, D., Kaltner, H., Dong, S., Gabius, H. J., and Brewer, C. F. (1996). A comparison of the fine saccharide-binding specificity of Dioclea grandiflora lectin and concanavalin A. *Eur. J. Biochem.* **242,** 320–326.

Iglesias, J. L., Lis, H., and Sharon, N. (1982). Purification and properties of a D-galactose/ N-acetyl-D-galactosamine–specific lectin from. *Erythrina cristagalli. Eur. J. Biochem.* **123,** 247–252.

Ko, K. S., Jaipuri, F. A., and Pohl, N. L. (2005). Fluorous-based carbohydrate microarrays. *J. Am. Chem. Soc.* **127,** 13162–13163.

Mrksich, M. (2004). An early taste of functional glycomics. *Chem. Biol.* **11,** 875–881.

Palma, A. S., Feizi, T., Zhang, Y., Stoll, M. S., Lawson, A. M., Diaz-Rodreguez, E., Campanero-Rhodes, M. A., Costa, J., Gordon, S., Brown, G. D., and Chai, W. (2005). Ligands for the beta-glucan receptor, Dectin-1, assigned using "designer" microarrays of oligosaccharide probes (neoglycolipids) generated from glucan polysaccharides. *J. Biol. Chem.* **21,** 5771–5779.

Park, S., and Shin, I. (2002). Fabrication of carbohydrate chips for studying protein-carbohydrate interactions. *Angew. Chem. Int. Ed.* **41,** 3180–3182.

Perkel J. M. (2002) Glycobiology goes to the ball. *The Scientist.* **16,** 32.

Schartz-Albiez, R., and Kniep, B. (2005). Perspectives for establishment and reorganization of carbohydrate-directed CD antibodies: Report of the carbohydrate section. *Cell Immunol.* **236,** 48–50.

Seeberger, P. H. (2005). Exploring life's sweet spot. *Nature.* **437,** 1239.

Sharon, N., and Lis, H. (2004). *In* "Lectins" (N. Sharon and H. Lis, eds.), 2nd ed., p. 470. Kluwer Academic Publishers, Dordrecht.

Shin, I., Park, S., and Lee, M. R. (2005). Carbohydrate microarrays: An advanced technology for functional studies of glycans. *Chemistry* **6,** 2894–2901.

Stevens, J., Blixt, O., Glaser, L., Taubenberger, J. K., Palese, P., Paulson, J. C., and Wilson, I. A. (2006). Glycan microarray analysis of the hemagglutinins from modern and pandemic influenza viruses reveals different receptor specificities. *J. Mol. Biol.* **355,** 1143–1155.

Taylor, M. E., and Drickamer, K. (2003). "Introduction to Glycobiology." Oxford University Press, Oxford.

van Vliet, S. J., van Liempt, E., Saeland, E., Aarnoudse, C. A., Appelmelk, B., Irimura, T., Geijtenbeek, T. B., Blixt, O., Alvarez, R., van Die, I., and van Kooyk, Y. (2005). Carbohydrate profiling reveals a distinctive role for the C-type lectin MGL in the recognition of helminth parasites and tumor antigens by dendritic cells. *Int. Immunol.* **17,** 661–669.

Varki, A., Cummings, R., Esko, J., Freeze, H., Hart, G., and Marth, J. (1999). "Essentials of Glycobiology." Cold Spring Harbor Press, Plainview, NY.

Wang, D., Liu, S., Trummer, B. J., Deng, C., and Wang, A. (2002). Carbohydrate microarrays for the recognition of cross-reactive molecular markers of microbes and host cells. *Nat. Biotechnol.* **20,** 275–281.

Willats, W. G. T., Rasmussen, S. E., Kristensen, T., Mikkelsen, J. D., and Knox, J. P. (2002). Sugar-coated microarrays: A novel slide surface for the high-throughput analysis of glycans. *Proteomics* **2,** 1666–1671.

[19]   High-Throughput Analysis of Lectin-Oligosaccharide
       Interactions by Automated Frontal
       Affinity Chromatography

By Sachiko Nakamura-Tsuruta,
Noboru Uchiyama, and Jun Hirabayashi

Abstract

Frontal affinity chromatography (FAC) is a quantitative method that enables sensitive and reproducible measurements of interactions between lectins and oligosaccharides. The method is suitable even for the measurement of low-affinity interactions and is based on a simple procedure and a clear principle. To achieve high-throughput and efficient analysis, an automated FAC system was developed. The system designated FAC-1 consists of two isocratic pumps, an autosampler, and a couple of miniature columns (bed volume, 31.4 μl) connected in parallel to either a fluorescence or an ultraviolet detector. By use of this parallel-column system, the time required for each analysis was reduced substantially. Under the established conditions, fewer than 10 hrs are required for 100 interaction analyses, consuming as little as 1 pmol pyridylaminated oligosaccharide for each analysis. This strategy for FAC should contribute to the construction of a lectin–oligosaccharide interaction database essential for future glycomics. Overall features and practical protocols for interaction analyses using FAC-1 are described.

Introduction

Substantial efforts have been made so far to understand some basic aspects of the sugar-binding specificities of carbohydrate-binding proteins (i.e., lectins) using various methods (Sharon and Lis, 2003; Van Damme et al., 1998). Extensive information has been accumulated about such sugar-binding specificities. However, in comparison with antibodies, lectins show much more diverse and lower binding affinity (e.g., a dissociation constant, K_d, of 10^{-3} to 10^{-6} M). It is therefore difficult to obtain precise affinity constants for lectins in a systematic manner, meaning that an appropriate system for quantitative analysis between lectins and glycans, which enables precise determination of even low-affinity constants (e.g., $>10^{-4}$ M), is thus desirable.

METHODS IN ENZYMOLOGY, VOL. 415   0076-6879/06 $35.00
                   DOI: 10.1016/S0076-6879(06)15019-3

Frontal affinity chromatography (FAC) is a quantitative affinity chromatography, which achieves sensitive and reproducible analyses even for low-affinity ($K_a < 10^4$ M^{-1}) interactions. As described previously (Hirabayashi et al., 2003), the principle is simple and clear. The system has been improved by incorporating high-performance liquid chromatography (HPLC) and pyridylaminated (PA) oligosaccharides for sensitive detection (Arata et al., 2001a; Hirabayashi et al., 2000). The method was originally developed by Kasai et al. in the 1970s (Kasai et al., 1986) and reinforced by Hirabayashi et al. (2002, 2003), who demonstrated that it was a powerful and comprehensive tool for the analysis of interactions between galectin and most commercially available fluorescently labeled glycans (Hirabayashi et al., 2002). Several tens of K_d values per day could be determined using a manual-injection system (Hirabayashi et al., 2003). To achieve improvements in terms of throughput and feasibility toward construction of a much larger database of lectin–oligosaccharide interactions (i.e., glycomics), an automated instrument for FAC was developed. This enabled us to analyze lectin–oligosaccharide interactions in a high-throughput manner (>100 analyses per day) (Hirabayashi, 2004; Nakamura et al., 2005). This chapter describes the overall features and protocols for interaction analysis using the automated FAC system, FAC-1.

Principle of FAC

Details of the principles of FAC are described in Hirabayashi et al. (2003). An excess volume of diluted fluorescently labeled glycan is continuously applied to a lectin-immobilized column (Fig. 1A). When the amount of the applied glycan exceeds the capacity of the column, leakage occurs, as shown by the appearance of an elution front at some elution point, and the concentration of glycan in the elution finally reaches a "plateau," where the concentration is equal to that of the initial solution (Fig. 1B). Glycans having some affinity for the immobilized lectin (see Fig. 1A, right) will be retained by the immobilized lectin, and their elution will be retarded, as shown in curve II (see Fig. 1B). The volume of the elution front (V) of each oligosaccharide is calculated according to the method originally described by Arata et al. (2001b). Retardation of the elution front relative to that of an appropriate standard glycan (see Fig. 1A, left, and B, curve I), that is, $V–V_0$, is then determined. K_d values for dissociation of lectin and glycans are obtained from $V–V_0$ and B_t, according to the basic equation of FAC, equation (1), where B_t is the effective ligand content (expressed in mol), and $[A]_0$ is the initial concentration of glycan. Equation (1) can be simplified

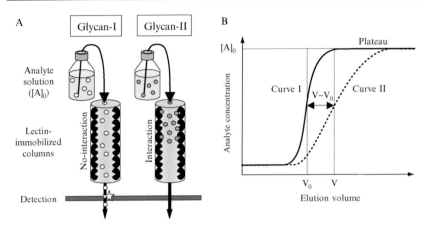

FIG. 1. Scheme of frontal affinity chromatography. (A) Analyte solution, the initial concentration of which is represented by $[A]_0$, is continuously applied to a lectin-immobilized column. The elution front of an analyte that does not interact with the immobilized lectin (glycan-I) is observed immediately, whereas that of an analyte that specifically interacts with the immobilized lectin (glycan-II) is retarded depending on the strength of the interaction. (B) The solid line indicates the elution profile of glycan-I, showing no interaction with the immobilized lectin; the dotted line, glycan-II, shows significant affinity for the immobilized lectin. The elution volumes of glycan-I and glycan-II are designated V_0 and V, respectively. When $[A]_0$ is negligibly small compared with K_d, the $V-V_0$ value is in inverse proportion to K_d.

to equation (2), where $[A]_0$ (e.g., $<10^{-8}$ M) is negligibly small compared with K_d (e.g., $>10^{-6}$ M).

$$K_d = B_t/(V - V_0) - [A]_0 \qquad (1)$$
$$K_d = B_t/(V - V_0), \text{if } K_d \gg [A]_0 \qquad (2)$$

For the determination of effective ligand content, B_t, "concentration-dependence analysis" and subsequent Woolf-Hofstee type plots are performed (Hirabayashi *et al.*, 2003), where various concentrations of labeled glycans (e.g., p-nitrophenyl [pNP] glycans) are applied to a lectin-immobilized column, and $V-V_0$ values are calculated. B_t and K_d values are determined from the intercept and the slope of Woolf-Hofstee–type plots, i.e., $(V-V_0)$ and $(V-V_0)[A]_0$, respectively.

By using B_t obtained by the previously mentioned "concentration dependence analysis" and $V-V_0$ values, K_d values are determined according to either equation (1) or equation (2).

Experimental

Automated FAC System

An automated system for FAC, designated "FAC-1" was developed by collaboration with Shimadzu (Kyoto, Japan). The system was equipped with a fluorescence (Shimadzu, RF10AXL) or ultraviolet light (UV) detector (Shimadzu, SPD-10A VP) and a personal computer (PC) workstation loaded with "LCsolution" software (Hirabayashi *et al.*, 2004; Nakamura *et al.*, 2005). A scheme for the total procedure is illustrated in Fig. 2A. The FAC-1 consists of two isocratic pumps (pump A for analysis, pump B for washing), an autosampling system (up to 210 samples), a column oven, and a couple of miniature columns connected in parallel to either a fluorescence or a UV detector. By use of the parallel-column system, the time for analysis is reduced, as, during analysis with one column, the other is being washed (Fig. 2B). Use of a capsule-type miniature column (inner diameter, 2 mm; length, 10 mm; bed volume, 31.4 μl) (Fig. 3A) allowed both injection and analysis times to be reduced.

For FAC analysis, the following three items should be prepared:

1. Lectin-immobilized resin (at least 50 to 100 μl)
2. A set of ligand solutions (2.5 or 5.0 n*M*, about 1 to 5 pmol per assay)
3. Running buffer (500 ml for overnight analysis).

Immobilization of Lectins

For FAC analysis, lectins or ligands (i.e., glycoproteins, glycopeptides, or glycans) are immobilized on resin. Considering the efficiency of the analysis, it is preferable to immobilize lectins to profile glycan specificities. Although immobilization of glycans or glycopeptides is also feasible (reverse FAC), only a limited glycan is practically applicable.

Preparing a suitable affinity adsorbent is essential for successful FAC (Hirabayashi *et al.*, 2003). Because it is difficult to chose an optimal protein concentration initially, the ratio of 3 to 5 mg protein per 1 ml resin is recommended in a trial experiment, on the assumption that K_a values are in the range of 10^4 to 10^5 M^{-1}. Since the bed volume of the miniature column is about 31.4 μl, 100 μl of resin is enough to prepare one column; that is, about 10 to 500 μg is required to make each column. Typically, 300 to 500 μg of purified lectin is dissolved in 500 μl to 1 ml of coupling buffer (10 m*M* NaHCO$_3$ buffer [pH, 8.3] containing 0.5 *M* NaCl) and coupled to 100 μl of NHS-activated Sepharose 4FF (Amersham) according to the manufacturer's instructions. After incubating for 0.5 to 2 hrs at room temperature with rotation, excess NHS groups are deactivated by 1 *M*

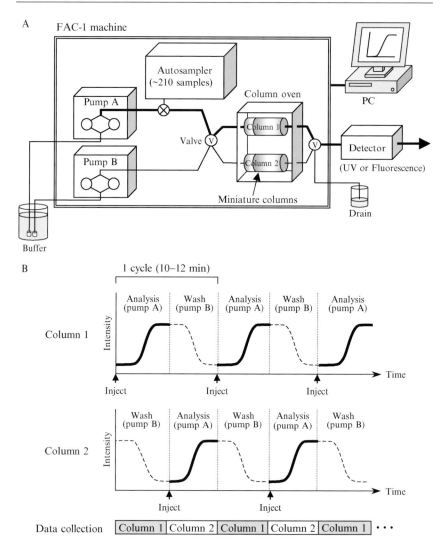

FIG. 2. Scheme of automated frontal affinity chromatography (FAC) system. (A) An automated system for FAC consists of an FAC-1 instrument, fluorescence or ultraviolet (UV) detector and a personal computer (PC) workstation. FAC-1 is equipped with two pumps, an autosampler, a column oven, and a couple of miniature columns connected in parallel. One of the pumps is used exclusively for analysis; the other is for regeneration of the columns. Analyte solutions are applied via the autosampler by pump A. Elution of analytes from the lectin-immobilized column is monitored by UV or fluorescence detectors. The bold line indicates the pathway when analysis is carried out with column 1. (B) Outline of the dual-column switching system. The two columns are used alternately to reduce the total analysis time by half. Data collection is carried out only for the "analyzing" column.

monoethanolamine. After washing with the coupling buffer and 10 mM
acetate buffer (pH, 4.0) containing 0.5 M NaCl, the lectin-agarose is
suspended in an equal volume of appropriate buffer (e.g., 10 mM Tris-
HCl [pH, 7.4] containing 1 mM CaCl$_2$ and 0.1 mM MnCl$_2$). Note that some
lectins are inactivated by washing at low pH.

Preparation of Miniature Column

Lectin-immobilized resin is packed into a miniature column as illustrated in Fig. 3B. A 50% slurry of the lectin-agarose is added to the
miniature column filled with appropriate buffer. To ensure complete pack-
ing, the column is slowly aspirated from the outlet with a syringe. After the
addition of sufficient resin, the inlet is closed with a cap-type filter (sepa-
rately manufactured by Shimadzu), with special care taken not to leave an
air space on the top of the resin. The resulting packed lectin column is
slotted into a stainless steel holder (Fig. 3C) and connected to the automated
FAC instrument (Fig. 3D).

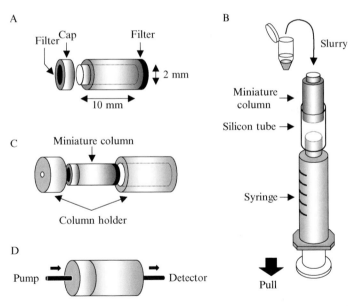

FIG. 3. Schematic diagram of the miniature column. Automated frontal affinity
chromatography (FAC) system adopting a capsule-type miniature column (A). Lectin-
immobilized resin is packed into a miniature column using a syringe (B). The resulting column
is slotted into column holder (C), and connected to FAC-1 (D).

Preparation of Ligand Solutions

In our detection system, either fluorescence-labeled (e.g., 2-aminopyridine, 2-aminobenzoic acid, and 4-methyl-umbelliferone) or UV-labeled (e. g., *p*-nitrophenyl) glycans can be used. Considering reasonable sensitivity, total absence of nonspecific interaction with resins, and ease of purification by HPLC as well as established two- or three-dimensional mapping system, PA glycans are the best for this purpose. Various PA glycans are commercially available from Takara Bio Inc. and Seikagaku Corp. Non-labeled glycans, which are available from several companies (e.g., Funakoshi Co., Dextra Laboratories, Ltd., and Calbiochem), can also be used after pyridylamination (Hase, 1994). PA glycans used in a recent study (Nakamura *et al.*, 2005) are shown in Fig. 4.

In the case of PA oligosaccharides, a 2.5-nM (*N*-glycans) or 5-nM (other glycans) solution is used for routine analysis. After 1 ml of running buffer is pipetted into glass vials, 2.5 or 5 μl of stock solution (1 pmol/μl) is added. Considering losses in the autosampling system, the injection volume plus 200 μl of PA-glycan solution should be prepared.

Operation of FAC

Lectin columns connected to FAC-1 are equilibrated with the appropriate buffer for the lectins. For example,

- 10 mM Tris-HCl buffer (pH, 7.4) containing 0.8% NaCl with or without 1 mM CaCl$_2$ (TBS)
- 10 mM phosphate buffer (pH, 7.0) containing 0.8% NaCl (PBS)
- 10 mM HEPES buffer (pH, 7.4) containing 0.8% NaCl with or without 0.1 mM CaCl$_2$ and 0.01 mM MnCl$_2$.

Because the FAC-1 uses a draw-and-push pumping system for both buffer flow and autosampling, detergent-containing buffer should be avoided. After 10 min of equilibration of the miniature columns, an excess volume (more than 300 μl) of labeled glycan solution dissolved in the running buffer is successively injected into a pair of lectin columns via an autosampling system (up to 210 samples). Serial injection of glycan solution is automatically performed according to programmed order on a batch file. Injection volume is also set in the same file, since it varies depending on the purpose of analysis and the strength of interactions (a large volume, e.g., 0.8 ml, is applied, when strong retardation is observed). Elution of PA oligosaccharides is monitored by fluorescence (excitation and emission wavelengths of 310 and 380 nm, respectively), whereas that of pNP-glycosides is detected by UV (280 nm). V–V$_0$ values are automatically calculated

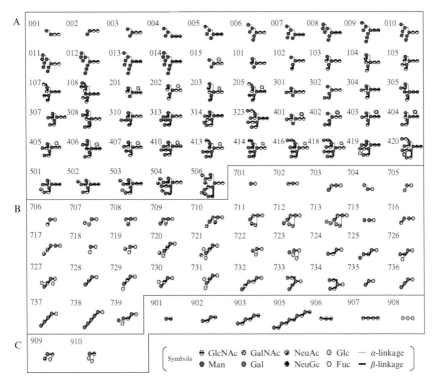

FIG. 4. Schematic representation of pyridylaminated glycan structures. The reducing terminal is pyridylaminated for frontal affinity chromatography analysis. Symbols used to represent pyranose rings of monosaccharides are shown at the bottom of glycan structures. Anomeric carbon (i.e., position 1) is placed at the right side, and 2, 3, 4, etc., are placed clockwise. (A) N-linked glycans. (B) Glycolipid-type glycans. (C) Others.

using the developed software "FAC analyzer," according to the method originally described by Arata *et al.* (2001b).

The overall procedure to operate the FAC-1 system is summarized below.

1. Connect a pair of columns to the FAC-1.
2. Equilibrate the columns with an appropriate buffer (e.g., TBS and PBS).
3. Set a series of sample vials on the autosampling system.
4. Prepare the time schedule (batch file) to direct the order of analysis and the volume of injection.
5. Run a batch file (start a sequence of analyses).
6. Summarize the resulting interaction data on a PC.

Initial Profiling Using Core 49 Glycans (Preliminary Analysis)

To check the quality of the prepared lectin column(s) and to obtain the overall features of their sugar-binding specificities, initial profiling is carried out using a fixed set of glycans. These represent various types of N-glycans and glycolipid-type glycans so that at least one is expected to have some affinity for each lectin. We selected 49 such glycans (designated "core 49") to cover almost all glycan epitopes for lectin recognition (Fig. 5, inset in graphs). Experimental conditions were as follows:

1. Lectin column: 1 column/lectin (3 to 5 mg/ml)
2. Injection volume: 300 μl (0.75 or 1.5 pmol/glycan/column)
3. Analysis time: 10 min/cycle (5 min × 2 columns)
4. Flow rate: 0.125 ml/min (analysis) or 0.250 ml/min (wash)
5. Detection: excitation/emission wavelength = 310 nm/380 nm.

Under these conditions, analysis time and ligand consumption are reduced. Analysis time is less than 10 hrs for 100 samples (2 lectin columns × 50 glycans), and the amount of each glycan used is less than 2 pmol. An example of preliminary analysis using *Erythrina cristagalli* agglutinin (ECA) (3 mg/ml) is shown in Fig. 5. Chromatograms of 49 glycans are shown with tentative retardation volumes (in microliters) and glycan structures. In 1 week, it is possible to analyze more than 700 interactions (i.e., 14 lectins × 50 glycans). An example of comprehensive analysis using 14 lectins toward 50 glycans (including 1 negative control, Rha) is shown in Fig. 6. Under the conditions used, however, elution of some ligand solutions may not reach plateau level completely. In such cases, it is practically impossible to obtain precise V–V_0 (or K_d) values. Nevertheless, much information on overall features of sugar-binding specificities may be derived.

Full Specificity Analysis Using More Than 100 Glycans

Based on the initial profiling mentioned previously, conditions for more detailed analyses were optimized. In the present FAC system, approximately 2 μl of experimental error in terms of V–V_0 value is included considering the data-collection interval (1 sec) and the flow rate (0.125 ml/min). Under the conditions, it is difficult to determine precisely the dissociation constants (K_d) of low-affinity glycans showing V–V_0 < 2 μl. In these cases, the lectin content must be increased by immobilizing more lectin protein on agarose. On the other hand, the maximum V–V_0 value that can be reliably determined is about 120 μl, due to limitations of the present autosampling system. When the observed V–V_0 exceeds this value, another column packed with agarose having a lower lectin concentration should be prepared.

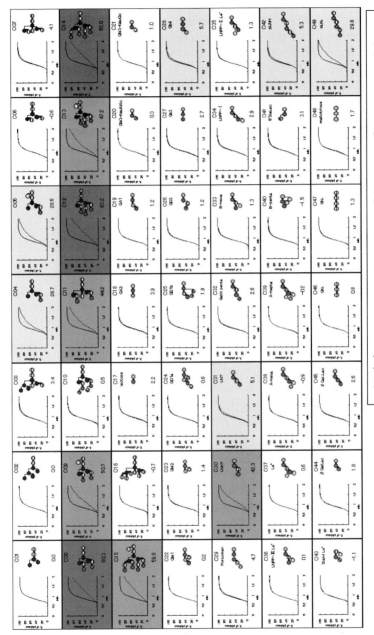

Typical conditions used for detailed analysis are as follows:

1. Number of columns: 1 to 3 columns with different immobilization contents/lectin (depending on the range of K_d)
2. Injection volume: 500 to 1000 μl (1.25 to 5 pmol/glycan/column, depending on affinity strength)
3. Analysis time: 10 to 12 min/cycle (5 to 6 min × 2 columns, depending on the injection volume)

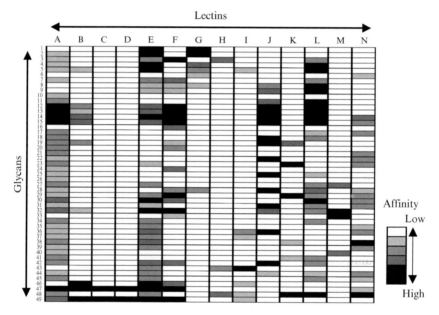

FIG. 6. Interaction data obtained within a week. With optimized conditions for preliminary analysis (injection volume, 300 μl; analysis time, 5 min), more than 700 interactions (14 lectins × 50 glycans) can be analyzed per week. Examples of interaction analysis are shown. The affinity strength of lectin–oligosaccharide interactions was assigned to six levels on the basis of V–V$_0$ values.

FIG. 5. Elution profiles of core 49 glycans with *Erythrina cristagalli* agglutinin (ECA) columns. Chromatograms of 49 glycans, selected to cover most glycan epitopes for lectins, are shown with tentative retardation volumes (in microliters) and glycan structures. For the sake of convenience, the elution pattern of each saccharide is overlaid with that of pyridylaminated Rha, which has no affinity for ECA (i.e., the negative control). Because the injection volume and protein concentration of the lectin column were fixed at 300 μl and 3 mg/ml, respectively, precise K_d values cannot be determined, but much information about the overall features of sugar-binding specificities of the lectin may be obtained rapidly. Time for 100 analysis (2 lectins × 50 glycans) is less than 10 hrs.

4. Flow rate: analysis 0.125 ml/min, wash 0.250 ml/min
5. Detection: excitation/emission wavelength of 310 nm/380 nm.

An example of detailed analysis using ECA and 102 PA glycan is shown in Fig. 7. For the analysis, 500 to 1000 μl of glycan solution was injected serially into a set of two ECA columns (3 and 0.6 mg/ml). $V-V_0$ values obtained using the latter column (0.6 mg/ml) were combined with those obtained using the former (3 mg/ml). In Fig. 7, the left vertical axis thus indicates the combined $V-V_0$ values obtained for the two columns.

Detailed FAC analysis indicates that ECA shows significant affinity for complex-type N-glycans, whereas it has no apparent affinity for high-mannose type and completely agalactosylated glycans (see Fig. 7). Since ECA is known to recognize the LacNAc unit (Iglesias et al., 1982), this result is quite reasonable. Notably, its affinity increases with the increase in the branching number, i.e., 323 (tetra-antennary) > 313 (tri-) > 307 (di-) >302 (mono-). These results are consistent with observations made by Kaladas et al. (1982). Judging from 401 vs. 402, and 403 vs. 404, branched positions are considered to be less important for ECA recognition. Since ECA prefers 313 (Galβ1-4GlcNAc $\times$ 3) to 314 (Galβ1-4GlcNAc $\times$ 2, Galβ1-3GlcNAc $\times$ 1) and the latter shows almost the same affinity for 307 (bi-antennary glycan, Galβ1-4GlcNAc $\times$ 2), it recognizes only Galβ1-4GlcNAc (type 2 chain), but not Galβ1-3GlcNAc (type 1). This finding is also confirmed by comparison between 734 (LNnT, C30 in Fig. 5) and 738 (LNT, C31 in Fig. 5). As observed for 307 vs. 308, and 405 vs. 406, the presence of bisecting GlcNAc (β1-4GlcNAc) somewhat diminishes, but does not totally eliminate, the affinity. On the other hand, addition of α1-6 Fuc increases the affinity of ECA. Similarly to N-glycans, ECA showed significant affinity for other-type glycans containing a nonmodified Galβ1-4GlcNAc unit; these include 724, 733, 734, 902, 903, and 905. The observation that affinity for 734 (Galβ1-4GlcNAc $\times$ 2) is significantly higher than that for 733 (Galβ1-4GlcNAc $\times$ 1, Galβ1-3GlcNAc $\times$ 1) again confirms the previously mentioned observation that ECA prefers a type 2 chain, whereas repetition of the Galβ1-4GlcNAc unit reduces affinity (902 > 903 > 905).

Concentration Dependence Analysis

After determination, $V-V_0$ values are converted into K_d values according to the basic equation of FAC (equation [1]). Under the usual conditions described previously, $[A]_0$ (2.5×10^{-9} or 5.0×10^{-9} M) is negligibly small compared with K_d (generally 10^{-3} to 10^{-6} M), and hence, equation (1) can be simplified to equation (2).

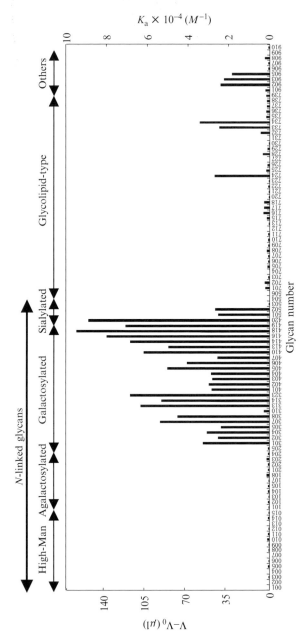

FIG. 7. Bar graph representation of affinity constants (K_a) of *Erythrina cristagalli* agglutinin toward 102 pyridylaminated glycans. By use of various injection volumes, depending on affinity strength and several columns with various lectin concentrations, K_a values for 102 glycans were determined. Glycan numbers correspond to glycan structures in Fig. 4.

$$(V - V_0)[A]_0 = -K_d(V - V_0) + Bt \qquad (3)$$

$$K_d = B_t/(V - V_0) \qquad (4)$$

To determine B_t, "concentration-dependence" analysis followed by Woolf-Hofstee–type plot is performed as described by Hirabayashi et al. (2003). From an economic viewpoint, pNP glycosides (obtainable from Sigma, Calbiochem, and Funakoshi Co.) are most suitable. In the case of ECA, various concentrations (8 to 50 μM) of pNP-lactose dissolved in TBS were applied to an ECA-immobilized column (3 mg/ml), and the elution was monitored by absorbance at 280 nm (Fig. 8A). By Woolf-Hofstee–type plots (equation [3]), B_t and K_d values were determined to be 1.72 nmol and 24.8 μM, respectively (Fig. 8B). For the series of experiments, 90 nmol (40 μg) of pNP-lactose was used, taking less than 100 min. Using the basic equation of FAC (equation [1]), K_d values for all other glycans were determined using the derived B_t value (1.72 nmol) (see Fig. 7, right vertical axis).

Concluding Remarks

In this article, we focused on interaction analysis between lectins and glycans. However, FAC is potentially applicable to many other purposes, for example, evaluation of enzyme–substrate interactions or receptor–ligand interactions. The automated FAC system should be a versatile tool for the characterization of biomolecular interactions by providing fundamental data. Applied to carbohydrate recognition events, FAC will assist in the development of the new field of "system glycobiology."

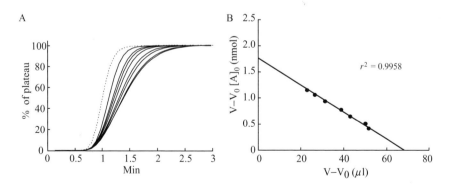

FIG. 8. Determination of B_t values. p-Nitrophenyl (pNP)-lactose was used for concentration-dependence analysis. For determination of B_t values for the immobilized ECA, pNP-lactose was diluted to various concentrations (8 to 50 μM) and applied to the column. The solid and dotted lines indicate elution profiles of pNP-lactose and control sugar (pNP-mannose), respectively (A). Woolf-Hofstee–type plots were made by using V–V$_0$ values (B).

Acknowledgments

This work was supported by New Energy and Industrial Technology Development Organization (NEDO) in Japan under the Ministry of Economy, Trade, and Industry (METI), Japan.

References

Arata, Y., Hirabayashi, J., and Kasai, K. (2001a). Sugar binding properties of the two lectin domains of tandem repeat-type galectin LEC-1 (N32) of *Caenorhabditis elegance*: Detailed analysis by an improved frontal affinity chromatography method. *J. Biol. Chem.* **276,** 3068–3077.

Arata, Y., Hirabayashi, J., and Kasai, K. (2001b). Application of reinforced frontal affinity chromatography and advanced processing procedure to the study of the binding property of a *Caenorhabditis elegans* galectin. *J. Chromatogr. A* **905,** 337–343.

Hase, S. (1994). High-performance liquid chromatography of pyridylaminated saccharides. *Methods Enzymol.* **230,** 225–237.

Hirabayashi, J. (2004). Lectin-based structural glycomics: Glycoproteomics and glycan profiling. *Glycoconj. J.* **21,** 35–40.

Hirabayashi, J., Arata, Y., and Kasai, K. (2000). Reinforcement of frontal affinity chromatography for effective analysis of lectin–oligosaccharide interactions. *J. Chromatogr. A* **890,** 261–271.

Hirabayashi, J., Arata, Y., and Kasai, K. (2003). Frontal affinity chromatography as a tool for elucidation of sugar recognition properties of lectins. *Methods Enzymol.* **362,** 353–368.

Hirabayashi, J., Hashidate, T., Arata, Y., Nishi, N., Nakamura, T., Hirashima, M., Urashima, T., Oka, T., Futai, M., Muller, W. E., Yagi, F., and Kasai, K. (2002). Oligosaccharide specificity of galectins: A search by frontal affinity chromatography. *Biochim. Biophys. Acta* **1572,** 232–254.

Iglesias, J. L., Halina, L., and Sharon, N. (1982). Purification and properties of a D-galactose/N-acetyl-D-galactosamine–specific lectin from *Erythrina cristagalli*. *Eur. J. Biochem.* **123,** 247–252.

Kaladas, P. M., Kabat, E. A., Iglesias, J. L., Lis, H., and Sharon, N. (1982). Immunochemical studies on the combining site of the D-galactose/N-acetyl-D-galactosamine specific lectin from *Erythrina cristagalli* seeds. *Arch. Biochem. Biophys.* **217,** 624–637.

Kasai, K., Oda, Y., Nishikawa, M., and Ishii, S. (1986). Frontal affinity chromatography: Theory, for its application to studies on specific interaction of biomolecules. *J. Chromatogr. A* **379,** 33–47.

Nakamura, S., Yagi, F., Totani, F., Ito, Y., and Hirabayashi, J. (2005). Comparative analysis of carbohydrate-binding properties of 2 tandem repeat-type Jacalin-related lectins, *Castanea crenata* agglutinin and *Cycas revoluta* leaf lectin. *FEBS J.* **272,** 2784–2799.

Sharon, N., and Lis, H. (2003). *In* "Lectins" pp. 63–103. Kluwer Academic Publishers, London.

Van Damme, E. J. M., Peumans, W. J., Barre, A., and Rougé, P. (1998). Plant lectins: A composite of several distinct families of structurally and evolutionary related proteins with diverse biological roles. *Crit. Rev. Plant Sci.* **17,** 575–692.

[20] Preparation of Neoglycolipids with Ring-Closed Cores *via* Chemoselective Oxime-Ligation for Microarray Analysis of Carbohydrate–Protein Interactions

By YAN LIU, WENGANG CHAI, ROBERT A. CHILDS, and TEN FEIZI

Abstract

Affinities of most oligosaccharide–protein interactions are so low that multivalent forms of ligand and protein are required for detecting interactions. The neoglycolipid (NGL) technology was designed to address the need for microscale presentation of oligosaccharides in a multivalent form for studying carbohydrate–protein interactions, and this is now the basis of a state-of-the-art carbohydrate microarray system. NGL technology involves conjugating oligosaccharides by reductive amination to the aminolipid 1,2-dihexadecyl-*sn*-glycero-3-phosphoethanolamine (DHPE). Other than ring-opening of the monosaccharide residues at reducing ends, oligosaccharides remain intact, and the NGLs derived from trisaccharides or larger oligosaccharides have performed well for the majority of carbohydrate-recognition systems that have the peripheral or backbone regions of oligosaccharides as recognition motifs. However, ring-opening of reducing end monosaccharides limits applicability to very short oligosaccharides (di- and trisaccharides) and, potentially, to *N*-glycans recognized by proteins such as *Pisum sativum* agglutinin (pea lectin) that require both intact core and backbone regions for strong binding. This chapter describes a method for preparing NGLs (designated AO-NGLs) from reducing oligosaccharides by chemoselective oxime-ligation to a new lipid reagent, *N*-aminooxyacetyl-DHPE. Microarray analyses of the AO-NGL derived from Lewis x (Le^x) trisaccharide probed with anti-Le^x antibodies indicate that a significant proportion of the core monosaccharide linked to lipid is in ring-closed form. Thus, AO-NGLs have broadened the applicability of NGLs as probes in studies of carbohydrate–protein interactions.

Overview

The diverse oligosaccharides that "decorate" glycoproteins, glycolipids, proteoglycans, and polysaccharides are potentially a vast source of information and could harbor a "glycocode" that is waiting to be deciphered in various contexts of biological and medical importance. It is desirable, therefore, to develop a knowledge base of biological systems that operate through

METHODS IN ENZYMOLOGY, VOL. 415
Copyright 2006, Elsevier Inc. All rights reserved.
0076-6879/06 $35.00
DOI: 10.1016/S0076-6879(06)15020-X

oligosaccharide recognition. Toward this goal, the neoglycolipid (NGL) technology was introduced in 1985 (Tang *et al.*, 1985) and further refined (Chai *et al.*, 2003; Feizi *et al.*, 1994; Stoll *et al.*, 1988). It involves conjugating oligosaccharides by reductive amination to the aminolipid 1,2-dihexadecyl-*sn*-glycero-3-phosphoethanolamine (DHPE). Tagging each oligosaccharide chain with the lipid tail confers an amphipathic property that enables immobilization in clustered display on solid matrices. This mode of display is crucial for detecting binding signals with the majority of carbohydrate-binding antibodies and receptors as the affinities of binding are low. A carbohydrate microarray platform has now been developed based on the NGL technology. It encompasses oligosaccharide probes that can, uniquely, be generated from naturally occurring sequences of glycoproteins, glycolipids, proteoglycans, and polysaccharides, as well as chemically synthesized oligosaccharides and glycolipids (Feizi and Chai, 2004; Fukui *et al.*, 2002; Galustian *et al.*, 2004; Palma *et al.*, 2006; Reddy *et al.*, 2004).

The NGL approach, originally developed using sequence-defined *O*-glycans for antibody binding studies (Tang *et al.*, 1985), has also been shown to be a powerful means of generating other types of oligosaccharide probes: from *N*-glycans (Mizuochi *et al.*, 1989), fragments of glycosamino-glycans (Fukui *et al.*, 2002; Leteux *et al.*, 2001) and polysaccharides (Palma *et al.*, 2006), and from diverse chemically synthesized oligosaccharides. Glycolipids, both naturally occurring and synthetic, are included in the repertoire of probes. A key development has been to combine carbohydrate–protein interaction studies with mass spectrometry in situ (Chai *et al.*, 1991, 2003; Feizi *et al.*, 1994) to gain information on monosac-charide composition, sequence, and branching pattern, as well as sulphate and phosphate substitution of the immobilized oligosaccharide probes at low picomole levels. NGL probes generated as mixtures from whole cells and tissues and probed with carbohydrate-recognizing proteins can be deconvoluted by thin-layer chromatography (TLC) combined with mass spectrometry (Fukui *et al.*, 2002; Osanai *et al.*, 1996).

The carbohydrate microarray system based on lipid-linked oligosaccha-ride probes currently has over 200 sequence-defined oligosaccharide probes and is continually expanding. It has served as the basis of "designer" microarrays generated from bacterial polysaccharides targeted by a novel carbohydrate-recognizing protein of the innate immune system, Dectin-1 (Palma *et al.*, 2006). Thus, this platform shows promise as a novel approach to surveying entire glycomes and proteomes for the molecular definition of carbohydrate-recognition systems.

The conjugation of oligosaccharides to DHPE occurs by reductive amination to form stable conjugates (referred to as DH-NGLs). Other than ring-opening of the monosaccharide residues at reducing ends,

oligosaccharides remain intact. DH-NGLs derived from trisaccharides or larger oligosaccharides have performed well for the majority of carbohydrate-recognition systems that have the peripheral or backbone regions of oligosaccharides as recognition motifs (Feizi and Chai, 2004; Feizi et al., 2003). However, ring-opening of reducing end monosaccharides may affect biological activities. For instance, the DH-NGL of Lewis x (Le^x) trisaccharide is not bound by anti-Le^x antibodies (Streit et al., 1996), and that of sialyl-Le^x tetrasaccharide is not bound by the selectins (Leteux et al., 1999). It is highly desirable to overcome this limitation and enhance the applicability of NGL probes derived from short oligosaccharides, which are the most accessible via chemical synthesis. In addition, certain plant lectins require both intact core and backbone regions of N-glycans to elicit strong binding (Leteux et al., 1998). To have a method of preparing NGLs widely applicable to short oligosaccharides in carbohydrate-recognition studies, also to cater for recognition systems that have a requirement for intact core monosaccharides as well as backbones of oligosaccharides, the aim has been to develop NGLs with ring-closed monosaccharide cores. This chapter describes the preparation of NGLs from reducing oligosaccharides by chemoselective oxime-ligation to an aminooxy-functionalized DHPE (AOPE). The Le^x-trisaccharide is used here to illustrate that its antigenicity is preserved when presented as an AO-NGL to two anti-Le^x antibodies, anti-L5 (Streit et al., 1996) and anti-SSEA-1 (Gooi et al., 1981).

Preparation of the Aminooxy-Functionalized Lipid AOPE

AOPE is prepared from commercially available DHPE in two steps, as illustrated in Fig. 1.

Materials

1. Phosphatidylethanolamine: 1,2-dihexadecyl-*sn*-glycero-3-phosphoethanolamine (DHPE, Fluka Chemicals, Dorset, UK)
2. Boc-aminooxyacetic acid (Novabiochem, Darmstadt, Germany)
3. 1-Ethyl-3(3-dimethyllaminopropyl)carbodiimide   (hydrochloride EDC; Aldrich, Dorest, UK)
4. Trifluoroacetic acid (TFA; Aldrich, Dorest, UK)
5. Silica cartridge (500 mg, from Waters, Milford, US)
6. Aluminium-backed high-performance TLC (HPTLC) plates (5 μm silica, from Merck Darmstadt, Germany)
7. Solvents of analytical reagent grade
8. Primulin staining reagent (Chai et al., 2003).

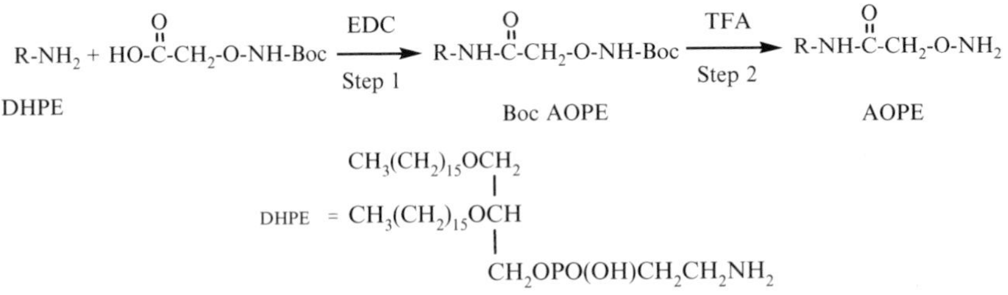

$$R\text{-}NH_2 + HO\text{-}\overset{O}{\underset{\|}{C}}\text{-}CH_2\text{-}O\text{-}NH\text{-}Boc \xrightarrow[\text{Step 1}]{\text{EDC}} R\text{-}NH\text{-}\overset{O}{\underset{\|}{C}}\text{-}CH_2\text{-}O\text{-}NH\text{-}Boc \xrightarrow[\text{Step 2}]{\text{TFA}} R\text{-}NH\text{-}\overset{O}{\underset{\|}{C}}\text{-}CH_2\text{-}O\text{-}NH_2$$

DHPE   Boc AOPE   AOPE

$$\text{DHPE} = \begin{array}{l} CH_3(CH_2)_{15}OCH_2 \\ \quad\quad\quad\quad\quad\quad | \\ CH_3(CH_2)_{15}OCH \\ \quad\quad\quad\quad\quad\quad | \\ \quad CH_2OPO(OH)CH_2CH_2NH_2 \end{array}$$

FIG. 1. Reaction scheme for preparation of aminooxy-functionalized 1,2-dihexadecyl-*sn*-glycero-3-phosphoethanolamine (AOPE). DHPE, 1,2-dihexadecyl-*sn*-glycero-3-phospho-ethanolamine; EDC, 1-ethyl-3(3-dimethyllaminopropyl)carbodiimide hydrochloride; TFA, trifluoroacetic acid.

Step 1: Preparation of N-Boc-aminooxyacetyl-DHPE (Boc-AOPE)

1. Dissolve DHPE (20 μmol, 13.3 mg) and Boc-aminooxyacetic acid (60 μmol, 11.5 mg) in 2 ml of chloroform.
2. Add 400 μl of freshly prepared EDC solution (20 μg/μl chloroform) to the solution previously mentioned at 0° (on ice). Stir the reaction mixture for 1 hr at 0° followed by 4 hrs at ambient temperature.
3. The reaction is monitored by HPTLC (developed with chloroform/ethanol/water, 50:50:1)[1] of an aliquot of the reaction mixture (~1 nmol of lipid). A major product can be revealed at R_f 0.4 (visualized under long-wave ultraviolet light after primulin staining [Chai *et al.*, 2003]); the starting material DHPE disappears.
4. After completion, the reaction mixture is washed twice with 2 ml of water; the organic layer is collected, and the solvent is evaporated under a nitrogen stream.
5. The product is purified using either a silica gel column eluted with dichloromethane/methanol/25% ammonium solution (100:10:1) as solvent or by semi-preparative TLC developed with chloroform/ethanol/water (50:50:1); yield is approximately 80%.
6. Matrix-assisted laser desorption/ionization mass spectrometry (MALDI-MS) analysis of the purified Boc-AOPE gives MNa$^+$ at m/z 860.

Step 2: Preparation of N-aminooxyacetyl-DHPE (AOPE)

1. Dissolve Boc-AOPE (10 μmol) in 500 μl of dichloromethane, and add 50 μl of TFA at 0° (on ice). Stir the reaction mixture for 5 hrs on ice.

---

[1] Ratios of mixed solvents used for reactions and chromatographies are by volume throughout.

2. Add toluene (50 μl) to the reaction mixture, and evaporate the solvents under a nitrogen stream. Repeat this evaporation three times for complete removal of TFA.

3. HPTLC analysis of an aliquot of the reaction mixture ($\sim$1 nmol) shows that the de-protected product, AOPE, is in quantitative yield (R_f0.1, developed with chloroform/ethanol/water, 50:50:1). The AOPE gives MNa^+ at m/z 760 in the MALDI mass spectrum.

4. Dissolve the dried residue in 2 ml of chloroform/methanol (1:1), and use this as a stock solution ($\sim$5 nmol/μl) for later conjugation with reducing oligosaccharides without purification. This stock solution is stored at $-20°$ and the AOPE is stable for at least 4 months.

Preparation of Oxime-Linked NGLs (AO-NGLs)

AO-NGLs can be prepared directly from reducing sugars by chemoselective oxime-ligation reaction with the lipid reagent AOPE, as depicted in Fig. 2. The solvent compositions and reaction times can be varied for conjugation of different oligosaccharides. Reactions are carried out in glass microvials sealed with Teflon-lined screw caps. Oligosaccharides are reacted with slight excess of AOPE without other reagents.

Materials

1. AOPE stock solution: 5 nmol/μl in chloroform/methanol (1:1) kept at $-20°$

2. Reducing oligosaccharides, 50 nmol. Examples illustrated are Le^a and Le^x trisaccharides and Le^x pentasaccharide:

Le^a-tri	Galβ-3GlcNAc
	Fucα-4
Le^x-tri	Galβ-4GlcNAcβ
	Fucα-3
Le^x-penta	Galβ-4GlcNAcβ-3Galβ-4Glc
	Fucα-3

3. Solvents of HPLC grade
4. Glass microvials with Teflon-lined caps (Chromacol, Herts, UK)
5. Aluminium-backed HPTLC plates
6. Orcinol staining reagent (Chai et al., 2003).

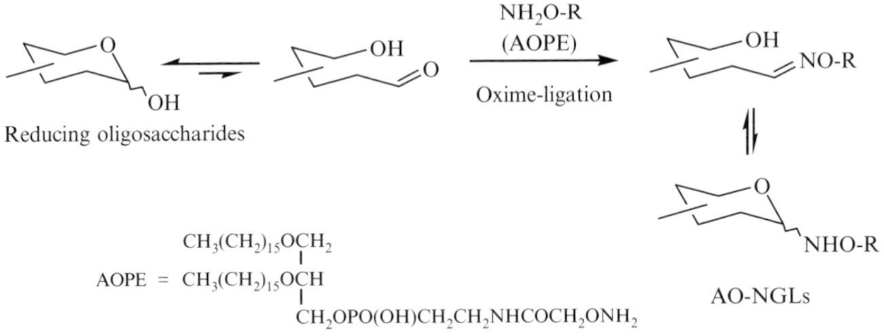

FIG. 2. Preparation of oxime-linked neoglycolipids (AO-NGLs) from reducing oligosaccharides. AOPE, aminooxy-functionalized 1,2-dihexadecyl-*sn*-glycero-3-phosphoethanolamine.

Preparation of NGLs from Le^a and Le^x Trisaccharides and Le^x Pentasaccharide

1. Add 20 μl of AOPE (~5 nmol/μl in chloroform/methanol, 1:1) to 50 nmol of oligosaccharide lyophilized in a reaction vial, and evaporate the mixture to dryness under a nitrogen stream.
2. Dissolve the reaction mixture in 50 μl of chloroform/methanol/water (10:10:1) and incubate at ambient temperature for 16 hrs.
3. Loosen the screw cap of the reaction vial and place it in a heating block at 60° to allow slow evaporation (over ~1 hr) to dryness.
4. Dissolve the residue in 100 μl of chloroform/methanol/water (25:25:8), apply 2 μl of the solution (containing ~1 nmol starting sugar) to a HPTLC plate, and develop the plate with chloroform/methanol/water (130:50:9).
5. Stain the chromatogram, first with primulin and then with orcinol reagent (Chai *et al.*, 2003). HPTLC of conjugation mixtures of Le^a-tri, Le^x-tri and Le^a-penta is shown in Fig. 3A.

Preparation of NGLs from Other Saccharides

The procedures for preparing AO-NGLs of monosaccharides and disaccharides are as described above for oligosaccharides up to pentasaccharides. As will be described elsewhere, for oligosaccharides larger than hexasaccharides, increased water content of the conjugation solvent is necessary to assist their dissolution. For example, chloroform/methanol/water at a ratio of 25:25:8 is used as conjugation solvent for complex-type and high-mannose–type *N*-glycans. For glucan oligomeric fragments (e.g., hepta- and tridecasaccharides from curdlan and pustulan), 10 molar excess of AOPE is applied and incubation is prolonged to 24 hrs under acidic conditions (chloroform/methanol/water/acetic acid, 25:25:8:1). Water content of the solvent used for TLC development

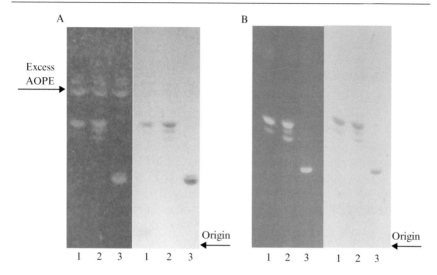

FIG. 3. HPTLC analyses of conjugation mixtures and of purified aminooxy-functionalized neoglycolipids (AO-NGLs) of Lewis a (Le^a) trisaccharide (lanes 1), Le^x trisaccharide (lanes 2), and Le^x pentasaccharide (lanes 3). Left panels show primulin staining, and right panels orcinol staining. Images are of the conjugation mixtures (A) and of the purified AO-NGLs from the corresponding mixtures (B). The duplex and triplex bands represent potential isomeric forms of the oxime-linked AO-NGLs as they gave the same molecular ions in matrix-assisted laser desorption/ionization mass spectra. AOPE, aminooxy-functionalized 1, 2-dihexadecyl-sn-glycero-3-phosphoethanolamine.

of larger oligosaccharides is also increased (e.g., chloroform/methanol/water 60:35:8 and 55:45:10), depending on the size of oligosaccharides as described by Chai *et al.* (2003). Other reaction procedures are as described above for the Le^a and Le^x analogs.

The water tolerance of oxime-ligation contrasts with reductive-amination where water is an inhibitor (Stoll *et al.*, 1988) and is a potential advantage for preparing NGLs of large and highly acidic oligosaccharides, which depend on the presence of water for solubilization.

Purification of AO-NGLs

Materials

1. Aluminium-backed HPTLC plates
2. Silica cartridges
3. Orcinol staining reagent
4. Solvents and solutions: water, methanol and chloroform are of HPLC grade; ammonium acetate solution (0.2 M in water); chloroform/methanol/water (C/M/W) and chloroform/ethanol/water (C/E/W).

Purification Procedure Using Silica Cartridge

AO-NGLs derived from trisaccharides and larger oligosaccharides (as well as acidic mono- and disaccharides) are purified using silica cartridges to remove the excess lipid reagent. The procedure is as follows:

1. Wash the column sequentially with 4 ml of methanol, 4 ml of water, 6 ml of ammonium acetate (0.2 M), 12 ml of water, 4 ml of methanol, and 6 ml of chloroform.
2. Dissolve the dried NGL conjugation mixture (<100 nmol of starting sugar) with up to 200 μl of C/M/W (130:50:9). Apply the solution onto the prewashed column and collect the fall through as fraction 1.
3. Wash the column four times with 300 μl of C/M/W (130:50:9) and collect the washings as fractions 2 through 5.
4. Elute the column four times with 300 μl of C/M/W (60:35:8) and four times with 300 μl of C/M/W (25:25:8); collect these as fractions 6 through 13.
5. Apply aliquots (1 to 2 μl) of each fraction for HPTLC to identify, by primulin and orcinol staining, fractions containing the NGLs.
6. Pool these fractions and dry under a nitrogen stream.

Semi-Preparative HPTLC Purification Procedure

Silica cartridge is not recommended for purifying AO-NGLs derived from neutral monosaccharides and disaccharides because they migrate near the excess lipid on TLC plates. These NGLs are purified by semi-preparative HPTLC and developed in one of the following solvents: C/E/W (50:50:1) for AO-NGL of GalNAc; C/M/W (50:25:1) for AO-NGLs of fucose (Fuc) and rhamnose (Rha); C/M/W (130:50:9) for AO-NGLs of Glc, Man, Gal, GlcNAc; and C/M/W (60:35:8) for AO-NGLs of disaccharides (e.g., lactose and N-acetyllactosamine). The loading of the sample depends on the required resolution; in general, it should be less than 20 nmol NGL per cm. The procedure progresses as follows:

1. Equilibrate the development solvent in a TLC tank for 1 hr at ambient temperature.
2. Apply NGL reaction mixture (<100 nmol of starting sugar) as a long band to an HPTLC plate (10 × 10 cm) 15 mm from the bottom edge, and allow 15 mm free at the right and left edges.
3. After drying, place the plate into the tank and develop to 5 mm below the top edge.
4. Air-dry the plate. The positions of AO-NGL bands can be identified and pencil marked after staining with primulin reagent at the edges (left and right) of the plate.

5. Cut out the required band as a strip of plate with a scalpel, and carefully scrape off the silica gel from the strip.
6. Pack the scraped silica gel as a mini-column and elute the NGLs three times with 500 μl of C/M/W (60:35:8) and three times with 500 μl of C/M/W (25:25:8).
7. Dry the eluate under a nitrogen stream.

Analysis and Quantitation of Purified AO-NGLs

Materials and Equipment

1. Primulin staining reagent
2. Densitometer (Shimadzu CS-9000 scanner)
3. MALDI time-of-flight mass spectrometer (Waters, Manchester, UK).

Procedures

Purified AO-NGLs are dissolved in C/M/W (25:25:8) to give a concentration of approximately 100 pmol/μl for analysis, quantitation, and storage (at $-20°$). HPTLC analysis of purified AO-NGLs of Le^a trisaccharides, Le^x trisaccharide, and Le^x pentasaccharide, with primulin- and orcinol-staining, is shown in Fig. 3. Densitometry is used to quantitate the AO-NGLs on HPTLC plates after TLC followed by primulin staining, as described for conventional DH-NGLs (Chai et al., 2003). The DH-NGL of maltopentaose applied at four levels (i.e., 500, 250, 100, and 50 pmol) is used as standard.

The purified AO-NGLs can be analyzed by MALDI-MS. For this, the NGLs are dissolved in chloroform/methanol/water (25:25:8) at a concentration of 10 to 20 pmol/μl; 0.5 μl is deposited on the sample target together with a matrix of 2-(4-hydroxyphenylazo)benzoic acid. Negative-ion MALDI spectra of the AO-NGLs of Le^a and Le^x trisaccharides and Le^x pentasaccharide are shown in Fig. 4.

Status of the Core Monosaccharide in AO-NGLs

Because the oxime ligation does not include reduction, it has been assumed that the ligated monosaccharide is in the ring-closed form. However, recent publications have variously suggested that the oxime-linked monosaccharide is in ring-opened form, or rather, in equilibrium with the ring-closed form (Ramsay et al., 2001), as depicted in Fig. 2. Phenylboronic acid (PBA)-derivatized silica gel is known, under controlled

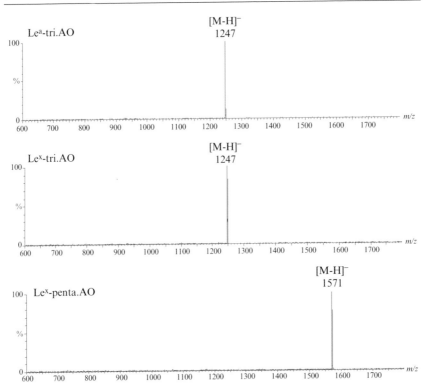

Fig. 4. Negative-ion matrix-assisted laser desorption/ionization mass spectra of the neoglyco-lipids (AO-NGLs) of Lewis a (Le^a) trisaccharide, Le^x trisaccharide, and Le^x pentasaccharide. The [M-H]$^-$ ions observed are in accord with their expected values.

conditions, to selectively react with the open-chain vicinal diols of reduced oligosaccharide alditols and of NGLs prepared by reductive amination (Chai *et al.*, 2003; Stoll and Hounsell, 1988). This has been shown previous-ly to provide a means of separating NGLs, which contain a vicinal diol in ring-opened core monosaccharide linked to lipid and are retained by the column, from other components in reaction mixtures that fall through (Chai *et al.*, 2003). In studies to be described elsewhere, we have carried out affinity chromatography on PBA cartridges to investigate the status of the core monosaccharide, using as models the AO- and DH-NGLs of galactose. The DH-NGL was, as predicted, fully retained on the PBA cartridge. In contrast, about 50% of the AO-NGL of galactose was in the fall-through fraction, and 50% was retained, consistent with there being an equilibrium of ring-closed and ring-opened forms of the monosaccharide.

It must be noted that the number of open-chain vicinal diols available and the acidity of the oligosaccharide influence the interaction with PBA, and hence the retention of the NGLs by PBA-silica cartridge. For example, the DH-NGL of Le^x trisaccharide, which has a disubstituted GlcNAc core, and DH-NGLs of sialyl oligosaccharides may not be fully retained by a PBA-cartridge. Therefore, this strategy is not always straightforward for showing the differentiation of DH- and AO-NGLs.

Evaluation of AO-NGLs in Carbohydrate–Protein Interaction Studies

Binding of NGLs by antibodies and lectins can be assayed by conventional enzyme-linked immunosorbent assay–type experiments in plastic microwells as described by Chai et al. (2003). Here, we have carried out microarray analyses of the interactions of the anti-Le^x antibodies anti-L5 and anti–SSEA-1, with the AO-NGLs of the Le^x and Le^a trisaccharides and Le^x pentasaccharide and of the corresponding DH-NGL analogues; the latter were prepared by protocols described by Chai et al. (2003). The NGLs were incorporated into liposomes for arraying, and two assay formats were used. In the first, the NGL probes were spotted onto nitrocellulose membranes, and binding of the anti-Le^x antibodies was detected using colorimetric biotin-streptavidin system. In the second format, the NGL probes were robotically arrayed onto nitrocellulose-coated glass slides, and binding was detected using a fluorescent biotin-streptavidin system.

Materials

1. Le^a-tri.DH, Le^a-tri.AO, Le^x-tri.DH, Le^x-tri.AO, Le^x-penta.DH, Le^x-penta.AO (100 pmol of each in chloroform/methanol/water [25:25:8])
2. Tris-buffered saline (TBS): 10 mM Tris-HCl (pH, 8.0), 150 mM NaCl
3. Phosphatidylcholine (PC) stock solution: 50 pmol/μl in methanol
4. Cholesterol (C) stock solution: 50 pmol/μl in methanol
5. Microtube (0.5 ml, Alpha, Hampshire, UK)
6. Sonic water bath (Branson 2510, Banbury, US)
7. Centrifuge (MSE Micro Centaur, UK)
8. Nitrocellulose membrane (Bio-Rad, Hemel, Hempstead, UK).

Preparation of Liposomes

1. Add 5 μl of methanol to a microtube, followed by 5 μl of PC, 3 μl of C, and 100 pmol NGL.
2. Dry at 37°. ·
3. Add 5 μl of TBS, vortex and sonicate for 10 min at 30°.
4. Centrifuge 13,000 rpm for 30 sec.

Procedures

In the first of the assay systems used here, the NGL probes incorporated into liposomes were spotted onto nitrocellulose membranes. Alternatively, the NGLs in chloroform/methanol/water (25:25:8) can be arrayed onto nitrocellulose membranes by jet spray as described by Chai *et al.* (2003) and Fukui *et al.* (2002). Binding of the two anti-Le^x antibodies, anti-L5 and anti–SSEA-1, was detected using biotin-conjugated rabbit anti-rat and goat anti-mouse immunoglobulins, respectively, and a colorimetric horseradish peroxidase-streptavidin system, exactly as described by Fukui *et al.* (2002). The DH-NGL of the Le^x-trisaccharide did not give binding signals with the two antibodies, whereas the AO-NGL was bound by both antibodies (Fig. 5), consistent with preservation of the ring of the *N*-acetylglucosamine linked to lipid. Both the DH- and AO-NGLs of the Le^x-pentasaccharide were bound by the antibodies in accord with earlier observations, and the NGLs of Le^a-trisaccharide served as negative controls.

In the second assay system, lipid-linked oligosaccharide probes were robotically arrayed onto nitrocellulose-coated glass slides using a non–contact-type arrayer. Details of the arraying, scanning, and analyses will be described elsewhere. In brief, each probe was printed at two levels (~2 and ~7 fmol per spot) in duplicate, with Cy3 dye as a marker to monitor sample application (Fig. 6A). In addition to the six AO- and DH-NGLs under investigation at positions 31 through 36, as shown in Fig. 6D, fifty eight lipid-linked oligosaccharide probes were arrayed. These will be described

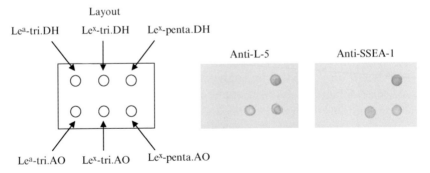

FIG. 5. Analyses of interactions of two anti-Le^x antibodies with DH-NGLs (by reductive-amination) and AO-NGL (by oxime-ligation) incorporated into liposomes and spotted onto nitrocellulose membranes. DH- and AO-NGLs of Lewis a (Le^a) trisaccharide, Le^x trisaccharide, and Le^x pentasaccharide were spotted on nitrocellulose membranes as shown in the layout (20 pmol per spot), overlaid with anti-L5 or anti-SSEA-1. Binding of the antibodies was detected calorimetrically as described by Fukui *et al.* (2002).

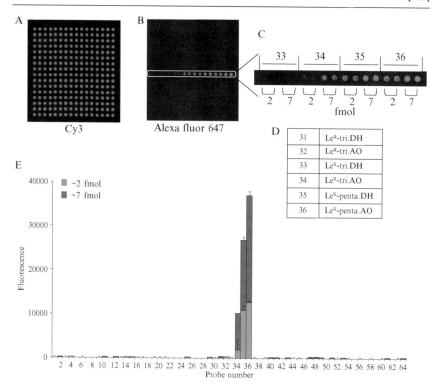

FIG. 6. Analyses of the interactions of rat anti–Lewis x (Le^x) antibody, anti-L5, overlaid onto a microarray of 64 lipid-linked oligosaccharide probes, each printed at approximately 2 and approximately 7 fmol/spot in duplicate on nitrocellulose-coated glass slides with Cy3 dye included as marker (green emission; A) and binding detected with biotinylated anti-rat immunoglobulins and Alexa Fluor 647-labeled streptavidin (red emission; B and C). (D) The positions of the Le^x- and Le^a-related NGLs. Numerical scores of the binding as measured by fluorescence intensity (means of duplicate spots with error bars) in blue for 2 fmol/spot and in red for 7 fmol/spot (E). (See color insert.)

elsewhere; suffice it to say that among them were AO- and DH-NGL pairs derived from several monosaccharides including fucose, acidic, and neutral oligosaccharides, including N-glycans with or without core fucose but lacking the Le^x sequence. Antibody binding was detected by a fluorescent (Alexa Fluor 647) biotin-streptavidin system as described by Campanero-Rhodes *et al.* (2006) and Palma *et al.* (2006). Here also, anti-L5 and anti–SSEA-1 gave binding to the AO-NGL of Le^x-trisaccharide and to both the AO- and DH-NGLs of the Le^x-pentasaccharide, but not to the DH-NGL of the Le^x-trisaccharide. There were no binding signals with any of the probes lacking the Le^x-sequence. Results with anti-L5 are shown in Fig. 6.

Perspectives

DH-NGLs derived from trisaccharides or larger oligosaccharides have been valuable probes for numerous carbohydrate-recognition systems that have the peripheral or backbone regions of oligosaccharides as recognition motifs (Feizi and Chai, 2004; Feizi and Childs, 1994; Feizi et al., 2003). However, as discussed in the Overview section, Le^x-antigenicity is abolished on the DH-NGL of the Le^x-trisaccharide (Streit et al., 1996), clearly a result of loss of the pyranose structure of the core N-acetylglucosamine and associated change in orientation of the fucose residue after the reductive amination. In contrast, on the AO-NGL, the antigenic activity is retained consistent with preservation of the ring-closed state in a significant proportion of the lipid-linked core monosaccharide. In studies to be described elsewhere, we have made additional observations on the advantages of AO-NGLs, namely for presenting trisaccharides (sialyllactoses) for interactions with Siglecs, and for presenting N-glycans to Pisum sativum agglutinin (pea lectin), which requires the nonreduced core regions of N-glycans for high-affinity binding. Thus, AO-NGLs have broadened the applicability of NGLs by providing access to short oligosaccharides such that they can be probed in parallel with long oligosaccharides in studies of carbohydrate–protein interactions. We will investigate whether N-methylation to force the equilibrium to maximize the ring-closed form will enhance the performance of AO-NGLs in carbohydrate-binding assays.

References

Campanero-Rhodes, M. A., Childs, R. A., Kiso, M., Komba, S., Le Narvor, C., Warren, J., Otto, D., Crocker, P. R., and Feizi, T. (2006). Carbohydrate microarrays reveal sulphation as a modulator of siglec binding. Biochem. Biophys. Res. Comm. 344, 1141–1146.

Chai, W., Cashmore, G. C., Carruthers, R. A., Stoll, M. S., and Lawson, A. M. (1991). Optimal procedure for combined high-performance thin-layer chromatography/high-sensitivity liquid secondary ion mass spectrometry. Biol. Mass Spectrom. 20, 169–178.

Chai, W., Stoll, M. S., Galustian, C., Lawson, A. M., and Feizi, T. (2003). Neoglycolipid technology: Deciphering information content of glycome. Methods Enzymol. 362, 160–195.

Feizi, T., and Chai, W. (2004). Oligosaccharide microarrays to decipher the glyco code. Nat. Rev. Mol. Cell Biol. 5, 582–588.

Feizi, T., and Childs, R. A. (1994). Neoglycolipids: Probes in structure/function assignments to oligosaccharides. Methods Enzymol. 242, 205–217.

Feizi, T., Lawson, A. M., and Chai, W. (2003). Neoglycolipids: Identification of functional carbohydrate epitopes. In "Carbohydrate-based Drug Discovery: From the Laboratory to the Clinic" (C.-H. Wong, ed.), Vol. 2, pp. 747–760. Wiley-VCH, Weinheim.

Feizi, T., Stoll, M. S., Yuen, C.-T., Chai, W., and Lawson, A. M. (1994). Neoglycolipids: Probes of oligosaccharide structure, antigenicity and function. Methods Enzymol. 230, 484–519.

Fukui, S., Feizi, T., Galustian, C., Lawson, A. M., and Chai, W. (2002). Oligosaccharide microarrays for high-throughput detection and specificity assignments of carbohydrate-protein interactions. *Nat. Biotechnol.* **20,** 1011–1017.

Galustian, C., Park, C. G., Chai, W., Kiso, M., Bruening, S. A., Kang, Y. S., Steinman, R. M., and Feizi, T. (2004). High and low affinity carbohydrate ligands revealed for murine SIGN-R1 by carbohydrate array and cell binding approaches, and differing specificities for SIGN-R3 and langerin. *Int. Immunol.* **16,** 853–866.

Gooi, H. C., Feizi, T., Kapadia, A., Knowles, B. B., Solter, D., and Evans, M. J. (1981). Stage specific embryonic antigen SSEA-1 involves a1–3 fucosylated type 2 blood group chains. *Nature* **292,** 156–158.

Leteux, C., Chai, W., Nagai, K., Herbert, C. G., Lawson, A. M., and Feizi, T. (2001). 10E4 Antigen of scrapie lesions contains an unusual nonsulfated heparan motif. *J. Biol. Chem.* **276,** 12539–12545.

Leteux, C., Childs, R. A., Chai, W., Stoll, M. S., Kogelberg, H., and Feizi, T. (1998). Biotinyl-L-3-(2-naphthyl)-alanine hydrazide derivatives of N-glycans: Versatile solid-phase probes for carbohydrate-recognition studies. *Glycobiology* **8,** 227–236.

Leteux, C., Stoll, M. S., Childs, R. A., Chai, W., Vorozhaikina, M., and Feizi, T. (1999). Influence of oligosaccharide presentation on the interactions of carbohydrate sequence-specific antibodies and the selectins: Observations with biotinylated oligosaccharides. *J. Immunol. Methods* **227,** 109–119.

Mizuochi, T., Loveless, R. W., Lawson, A. M., Chai, W., Lachmann, P. J., Childs, R. A., Thiel, S., and Feizi, T. (1989). A library of oligosaccharide probes (neoglycolipids) from N-glycosylated proteins reveals that conglutinin binds to certain complex type as well as high-mannose type oligosaccharide chains. *J. Biol. Chem.* **264,** 13834–13839.

Osanai, T., Feizi, T., Chai, W., Lawson, A. M., Gustavsson, M. L., Sudo, K., Araki, M., Araki, K., and Yuen, C. T. (1996). Two families of murine carbohydrate ligands for E-selectin. *Biochem. Biophys. Res. Commun.* **218,** 610–615.

Palma, A. S., Feizi, T., Zhang, Y., Stoll, M. S., Lawson, A. M., Diaz-Rodríguez, E., Campanero-Rhodes, A. S., Costa, J., Brown, G. D., and Chai, W. (2006). Ligands for the beta-glucan receptor, Dectin-1, assigned using 'designer' microarrays of oligosaccharide probes (neoglycolipids) generated from glucan polysaccharides. *J. Biol. Chem.* **281,** 5771–5779.

Ramsay, S. L., Freeman, C., Grace, P. B., Redmond, J. W., and MacLeod, J. K. (2001). Mild tagging procedures for the structural analysis of glycans. *Carb. Res.* **333,** 59–71.

Reddy, S. T., Chai, W., Childs, R. A., Page, J. D., Feizi, T., and Dahms, N. M. (2004). Identification of a low affinity mannose 6-phosphate–binding site in domain 5 of the cation-independent mannose 6-phosphate receptor. *J. Biol. Chem.* **279,** 38658–38667.

Stoll, M. S., and Hounsell, E. F. (1988). Selective purification of reduced oligosaccharides using a phenylboronic acid bond elut column: Potential application in HPLC, mass spectrometry reductive amination procedures and antigenic/serum analysis. *Bio. Med. Chromatog.* **2,** 249–253.

Stoll, M. S., Mizuochi, T., Childs, R. A., and Feizi, T. (1988). Improved procedure for the construction of neoglycolipids having antigenic and lectin-binding activities from reducing oligosaccharides. *Biochem. J.* **256,** 661–664.

Streit, A., Yuen, C.-T., Loveless, R. W., Lawson, A. M., Finne, J., Schmitz, B., Feizi, T., and Stern, C. D. (1996). The Le^x carbohydrate sequence is recognized by antibody to L5, a functional antigen in early neural development. *J. Neurochem.* **66,** 834–844.

Tang, P. W., Gooi, H. C., Hardy, M., Lee, Y. C., and Feizi, T. (1985). Novel approach to the study of the antigenicities and receptor functions of carbohydrate chains of glycoproteins. *Biochem. Biophys. Res. Commun.* **132,** 474–480.

[21] Development of a Lectin Microarray Based on an Evanescent-Field Fluorescence Principle

By NOBORU UCHIYAMA, ATSUSHI KUNO, SHIORI KOSEKI-KUNO, YOUJI EBE, KOJI HORIO, MASAO YAMADA, and JUN HIRABAYASHI

Abstract

To investigate protein–carbohydrate interactions in a comprehensive and high-throughput manner, carbohydrate biosensors including microarrays have recently attracted increased attention. In this context, carbohydrate and lectin microarrays are emerging as techniques to meet such requisites. However, most of these methods adopt a conventional immuno-detection system, which requires repetitive washing steps before detection. Since lectin–carbohydrate interactions are relatively weak compared with those between antigens and antibodies, a more precise analytical method, which does not require any washing step, is desirable. We describe here a novel platform for lectin microarray that enables direct observation of lectin–carbohydrate interactions under equilibrium conditions, on the basis of an evanescent-field fluorescence-assisted detection principle. This method allows the analysis of a panel of glycoproteins (glycopeptides) in an extremely sensitive manner. The system also allows real-time observation of lectin–glycoprotein interactions in an aqueous phase. No washing procedures are required, thus relatively weak interactions are detectable. The described lectin microarray is expected to be useful for various fields of glycomics requiring high-throughput analysis of not only purified glycoproteins but also of crude samples.

Introduction

Carbohydrates encode enormous amounts of information essential for a variety of biological phenomena including cell differentiation, development, fertilization, and inflammation (Crocker and Varki, 2001; Feizi, 2000; Helenius and Aebi, 2001). In the context of post-genome or proteome sciences, key technologies to facilitate comprehensive carbohydrate analysis (i.e., "glycomics"), which enable large-scale analysis in a high-throughput and sensitive manner, are required. To meet such requisites, various microarray platforms were proposed in a recent few years. For example, glycan arrays, first described in 2002 (Fazio *et al.*, 2002; Fukui *et al.*, 2002; Houseman and Mrksich, 2002; Wang *et al.*, 2002; Willats *et al.*, 2002), have now been widely used for screening and characterization of carbohydrate-binding

METHODS IN ENZYMOLOGY, VOL. 415
0076-6879/06 $35.00
DOI: 10.1016/S0076-6879(06)15021-1

proteins, that is, lectins. More recently, an alternative method termed *lectin microarray* was introduced to profile complex glycosylation features of various glycoproteins (Angeloni *et al.*, 2005; Fromell *et al.*, 2005; Pilobello *et al.*, 2005; Zheng *et al.*, 2005). However, lectin–glycan interactions are relatively weak—for example, in the range of 10^{-4} to 10^{-7} M in terms of dissociation constant (K_d)—compared with antigen–antibody interactions (10^{-8} to 10^{-12} M). In conventional array systems, which require repeated washing procedures before scanning, dissociation of lectin–glycan complexes occurs. To realize interaction analysis without washing procedures, an evanescent-field fluorescence-assisted detection principle (Kuno and Uchiyama *et al.*, 2005), by which direct observation of even weak lectin–carbohydrate interactions is possible under equilibrium conditions (i.e., real-time imaging) (Lehr *et al.*, 2003; Pawlak *et al.*, 2002; Stimpson *et al.*, 1995; Xu and Yeung, 1998), was adopted.

Experimental Methods

Materials

Chemicals for buffer preparation were purchased from Wako (Osaka, Japan). Bovine serum albumin (BSA, >99%), asialofetuin (ASF), fetuin (FET), and RNase A and B were from Sigma-Aldrich (St. Louis, MO, USA). Concanavalin A (Con A), *Ricinus communis* agglutinin 120 (RCA120) were from Seikagaku Kogyo (Tokyo, Japan). *Sambucus nigra* lectin (SNA) was from Vector Lab (Burlingame, CA, USA). Antibodies (anti-FET and anti-RNase) were from Biogenesis Ltd. (Poole, England, UK). Tween-20 was purchased from Bio-Rad Lab (Hercules, CA, USA). 3-glycidoxypropyltrimethoxysilane was from Shin-Etsu Silicones (Tokyo, Japan). The DNA microarray-grade slide was purchased from Matsunami Glass Ind., Ltd. (Osaka, Japan).

Evanescent-Field Fluorescence-Assisted Microarray System

An evanescent field, created at the surface of the glass slide by internal reflection of the excitation light, excites the fluorescence probes when they bind selectively to the proteins on the slide surface. This surface-confined excitation has a limited penetration depth of about 200 nm into the adjacent probe solution (Fig. 1A). The instrumental scheme is shown in Fig. 1B. The system adopts a metal halide lamp as an excitation light, and appropriate optical filters (excitation filter: HQ535/50, emission filter 570DF10) purchased from Chroma Technology (Rockingham, VT, USA) and Omega Optical Inc. (Brattleboro, VT, USA), respectively.

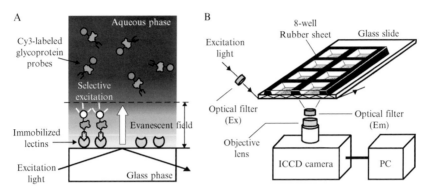

FIG. 1. Schematic illustration of lectin microarray system. (A) Principle of the evanescent-field fluorescence detection system. (B) Schematic illustration of the optical scheme of the detection system. Excitation light is injected into the edge of a glass slide, propagating with total internal reflection, and the fluorescence signals generated in the near-optical field (i.e., evanescent field, <200 nm) are captured through an intensified charge-coupled device (ICCD) camera located below the glass slide. Eight wells on the glass slide, which are fabricated by a rubber sheet, allow reaction with eight different probes simultaneously. A suitable probe volume for each well is approximately 100 μl. PC, personal computer.

Fabrication of Lectin Microarray

To observe weak interactions on a lectin microarray, high-density lectin immobilization is the most important requisite. Among various kinds of surface-coated glass slides tested for protein immobilization, such as epoxy- (Kusnezow et al., 2003), polylysine, and amino-coated (Angenendt et al., 2002), aldehyde-coated and hydrogel-coated (Kiyonaka et al., 2004) glass slides, the epoxysilane-coated slide showed the best performance with regard to signal-to-noise ratio. The epoxysilane surface allows random orientation upon immobilization of lectin proteins. It is simple to produce, showing a high cost–performance ratio as well as low background. It also shows sufficient capacity for immobilization of a variety of proteins (Mateo et al., 2000; Zhu et al., 2000). As a silane-coupling reagent, 3-glycidoxypropyltrimethoxysilane, which contains epoxy groups adequate for immobilization of various kinds of lectins on the glass slide, was chosen. Each lectin sample was diluted into a spotting solution at a final concentration of 1 mg/ml, except for RCA (0.25 mg/ml) and antibodies (0.1 mg/ml). The diluted proteins were spotted onto epoxy-coated slides using a spotting robot (STAMPMAN; Nippon Laser & Electronics Lab, Nagoya, Japan) with spot diameter sizes of 400 μm spaced at 375 μm intervals. During the array printing, the humidity of the internal space of the spotting robot was maintained between 60% and 80% to prevent sample drying. After spotting,

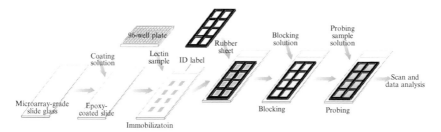

FIG. 2. Experimental scheme of the lectin microarray.

excess amounts of nonimmobilized lectins were washed out with phosphate-buffered saline (PBS) (pH, 7.4) containing 0.05% (v/v) Tween-20. The glass slide was incubated in a chamber (>80% humidity) at room temperature for 3 hrs to stabilize lectin immobilization. After incubation, a silicone rubber sheet with eight wells was carefully attached to the glass slide, and to each well was added 100 μl of blocking solution (1% BSA in PBS). Blocking was performed at room temperature for 1 hr (Fig. 2).

Preparation of Cy3-Labeled Glycoprotein Probes

Target glycoproteins were labeled with Cy3 Mono Reactive Dye (Amersham Biosciences, NJ) dissolved in 0.1 M sodium bicarbonate buffer (pH, 9.3) for 1 hr. Residual dye was removed by gel filtration on a Sephadex G-25 column (Amersham Biosciences). The dye/protein ratio determined by measuring absorbance of the dye-glycoprotein conjugate was 2.0 mol to 3.0 mol. For analysis, the concentration of each Cy3-labeled glycoprotein probe was adjusted to 0.67 nM with respect to Cy3 groups. The protein concentration of each probe was as follows: Cy3-ASF, 109 ng/ml; Cy3-FET, 15 ng/ml; Cy3-RNase A, 43 ng/ml; Cy3-RNase B, 68 ng/ml; and Cy3-BSA, 60 ng/ml.

Scanning and Data Analysis

One hundred microliters of Cy3-labeled glycoprotein solution dissolved in the probing buffer (TBS containing 0.05% Tween-20) was applied to each well on the lectin microarray slide, which was then incubated at room temperature until the binding reaction reached equilibrium. After the binding of the Cy3-labeled glycoprotein probes to the immobilized lectins, a fluorescence image of the array was continuously acquired using an evanescent-field fluorescence scanner, GTMASScan III (Nippon Laser & Electronics Lab). In all experiments, scanning conditions of the intensified

charge-coupled device (ICCD) camera (i.e., resolution, 5 μm), number of times for integration (8 times), and exposure time (110 msec) were fixed, whereas the microchannel plate multiplier gain was adjusted to maximize the dynamic range without causing saturation of the signal (not exceeding 40,000, which corresponds to 61% of a 16-bit signal). All data obtained were imported in TIFF format to a personal computer and were analyzed with the Array Pro analyzer Ver. 4.5 (Media Cybernetics, Inc., MD). The net intensity value for each spot was calculated on the basis of signal intensity minus the background value. Five replicates of net signal intensity values were averaged.

Results

Performance Evaluation of the Detection System

To confirm the range of linearity in signal intensity, 1.0 μl of diluted Cy3-labeled Tris solution (1 to 100 nM) was serially spotted (about 3.1 mm^2 in size) on a glass slide. As a result of scanning with the evanescent-field fluorescence scanner (Fig. 3A), it was found that the signal intensity vs. amount of dye applied was linear for a wide range of gains ($\times100$ to $\times1000$), and it displayed a high sensitivity at low dye concentrations (each spot containing 0 to 3.2 zmol/μm^2 dye, Fig. 3B). The dose–response curve also indicated good linearity in terms of net intensity, that is, in the range between 100 and 40,000. From the inclination of the standard curve obtained at a gain of $\times1000$, 1 zmol/μm^2 of Cy3 molecule, which gives 31,530 units of net intensity, the detection limit of Cy3 molecule was calculated to be 32 ymol/μm^2 (around 5 molecules/μm^2). These results suggest that this detection system should be applicable for fully quantitative and high-sensitivity analysis of low amounts of probe (glycoprotein) molecules.

Validation of Specific Binding Between Oligosaccharides and Immobilized Lectins

To confirm glycan detection ability and validate binding specificity with the previously mentioned lectin microarray, RCA120 (Lac/LacNAc binder; Baenziger and Fiete, 1979b) and ConA (Man binder; Baenziger and Fiete, 1979a) were chosen as model immobilized lectins.

First: 50 ng/ml (corresponding to approximately 1 nM) of Cy3-labeled ASF was used as a specific probe to RCA120. ASF is known to be a good ligand for RCA120, having 3 asialo-, bi- or tri-antennary complex-type *N-glycans* per molecule (Edge and Spiro, 1987). To preserve the equilibrium conditions, scanning was performed without removing the probe, and

A   B

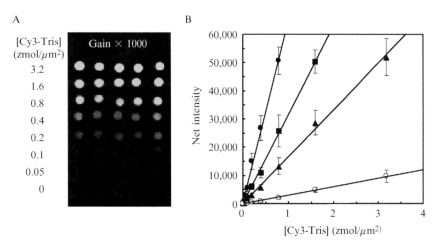

FIG. 3. Evaluation of the range of linearity in signal intensity. (A) Scan image of spots corresponding to nine different amounts of Cy3-Tris in the range 0 to 3.2 zmol/μm^2 on the glass slide. (B) Standard curves obtained by the evanescent field fluorescence detector when microchannel plate multiplier gain values were set at ×4000 (solid circle), ×1000 (solid square), ×500 (solid triangle), and ×100 (open circle). Data points were taken from the mean of net intensities (signal intensity minus background intensity) of every five spots. The symbols represent the mean of net intensity from replicate spots, and error bars represent the standard deviation of these replicate spots. Straight lines represent the linear regression fitting.

washing procedures were omitted. We observed sufficiently intense signals on the RCA120 spot probed by Cy3-ASF, and this signal was inhibited by the presence of a competitor saccharide, lactose (IC$_{50}$ = 94 μM; Fig. 4A).

Second: ConA was probed by Cy3-RNAse B, which is a well-known ligand for ConA, having a high-mannose–type *N*-glycan on the core protein (Avey *et al.*, 1967). In this case, the observed signal was inhibited by a competitor saccharide, mannose, but less strongly (IC$_{50}$ = 1.6 mM; Fig. 4B). These results clearly indicate that the fluorescence signals on the glass slide are based on specific lectin affinity to target glycans.

Real-Time Scanning of Binding Reactions

One of the greatest advantages of the evanescent-field fluorescence-assisted scanning method is the feasibility of "real-time" detection (Stimpson *et al.*, 1995). On this basis, we next investigated whether the binding reaction between immobilized lectins and probe glycans (glycoproteins) could be observed in real time by using RCA120 and Cy3-ASF. To achieve this, scanning was carried out continuously every minute after injection of

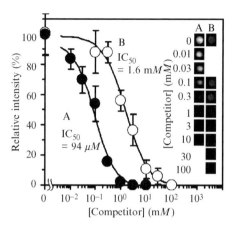

FIG. 4. Competition assay with the lectin microarray. (A) Competition assay using 0 to 10 mM lactose. (B) Competition assay using 0 to 100 mM mannose. Competitors were coincubated with 100 ng/ml of an appropriate probe (Cy3-ASF or Cy3-RNase B) for 3 hrs, and fluorescent signals were detected by scanning the slide. The error bars represent the standard deviations of five replicate spots.

the glycoprotein probe. The resulting signal intensities of RCA120 increased in a time-dependent manner and eventually became saturated within 120 min (Fig. 5). To observe a dissociation process between RCA120 and Cy3-ASF, we displaced the Cy3-ASF probe with a washing buffer at 125 min, and the signal intensity was continuously monitored. It was found that the signal intensity decreased in a time-dependent manner. This observation indicates that the lectin–glycan interaction is easily dissociated by incorporation of a washing procedure, as expected. The system is unique in its ability to analyze lectin–oligosaccharide interactions in a precise manner and has potential to accelerate studies dealing with a variety of lectin–glycan interactions.

Application of Glycoprotein Probes on the Lectin-Antibody Hybrid-Array Slide

To validate the system for comprehensive analysis of glycans, two sets of glycoproteins (RNase A vs. RNase B, and ASF vs. FET) whose glycan structures have been well characterized were chosen as model glycoprotein probes. In both cases, the core protein structures are identical, whereas the glycan structures are markedly different (Fig. 6). The microarray used for this purpose comprised six plant lectins: RCA120, ECA, ConA, GNA, SSA, and SNA. In addition to the previously mentioned lectins, BSA (negative

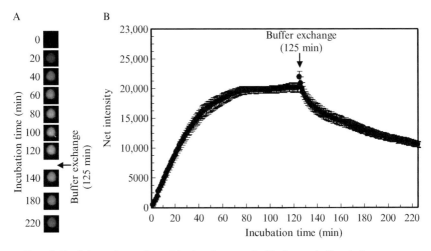

FIG. 5. Real-time observation of lectin–glycoprotein binding and dissociation processes. (A) Scanning image of the glass slide, on which RCA120 was immobilized. Scanning was initiated by the addition of 100 ng/ml of Cy3-ASF. (B) Illustration of the time course of the RCA120-ASF interaction. The error bars represent the standard deviations of five replicate spots.

control) and antibodies specific for the core proteins of both RNase and FET were immobilized on the same glass. The first set of glycoprotein probes comprised glycoprotein (RNase B) and non-glycoprotein (RNase A). When the microarray was probed with Cy3-RNase B, which has high-mannose type N-glycans at a single glycosylation site, intense signals were observed on the spots corresponding to ConA and GNA, whereas less intense but significant signals were observed on the anti-RNase spots. In the case of RNase A, which completely lacks glycan, the signal was detected only on the spots of anti-RNase at the same intensity as that of RNase B (see Fig. 6). In the second set of glycoprotein probes, Cy3-FET and Cy3-ASF, quite different signal patterns were obtained, which could be fully rationalized as the result of removal of terminal sialic acids from FET. Intense fluorescence signals were observed for Cy3-ASF on the spots corresponding to RCA120, ECA, and GNA, plus less intense but significant ones on ConA and anti-FET antibody spots. In the case of Cy3-FET, the signals on RCA120 and ECA spots dramatically decreased, whereas enhanced signals on SSA and SNA spots were observed (see Fig. 6). These results clearly reflect the established differences in glycan structures for ASF and FET.

Concluding Remarks

In this chapter, we described a novel microarray method for detecting lectin–glycoprotein interactions under equilibrium conditions on the basis

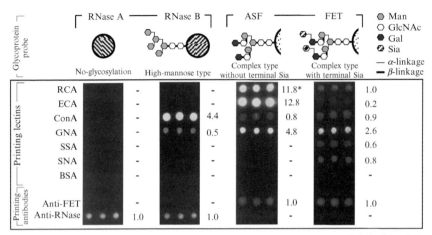

Relative intensity*

FIG. 6. Quantitative scanning with a lectin-antibody hybrid microarray for comprehensive analysis. The hybrid microarray platform comprising six plant lectins (RCA120, ECA, ConA, GNA, SSA and SNA), BSA (negative control), and two antibodies (anti-FET and anti-RNase) with five replicate spots per each analyte were probed with either Cy3-RNase A, Cy3-RNase B, Cy3-ASF, or Cy3-FET. The amounts of Cy3 probes were between 10 to 80 ng/well. Relative intensities determined by the ratio of each signal's intensity to that of an appropriate antibody (anti-RNase for RNase A and RNase B, and anti-FET for ASF and FET) are shown.

of an evanescent-field–assisted fluorescence detection principle. The system enables not only simplified structural profiling of complex glycans without washing steps, but also the detection of a wide range (i.e., weak-to-strong) of interactions. Since these features had not been covered by previous technologies, our system should contribute greatly to future studies that require high-throughput and sensitive analyses of complex glycans for various purposes; for example, prescreening of appropriate lectins for detection/purification of target glycoproteins (search for glycan-related biomarkers) and rapid profiling of glycoforms of particular glycoproteins (quality control of glycoprotein drugs). The system has also proved of use as a "protein array." In this context, combined use with antibodies and other binding proteins is also promising.

Acknowledgments

We thank S. Nakamura from AIST and K. Kasai from Teikyo University for critical discussion about the lectin–carbohydrate interactions. We also thank Y. Kubo for technical assistance. This work was supported in part by New Energy and Industrial Technology Development Organization (NEDO) in Japan.

References

Angeloni, S., Ridet, J. L., Kusy, N., Gao, H., Crevoisier, F., Guinchard, S., Kochhar, S., Sigrist, H., and Sprenger, N. (2005). Glycoprofiling with micro-arrays of glycoconjugates and lectins. *Glycobiology* **15**, 31–41.

Angenendt, P., Glokler, J., Murphy, D., Lehrach, H., and Cahill, D. J. (2002). Toward optimized antibody microarrays: A comparison of current microarray support materials. *Anal. Biochem.* **309**, 253–260.

Avey, H. P., Boles, M. O., Carlisle, C. H., Evans, S. A., Morris, S. J., Palmer, R. A., Woolhouse, B. A., and Shall, S. (1967). Structure of ribonuclease. *Nature* **213**, 557–562.

Baenziger, J. U., and Fiete, D. (1979a). Structural determinants of concanavalin A specificity for oligosaccharides. *J. Biol. Chem.* **254**, 2400–2407.

Baenziger, J. U., and Fiete, D. (1979b). Structural determinants of Ricinus communis agglutinin and toxin specificity for oligosaccharides. *J. Biol. Chem.* **254**, 9795–9799.

Crocker, P. R., and Varki, A. (2001). Siglecs in the immune system. *Immunology* **103**, 137–145.

Edge, A. S., and Spiro, R. G. (1987). Presence of an O-glycosidically linked hexasaccharide in fetuin. *J. Biol. Chem.* **262**, 16135–16141.

Fazio, F., Marian, C., Bryan, O. B., James, C. P., and Chi-Huey, W. (2002). Synthesis of sugar arrays in microtiter plate. *J. Am. Chem. Soc.* **124**, 14397–14402.

Feizi, T. (2000). Progress in deciphering the information content of the 'glycome': A crescendo in the closing years of the millennium. *Glycoconj. J.* **17**, 553–565.

Fromell, K., Andersson, M., Elihn, K., and Caldwell, K. D. (2005). Nanoparticle decorated surfaces with potential use in glycosylation analysis. *Colloids Surf. B: Biointerfaces* **46**, 84–91.

Fukui, S., Feizi, T., Galustian, C., Lawson, A. M., and Chai, W. (2002). Oligosaccharide microarrays for high-throughput detection and specificity assignments of carbohydrate-protein interactions. *Nat. Biotechnol.* **20**, 1011–1017.

Helenius, A., and Aebi, M. (2001). Intracellular functions of *N*-linked glycans. *Science* **291**, 2364–2369.

Houseman, B. T., and Mrksich, M. (2002). Carbohydrate arrays for the evaluation of protein binding and enzymatic modification. *Chem. Biol.* **9**, 443–454.

Kiyonaka, S., Sada, K., Yoshimura, I., Shinkai, S., Kato, N., and Hamachi, I. (2004). Semi-wet peptide/protein array using supramolecular hydrogel. *Nat. Mater.* **3**, 58–64.

Kuno, A., Uchiyama, N., Koseki-Kuno, S., Ebe, Y., Takashima, S., Yamada, M., and Hirabayashi, J. (2005). Evanescent-field fluorescence-assisted lectin microarray: A new strategy for glycan profiling. *Nat. Methods* **2**, 851–856.

Kusnezow, W., Jacob, A., Walijew, A., Diehl, F., and Hoheisel, J. D. (2003). Antibody microarrays: An evaluation of production parameters. *Proteomics* **3**, 254–264.

Lehr, H. P., Reimann, M., Brandenburg, A., Sulz, G., and Klapproth, H. (2003). Real-time detection of nucleic acid interactions by total internal reflection fluorescence. *Anal. Chem.* **75**, 2414–2420.

Mateo, C., Fernandez-Lorente, G., Abian, O., Fernandez-Lorenta, R., and Guisan, J. M. (2000). Multifunctional epoxy supports: A new tool to improve the covalent immobilization of proteins. The promotion of physical adsorptions of proteins on the supports before their covalent linkage. *Biomacromolecules* **1**, 739–745.

Pawlak, M., Schick, E., Bopp, M. A., Schneider, M. J., Oroszlan, P., and Ehrat, M. (2002). Zeptosens' protein microarrays: A novel high performance microarray platform for low abundance protein analysis. *Proteomics* **2**, 383–393.

Pilobello, K. T., Krishnamoorthy, L., Slawek, D., and Mahal, L. K. (2005). Development of a lectin microarray for the rapid analysis of protein glycopatterns. *Chembiochem* **6,** 985–989.

Stimpson, D. I., Hoijer, J. V., Hsieh, W. T., Jou, C., Gordon, J., Theriault, T., Gamble, R., and Baldeschwieler, J. D. (1995). Real-time detection of DNA hybridization and melting on oligonucleotide arrays by using optical wave guides. *Proc. Natl. Acad. Sci.* **92,** 6379–6383.

Xu, X. H., and Yeung, E. S. (1998). Long-range electrostatic trapping of single-protein molecules at a liquid-solid interface. *Science* **281,** 1650–1653.

Wang, D., Liu, S., Trummer, B. J., Deng, C., and Wang, A. (2002). Carbohydrate microarrays for the recognition of cross-reactive molecular markers of microbes and host cells. *Nat. Biotechnol.* **20,** 275–281.

Willats, W. G., Rasmussen, S. E., Kristensen, T., Mikkelsen, J.D, and Knox, J. P. (2002). Sugar-coated microarrays: A novel slide surface for the high-throughput analysis of glycans. *Proteomics* **2,** 1666–1671.

Zheng, T., Peelen, D., and Smith, L. M. (2005). Lectin arrays for profiling cell surface carbohydrate expression. *J. Am. Chem. Soc.* **127,** 9982–9983.

Zhu, H., Klemic, J. F., Chang, S., Bertone, P., Casamayor, A., Klemic, K. G., Smith, D., Gerstein, M., Reed, M. A., and Snyder, M. (2000). Analysis of yeast protein kinases using protein chips. *Nat. Genet.* **26,** 283–289.

Author Index

A

Aarnoudse, C. A., 294
Aberg, P. M., 143
Abian, O., 343
Aboul-ela, F., 183, 184, 194, 195
Accavitti, M. A., 114, 115, 116
Adachi, M., 138
Adams, E. W., 269, 270, 271, 310
Adhikari, S. S., 183
Aebersold, R., 124, 203, 254, 260
Aebi, M., 5, 21, 28, 31, 341
Agard, N. J., 230, 231, 238, 240, 243, 260
Agrawal, R. K., 191
Agris, P. F., 193
Aguilar, A., 155, 162
Ahn, J., 47, 48, 50, 52
Akamatsu, T., 164, 165
Akiyama, S., 27
Akshay, S., 195, 196, 197
Albers, M. W., 216
Albersheim, P., 73
Albrecht, R., 183
Alexander, S., 47, 89
Alhussaini, M., 176
Allende, M. L., 105
Allin, K., 138, 139, 143, 147, 149
Alm, R. A., 169
Alstock, R. T., 310
Altmann, F., 89
Altraja, S., 165
Altstock, R. T., 270
Alvarez, R., 138, 270, 292, 293, 294, 297, 298,
    301, 307, 308
Aly, M. R. E., 145
Amara, J. F., 216
Amsili, S., 213
Anderson, P. W., 154, 155
Anderson, R., 158, 161
Anderson, T. R., 244
Andersson, M., 342
Angara, K., 105, 107
Angeloni, S., 293, 342

Angenendt, P., 343
Angstrom, J., 165
Angulo, J., 278
Apodaca, J., 47, 48, 50, 52
Appelmelk, B., 294
Araki, K., 327
Araki, M., 327
Arata, Y., 87, 312, 313, 314, 318, 324
Argade, S., 59, 242
Argov, Z., 213
Arison, B., 154
Arndt, S., 114, 128, 129
Arnqvist, A., 165
Aruffo, A., 168
Asaka, M., 166
Asara, J. M., 21
Ashwell, G., 28, 158
Asou, N., 149
Astronomo, R. D., 270
Athanassiadis, A., 31, 32, 34
Atsumi, S., 28
Auerbach, T., 183, 184, 185, 187, 189
Auge, C., 138, 145
Austin, D. J., 216
Averani, G., 155, 158
Avey, H. P., 346
Avila, H., 183, 184, 185, 187, 189
Awaya, J., 87
Ayida, B. K., 198
Azuma, T., 166

B

Badet, J., 5
Baenziger, J. U., 171, 345
Bafna, V., 75
Baird, C. L., 42
Bairoch, A., 31
Balanzino, L., 104
Baldeschwieler, J. D., 342, 346
Baldwin, M. A., 74
Baly, A., 155, 162
Ban, N., 181

353

U

Uchikawa, M., 171
Uchimura, K., 90, 139, 150
Uchiyama, N., 342
Ujita, M., 145
Umeda, Y., 97
Underhill, C. B., 168
Urano, T., 171
Urashima, T., 312
Urata, T., 166, 167
Uria-Nickelsen, M., 169
Uzawa, H., 145

V

Valdés, Y., 155, 161, 162
Valiando, J., 21
Vallee, F., 33, 34, 35, 42
Van Beeumen, J., 34
Van Damme, E. J. M., 311
van de Heyning, P. H., 280
van den Eijnden, D. H., 104, 105, 107, 217
van den Nieuwenhof, I. M., 138, 295
Vandersall-Nairn, A. S., 32, 33
Vandersteen, D. P., 165
van Die, I., 138, 270, 293, 294, 298, 301, 307, 308
van Duin, M., 217
van Gug-Van der Toorn, C. J. G., 246
van Halbeek, H., 59, 243
van Kessel, A. G., 139
van Kooyk, Y., 294
van Liempt, E., 294
Van Petegem, F., 34
van Vliet, C., 213
van Vliet, S. J., 294
Varki, A., 59, 213, 230, 253, 293, 341
Varki, N. M., 230
Vasella, A., 195, 196, 197, 261
Vasiliu, D., 138, 139, 143, 147
Vazquez, D., 181
VedBrat, S., 270
Vella, P. P., 154
Venerando, B., 203, 209
Verdine, G. L., 216
Verdonk, G., 217
Verdugo, D., 214
Vérez-Bencomo, V., 155, 161, 162
Vervoort, J., 246

Vestweber, D., 213
Vicens, Q., 183, 184, 191, 192, 193, 194, 195, 196, 197, 198, 280
Vijay, I. K., 257
Villar, A., 155, 162
Vimr, E., 209
Vink, M. K. S., 243
Virtanen, A., 280
Viti, S., 155
Vo, L., 215
Vocadlo, D. J., 114, 131, 231, 253, 259, 260
Vogelman, J. H., 165
Volker, C., 34
von der Lieth, C. W., 75
Von Figura, K., 28
von Nicolai, H., 60
Vonrhein, C., 181, 198
Vorozhaikina, M., 328
Vosseller, K., 114, 115, 116, 124, 128
Vourloumis, D., 184, 193, 198
Vouros, P., 73
Vovis, G. F., 169
Voynow, J. A., 250

W

Wada, I., 31, 33
Wada, T., 205
Wada, Y., 97
Wadstrom, T., 165
Waechter, C. J., 4
Wakarchuk, W., 138, 139, 143, 147
Wakatsuki, S., 203, 209
Walijew, A., 343
Walker, P., 48
Wall, D., 184, 193
Walsh, C., 280
Walter, F., 280
Wandless, T. J., 216, 218, 227
Wang, A., 293, 341
Wang, B. C., 33, 34, 35, 36, 37, 38, 39, 40, 42
Wang, D., 293, 310, 341
Wang, G., 170
Wang, P. G., 105
Wang, R., 310
Wang, S., 230
Ward, C. L., 48
Warnecke, D., 172

Subject Index

A

N-Acetylglucosamine
  azido sugar, *see* Azido sugars
  chemoenzymatic enrichment of *O*-linked
    proteins, 128–130
  function in *O*-linked proteins, 114, 130
  immunoblot detection
    blotting, 116–117
    gel electrophoresis, 116
    materials, 115–116
    overview, 115
  mucin protection against *Helicobacter
    pylori*, *see* Helicobacter pylori
  prospects for study, 130–131
  site mapping in proteins
    β-elimination and Michael addition with
      dithiothreitol
      materials, 124–125
      mild BEMAD treatment, 126–127
      performic acid oxidation, 126
      phosphatase treatment, 126
      principles, 123–124
      protein digestion, 125–126
      strong BEMAD treatment, 127
      thiol affinity chromatography, 127–128
    controls
      hexosaminidase digestion, 121–122
      peptide:N-glycosidase F digestion, 121
      Edman degradation, 122–123
    galactosyltransferase
      glycan blocking, 117–118
      radiolabeling, 118–120
    overview, 114–115
    principles, 117
Activity-based affinity probes, *see*
  Glycosylating enzyme proteomics
Aminoglycosides, *see* Carbohydrate
  microarray; Ribosome
Antibiotics, *see* Carbohydrate microarray;
  Helicobacter pylori; Ribosome
Apramycin, ribosome binding studies,
  193–194

Azido sugars
  carbohydrate detection, 244–245
  metabolic oligosaccharide engineering
    cell culture labeling, 238–239
    glycan enrichment and identification,
      241–242, 245–247
    labeling reagent reaction with azido
      sugars, 239–241
    materials, 233, 234
    mouse labeling, 239, 241
    optimization parameters, 242–243
    overview, 231, 233
    prospects, 248
  phosphine-histidine-tagged FLAG
    synthesis, 238
  synthesis
    Ac4GalNaz, 235
    Ac4GlcNaz, 235
    Ac4ManNAz, 234–235
    cyclooctyne scaffold, 243–244
    ManNAz, 234
    1-methyl-2-
      diphenylphosphinoterephthalate, 237
    1-methyl-2-iodoterephthalate, 237
    phosphine-PFP, 237–238

B

BEMAD, *see* N-Acetylglucosamine
Biacore, *see* Surface plasmon resonance

C

Carbohydrate microarray
  aminoglycoside microarrays
    applications, 280
    materials, 280–281
    preparation, 281, 283
    protein hybridization, 285–286
    RNA
      binding candidate synthesis, 283–274
      hybridization, 284–285
    applications, 269–270, 341–342

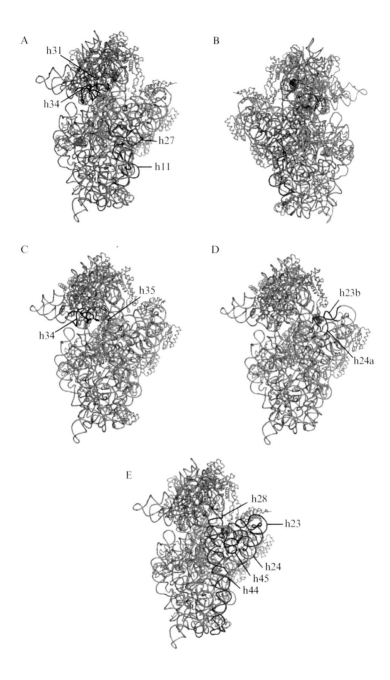

WIRMER AND WESTHOF, CHAPTER 12, FIG. 2. Three-dimensional structures of antibiotics bound to the 30S ribosomal particle. RNA is colored grey, black, and blue; proteins are colored green; antibiotics are indicated as spheres. (A and B), Tetracycline (1I97.pdb (Pioletti *et al.*, 2001)); the six tetracycline molecules binding to the subunit are shown (tet1, red; tet2, cyan; tet3, blue; tet4, magenta; tet5, yellow; tet6, orange). Helices in contact with tet1 and tet5 are indicated (h31, blue; h34, black; h27, blue; h11, black). (C) Spectinomycin (1FJG.pdb (Carter *et al.*, 2000)); helices in contact with the antibiotic are indicated (h34, black; h35, blue). (D) Pactamycin (1HNX.pdb (Brodersen *et al.*, 2000)) h23b (black) and h24a (blue) are indicated. (E) Edeine (1I95.pdb (Pioletti *et al.*, 2001)); interacting helices are indicated (h28, black; h44, blue; h45, black; h24, blue; h23, black).

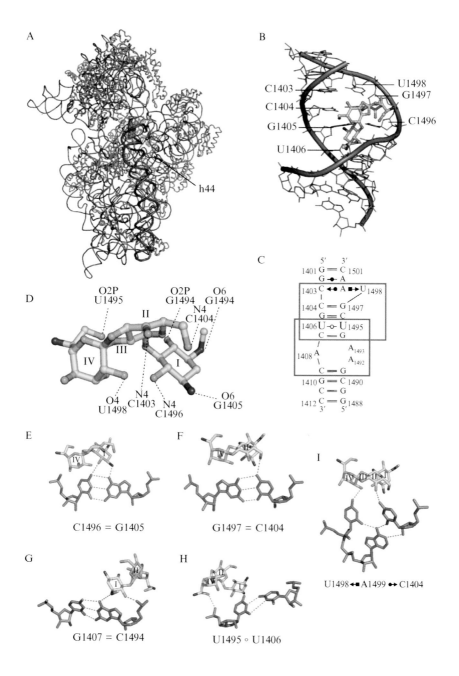

A

B

C1403
C1404
G1405
U1406

U1498
G1497
C1496

C
5'   3'
1401 G ═══ C 1501
        G ─●─ A
1403 C ◄─●─ A ─■─► U 1498
        │
1404 C ═══ G 1497
        G ═══ C
1406 U ─○─ U 1495
        C ═══ G
1408 A   A 1493
              A 1492
        C ═══ G
1410 G ═══ C 1490
        C ═══ G
1412 C ═══ G 1488
        3'   5'

D
              O2P   O2P   O6
              U1495   G1494 G1494
                        N4
                    II C1404
                III
        IV   I
        O4   N4   N4   O6
        U1498 C1403   C1496   G1405

E

IV   I

C1496 = G1405

F

IV   II

G1497 = C1404

G

I   II

G1407 = C1494

H

IV   II

U1495 ∘ U1406

I

IV III II I

U1498 ◄─■ A1499 ─►► C1404

WIRMER AND WESTHOF, CHAPTER 12, FIG. 3. (A) 3D structure of hygromycin bound to the 30S ribosome (1HNZ.pdb (Brodersen *et al.*, 2000)). Hygromycin is shown as red spheres. (B) Close-up view of the hygromycin binding site (helix h44 residues 1401–1410 and 1490–1501). Hygromycin is shown as ball and sticks. (C) Secondary structure diagram of parts of h44 that interact with hygromycin (red boxed area) and paromomycin (green boxed area). Tertiary interactions are indicated using the Leontis-Westhof nomenclature (Leontis and Westhof, 2001). (D through I) Interactions of hygromycin with the 30S particle. Hydrogen bonds with donor-acceptor distance of less than 3.5 Å are shown.

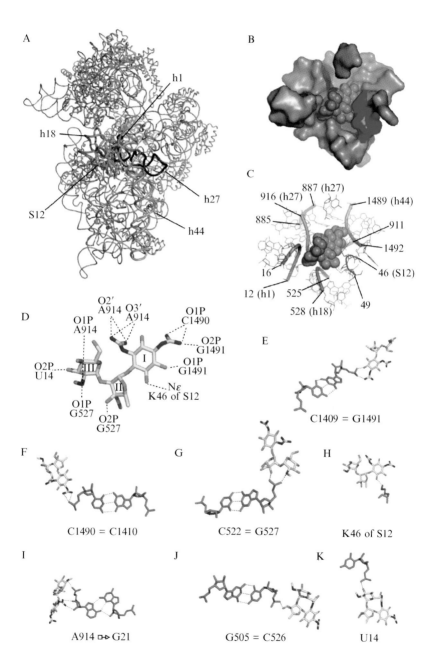

A

h1

h18

S12

h27

h44

B

C

887 (h27)

916 (h27)

1489 (h44)

885

911

1492

16

46 (S12)

12 (h1)   525

49

528 (h18)

D

O2′
A914   O3′
A914

O1P
A914

O1P
C1490

O2P
G1491

O2P
U14

III

I

O1P
G1491

II

Nε
K46 of S12

O1P
G527

O2P
G527

E

C1409 = G1491

F

C1490 = C1410

G

C522 = G527

H

K46 of S12

I

A914 ▭▸ G21

J

G505 = C526

K

U14

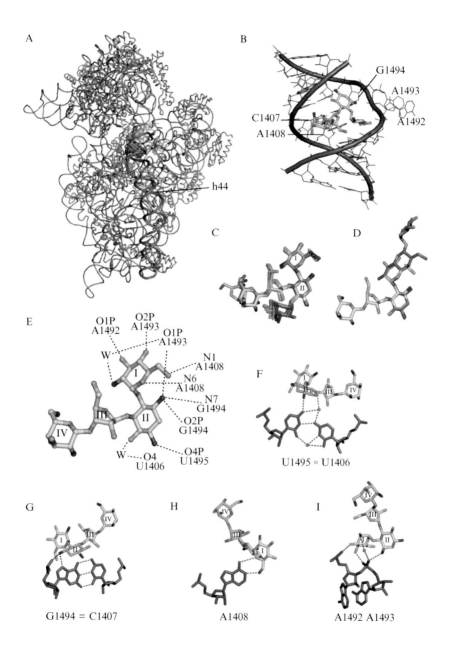

A

h44

B

G1494

A1493

C1407

A1408

A1492

C

I

II

D

E

O1P
A1492

O2P
A1493

O1P
A1493

W

I

N1
A1408

N6
A1408

III

II

N7
G1494

O2P
G1494

IV

W

O4
U1406

O4P
U1495

F

I

II III IV

U1495 ∘ U1406

G

I

II III

IV

G1494 = C1407

H

IV

III

I

A1408

I

IV

III

II

I

A1492 A1493

WIRMER AND WESTHOF, CHAPTER 12, FIG. 5. (A) Three-dimensional structure of paromomycin bound to the 30S ribosome (1FJG.pdb (Carter *et al.*, 2000)). Paromomycin is shown as red spheres, proteins are green, and RNA is grey. (B) paromomycin binding site taken from a structure presenting the minimal A-site (1J7T.pdb (Vicens and Westhof, 2001)). (C) Superposition of aminoglycoside antibiotics containing the neamine moiety onto paromomycin (yellow). Color coding: Geneticin (cyan), gentamicin A (magenta), kanamycin A (green), lividomycin (deep-salmon), neamine (grey), neomycin B (slate-blue), ribostamycin (orange), and tobramycin (marine). For pdb codes and references, see Table I. (D) Superpositon of apramycin (1YRJ.pdb (Han *et al.*, 2005), blue-green) onto paromomycin (yellow). (E through I) Common interactions of ring I and II of neamine-containing aminoglycosides; paromomycin is shown as example. Common hydrogen bonds with donor-acceptor distance of less than 3.5 Å are shown.

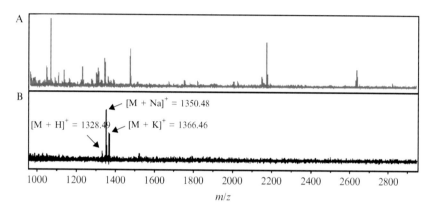

Hɪɴᴏᴜ *ᴇᴛ ᴀʟ.*, Cʜᴀᴘᴛᴇʀ 13, Fɪɢ. 4. MALDI-TOF mass spectra of fluorescence-labeled peptides. Peptide fragments produced by digestion with pepsin were analyzed. (A) Mixtures of peptides derived from labeled enzyme; (B) Peptide isolated by using anti-dansyl antibody column chromatography.

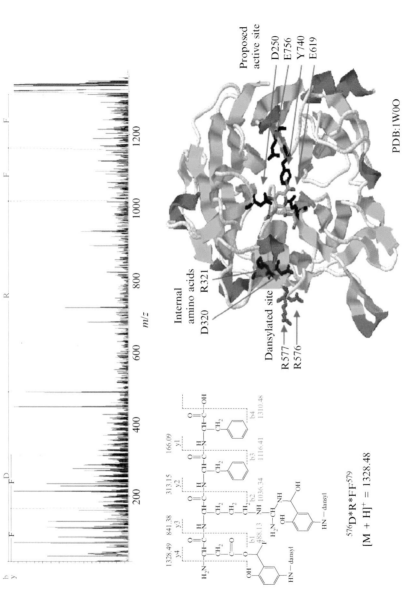

HINOU ET AL., CHAPTER 13, FIG. 5. MALDI-TOF/TOF mass spectrum of the dansylated peptide fragment (m/z 1328.49) and topological assignment of the identified sequence $^{576}DRFF^{579}$, which contains dansylated Asp576 and Arg577.

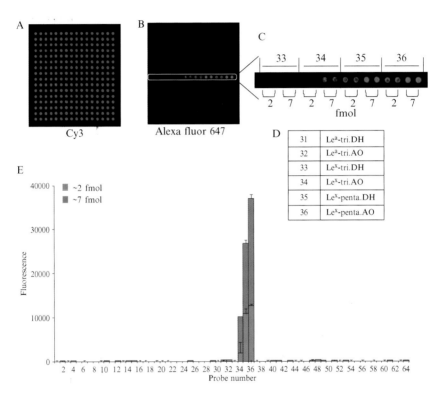

LIU *ET AL.*, CHAPTER 20, FIG. 6. Analyses of the interactions of rat anti–Lewis x (Le^x) antibody, anti-L5, overlaid onto a microarray of 64 lipid-linked oligosaccharide probes, each printed at approximately 2 and approximately 7 fmol/spot in duplicate on nitrocellulose-coated glass slides with Cy3 dye included as marker (green emission; A) and binding detected with biotinylated anti-rat immunoglobulins and Alexa Fluor 647-labeled streptavidin as described by Campanero-Rhodes *et al.*(2006) and Palma *et al.* (2006) (red emission; B and C). (D) The positions of the Le^x- and Le^a-related NGLs. Numerical scores of the binding as measured by fluorescence intensity (means of duplicate spots with error bars) in blue for 2 fmol/spot and in red for 7 fmol/spot (E).